VOLUME FOUR TWENTY SIX

METHODS IN
ENZYMOLOGY
Integrins

METHODS IN ENZYMOLOGY

Editors-in-Chief

JOHN N. ABELSON AND MELVIN I. SIMON

Division of Biology
California Institute of Technology
Pasadena, California

Founding Editors

SIDNEY P. COLOWICK AND NATHAN O. KAPLAN

VOLUME FOUR TWENTY SIX

METHODS IN
ENZYMOLOGY
Integrins

EDITED BY

DAVID A. CHERESH
Department of Pathology
Moores Cancer Center
University of California
San Diego, La Jolla

QP601
C71
v.426
2007

AMSTERDAM • BOSTON • HEIDELBERG • LONDON
NEW YORK • OXFORD • PARIS • SAN DIEGO
SAN FRANCISCO • SINGAPORE • SYDNEY • TOKYO
Academic Press is an imprint of Elsevier

Academic Press is an imprint of Elsevier
525 B Street, Suite 1900, San Diego, California 92101-4495, USA
84 Theobald's Road, London WC1X 8RR, UK

This book is printed on acid-free paper.

Copyright © 2007, Elsevier Inc. All Rights Reserved.

No part of this publication may be reproduced or transmitted in any form or by any means, electronic or mechanical, including photocopy, recording, or any information storage and retrieval system, without permission in writing from the Publisher.

The appearance of the code at the bottom of the first page of a chapter in this book indicates the Publisher's consent that copies of the chapter may be made for personal or internal use of specific clients. This consent is given on the condition, however, that the copier pay the stated per copy fee through the Copyright Clearance Center, Inc. (www.copyright.com), for copying beyond that permitted by Sections 107 or 108 of the U.S. Copyright Law. This consent does not extend to other kinds of copying, such as copying for general distribution, for advertising or promotional purposes, for creating new collective works, or for resale. Copy fees for pre-2007 chapters are as shown on the title pages. If no fee code appears on the title page, the copy fee is the same as for current chapters. 0076-6879/2007 $35.00

Permissions may be sought directly from Elsevier's Science & Technology Rights Department in Oxford, UK: phone: (+44) 1865 843830, fax: (+44) 1865 853333, E-mail: permissions@elsevier. com. You may also complete your request on-line via the Elsevier homepage (http://elsevier.com), by selecting "Support & Contact" then "Copyright and Permission" and then "Obtaining Permissions."

For information on all Elsevier Academic Press publications
visit our Web site at www.books.elsevier.com

ISBN: 978-0-12-373924-7

PRINTED IN THE UNITED STATES OF AMERICA
07   08   09   10   9   8   7   6   5   4   3   2   1

Working together to grow
libraries in developing countries

www.elsevier.com | www.bookaid.org | www.sabre.org

ELSEVIER   BOOK AID
               International   Sabre Foundation

CONTENTS

12. Analysis of Integrin Functions in Peri-Implantation Embryos, Hematopoietic System, and Skin   239

Eloi Montanez, Aleksandra Piwko-Czuchra, Martina Bauer,
Shaohua Li, Peter Yurchenco, and Reinhard Fässler

13. Identification and Molecular Characterization of Multiple Phenotypes in Integrin Knockout Mice   291

Chun Chen and Dean Sheppard

14. Purification, Analysis, and Crystal Structure of Integrins   307

Jian-Ping Xiong, Simon L. Goodman, and M. Amin Arnaout

19. Analysis of Integrin Signaling in Genetically Engineered Mouse Models of Mammary Tumor Progression

439

Yuliya Pylayeva, Wenjun Guo, and Filippo G. Giancotti

20. Design and Chemical Synthesis of Integrin Ligands

463

Dominik Heckmann and Horst Kessler

21. Evaluating Integrin Function in Models of Angiogenesis and Vascular Permeability   505

Sara M. Weis

CONTRIBUTORS

Brian D. Adair
Department of Cell Biology, The Scripps Research Institute, La Jolla, California

Alireza Alavi
Department of Pathology, School of Medicine, Moores Cancer Center, University of California at San Diego, La Jolla, California

M. Amin Arnaout
Structural Biology Program, Leukocyte Biology and Inflammation Program, Nephrology Division, Massachusetts General Hospital and Harvard Medical School, Charlestown, Massachusetts

Richard K. Assoian
Department of Pharmacology, University of Pennsylvania School of Medicine, Philadelphia, Pennsylvania

Martina Bauer
Max Planck Institute of Biochemistry, Department of Molecular Medicine, Martinsried, Germany

Laurent Bodin
Division of Hematology-Oncology, Department of Medicine, University of California, San Diego, La Jolla, California

David Boettiger
Department of Microbiology, University of Pennsylvania, Philadelphia, Pennsylvania

David A. Calderwood
Department of Pharmacology and Interdepartmental Program in Vascular Biology and Transplantation, Yale University School of Medicine, New Haven, Connecticut

Paola Castagnino
Department of Pharmacology, University of Pennsylvania School of Medicine, Philadelphia, Pennsylvania

Keefe T. Chan
Department of Molecular and Cellular Pharmacology, University of Wisconsin, Madison, Wisconsin

Chun Chen
Lung Biology Center, Department of Medicine, University of California, San Francisco, San Francisco, California

Shu Chien
Departments of Bioengineering and Medicine, Whitaker Institute of Biomedical Engineering, University of California at San Diego, La Jolla, California

Matthew W. Conklin
Department of Pharmacology, University of Wisconsin Medical School, University of Wisconsin–Madison, Madison, Wisconsin

Christa L. Cortesio
Department of Biomolecular Chemistry, University of Wisconsin, Madison, Wisconsin

Lance A. Davidson
Department of Bioengineering, University of Pittsburgh, Pittsburgh, Pennsylvania

Sumiko Denda
Shiseido Research Center 2, Kanazawa-ku, Yokohama, Japan

Douglas W. DeSimone
Department of Cell Biology, and Morphogenesis and Regenerative Medicine Institute, School of Medicine, University of Virginia, Charlottesville, Virginia

Bette Dzamba
Department of Cell Biology, School of Medicine, University of Virginia, Charlottesville, Virginia

Reinhard Fässler
Max Planck Institute of Biochemistry, Department of Molecular Medicine, Martinsried, Germany

Mark Fuster
Department of Medicine, University of California, San Diego, California

Barbara Garmy-Susini
Moores UCSD Cancer Center, University of California, San Diego, La Jolla, California

Scott Gehler
Department of Pharmacology, University of Wisconsin Medical School, University of Wisconsin–Madison, Madison, Wisconsin

Filippo G. Giancotti
Sloan-Kettering Division, Weill Graduate School of Medical Sciences, Cornell University, New York, New York

Simon L. Goodman
Preclinical Oncology Research, Merck KGaA, Damstadt, Germany

Wenjun Guo
Whitehead Institute for Biomedical Research, Cambridge, Massachusetts

David S. Harburger
Department of Pharmacology and Interdepartmental Program in Vascular Biology and Transplantation, Yale University School of Medicine, New Haven, Connecticut

Dominik Heckmann
Department of Chemistry, Technical University München, Garching, Germany

B. G. Hoffstrom
Department of Biological Sciences, Columbia University, New York, New York

Anna Huttenlocher
Departments of Pediatrics and Molecular and Cellular Pharmacology, University of Wisconsin, Madison, Wisconsin

Hisashi Kato
Division of Hematology-Oncology, Department of Medicine, University of California, San Diego, La Jolla, California

Patricia J. Keely
Department of Pharmacology, University of Wisconsin Medical School, University of Wisconsin–Madison, Madison, Wisconsin

Horst Kessler
Department of Chemistry, Technical University München, Garching, Germany

Eric A. Klein
Department of Pharmacology, University of Pennsylvania School of Medicine, Philadelphia, Pennsylvania

Devashish Kothapalli
Department of Pharmacology, University of Pennsylvania School of Medicine, Philadelphia, Pennsylvania

Yatish Lad
Department of Pharmacology and Interdepartmental Program in Vascular Biology and Transplantation, Yale University School of Medicine, New Haven, Connecticut

Shaohua Li
Department of Surgery, Robert Wood Johnson Medical School, New Brunswick, New Jersey

Milan Makale
Moores UCSD Cancer Center, University of California, San Diego, La Jolla, California

Denise K. Marciano
Department of Medicine, Division of Nephrology and Neuroscience Program, Department of Physiology and Howard Hughes Medical Institute, University of California, San Francisco, San Francisco, California

Eloi Montanez
Max Planck Institute of Biochemistry, Department of Molecular Medicine, Martinsried, Germany

Aleksandra Piwko-Czuchra
Max Planck Institute of Biochemistry, Department of Molecular Medicine, Martinsried, Germany

Suzanne M. Ponik
Department of Pharmacology, University of Wisconsin Medical School, University of Wisconsin–Madison, Madison, Wisconsin

Nicolas Prévost
Division of Hematology–Oncology, Department of Medicine, University of California, San Diego, La Jolla, California

Paolo P. Provenzano
Department of Pharmacology, University of Wisconsin Medical School, University of Wisconsin–Madison, Madison, Wisconsin

Yuliya Pylayeva
Cell Biology Program, Memorial Sloan-Kettering Cancer Center, and Sloan-Kettering Division, Weill Graduate School of Medical Sciences, Cornell University, New York, New York

Louis F. Reichardt
Department of Physiology, Neuroscience Program, Howard Hughes Medical Institute, University of California, San Francisco, San Francisco, California

Sanford J. Shattil
Division of Hematology–Oncology, Department of Medicine, University of California, San Diego, La Jolla, California

Dean Sheppard
Lung Biology Center, Department of Medicine, University of California, San Francisco, San Francisco, California

Dwayne G. Stupack
Department of Pathology, School of Medicine, Moores Cancer Center, University of California at San Diego, La Jolla, California

Judith A. Varner
Moores UCSD Cancer Center, University of California, San Diego, La Jolla, California

Yingxiao Wang
Department of Bioengineering and Molecular & Integrative Physiology, Neuroscience Program, Center for Biophysics and Computational Biology, Beckman Institute for Advanced Science and Technology, University of Illinois, Urbana-Champaign, Urbana, Illinios

E. A. Wayner
Antibody Development Laboratory, Fred Hutchinson Cancer Research Center, Seattle, Washington

Sara M. Weis
Moores UCSD Cancer Center, University of California, San Diego, La Jolla, California

Jian-Ping Xiong
Structural Biology Program, Leukocyte Biology and Inflammation Program, Nephrology Division, Massachusetts General Hospital and Harvard Medical School, Charlestown, Massachusetts

Mark Yeager
Department of Cell Biology, The Scripps Research Institute, and Division of Cardiovascular Diseases, Scripps Clinic, La Jolla, California

Yuval Yung
Department of Pharmacology, University of Pennsylvania School of Medicine, Philadelphia, Pennsylvania

Peter Yurchenco
Department of Pathology, Robert Wood Johnson Medical School, Piscataway, New Jersey

Preface

Integrins are a family of cell surface receptors that mediate contact between a cell and its surrounding extracellular matrix and microenvironment. Integrins not only regulate cell attachment (Chapter 1), but also act to transmit signals across the plasma membrane (2, 6, 9). They also serve as mechanotransducers (8) that, in turn, influence the cells' cytoskeleton assembly (4), leading to changes in cell migration/invasion (3, 16) and ultimately influencing cell survival (5) and cell cycle control (8). The techniques associated with understanding how cells respond to integrin ligation in two dimensional setting (1) and in three dimensional microenvironments (2, 5) have greatly advanced the field and our basic understanding of biology and disease. As such, integrins have been shown to play a key role in embryonic development (10, 12), immune recognition, tissue homeostasis, and wound repair (13). In addition, integrins have been shown to regulate various pathological conditions such as cancer, inflammation, and cardiovascular disease (6, 12, 13, 18, 21). In fact, integrin mutations or disregulation of integrin function are responsible for diseases associated with defective platelet aggregation and clotting, altered immune function, and altered tissue morphogenesis (6, 12, 18, 21). Integrin ligands (11) are typically found in the extacellular matrix and basement membrane and include proteins such as collagens, fibronectins, and laminins. However, during tissue remodeling, cancer, and angiogenesis (21), some specific integrins expressed on invasive cells recognize provisional matrix proteins including fibronectin, vitronectin, fibrin, and osteopontin, among others. For this reason it is important to have techniques available to understand how integrins function to promote cell adhesion to and invasion of the extracellular matrix (1–4). In some cases, integrins can recognize ligands on the surface of other cells. This is particularly true among hemaptoietic cells, those in the blood stream, and in the lymphatic system (6, 18, 21).

Recently, integrins have been recognized as important drug targets. In fact, there has been considerable effort in establishing technology to design integrin antagonists for use in treating various disease conditions (7, 20). For example, inhibitors (antagonists) of the platelet integrin $\alpha IIb\beta 3$ are used to suppress clot formation in patients with thrombotic disorders (6). Other integrin antagonists suppress immune recognition and thereby regulate inflammatory disease and/or autoimmune diseases such as multiple sclerosis. More recently, clinical trials have established that alpha V integrin antagonists can be used to treat or diagnose human cancer (20, 21). These alpha V integrin antagonists have been shown to directly suppress tumor growth and invasion and/or suppress the process of tumor angiogenesis. There are three forms of integrin antagonists: antibodies, peptides, and small organic peptidomimetics. Anti-integrin antibodies can directly compete for ligand binding or act as allosteric inhibitors. In general, integrin antibody antagonists tend to be more specific than peptide or peptidomimetic antagonists. The development of specific, function-blocking antibodies to integrins (7) has provided the most important tool for the biologist to understand how integrins function in the context of cells and the intact organism. However, integrin antibodies are generally produced in mice (7) and then subjected to humanization prior to being developed as clinical candidates. The development of integrin inhibitors has been aided by scientific approaches to studying how integrins function on cells (1–6, 8, 18, 21) and how they structurally interact with their ligands (14, 15, 20). In addition, the use of genetic models of mice lacking integrins or expressing mutant integrins have been absolutely critically in understanding the role that integrins play in the intact organism (12, 13, 17, 19). However, in a number of cases integrin knockout mice can have a different phenotypes than what one observes when treating mice with specific integrin antagonists. For example, mice deficient in alpha V integrins can develop with a normal-looking vasculature, yet animals treated with alpha V integrin antagonists show a disrupted angiogenic response (21). This may be due to compensatory changes that occur in response to the genetic knockout or due to molecular redundancy. In either event, it is difficult to compare the phenotypes of wild-type mice treated with integrin antagonists to mice entirely lacking an integrin to begin with.

The integrin field has not only made a significant impact on our understanding of basic cell biology, but it has provided important insight into tissue remodeling in the embryo and the adult. The structural, molecular, and biological techniques have combined to elucidate the role that integrins play in these processes and in the development of a wide array of pathological conditions. The field has now progressed to the point where new therapeutic strategies have been developed or are under development to treat everything from cardiovascular disease to inflammatory disease and cancer. The techniques outlined in this volume provide a complete guide to understanding the structure, function, and biological properties of integrins.

I would like to thank the authors of this volume for agreeing to participate in this project as these individuals, having made many of the key contributions to our understanding of the structure, function, and biology of integrins, represent the leaders in this field. I am particularly grateful to Cindy Minor and Jamey Stegmaier and their project management efforts in making this volume possible.

David A. Cheresh

METHODS IN ENZYMOLOGY

VOLUME 419. Adult Stem Cells
Edited by IRINA KLIMANSKAYA AND ROBERT LANZA

VOLUME 420. Stem Cell Tools and Other Experimental Protocols
Edited by IRINA KLIMANSKAYA AND ROBERT LANZA

VOLUME 421. Advanced Bacterial Genetics: Use of Transposons and Phage for Genomic Engineering
Edited by KELLY T. HUGHES

VOLUME 422. Two-Component Signaling Systems, Part A
Edited by MELVIN I. SIMON, BRIAN R. CRANE, AND ALEXANDRINE CRANE

VOLUME 423. Two-Component Signaling Systems, Part B
Edited by MELVIN I. SIMON, BRIAN R. CRANE, AND ALEXANDRINE CRANE

VOLUME 424. RNA Editing
Edited by JONATHA M. GOTT

VOLUME 425. RNA Modification
Edited by JONATHA M. GOTT

VOLUME 426. Integrins
Edited by DAVID CHERESH

VOLUME 427. MicroRNA Methods (in preparation)
Edited by JOHN J. ROSSI

VOLUME 428. Osmosensing and Osmosignaling (in preparation)
Edited by HELMUT SIES AND DIETER HAUSSINGER

VOLUME 429. Translation Initiation: Extract Systems and Molecular Genetics (in preparation)
Edited by JON LORSCH

VOLUME 430. Translation Initiation: Reconstituted Systems and Biophysical Methods (in preparation)
Edited by JON LORSCH

VOLUME 431. Translation Initiation: Cell Biology, High-Throughput, and Chemical-Based Approaches (in preparation)
Edited by JON LORSCH

VOLUME 432. Lipidomics and Bioactive Lipids: Mass Spectrometry Based Lipid Analysis (in preparation)
Edited by H. ALEX BROWN

VOLUME 433. Lipidomics and Bioactive Lipids: Specialized Analytical Methods and Lipids in Disease (in preparation)
Edited by H. ALEX BROWN

VOLUME 434. Lipidomics and Bioactive Lipids: Lipids and Cell Signaling (in preparation)
Edited by H. ALEX BROWN

CHAPTER ONE

QUANTITATIVE MEASUREMENTS OF INTEGRIN-MEDIATED ADHESION TO EXTRACELLULAR MATRIX

David Boettiger

Contents

Abstract

Integrin-mediated adhesion is based on the binding of integrins to immobilized ligands either in the extracellular matrix or on the surface of adjacent cells. The strength of adhesion is determined primarily by the number of adhesive bonds that form. Integrins have also been described as signaling receptors. Like adhesion, signals from integrins receptors can depend on the number of integrins that are bound to substrate attached ligands. The common methods for measuring cell adhesion are only capable of measuring initial interactions because the multivalent nature of adhesive bonds when the cell is considered as a unit, these assays reach plateau levels. To measure adhesive integrin-ligand bonds, a spinning disc device is described in which the mean cell detachment force is proportional to the

Department of Microbiology, University of Pennsylvania, Philadelphia, Pennsylvania

Methods in Enzymology, Volume 426

ISSN 0076-6879, DOI: 10.1016/S0076-6879(07)26001-X

© 2007 Elsevier Inc.

All rights reserved.

number of adhesive bonds. Particularly as the field has moved from identification of integrins and their ligands into the analysis of adhesion and its relationship to cell signaling, it is important to shift to assays that relate the extracellular binding to the intracellular signals. Specific plans, methods, and analytic strategies are provided to apply this technology to current problems in integrin biology. At present, this is the only published approach that will provide good measures of the number of adhesive bonds that can be generally applied.

1. INTRODUCTION

The process of cell adhesion can be divided into two basic steps: initial attachment of the cell to a surface and the increase in the number of adhesive bonds that follows from the initial attachment. Initial attachment is relatively easy to measure because non-adherent cells can be physically separated from the adherent cells and the proportion of adherent cells determined. This is the operation of the common adhesion assays that involve plating the cells, incubating, washing off nonadherent cells and counting the remaining cells. These plate-and-wash type assays are good for identifying adhesion receptors and determining their specificity (Humphries, 1998; Pierschbacher and Ruoslahti, 1984). However, the quantification involves separating cells into adherent and non-adherent groups and hence effectively partitions the cells into two categories against a wide range of adhesive strengths. This range becomes more obvious when you consider that the average fibroblast has 1 to 2×10^5 cell-surface $\alpha5\beta1$ integrins and if only 10% are bound this still gives 10,000 to 20,000 bonds per cell. Measurements of the forces required to break single integrin–ligand bonds are in the 50 to 100 pN range (Litvinov et al., 2002; Li et al., 2003; Weisel et al., 2003). The forces applied in the plate-and-wash type assays can be less than 100 pN/cell, if the plate is just inverted to drain the medium and then gently refilled, up to several 100 pN for more vigorous washing. The actual force required to detach a spread fibroblast is about 1000-fold higher. Therefore, it is easy to see that the plate-and-wash type assays only function to measure the initial adhesion event and very soon are out-of-measurement-range. This adhesion strengthening to "out-of-range" values was originally described for a centrifugation assay (Lots et al., 1989). The published assays use microtiter plates using forces of 50 to 250 g and the force per cell or force per cell from 30 to 450 pN (Redick et al., 2000; Reyes and Garcia, 2003). To detach a spread fibroblast would require ~300,000 g. Hence, this approach is not that practical as a cell detachment assay and the inability of this method to detach cells that are attached by more than ~10 bonds means that the data look very like the plate-and-wash assays.

Therefore, in order to measure cell detachment forces, a method that applied more force was required. The method we chose was hydrodynamic shear using a spinning disc device (Garcia *et al.*, 1997). The advantage of hydrodynamic shear is that it can apply much higher forces, and the forces are to a large extent distributed over the cell surface rather than applied at a point. This is advantageous if one wishes to measure the detachment forces for the individual cell rather than a subregion of the cell. There are many configurations to deliver hydrodynamic shear (Boettiger, 2006). The spinning disc has the advantages that it used a well-understood model for the distribution of shear forces and that shear stress is a linear function of radial position (distance from the axis of rotation). The linear shear gradient allows you to plot shear stress as a function of cell detachment and determine the mean shear stress for cell detachment in a cell population from a single disc. It will require the construction of a spinning disc device and a controller. Plans and parts lists are given in the appendix but the construction requires a skilled machinist and someone capable of building electronic circuits. To obtain acceptable statistics, we count 61 10× microscope fields, or about 12% of the total surface area for 25-mm discs (typically 5,000 to 15,000 cells). This requires a fluorescent microscope with a computer-driven stage and image processing software for counting the discs.

2. SPINNING DISC MEASUREMENTS

The spinning disc measures the mean shear stress required for cell detachment. For many cell types and conditions, we have found that the mean cell detachment force is proportional to the number of adhesive bonds (Boettiger *et al.*, 2001; Garcia and Boettiger, 1999; Garcia *et al.*, 1998; Shi and Boettiger, 2003). It has been argued that the cell would detach by a peeling mechanism. The adhesive bonds would detach sequentially, and hence there would not be a direct relationship between detachment force and the number of adhesive bonds per cell. The peeling model does apply when the detachment force is applied locally using a micropipette. The application of the peeling model to cell detachment by hydrodynamic shear stress assumes that the cell is relatively stiff and inert. However, the cell is not inert and responds to force application by strengthening attachments, accumulation of vinculin at stress points, and redistribution of forces through the cytoskeleton (Choquet *et al.*, 1997; Ingber, 2003; Riveline *et al.*, 2001). Our preliminary data show increased vinculin staining at the cell edge facing the flow where the highest force would be. The relatively slow acceleration and the distribution of the force over the cell cooperate with effects on the cytoskeleton to redistribute forces so that the adhesive bonds are reasonable equally stressed. The experimental data fit the simpler model in which the cell detaches as a unit, implying that all integrins are similarly stressed at the breaking point. Therefore, applying Occam's razor, we use the simpler model until it is proven inadequate.

In measuring the mean cell detachment force, we normalize to the center position on the spinning disc where minimal force is applied. Any cells that do not attach do not affect the assay. Of the cells that attach it measures the mean cell detachment force. This is in part why the adhesion measurements made using the plate-and-wash assays and the spinning disc assay do not measure the same parameter. If one views cell adhesion in terms of the strength of adhesion or adhesion-receptor–mediated signaling, then the relevant measure is the number of adhesive bonds. The spinning disc represents the most thoroughly documented method to obtain these values. The strength of adhesion is proportional to the number of adhesive bonds. Likewise, the strength of downstream signaling is related to the number of bonds (Shi *et al.*, 2003). It is common in the literature to see the plate-and-wash type assay used as evidence for a connection between some signaling event and cell adhesion. Do we expect the signaling to be related to the number of adhesive bonds or to the initial attachment rate of the cell? The initial attachment rate is the rate-limiting step, and hence important for the exit of platelets and leukocytes from flowing blood to perform their defense and repair functions. However, it is less important for other cells in the body that do not circulate. As we move to a more mechanistic understanding of adhesion-mediated signaling, the ability to measure the number of adhesive bonds that could be involved in the signaling will be critical. Since the spinning disc uses shear stress for the measurements, it can in principle be applied to any adhesive bonds. Understanding the mechanisms of cell adhesion involves a balance of many elements so quantification is key.

3. PROTOCOLS

3.1. Preparation of the substrate

3.1.1. Treating glass

Much of the adhesion literature does not pay much attention to preparation of the substrate layer. Protein adsorption to glass or plastic substrates can vary in both efficiency and in the conformation of the adsorbed protein (Garcia *et al.*, 1999; Miller and Boettiger, 2003). Glass coverslips are most convenient for the later counting of cells under a fluorescent microscope, the plastic coverslips are more uniform in performance. (Our sources for both are given in Appendix 1.) Glass coverslips have a surface oxidized layer that differs by manufacturer and batch of coverslips. The main problem is uniformity of the oxidized layer. The oxidized layer increases protein adsorption and affects its adsorbed conformation. After trying both ways, we decided to retain the oxidized layer. To remove oils from the manufacture, we soak coverslips in coverslip racks in 95% ethanol overnight or longer and wash them with PBS–ABC 3× before adding protein.

Protein adsorption Adsorption of fibronectin to glass coverslips occurs in two phases shown as different slopes of the adsorption profile (Fig. 1.1). The initial phase is dominated by adsorption of fibronectin to the glass and the second by adsorption of fibronectin to fibronectin, which is less efficient (note the break in the slope of the adsorption profile above 10 μg/ml fibronectin used for adsorption). We prefer to remain in the first phase and use two-thirds saturation for most experiments (adsorption from a 5 μg/ml solution for fibronectin). We use a strict regime for adsorption so that we can use previously generated adsorption profiles. Fibronectin is adsorbed in PBS-ABC for 30 min at room temperature, the excess removed, and the plate washed 1× with PBS-ABC. The coverslips are blocked for 30 min with 1% BSA (heat inactivated at 56° for 30 min), washed 3× with PBS-ABC or HEPES adhesion buffer and stored in adhesion buffer up to 3 hours. Other extracellular matrix proteins will have different adsorption profiles but are roughly in the same range as fibronectin (approximately threefold).

Surface roughness Levels of protein adsorption are affected by surface roughness (Lee *et al.*, 2006). Plastic is rougher than glass and adsorbs about twice the level of protein (Fig. 1.2). However, the increase in adsorbed fibronectin on the polystyrene culture dish surfaces does not increase cell adhesion strength, relative to using the same adsorption concentration for glass coverslips, suggesting that it does not lead to increased number of exposed receptor-binding domains.

Protocol for proteins with weak adsorption Some proteins, like immunoglobulins, adsorb to glass but the strength of this adsorption is not sufficient to withstand forces applied by the spinning disc when cells are

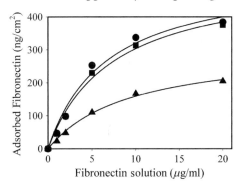

Figure 1.1 Adsorption of fibronectin. Using the protocol given, these plots can be used to determine the level of adsorbed fibronectin on untreated polystyrene (■), tissue culture-treated polystyrene (●), and glass (▲). Reproduced with permission from Garcia *et al.* (1999). *Mol. Biol. Cell* **10**, 785.

Figure 1.2   Spinning disc device. (A) Detail of the spinning disc platform. The disc fits in the center of the platform; there is a 2-mm rim around the disc and a hole in the center to apply vacuum. (B) Shows inverting the spinning platform into the chamber. (C) Shows both the spinning disc and our controller ready to begin the spin. Note that the inverted design shown here is easier to use and maintain than the upright design, but it is not suitable for weak adhesion. This system is optimized to cover the range of adhesion strengths found for solid tissue cells.

attached to the protein. An alternative method to attach these proteins is through nitrocellulose. First a layer of gelatin or collagen is dried on the coverslip (100 μg/ml solution added for 5 min, the excess is removed, and the surfaces are air dried). Add nitrocellulose solution 300 μl/25-mm– diameter coverslip, and allow to dry. These surfaces can be kept for months at room temperature. For use, add I$_g$G (or similar protein) at 10 μg/ml for 10 min. For immuno-globulin, $\sim$10^4 molecules are adsorbed/μm^2 with $\sim$3.3 $\times$ 10^3 available in epitope-binding domains (Lee *et al.*, 2004).

3.2. Preparation of the cells

3.2.1. EDTA

The ideal preparation of the cells involves using EDTA to remove them from the growth culture plates and minimal centrifugation. The exact method will differ according to cell type, but passaging the cells the day before assay and serum starving the cells tends to make the dissociation with EDTA more effective. Remember that a single cell suspension is important for this assay.

3.2.2. Trypsin

Trypsin can be used to dissociate cells. After trying several trypsin inhibitors, we returned to the use of serum to neutralize the trypsin. This requires centrifugation and washing to remove the serum. When using trypsin inhibitors (TLCK, soybean, and egg white), we found that they needed to be carefully titrated for each batch because insufficient neutralization will result in trypsinization of the adhesive substrate, while too much inhibitor also reduced cell adhesion in a dose-dependent manner by unknown mechanisms.

3.3. Setup of the spinning disc device

3.3.1. Seeding the cells

Prepared 25-mm, round coverslips are placed in 35-mm culture dishes, and 400 μl of cells in the appropriate adhesion buffer are seeded with a 1-ml micropipette using a spiral pattern beginning at the outside of the disc and moving toward the center. With practice this will give a reasonably uniform cell distribution. It can be checked by allowing cells to attach and then counting the disc without spinning. The number of cells seeded will depend on cell size, but ideally the cell should be separated by more than one cell diameter to reduce interference with the flow. Plating 1 to 2 $\times$ 10^5 cells/coverslip works for many cell lines. Cells can be analyzed from 5 min to several days after seeding.

3.3.2. Preparing the spinning disc device

Table 1.1 shows the relationship between the spinning speed (rpm), and the wall shear stress for a range of radial positions across the disc (from the axis of rotation).

1. Fill chamber ($\sim$1100 ml) with spinning buffer that is prewarmed to room temperature (25°), or as required. Set the speed and time. To set speed, place the spinning plate in the chamber and adjust speed in a trial run. We use 5 min as our standard run time.

Table 1.1 Shear stress in pN/μm² for selected speeds

RPM	2 mm	4 mm	6 mm	8 mm	10 mm	12 mm
Spinning in adhesion buffer						
500	0.54	1.08	1.62	2.17	2.71	3.25
1000	1.53	3.06	4.6	6.13	7.66	9.19
1500	28.1	5.63	8.44	11.26	14.07	16.88
2000	4.33	8.67	13.0	17.33	21.66	26.00
2500	6.05	12.11	18.16	24.22	30.27	36.33
3000	7.96	15.92	23.88	31.84	39.80	47.76
3500	10.03	20.06	30.09	40.12	50.15	60.18
4000	12.25	24.51	36.76	49.02	61.27	73.53
4500	14.62	29.25	43.87	58.49	73.11	87.74
5000	17.13	34.25	51.38	68.50	85.63	102.76
5500	19.76	39.52	59.27	79.03	98.79	118.55
6000	22.51	45.03	67.54	90.05	112.56	135.08
Spinning in 5% Dextran buffer						
500	1.6	3.2	4.80	6.4	8.01	9.61
1000	4.53	9.06	12.58	18.11	22.64	27.17
1500	8.32	16.64	24.96	33.28	41.60	49.91
2000	12.81	25.62	38.42	51.23	64.04	76.85
2500	17.90	35.80	53.70	71.60	89.50	107.40
3000	23.53	47.06	70.59	94.12	117.65	141.18
3500	29.65	59.30	88.95	118.60	148.25	177.91
4000	36.23	72.45	108.68	144.91	181.13	217.36
4500	43.23	86.45	129.68	172.91	216.14	259.36
5000	50.63	101.26	151.88	202.51	253.14	303.77
5500	58.41	116.82	175.23	233.64	292.05	350.45
6000	66.55	133.10	199.66	266.21	332.76	399.31
6500	75.04	150.08	251.13	300.17	375.21	450.25
7000	83.87	167.73	251.60	335.46	419.33	503.19
7500	93.01	186.02	279.03	372.04	465.05	558.06
8000	102.46	204.93	307.39	409.86	512.32	614.78
8500	112.22	224.44	336.66	448.87	516.09	733.59

Note: Data in Figs. 1.2 and 1.4 are given in dynes/cm². 10 dynes/cm² = 1pN/mm². We have switched to the latter values as they are easier to relate to the cell length and force scales.

2. Apply a thin layer of vacuum grease to the spinning platform. Place coverslip on spinning plate and center, apply vacuum to the port on the shaft to such coverslip tight on the spinning plate. Figure 1.2A shows details of the spinning disc platform. Invert the spinning plate and supporting structure into the chamber slowly to minimize disturbances to the cells, particularly by the buffer surface tension (Fig. 1.2B).

3. Start the machine. Figure 1.2C shows the device ready to spin. Circulating bubbles in the chamber are acceptable up to a point. If the chamber is not filled to the top baffle, there will be more turbulence and this should be avoided.
4. At the end of the spin, remove the spinning plate from the spinning chamber while being careful to avoid buffer dripping on the bearings as this will reduce their life.
5. Remove disc with fine forceps into a 35-mm dish for fixation and staining. It is sometimes useful to view the cells on an inverted microscope to check speed and other experimental parameters.

3.3.3. Maintenance of the spinning disc device

To clean, fill the chamber with deionized water, wipe the vacuum grease off of the disk, and invert the shaft into the chamber. Spin for several minutes, and then wipe the shaft and disk dry. Invert the chamber on the metal supports to dry. Bearing replacement requires a machine shop. Adding 3-in-1 oil to the bearings occasionally will prolong the life of the bearings. Add about 50 μl of oil to the bearing while it is spinning (slowly).

3.4. Counting the discs

3.4.1. Staining cells

We have used two main staining procedures. The first is fixing the cells with 95% ethanol for 10 min and staining with ethidium homodimer for 10 min. The second is fixing the cells in 3.7% formalin for 10 min, permeabilizing in PBS–ABC containing 0.5% triton X-100, and staining with DAPI. Coverslips are mounted in Gel/Mount (Biomedia) and sealed with nail polish for longer storage. Once mounted firmly the vacuum grease can be removed with ethanol swabs. Generally we allow overnight for this process.

3.4.2. Counting

Ethidium homodimer is counted using tetramethylrhodamine isothiocyanate (TRITC) filters. DAPI uses an ultraviolet fluorescence filter set. These stains generally will give sufficient contrast so that slight differences in focus during the scan are not a problem and automated counting can be implemented. Other strategies are to use software discrimination for nuclear size and shape, or a z-axis motor and software-driven refocus across the scan.

To keep track of radial position and to automate the counting process, we use ImagePro software and its microscope-driving extension, ScopePro (Media Cybernectics). To orient the counting, the center of the disc must be located. This is done by determining the center points of two perpendicular cords (in the x and y directions) across the coverslip. Locate one edge point by centering the edge across the field, note coordinates, move in the x-direction to opposite edge, note coordinates, move in y direction to the next edge and note coordinates. The x coordinate of the center is the mid-point on the x-axis cord and the y coordinate is the mid-point of the y-axis cord. This can be automated with macros in most imaging/microscope software.

On the Nikon Optiphot microscope, the field diameter for the $10\times$ objective is 1 mm. A Ludl automated XY stage driven by the Scope-Pro extension of ImagePro (Media Cybernetics) is used to scan the stage in 1-mm increments left to right and top to bottom, and diagonals over 1 mm and up 1 mm. These scans are used to generate a pattern of 60 fields with 4 at each radial distance plus a center count that is used to normalize counts in the other fields.

Thresholding usually needs to be set for each disc. ImagePro automatically counts each field based on the threshold and size/shape parameters (if used), transfers the count to a Microsoft Excel spreadsheet, and moves to the next field. Each field is displayed on the screen as it is counted, so the quality of the counting can (and should) be monitored.

3.5. Generating adhesion plots using SigmaPlot

ImagePro will dump the counts for each field into a Microsoft Excel spreadsheet column, which is then transferred to a SigmaPlot template. This generates a plot of shear stress versus the proportion of cells that remain attached following the spin, and then a curve-fit that reports a value for the mean shear stress for cell detachment or τ_{50}. This form of quantification is based only on cells that have attached in the allotted time, and remaining nonadherent cells have no influence on the measured values.

3.5.1. Mathematical analysis

Our counting regime returns values for 15 radial positions with four data points per position plus a center value that is used to normalize. The shear stress for each radial position is calculated from the formula,

$$\tau = 0.800 \; r \; \sqrt{(\rho\mu\omega^3)} \tag{1}$$

where τ is the shear stress, r is the radial distance from the center of rotation, ρ is the fluid density, μ is fluid viscosity, and ω is rotational velocity in radians per minute.

The data are fit to the sigmoid probabilistic model,

$$f = 1/\left(a - \exp[b(\tau - c)]\right), \tag{2}$$

where normalized a is the asymptote and constrained to be $0.95 < a < 1.05$, c is the inflection point and equal to the point for 50% cell detachment, and

b is the slope at the inflection point. We use the c value to represent the mean shear stress for cell detachment, or τ_{50}. The data points will show a normal distribution at about τ_{50}, and the spread will represent the heterogeneity of the cell population with respect to adhesive strength as well as variations due to experimental issues.

3.5.2. Using sigmaplot transforms to curve-fit the data

1. Open SigmaPlot Data sheet. In Column 1, Row 12, enter speed in RPM. In Column 1, Row 16 enter dynamic viscosity (μ); the viscosity for buffer at $25°$ is 0.01 Poise, and for 5% dextran 0.07 Poise. Column 2 contains the radial positions for each successive microscope field counted. This is the basic template. Copy count data to Column 4. The transform *Adhesion.xfm* below uses information from Columns 1 and 2 and the data from Column 4 to compute shear stress for each radial position, places results in Column 3; and normalizes the data in Column 4 dividing by the center position (in our template, Row 12).

Adhesion.xfm

```
;calculate shear stress t
for i = 1 to 61 do
    cell(3,i)=0.8*(cell(2,i)/10)*sqrt(cell(1,16)*(cell(1,12)*0.1048)^3)
end for
;compute cell fractions
for j = 1 to 61 do
    cell(5,j) = cell(4,j)/cell(4,12)
end for
```

2. Create scatter plot with x = Column 3 and y = Column 5.

3. Fit curve to user–defined fit *ADFIT*. Select Statistics; Regression Wizard; User Defined; New. enter the following:

ADFIT

Equation: $f = a/(1 + \exp(b*(x - c)))$
         fit f to y
Variables: $x = \text{col}(3)$
         $y = \text{col}(5)$.
Initial Parameters: $a = 1$
              $b = 0.02$
              $c = 400$
Constraints: $a < 1.05$
           $a > 0.95$
           $b > 0.001$
           $c > 0$

Running this analysis will generate the curve-fit to the data plot and report values for the fit variables a, b, and c. Variable c is the mean cell detachment shear stress, or τ_{50}. It also reposts statistical measures for the curve-fit (R^2).

4. Data Analysis

4.1. Interpretation and sources of error

The emphasis is on quantitative measures that can be fit to expected functional relationships for chemical and physical interactions. Examples of plots obtained from the spinning disc analysis are shown in Fig. 1.3. The data points represent normalized cell densities for the 61 fields counted, and the lines represent the curve-fit. The critical parameters to focus on are τ_{50} and R^2, which are obtained from the curve-fit, rotation speed, and center count used for normalization.

1. The x-axis has been converted from radial distance to shear stress based on Eq. 1 as part of the data transform. $X = 0$ is the center with minimal shear stress. Increasing the rate of rotation increases the slope of the shear gradient. There is also generally a decrease in density at the edge even without spinning due to surface tension. Hence, values of τ_{50} that fall in the outer 25% of the gradient should be treated as suspect, and duplicate discs should be spun at a higher speed to validate the values obtained (Fig. 1.3I). By using the data in Table 1.1 and the measured values for τ_{50}, it is easy to determine whether the curve-fit meets the criteria. It is possible to see whether the data are likely to create a good plot (such as in Fig. 3A–F) by viewing the cells on an inverted phase microscope immediately after spinning.

2. The y-axis has been converted to a normalized cell density based on the center count. Sometimes this count may be lower than the neighboring areas due to random plating artifacts, low overall cell counts, inaccurate centering, or a bad spot on the disc. In these cases, the average of the nearest set of four points can be used as an alternative. It is easiest to replace the center value with this average and rerun the curve-fit. If you also record the center count, it will provide an indication of less certainty in the measurement. This will also be apparent by observing the plot (see Fig. 1.3H).

3. The largest source of variability has been in the uniformity of the coverslip surface due to variations in the oxide layer as discussed above. This will result in areas of poor adhesion and points accumulating along the x-axis (Fig. 1.3G). Often the variations will not be as severe as the example in Fig. 1.3G, with fewer points close to the x-axis and others that follow a proper distribution for the plot. The low points can be deleted and a new

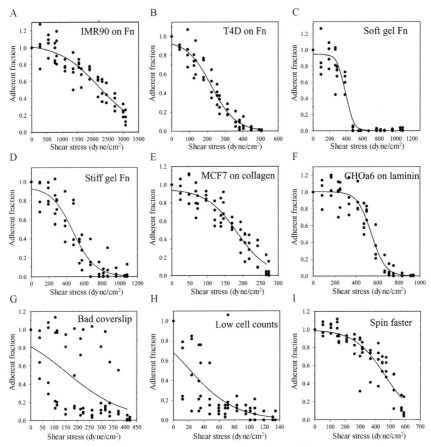

Figure 1.3 (A) High shear stress for spread IMR90 fibroblasts on fibronectin (160 ng/cm²). Spin at 5240 RPM in 5% Dextran, $\tau_{50} = 2239$ dynes/cm², $R^2 = 0.83$. (B) Intermediate adhesion, same conditions with mammary tumor cells T24D. Spin at 3000 RPM, $\tau_{50} = 222$ dynes/cm², $R^2 = 0.94$. (C) Cells attached to stiff polyacrylamide gels with fibronectin substrate. NIH3T3α5 cells spin at 4995 RPM, $\tau_{50} = 389$ dynes/cm², $R^2 = 0.93$. (D) Same as C using stiff gels. NIH3T3α5 cells. Spin at 4995 RPM, $\tau_{50} = 474$ dynes/cm², $R^2 = 0.87$. (E) Using collagen substrate (dried 0.1%), MCF7 cells. Spin at 4541 RPM, $\tau_{50} = 541$ dynes/cm², $R^2 = 0.93$. (F) Using laminin substrate adsorbed at 5 μg/ml. CHOa6 cells. Spin at 1995 RPM $\tau_{50} = 183$ dynes/cm², $R^2 = 0.87$. (G) Example of a bad disc due to poor oxide coating; cells failed to adhere to regions of the disc. IMR90 cells on fibronectin (75 ng/cm²) spin at1390 RPM in 5% Dextran, $\tau_{50} = 154$ dynes/cm², $R^2 = 0.31$. (H) Low counts for background adhesion on BSA are sometimes unreliable. IMR90 cells. Spin at 650 RPM in 5% dextran, $\tau_{50} = 19$ dynes/cm², $R^2 = 0.45$. (I) Not spun fast enough. τ_{50} is within the last 25% of the disc and could be due to fewer cells seeding. This material requires faster disc spin to test whether it is real or an artifact, even though the R^2 is in the acceptable range. C2C12 cells on fibronectin (160160 ng/cm²). Spin at 3330 rpm, $\tau_{50} = 458$ dynes/cm², $R^2 = 0.82$.

curve-fit generated to get a better measurement or for confirmation. Counting artifacts generally result in counts that are too low. It is important to examine the plots while keeping in mind that an estimated 20 to 30% of the glass coverslips show some lack of uniformity.

4. From the curve-fit, values for a (the plateau maximum), b (the slope of the plot at the inflection point), and c is τ_{50}. Variable a is constrained as follows: $0.95 < a < 1.05$. Variable b reflects the distribution. τ_{50} is the mean shear stress required for cell detachment and is proportional to the number of adhesive bonds.

5. The R^2 value for the curve-fit provides a basis for accepting or rejecting individual discs. The use of 61 fields provides a large number of points to fit that will result in less perfect fits, so we tolerate lower R^2 values for the initial fit. Generally, R^2 values less than 0.7 should be rejected, but the plot should also be examined because R^2 is only one measure of how well the data fit the theory. Compare values for Fig. 1.3A through F with Fig. 1.3G and H.

6. Figure 1.3A through F show a range of conditions and plot shapes obtained. The range of τ_{50} values for adherent cells shows a high for spread fibroblasts (Fig. 1.3A) to lower values for epithelial derived cells and different substrates. Some curves show relatively sharp transitions (Fig. 1.3C and F), while others show a more gradual transition (e.g., Fig. 1.3A). This is presumably related to differences in the heterogeneity of the cell population. The data show specific adhesion to fibronectin, laminin, and collagen, using a variety of cell lines. Because of the effect of substrate compliance on the forces generated by cells and on cell signaling (Discher *et al.*, 2005; Paszek *et al.*, 2005), it was interesting to examine the effect of adhesion. The polyacrilimide gel system involves covalent cross-linking of the gel to the disc and or the fibronectin to the gel. Detachment profiles for both soft and stiff gels are shown (Fig. 1.3C and D).

4.2. Quantification of other parameters

To relate the spinning disc measurement to adhesive bond formation, additional information is required. This includes ligand density, cell–surface integrin density, integrin activation state, the size of the cell–substrate contact zone, and the cell detachment mechanism. These parameters have been constrained by experimental analyses; however, these determinations involve the following initial assumptions:

1. Ligand density. The surface density of the ligand molecule can be determined from adsorption, except in the case of the gels. Adsorption to a surface is usually not specific in terms of molecular orientation, and can involve some protein denaturation will occur on hydrophobic surfaces

like polystyrene. This makes the calculation of the actual number of receptor-binding domains available to the cell surface integrins less certain. Our estimate for fibronectin, based on several approaches, is that 15 to 25% of the total cell binding domains are available for fibronectin adsorbed to glass surfaces and 7 to 14% on polystyrene surfaces.

2. Integrin surface density. Integrin surface density is measured by saturation binding of soluble ligand. For analysis of $\alpha 5 \beta 1$ integrin levels, K562 cells provide a convenient standard with a stable surface level of $\sim 10^5$/cell and can be used with FACS analysis to determine values for other cells (Faull et al., 1993; Garcia et al., 1998). To determine average integrin density, a measure of surface area is needed. Cell size and shape measurements can be used to obtain approximate surface areas, but these do not take into account invaginations of the membrane. As a first approximation, we ignore this issue and assume that all surface integrins are available. Another issue is recruiting of integrins from areas where there is no substrate contact into the areas of cell substrate adhesion. Although this may occur for leukocytes (Michl et al., 1983), we do not find this redistribution for HT1080 cells (Lee et al., manuscript in preparation).

3. Integrin activation state. Many members of the integrin family of adhesion receptors undergo a conformation switch between inactive and active states (Liddington and Ginsberg, 2002; Takagi et al., 2002). Conversion to the active state is thought to be a prerequisite for efficient ligand binding. The data to support the model are based on the binding of soluble ligand. The role of integrin activation in the formation of adhesive bonds has yet to be experimentally determined. In the HT1080 model system, integrin activation does not appear to affect the formation of adhesive bonds, although it does affect the initial attachment of the cell to a ligand–coated substrate (Lee et al., manuscript in preparation).

4. Cell–substrate contact zone. For the contact area we use the full interface between cell and substrate as the total area in which adhesive bonds could conceivably form. Others have used the area of close contact determined by interference reflection microscopy (IRM) (DePasquale and Izzard, 1987). Initial bonds must form in areas in which IRM would not detect small areas of close apposition. The adhesion structures are dynamic, so that new ones are forming within the larger area where there is a cell–substrate interface. The IRM distance of 15 nm is somewhat arbitrary and perhaps inconsistent with the 20-nm length of the extended integrin (Xiong et al., 2003), and the 3- to 10-nm extension of the cell-binding domain on fibronectin. The full interface gives a better baseline, and one can consider separately what is happening within this region. Hence using the full area of the cell–substrate interface as the contact zone provides a more useful computational model.

5. Mechanism of cell detachment. It has been widely argued that the application of more force to the side of the cell that faces the flow

would lead to a progressive detachment (the peeling or Velcro model for cell detachment). This form of cell detachment has been demonstrated for pulling on an attached cell with a micropipette (Griffin *et al.*, 2004). It was also expected in hydrodynamic shear-induced detachment because the edge of the cell facing the flow would experience a higher shear stress. However, the drag forces due to the shear are distributed over the cell surface, and cells respond to asymmetric application of force by reinforcing adhesive connections (Choquet *et al.*, 1997) that may be due to vinculin recruitment to sites of stress (Riveline *et al.*, 2001). For the cells we have examined, the cell detachment measurements are a function of ligand density, and receptor density plot fit adheres to the simpler assumption that the cells detach in a single step at a critical shear stress. Moreover, conversion of the mean shear stress for cell detachment to the force applied per cell (described below) and dividing by the number of adhesive bonds (based on binding equilibrium analysis), gives a value of $\sim$30 pN/bond for the $\alpha 5\beta 1$-integrin fibronectin adhesive bond strength, which is within a factor of 2 of the single molecule measurements. Since the peeling model involves detachment of successive groups of bonds, it would require much less total force. Peeling detachment may occur in some cases, but it does not appear to be the dominant mode in our spinning disc experiments.

4.3. Equilibrium and rates

1. Equilibrium constants can be measured for the interaction of proteins in solution by saturation binding. Under receptor-limiting conditions, a plot of the number of bonds that form as a function of ligand density is represented by the hyperbolic function $R_B = R_T L/(K_D + L)$, where R_B is the number of bonds, R_T is the total receptors available, L is ligand concentration (density), and K_D is the dissociation constant. The process of cell adhesion involves complex regulation of the adhesive bonds by interaction of the receptors with cytoplasmic proteins (Kim *et al.*, 2003; Liddington *et al.*, 2002), This means that the equilibrium point (or energy minimum) will be controlled by reactions other than ligand binding itself, and makes the equilibrium a moving target. The τ_{50} measured with the spinning disc gives a snapshot of the number of bonds and hence can be used as a quasi-equilibrium analysis. This assumes that the exchange rate for the integrin ligand bond is faster than the other reactions controlling the binding (e.g., focal complex formation), which results in the ligand binding reaction being close to equilibrium. Figure 1.4A shows the mean shear stress required for cell detachment (τ_{50}) as a function of ligand density for $\alpha 5\beta 1$ on HT1080 cells binding to fibronectin. When K_D is $\gg$ L, the plot will be linear as shown. Both linear and hyperbolic fits of the data give R^2 values of 0.994. Figure 1.4B shows

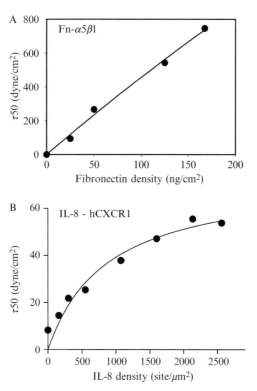

Figure 1.4 (A) HT1080 cells were plated on fibronectin 60 min and analyzed in the spinning disc device. Curve fit for linear regression $R^2 = 0.9938$; for hyperbolic fit $R^2 = 0.9945$; and the apparent K_D would be 3400 ng/cm². (B) 300.19 cells expressing CXCR1 were plated on an IL-8 surface and adhesion was analyzed using the spinning disc. B is reproduced with permission from Le *et al.* (2004).

a similar analysis for the interaction of a form of IL-8 that has been linked to a surface with its cell receptor CXCR1 with an evident hyperbolic shape. Therefore, the spinning disc approach is able to generate the expected hyperbolic curves that are distinct from linearity when sufficiently high ligand density can be achieved. The apparent linearity of Fig. 1.4A is actually a portion of the hyperbolic plot when the ligand density is much less than the apparent K_D of the reaction. While the inability to generate binding data for ligand densities greater than the K_D, limits the accuracy of the estimate of the apparent K_D, it does constrain the possible values. These estimates for the apparent K_D provide a means of moving from relative measures of bond number to actual estimates of adhesive bonds/cell.

2. Another form of the analysis that can be informative is studying the rate of bond formation. This is obtained by making the measurements of τ_{50} as a function of time. Since this generally involves a change in cell shape, and changes in cell shape will affect the relationship between the

hydrodynamic shear stress and the force applied per cell, it is necessary to consider transforming the data from shear stress to force as discussed ion the next section. The plot of mean cell detachment forces as a function of time can be fit to the first-order rate equation, $R_B = R_T(1 - e^{-bt})$, where R_T is the maximum bonds/cell, b is a rate fitting parameter, and t is time. The time for 50% R_T bonds is ln2/b.

4.4. Making comparisons

In order to compare different cell lines or even the same cell under different treatment conditions that affect cell morphology, it is necessary to convert hydrodynamic shear stress values to applied force per cell. Hydrodynamic shear applies force to the cell through two mechanisms: (1) pushing against the cell on the face exposed to flow directly that tends to push the cell over causing a rolling component, which is called torque (T), and (2) the friction or drag of the medium as it flows over the surface of the cell (F_D). Taking into account the complexities of cell morphology would be too mathematically complex; hence we use geometric approximations of cell shape. The best approximation (compromise) we have found is a model that approximates cell shape as a hemispherical cap (a sphere that has been cut by a plane which represents the substrate interface) (Truskey and Proulx, 1993). For this model:

$$F_D = 4.33\pi r^2 \tau \, (\text{drag component}) \tag{3}$$

$$T = 2.44\pi r^2 h \tau \, (\text{torque component}) \tag{4}$$

$$F_y = (0.75)/a(T + F_D h)(\text{force perpendicular to the plane of flow}) \tag{5}$$

$$F_T = (F_D^2 + F_y^2)(\text{total force is the vector addition of forces}) \tag{6}$$

F_D is drag, r is the radius of the sphere, τ is shear stress, T is torque, h is height of the cap, a is the radius of the contact area where the plane cuts the sphere, and F_T is the total force. In cells we have examined, by 15 min, $a = r$, that is, the contact area is equal to the projected area. When $a < r$, the torque dominates; when $a = r$ as the cell spreads, the drag component becomes dominant (a cannot be greater than r). The drag is proportional to the exposed surface area; hence specific morphology of the individual cell has less influence and the method provides a reasonable approximation.

To see what the range of forces looks like, K562 CML cells were compared with HT1080 fibroblastic tumor cells. K562 cells have a radius (r) of 7.7 μm and spread only until radius of the contact area (a) is also ~7.7 μm. For HT1080 cells, $r = 10$ μm, and for the spread cell, it reaches 19.5 μm. For K562, $F_T = 4573\tau$, and for spread HT1080 cells, $F_T = 8537\tau$. Taking the measurements of τ_{50} for these cell population and substituting for τ, K562 = (4523)(9) pN = 40.7 nN/cell; HT1080 = (8537)(80) = 683 nN/cell.

(Note that pN is piconewtons, nN is nanonewtons, units for length are in micrometers, and shear stress is in piconewtons per micrometers squared [1 pN/μm^2 = 10 dyne/cm^2]). A similar cell detachment force for K562 cells was measured using a high-speed centrifugation technique, showing that the hemispherical cap model does give a reasonable measurement of applied force. If the (breaking) strength of a single $\alpha 5\beta 1$-fibronectin bond is 60 pN (Li *et al.*, 2003), then 11,400 bonds would be required to give 683 nN. Since the breaking strength of single bonds is measured with an atomic force microscope, the rate of pulling on the bond will be much faster than for the cell or the spinning disc, and hence the 60pN value would be an overestimate of the bond strength under the conditions discussed here (Evans, 2001).

Comparisons between cells also need to take into account the levels of surface integrin receptor that could be involved in the binding. FACS analyses can be used to compare relative levels of surface expression for the specific integrin receptor involved. Also, as the cells change shape there may be changes in the proportion of the cell surface that is in contact with the ligand surface. This would also generate a proportional change in the number of integrin receptors per cell that could participate in binding. Finally, the redistribution of integrins into the contact area has been documented in some T-cell experiments, although we do not see this occurring for HT1080 fibroblasts (Michl *et al.*, 1983; *et al.*, manuscript in preparation).

ACKNOWLEDGMENTS

Substantial contributions to the spinning disc technology were made by Andres Garcia, Laura Lynch, Francois Huber, Mark Lee, Andriban Datta, Todd Miller, Marina Guvakova, May Chan, Qi Shi, and Deshan Yu. The research was supported by the National Cancer Institute (grant RO1 CA 16502) and General Medical Sciences (grants GM 49866 and GM 57388).

APPENDIX 1. REAGENTS

PBS–ABC (Dulbecco's PBS) (8 parts A 1 part B 1 part C). Each part can be autoclaved before mixing.
PBS-A (5 liters):
NaCl, 50.0 g
KCl, 1.25 g
Na$_2$HPO$_4$.7H$_2$O (dibasic), 7.23 g
KH$_2$PO$_4$ (monobasic), 1.25 g
PBS-B CaCl$_2$.2H$_2$O (to make 500 ml) 0.67 g
PBS-C MgCl2.6H$_2$O (to make 500 ml) 1.0 g
PBS-A + EDTA
0.1 mM EDTA; 2 mM glucose; 8 parts PBS-A, 2 parts H$_2$O

HEPES Adhesion Buffer (1 liter, filter sterilize)
$1M$ HEPES pH 7.4, 24 ml
5 M NaCl, 27.4 ml
0.1 M KCl, 27 ml
200 mM glucose, 10 ml
1 mM MgCl$_2$, or 1 mM MgCl$_2$

Dextran Adhesion Buffer (1 liter) (Dextran Sigma catalog # D1037)
50 g/liter of adhesion buffer (PBS-ABC, or HEPES adhesion buffer).
Add dextran slowly; it takes several hours to dissolve. Filter sterilize.

Glass Coverslips
Bellco Glass, 25-mm round #1, catalog # 1943–10025.

Plastic Coverslips
Sarstedt tissue culture–treated 25 mm coverslips, catalog # 88.1840

Human plasma fibronectin (Fn)
(Invitrogen, catalog # 33016–015). Reconstitute to 1 mg/ml in ddH$_2$O.
Allow solution to dissolve for 30 min; if solution is cloudy or protein
does not dissolve completely (some gentle shaking might be necessary),
discard the solution. Aliquot into microcentrifuge tubes and flash-freeze
in liquid nitrogen. Store at −80°.

1% Bovine serum albumin (BSA)(Fraction V; Sigma catalog # A7030)
Dissolve 1 g of BSA in 100 ml of complete PBS. Heat-inactivate for
30 min at 56° and filter sterilize.

Collagen (Vitrogen Bovine Collagen, Cohesion Technologies, catalog #
FXP-019)
Dilute 1:30 in sterile ddH$_2$O and use immediately.

Nitrocellulose solution
Use Millipore Immobilon transfer membrane (catalog # 1PVH00010),
and dissolve 1 cm^2/ml in methanol. Use 300 μl/25-mm coverslip.

Ethidium homodimer (Molecular Probes, catalog # E1169)
Dissolve in DMSO at 1 mg/ml and store at −20°. Use at 1:500 in PBS-A.

DAPI
300 nM in PBS-A

Formaldehyde
3.7% v/v in PBS-A

Gel mount (Fisher Scientific, BioMeda, BM-M01)

Silicone grease (Fisher Scientific, Corning High-Vacuum Grease, 14–635–5D)

 ## APPENDIX 2. SPINNING DISC DEVICE

The device design given below will deliver a linear shear gradient up
to 6000 RPM in the PBS-ABC or HEPES adhesion buffer and up to
~8500 RPM in the 5% Dextran buffer. It represents a design we have
optimized for analysis of adherent cells. The shaft (blue) and support

(yellow) including the bearings (grey) will require a good machine shop and machinist to construct as alignments are critical (See color inset for color coding). The spinning chamber (green) is relatively easy to assemble for a person accustomed to working with plexiglass.

Chamber Parts List

See Fig. 1.5.
    Chamber 5″ diameter plexiglass tube 7.5″ long; base 8″ square, 3/16″ plexiglass; 4 1/8″ plexiglass baffles, 1″ × 3–1/2″

Spinning platform – motor and shaft support

Plexiglass circles: 1/8″ plexiglass 4–7/8″ diameter with 3/4″ center hole and holes to mount 3 posts; 11/16″ plexiglass, 4–7/8″ diameter with bearing mount recess and holes for mounting 6 posts (3 up and 3 down); 11/16″ plexiglass, 6″ diameter with bearing mount and holes for 6 posts (3 up and 3 down); 3/16″ plexiglass, 5″ diameter, 3/4″ hole in center, and holes for motor mount and 3 posts.

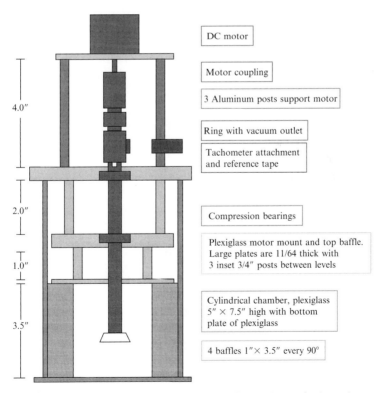

DC motor	
Motor coupling	
3 Aluminum posts support motor	
Ring with vacuum outlet	
Tachometer attachment and reference tape	
Compression bearings	
Plexiglass motor mount and top baffle. Large plates are 11/64 thick with 3 inset 3/4″ posts between levels	
Cylindrical chamber, plexiglass 5″ × 7.5″ high with bottom plate of plexiglass	
4 baffles 1″× 3.5″ every 90°	

4.0″
2.0″
1.0″
3.5″

Figure 1.5   Diagram of the spinning disc device. (See color insert.)

Posts: mounted to separate circles, Bottom (3) 11/16″ diameter X 1 3/16″ threaded plexiglass posts; middle (3) 5/8″ diameter × 2 1/4 threaded Derlin posts recessed 3/32″ each end into the plexiglass circles; upper (3) 1/2″ diameter × 4″ threaded aluminin posts.

Optional aluminum support for chamber and device is shown in Fig. 1.2C.

Shaft

9″ Derlin shaft stock, 3/4″ diameter, needs to be machined with a center hole directed to the side at the motor end to apply a vacuum, and machined to fit the bearings. Our shafts are tapered to 3/8″ and threaded to fit a spinning platform (Fig. 1.2A).

Bearings

Koyo 7201BGC3 S9903 (Pruyn Bearings Co., 1324–28 Frankfors Ave., Philadelphia, PA 19125)

Note: The dimensions of the chamber with baffles are critical to get laminar flow over the RPM range. This design has been validated using electro chemical transport. For the method, see Garcia *et al.* (1997). The objective of the motor mount is to provide a stable motor with minimal vibration. A center hole in the shaft is important to apply vacuum to the disc so that it stays on during the spin. Variations on the design are possible.

Spinning Disc Controller Unit Parts List

For circuit diagram see Fig. 1.6.

Motor and control unit (B.E.I./Kimco Magnetics Division, 804-A Roncheros Dr., San Marcos, CA 92069)

BLDC motor	BEI KIMCO	DIH23-16-BBNA
Commutator	BEI KIMCO	ECH05-30-001CL
D–sub 9-pin female connector	DIGIKEY	A2047-ND
Cable assembly 9-pin M-M	DIGIKEY	AE1019-ND

Power supply and voltage control

Power Supply Cosel RMB50U-2	ALLIED	800–1045
SPST rocker switch, red (run)	DIGIKEY	SW344-ND
SPST rocker switch, white (power)	DIGIKEY	SW345-ND
SPST rocker switch, blue	DIGIKEY	SW347-ND
Red panel LED	RADIO SHACK	
Green panel LED	RADIO SHACK	
Ampmeter 0–3A	ALLIED	701–0582
C1 Capacitor 2200 uF	DIGIKEY	P1160-ND
R1 10K 1/4-W resistor		
R2 50K trimmer pot	DIGIKEY	201XR503B-ND
R3 5.1 KΩ resistor	DIGIKEY	5.1K 1.0W-1-ND
R4 25K pot Clarostat 10-turn	ALLIED	753–0345
Knob 10-turn calibrated	DIGIKEY	411KL-ND

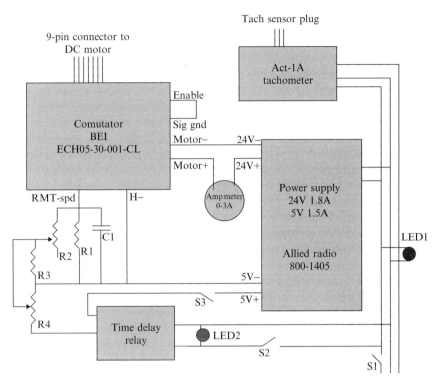

Figure 1.6 Wiring diagram for motor controller and tachometer. For specific components, see parts list. (See color insert.)

Timer (time to switch-off)

Timer 1–10 min	DIGIKEY	Z931-ND
Timer panel mount	DIGIKEY	Z885-ND
Timer socket	DIGIKEY	Z883-ND

Tachometer (Monarch Instrument, 15 Columbia Dr., Amherst, NH 03031)

Digital tachometer ACT-1A/ 115	Servo Systems	6140–010
Remote optical sensor	Servo Systems 6180– 015	
3 conductor plug	DIGIKEY	SC1083-ND
Jack for 3 conductor plug	DIGIKEY	SC1107-ND

This system will provide good speed control and calibration (Fig. 1.2C). A mounting box is desirable. Front controls include speed (10-turn potentiometer), three switches, two LED, ampmeter (if used, the current draw will rise as the bearings become worn and need replacement), timer, and connections for motor and tach sensor. Simpler DC motor systems are possible, but the tachometer is particularly recommended and easy to set up. The DC motor is important for speed regulation and control. The BEI motor requires the commutator. BEI engineers are helpful with wiring issues. The estimated total cost of parts is $800 with motor timer and tachometer.

REFERENCES

Boettiger, D. (2006). Use of hydrodynamic shear stress to analyze cell adhesion. In "Understanding the Biomolecular Interface" (M. King, ed.). Academic Press, New York.

Boettiger, D., Lynch, L., Blystone, S., and Huber, F. (2001). Distinct ligand-binding modes for integrin $\alpha(v)\beta(3)$-mediated adhesion to fibronectin versus vitronectin. J. Biol. Chem. **276,** 31684–31690.

Choquet, D., Felsenfield, D. P., and Sheetz, M. P. (1997). Extracellular matrix rigidity causes strengthening of integrin-cytoskeletal linkages. Cell **88,** 39–48.

DePasquale, J. A., and Izzard, C. S. (1987). Evidence for an actin-containing cytoplasmic precursor of the focal contact and the timing of incorporation of vinculin at the focal contact. J. Cell Biol. **105,** 2803–2809.

Discher, D. E., Janmey, P., and Wang, Y. L. (2005). Tissue cells feel and respond to the stiffness of their substrate. Science **310,** 1139–1143.

Evans, E. (2001). Probing the relation between force—lifetime—and chemistry in single molecular bonds. Annu. Rev. Biophys. Biomol. Struc. **30,** 105–128.

Faull, R. J., Kovach, N. L., Harlan, J. M., and Ginsberg, M. H. (1993). Affinity modulation of integrin $\alpha5$ $\beta1$: Regulation of the functional response by soluble fibronectin. J. Cell Biol. **121,** 155–162.

Garcia, A. J., and Boettiger, D. (1999). Integrin-fibronectin interactions at the cell-material interface: Initial integrin binding and signaling. Biomaterials **20,** 2427–2433.

Garcia, A. J., Ducheyne, P., and Boettiger, D. (1997). Quantification of cell adhesion using a spinning disc device and application to surface-reactive materials. Biomaterials **18,** 1091–1098.

Garcia, A. J., Huber, F., and Boettiger, D. (1998). Force required to break $\alpha5\beta1$ integrin-fibronectin bonds in intact adherent cells is sensitive to integrin activation state. J. Biol. Chem. **273,** 10988–10993.

Garcia, A. J., Vega, M. D., and Boettiger, D. (1999). Modulation of cell proliferation and differentiation through substrate-dependent changes in fibronectin conformation. Mol. Biol. Cell **10,** 785–798.

Griffin, M. A., Engler, A. J., Barber, T. A., Healy, K. E., Sweeney, H. L., and Discher, D. E. (2004). Patterning, prestress, and peeling dynamics of myocytes. Biophys. J. **86,** 1209–1222.

Humphries, M. J. (1998). Cell Adhesion. In "Current Protocols in Cell Biology" (J. S. Bonifacino, M. Dasso, J. B. Harford, J. Lippincott-Schwartz, and K. M. Yamada, eds.). John Wiley & Sons Inc., New York.

Ingber, D. E. (2003). Tensegrity II. How structural networks influence cellular information processing networks. J. Cell Sci. **116,** 1397–1408.

Kim, M., Carman, C. V., and Springer, T. A. (2003). Bidirectional transmembrane signaling by cytoplasmic domain separation in integrins. Science **301,** 1720–1725.

Lee, F. H., Haskell, C., Chario, I. F., and Boettiger, D. (2004). Receptor-ligand binding in the cell-substrate contact zone: A quantitative analysis using CX3CR1 and CXCR1 chemokine receptors. Biochemistry **43,** 7179–7186.

Lee, M. H., Ducheyne, P., Lynch, L., Boettiger, D., and Composto, R. J. (2006). Effect of biomaterial surface properties on fibronectin-$\alpha5\beta1$ integrin interaction and cellular attachment. *Biomaterials* **27**, 1907–1916.

Li, F., Redick, S. D., Erickson, H. P., and Moy, V. T. (2003). Force measurements of the $\alpha(5)\beta(1)$ integrin-fibronectin interaction. *Biophys. J.* **84**, 1252–1262.

Liddington, R. C., and Ginsberg, M. H. (2002). Integrin activation takes shape. *J. Cell Biol.* **158**, 833–839.

Litvinov, R. I., Shuman, H., Bennett, J. S., and Weisel, J. W. (2002). Binding strength and activation state of single fibrinogen-integrin pairs on living cells. *Proc. Natl. Acad. Sci. USA* **99**, 7426–7431.

Lots, M. M., Burdsal, C. A., Erickson, H. P., and McClay, D. R. (1989). Cell adhesion to fibronectin and tenascin: Quantitative measurements of initial binding and subsequent strengthening response. *J. Cell Biol.* **109**, 1795–1805.

Michl, J., Pieczonka, M. M., Unkeless, J. C., Bell, G. I., and Silverstein, S. C. (1983). Fc receptor modulation in mononuclear phagocytes maintained on immobilized immune complexes occurs by diffusion of the receptor molecule. *J. Exp. Med.* **157**, 2121–2139.

Miller, T., and Boettiger, D. (2003). Control of intracellular signaling by modulation of fibronectin conformation at the cell-materials interface. *Langmuir* **19**, 1730–1737.

Paszek, M. J., Zahir, N., Johnson, K. R., Lakins, J. N., Rozenberg, G. I., Gefen, A., Reinhart-King, C. A., Margulies, S. S., Dembo, M., Boettiger, D., Hammer, D. A., and Weaver, V. M. (2005). Tensional homeostasis and the malignant phenotype. *Cancer Cell* **8**, 241–254.

Pierschbacher, M. D., and Ruoslahti, E. (1984). Cell attachment activity of fibronectin can be duplicated by small synthetic fragments of the molecule. *Nature* **309**, 30–33.

Redick, S. D., Settles, D. L., Briscoe, G., and Erickson, H. P. (2000). Defining fibronectin's cell adhesion synergy site by site-directed mutagenesis. *J. Cell Biol.* **149**, 521–527.

Reyes, C. D., and Garcia, A. J. (2003). A centrifugation cell adhesion assay for high-throughput screening of biomaterial surfaces. *J. Biomed. Mater. Res.* **67A**, 328–333.

Riveline, D., Zamir, E., Balaban, N., Kam, Z., Geiger, B., and Bershadsky, A. (2001). Focal contact as a mechanosenser: Directional growth in response to local strain. *Mol. Biol. Cell* **10**(suppl.), 341a.

Shi, Q., and Boettiger, D. (2003). A novel mode for integrin-mediated signaling: Tethering is required for phosphorylation of FAK Y397. *Mol. Biol. Cell* **14**, 4306–4315.

Takagi, J., Petre, B. M., Walz, T., and Springer, T. A. (2002). Global conformational rearrangements in integrin extracellular domains in outside-in and inside-out signaling. *Cell* **110**, 599–611.

Truskey, G. A., and Proulx, T. L. (1993). Relationship between 3T3 cell spreading and the strength of adhesion on glass and silane surfaces. *Biomaterials* **14**, 243–254.

Weisel, J. W., Shuman, H., and Litvinov, T. I. (2003). Protein–protein unbinding induced by force: Single-molecule studies. *Curr. Opin. Struct. Biol.* **13**, 227–235.

Xiong, J. P., Stehle, T., Goodman, S. L., and Arnaout, M. A. (2003). New insights into the structural basis of integrin activation. *Blood* **102**, 1155–1159.

CHAPTER TWO

INVESTIGATING INTEGRIN REGULATION AND SIGNALING EVENTS IN THREE-DIMENSIONAL SYSTEMS

Patricia J. Keely, Matthew W. Conklin, Scott Gehler, Suzanne M. Ponik, *and* Paolo P. Provenzano

Contents

Department of Pharmacology, University of Wisconsin Medical School, University of Wisconsin–Madison, Madison, Wisconsin

Methods in Enzymology, Volume 426
ISSN 0076-6879, DOI: 10.1016/S0076-6879(07)26002-1

© 2007 Elsevier Inc.
All rights reserved.

Abstract

There has been much recent interest in working with cells cultured in three-dimensional (3D) matrices to better model the properties of the extracellular matrix environment found *in vivo*. However, working within 3D matrices adds several difficulties to experiments that have become routine in two-dimensional (2D) culture systems. Biochemical approaches are made difficult by the presence of milligram quantities of matrix protein, while cell biology approaches are more difficult to assess and image. Moreover, 3D imaging adds complexity to fluorescence studies, including the inherent challenge of a 3D volume as opposed to a 2D image, increased depths of field, and problems of light scatter. The purpose of this chapter is to provide a few overall strategies for working within 3D culture systems, focusing on biochemical and molecular imaging approaches.

1. INTRODUCTION

Integrin localization and function has been historically studied on rigid two-dimensional surfaces coated with various extracellular matrix (ECM) proteins. Early identification of integrins as key architects of focal adhesions relied on their immunolocalization within cells cultured in this manner (Burridge *et al.*, 1992). Indeed, the rapid advances in the field of integrin signaling, imaging of molecular events underlying cell migration, and understanding of focal adhesions as signaling scaffolds would not have been possible without these two dimensional (2D) culture systems.

Despite the value of classic culture approaches, it is becoming increasingly apparent that the response of cells cultured in three-dimensional (3D) matrices is not completely consistent with responses on coated 2D surfaces (Weaver *et al.*, 2002; Wozniak *et al.*, 2003). Cellular responses in 3D systems are driven in no small part by contractility-based signaling linked to the Rho small GTPase (Wozniak *et al.*, 2003). Given the extreme rigidity of a 2D surface, regulation by contractile events is likely to be quite different in 3D systems versus 2D. Indeed, it appears that the "compliance" of the 3D matrix is a key aspect of cellular response and signaling regulation (Paszek *et al.*, 2005; Wozniak *et al.*, 2003). Moreover, while cells make adhesive structures in 3D matrices, it is clear that these structures are smaller and differ from their 2D counterparts, and thus have been termed "matrix adhesions" rather than "focal adhesions" (Cukierman *et al.*, 2001).

Working within 3D matrices adds several difficulties to experiments that have become routine in 2D culture systems. Biochemical approaches are made difficult by the presence of milligram quantities of matrix protein, which get in the way of cellular proteins during routine analysis. Cell biology is more difficult to assess and image. Moreover, 3D imaging adds complexity to fluorescence studies, including the inherent challenge of a 3D volume as opposed to a 2D image, increased depths of field, and problems of light scatter. The purpose of this chapter is to provide a few overall strategies for working within 3D culture systems, as well as some specific approaches in use in our laboratory. This is not to say that 2D systems are not important complements to the 3D approach, and indeed we find that some of our studies are still best accomplished under 2D culture systems (see section 6). In this chapter, we will address the issues of performing biochemistry, cell biology studies, and imaging within the context of 3D collagen matrices. Included are some basic nonlinear optical approaches for imaging within 3D matrices using multiphoton microscopy.

1.1. Types of 3D matrices

A few different types of three-dimensional (3D) ECM culture systems have been developed to great advantage. Commonly used are 3D gels composed of rBM or matrigel, a purification of the basement membrane components made in abundance by the EHS tumor and containing laminin, collagen IV, pro- teoglycans, and several other more minor components (Kleinman *et al.*, 1982). Matrigel or rBM is commercially available, or can be produced in one's own lab, but with particular resources needed to generate and harvest tumors and purify the basement membrane fraction. While rBM has several applications for the study of epithelial structure (e.g., see Barcellos-Hoff *et al.*, 1989; Muthuswamy *et al.*, 2001), gelling properties vary tremendously lot to lot, and the commercial supply is sometimes inconsistent. Moreover, for our purposes, we find that these gels are too compliant for addressing our particular questions related to mechanical signal transduction, and do not seem to adequately mimic certain *in vivo* tissue environments.

Gels composed of Type I collagen, derived from rat tails, are a less expensive and more consistent source of 3D matrix. Collagen gels can be cast in concentrations ranging from ∼0.7 mg/ml up to 5 mg/ml (if one has a sufficiently concentrated source of collagen I), which makes them readily customized to the specific structural needs of each individual cell type and the scientific question at hand (see "Collagen Gels of Varying Densities," below). Thus, collagen gels can be created that cross a range of elastic moduli appropriate to mimic the physical properties of various tissue environments.

The disadvantage of collagen gels is that they do not specifically reproduce the biochemical components of the basement membrane in the way that rBM/matrigel does. However, we have found that after a period of time (≥ 7 days, usually), "normal" mammary epithelial cells will deposit a thin basement membrane, and thus provide a decent mimic of the 0.2-μm thick basement membrane found *in vivo*. Alternatives are to make mixed gels of matrigel/collagen I, or to "precoat" clumps of cells by incubating them for 10 min in ungelled matrigel, prior to casting them into collagen I gels. It should be noted, however, that mixing matrigel with collagen gels will alter the physical properties of the resultant gel.

An alternative approach to such "exogenous" matrices is to allow cells to deposit their own 3D matrix. The most commonly used are fibroblasts, already specialized for making and assembling 3D matrices. Once a matrix is created, the cells are extracted and the resulting cell-derived matrix used in subsequent experiments. This approach will not be specifically discussed further here, but the reader is encouraged to see Cukierman *et al.* (2001) for more information.

1.2. Challenges of working in 3D matrices

Several challenges must be overcome to obtain consistent and satisfying results from experiments performed with cells cultured in 3D matrices. Standard biochemical experiments, such as immunoprecipitations, pull-down assays, and molecular signaling events (see sections 4 and 5) can be complicated by the presence of milligram quantities of ECM protein in the lysate. Following are examples of issues that we have encountered:

- **2D versus 3D.** In comparisons of samples from cells cultured in 2D versus a 3D matrix, the presence of contaminant ECM protein is so abundant in the 3D sample that this sample will run aberrantly on SDS-PAGE compared to the 2D sample. To solve this, we create an equal-volume "dummy" gel/matrix that is added to the 2D sample upon lysis.
- **Releasing cells from 3D gels with collagenase (or other proteases).** It is possible to recover cells cultured in 3D collagen gels by treating the gels with collagenase for about 15 to 30 min, according to the specific activity of the collagenase purchased and the concentration of the gel. This approach is acceptable for a subset of experiments, such as cellular growth curves, in which disrupting the cell–matrix interactions will not alter the outcome of the experiments. However, this approach is not appropriate for signaling studies, gene expression studies, or even protein expression, as 15 to 30 min in a protease is enough time for the cell to completely change the response being investigated. Thus, we lyse cells in residence within the 3D collagen gel.
- **Protein assays are not recommended.** Unfortunately, there is far too much contaminating ECM protein, present in milligram quantities, to ever get an accurate read by any standard protein assay.
- **Differences in cell growth rates.** When comparing different conditions in 3D over several days prior to cell lysis, a complicating factor is that sometimes the growth rates differ (Wozniak *et al.*, 2003), such that one is not comparing equal cell numbers by the time the experiment is performed. Because it is not possible to determine equal protein levels in the samples, alternative approaches must be used. One solution is to culture a parallel set of gels from which cells will be released by collagenase treatment and counted, or that will be stained with DAPI and cell number counted via nuclei. This allows adjustment of sample volumes either at lysis or when loading SDS-PAGE gels.

- **Key to all experiments is inclusion of loading controls, such as parallel blot for GAPDH.** We do not favor actin as a reference protein, as we find changes in actin levels across cells cultured in 2D versus 3D.
- **RNA isolation remarkably easier than protein work in 3D matrices.** We find that using standard RNA reagents, such as Trizol (InVitrogen) and following the manufacturer's suggestion for isolation of RNA from tissue, rather than from cell culture, works quite well.

Cell biology and imaging approaches are also more difficult in 3D matrices, but there are approaches to work around these difficulties.

- Phase-contrast images of cells in 3D matrices complicated by lack of single focal plane, making satisfactory images difficult. This is particularly true when working with floating gels, which do not stay still long enough to acquire an adequate image. To assist in this, we find that removing the liquid from the culture, and allowing the 3D gel to "sit" on the bottom of the culture dish aids in obtaining better phase-contrast images.
- Immunostaining. Satisfying results can be obtained immunostaining within 3D matrices, without the need to section (Wozniak *et al.*, 2003). We have previously developed and published protocols for this (Wozniak and Keely, 2005). The key point is that the gels must be handled delicately, and washed much more extensively, than when working on 2D surfaces.
- Fluorescence microscopy. There are inherent issues imaging a cell that has significant 3D volume compared to one that is relatively 2D. Imaging of immunofluorescent staining, or GFP-labeled proteins, by standard epifluorescence is vastly improved by the use of 3D deconvolution software or the use of a confocal microscope. We find identical results whether we use confocal microscopy or standard epifluorescence combined with 3D deconvolution analysis; the latter approach is used, as it is resident within our laboratory.
- Multiphoton laser scanning microscopy (MPLSM). In addition to the inherent challenge of a 3D volume as opposed to a 2D image, low signals and light scatter create challenges when imaging 3D samples. The use of MPLSM mitigates much of the difficulties of working within 3D samples, and also has the added benefit of allowing the simultaneous imaging of unstained collagen matrices (see "Imaging Cell–Matrix Interactions in 3D Collagen Gels" below).

2. COLLAGEN MATRICES OF DIFFERENT DENSITIES

For the purposes of our particular set of studies, in which we are investigating the effects of collagen density on cell behavior, we often compare cells cultured in collagen gels of varying densities. Gels composed of rat tail collagen can be poured in concentrations from ~0.7 mg/ml to several milligrams per milliliter. In practical terms, we have not exceeded 5 mg/ml of collagen, as it is necessary to have a concentrated stock from which to make these gels. It is our experience that the physical properties, or elastic modulus, of these gels range in a predictable way that matches published data (Roeder *et al.*, 2002). The ability to cast gels in a wide range of concentrations, and thus achieve a range of elastic moduli, is one of the advantages of working with gels composed of Type I collagen.

We find that different cells vary in their contractile behavior, and thus the optimum concentration of collagen for a desired cellular behavior varies as well, and may need to be determined empirically. For example, T47D breast carcinoma cells, which are not very contractile, undergo tubulogenesis at an optimum collagen concentration of ~1.0 to 1.3 mg/ml. In contrast, this concentration is a bit too compliant for the more contractile normal murine mammary cell line, NMuMG, which undergoes optimum tubulogenesis ~3.0 mg/ml. Fibroblasts are generally more contractile as well, and prefer gels of higher collagen concentrations.

As collagen is acid soluble, sterile solutions are usually commercially available in dilute acid (i.e., BD Biosciences, San Jose, CA provides type I collagen in 0.03 N acetic acid at concentrations of 3 to 5 mg/ml or 8 to 11 mg/ml). Collagen will gel spontaneously upon neutralization at 37°. To accomplish this, we neutralize collagen by the addition of an equal volume of 2× PBS/100 mM HEPES at pH 7.3. To make more concentrated gels, one-half volume of 4× PBS/200 mM HEPES can be used. Gels can be cast into tissue culture dishes of various sizes, but typically we use six-well plates and cast a total volume of 1 ml per well. Thus, to cast a 2-mg/ml collagen gel from a 5 mg/ml stock solution, one would use 0.4 ml of collagen, 0.4 ml 2× PBS/100 mM HEPES, and the needed number of cells in 0.2 ml of medium. For tubulogenesis assays with mammary epithelial cells, we generally cast 100,000 to 125,000 cells/ml of gel. Depending on the specific experiment, cell concentrations up to 10×10^6 cells/ml have been used.

When casting collagen gels, it is important to keep the components cold or to work quickly, to mix thoroughly being careful not to introduce air bubbles, and to add the cells last to the mixture. The viscous solution of neutralized collagen plus cells is immediately pipetted into individual wells, and the tissue culture dish placed in a 37° incubator for 60 min to overnight to allow the gel to polymerize. For floating gels, the polymerized gels are gently released by edging the well with a Pasteur pipette, adding culture medium (2 ml for a 1-ml gel), and gently shaking the gels to release them from the sides and bottom of the dish. For more detail, an excellent protocol for pouring collagen gels has already been published (Wozniak and Keely, 2005).

3. GEL CONTRACTION AS A MEASURE TO QUANTIFY CELL CONTRACTILITY

Because much of the response of epithelial cells to 3D matrices, and the ability of the cells to sense the density of the matrix appears to involve signaling through Rho-mediated contractile events (Wozniak *et al.*, 2003), measurements of collagen gel contraction is a useful approach to assess the response of cells to a particular matrix. Moreover, collagen gel contraction allows the assessment of the contractility of a population of cells relative to another set of cells, or relative to a different set of culture conditions.

Collagen gels are cast in a six-well plate as described above for morphogenesis assays (e.g., 100,000 T47D cells/1 ml of collagen gel). Denser gels will be contracted to a lesser extent, and this should be taken into consideration when deciding the purpose of the experiment. Gels are polymerized overnight at 37°, and released the next day. The day at which the gels are rendered floating is considered Day 0. For example, if the gels are poured in a six-well plate, the diameter of a gel at Day 0 would be 35 mm. The gel diameter should be measured every 24 h for up to 10 days. Generally, the culture medium is changed every 4th day. Cells that are more highly contractile should be followed for shorter intervals. Measurements may be taken using a small ruler or using imaging software on gel images acquired from a dissecting microscope. In general, T47D breast epithelial cells will contract a 1.3-mg/ml collagen gel ~7.5 mm in 8 days, while the more contractile MCF10A cell line will contract a similar gel ~10 mm (Wozniak *et al.*, 2003).

4. RHO ACTIVITY ASSAY FROM CELLS CULTURED IN 3D COLLAGEN GELS

Recently we have been able to optimize the G-LISA RhoA activation assay (luminescence based) from Cytoskeleton, Inc. (Denver, CO) to analyze RhoA activity using breast epithelial cells cultured within 3D collagen gels. In the past the only way to analyze the activity of RhoGTPase was to perform a very labor-intensive Rho pull-down assay. While a very reproducible Rho pull-down protocol was developed in our lab to assay breast epithelial cells in 3D collagen gels (Wozniak and Keely, 2005), the assay had several limitations. In order to produce a detectable signal, the pull-down assay required the use of approximately 10 to 20 million cells per 1 ml of gel. A concern with this approach was the effects on cellular signaling due to a cell density far greater than that normally used for these 3D collagen-tubulogenesis assays. Due to the high cell number needed for this assay, analysis was restricted to time

points that could be completed within a few hours and could not be carried out
to 10 days (the time needed for tubulogenesis to occur). Using the G-LISA
RhoA assay, not only can significantly lower cell numbers be used for short-
term experiments (i.e., 1 h), but importantly the cells can be cultured in 3D
matrices for 10 or more days and Rho activation can be measured at the same
time point that tubulogenesis is assessed.

4.1. Cell culture

The example given here uses T47D breast carcinoma cells, but can be
adapted for several other types of cells. For the 1-h protocol, cells are
harvested in 0.5 m-M EDTA in PBS, resuspended in RPMI media supplemented with 5 mg/ml of fatty acid–free BSA, and just 2 million cells per
1 ml of gel (cultured in a six-well plate) are used for each condition to be
analyzed. The collagen gels are allowed to polymerize for 1 h at 37° after
which 1 ml of RPMI/BSA is added, and the gels rendered floating by gently
displacing their attachment to the dish, or left attached. Following 1 h of
additional incubation at 37° the gels are ready to be assayed.

   To measure RhoA activity after 10 days of culture in 3D collagen gels,
cells are cast into 3D collagen gels using the same number of cells that we
typically use for tubulogenesis endpoints: 100,000 cells per 1 ml of gel. Using
full growth media (RPMI with 10% heat-inactivated FBS and 8 mg/ml of
insulin), 1-ml gels are poured and allowed to polymerize for 8 to 16 h prior
to rendering them floating or leaving them attached in 2 ml of full media.
After 10 days in culture, replacing 1 ml of full media every 4 days, the gels are
ready to be assayed.

4.2. G-LISA assay

The protocol provided in the G-LISA RhoA activity assay (Cytoskeleton,
Inc., Denver, CO) is generally followed with a few modifications. Briefly, the
gel is carefully removed from the well with forceps and quickly rinsed in
approximately 2 ml of ice cold PBS. Residual PBS is removed by carefully
blotting the edge of the gel on a Kim wipe, and then the gel is promptly lysed
on ice in a microfuge tube containing 400 μl of the provided lysis buffer, ice
cold with protease inhibitors. The gels are quickly sheared through an
18-gauge needle and then the sample is spun at 14,000 rpm at 4° for 2 min.
Remove 80 μl of supernatant and mix it with 80 μl of binding buffer to load
75 μl per well of a G-LISA plate in duplicate (transfer the remainder of the
supernatant to a fresh tube). The negative control is also made up so that 75 μl
can be loaded per well; however, the positive control is added as recommended in the kit protocol (50 μl per well) due to the fact that increasing
the volume can result in values outside the detectable range. The remainder of
the assay is preformed as described in the protocol provided with the kit.
Briefly, the plate is incubated at 4°, shaking at a minimum of 200 rpm, for

exactly 30 min. After washing the wells, anti-RhoA primary is added and the plate is incubated for 45 min while shaking at room temperature. Again the wells are washed and then the secondary antibody is added for an additional 45-min incubation followed by the addition of HRP detection reagent. The plate is then read on a luminometer.

Although the cell number is counted when loading the gels, we find that the reproducibility of the assay is optimal when relative RhoA levels are quantitated. For quantitation of relative RhoA activity, the total Rho level in each sample also needs to be determined. Total Rho can be measured by loading a constant volume of supernatant (30 μl of supernatant mixed with 10 μl of 3× Laemmli sample buffer) on a 6% SDS page gel followed by Western blot using anti-Rho antibody (Transduction Laboratories). Densitometry is performed and active RhoA (luminescence values)/total Rho (densitometry) values are calculated. The results can then be expressed as relative RhoA activity.

5. Co-Immunoprecipitation of Integrin-Associated Proteins from Cells Cultured in 3D Collagen Gels

Co-immunoprecipitation has emerged as a useful approach to elucidate signaling complexes, and to understand the molecular interactions among signaling molecules in cells. A few modifications to classic immunoprecipitations are necessary to obtain satisfactory results from 3D matrix cultures. The example given here allows for the analysis of integrin-associated signaling molecules, or other early signaling events associated with changes in the physical properties (density) of the ECM in breast epithelial cells.

For both T47D and NMuMG cells, approximately 10 million cells are required for each co-immunoprecipitation sample, and cast into a 1-ml gel. Cells are mixed with the appropriate concentration of collagen I and then allowed to polymerize for 1 h. After 1 h, gels may either be released from the culture dish (floating) or maintained in the attached state. After 1 h more, cells are lysed using 1 ml of 2× lysis buffer (50 mM Tris-HCl, pH 7.4, 100 mM NaCl, 10 mM EDTA, 100 mM NaF, 0.2% BSA, 2% Triton X-100, 1 mM sodium pervanadate, and protease inhibitors). It is important to break up the collagen gel by carefully triturating cells with a pipettor until the gel is a suspension uniform in consistency. To ensure adequate lysis, the lysate is incubated for 15 min at 4° on a rotating wheel, followed by centrifugation for 10 min at 4° to clear the lysate, which is transferred to a new tube. For immunoprecipitations, add 3 to 4 μg of antibody plus 30 μl GammaBind G-sepharose (Amersham Biosciences) and incubate lysate for 2 h to overnight at 4° on a rotating wheel. The length of incubation depends on the antibody used for the immunoprecipitation and the proteins that are being co-immunoprecipitated.

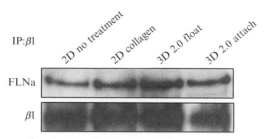

Figure 2.1 Co-immunoprecipitation of β1 integrin and filamin A from T47D cells cultured in 3D collagen gels. Cells plated on collagen-coated coverslips (2D collagen) undergo enhanced filamin A–β1 integrin interactions relative to uncoated coverslips (2D no treatment). However, under conditions in which cells are cultured in a high-density 3D collagen matrix that is not readily contracted, filamin A–β1 integrin interactions are elevated in both floating and attached gels.

After immunoprecipitation, briefly centrifuge the lysate at 4° at low speed in a microfuge to pellet the sepharose beads, discard the supernatant, and wash the beads with 1× lysis buffer, resuspending and centrifuging after each wash, for a total of three washes. Remove excess lysis buffer using a syringe attached to a 27-gauge needle, leaving damp beads without any remaining supernatant. Immunoprecipitated proteins are released by adding 30 μl of 1× Laemmli sample buffer to the rinsed beads and heating the samples for 3 min at 95°. The samples can then be analyzed using standard SDS-PAGE and immunoblotting approaches (for an example, see Fig. 2.1).

6. β_1-INTEGRIN ENDOCYTOSIS ASSAYS

Because cell adhesion is affected by the amount of adhesive receptors on the surface of the cell, the mechanism and efficiency of β_1-integrin endocytosis will affect adhesion and migration. We hope at some point to adapt these approaches for comparisons to cells cultured in 3D matrices, but for now have worked out approaches in 2D culture.

6.1. Time-lapse imaging of integrin trafficking in cells cultured on 2D surfaces

Considering recent interest and abilities in imaging cellular signaling events during cell adhesion and migration, the ability to also visualize integrin endocytosis and trafficking processes in live cells would prove to be highly informative. The use of commercially available kits that fluorescently label antibodies make this more readily possible. Direct labeling of β_1-integrin antibody (sc-9970, mouse monoclonal IgG$_1$ [4B7R], Santa Cruz) is carried out using a Zenon Alexa Fluor-555 mouse IgG$_1$ labeling kit from Molecular Probes (Eugene, OR), according to the manufacturer's instructions.

Cos-7 cells, obtained as a generous gift from Richard Anderson (Madison, WI) and maintained in DMEM (containing high glucose, L-glutamine, sodium pyruvate and pyridoxine hydrochloride) plus 10% fetal bovine serum at 5% CO_2, are plated into poly-L-lysine coated glass bottomed dishes (MatTek Ashland, MA) for 24 h at low confluency to minimize cell–cell interactions. Labeled antibodies are applied to cells for 30 min at 25°. Excess antibody is then washed out with PBS, followed immediately by live cell imaging.

6.1.1. Microscopy

Images were acquired using a CoolSnap FX (Roper) CCD camera mounted to an inverted microscope working in the epifluorescence mode. The emission intensity of a 100-W mercury arc bulb was attenuated by the appropriate excitation filter (545/30 for red fluorescence), directed to the cells with a dichroic mirror (565 DCLP), and fluorescence emission filtered (620/60) before reaching the camera. Exposure times ranged between 100 and 500 msec and the oil immersion objective used was $40\times$ N.A. = 1.3. We find it useful to perform no-neighbors deconvolution to remove out-of-focus fluorescence from the image. Digitized images can be exported to Image J for further analysis.

Figure 2.2A is a deconvolved epifluorescent image of a cell incubated with AlexaFluor555-tagged β_1-integrin antibody (4B7R, Santa Cruz). The staining was dimly fluorescent around the cell periphery where, in addition, higher-intensity microdomains of staining were also observed. Interior regions of the cell contained punctate and vesicular staining, which likely were trafficking vesicles and endosomal compartments. Figure 2.2A was the first image acquired in a series, and therefore some internalization of surface bound antibody has already occurred. This was necessary, as this technique required a 30-min incubation period with the fluorescent antibody followed by washout; otherwise cells would be unobservable during the incubation period due to high fluorescence, but at the same time require a proper amount of labeling. Compare the image in panel A with that of a cell stained using conventional immunocytochemistry techniques (panel B). The fixed cell did retain the phenotype of plasma membrane microdomain staining of β_1-integrin (arrow), and because the cell was permeabilized, there was accumulation of the antibody in the perinuclear region and labeling of intracellular, trafficked β_1-integrin. The advantage of potentially increased sensitivity obtained using immunocytochemistry, as witnessed by a greater number of small puncta of labeled β_1-integrin, is offset by the doubt raised concerning the specificity of staining. However the main advantage of avoiding the use of a fixed cell lies in the ability to perform time-lapse imaging experiments to look at β_1-integrin dynamics. Figure 2.2C shows a series of images of the boxed region in A acquired at approximately 2-min intervals. In this manner, one could observe how endocytosis of the microdomain of β_1-integrin occurred not through invagination of the membrane

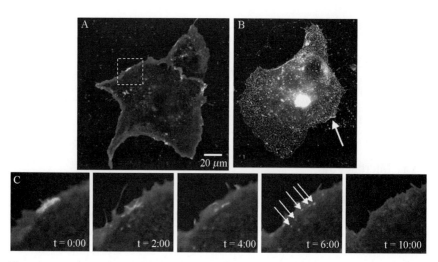

Figure 2.2   Direct fluorescent labeling of β_1-integrin antibody permits the visualization of β_1-integrin dynamics. (A) In live Cos-7 cells, β_1-integrin fluorescent microdomains were observed at the plasma membrane. This phenotype is preserved in a fixed, permeabilized cell (B, arrow) stained using conventional immunocytochemistry along with punctate intracellular staining that may be nonspecific to endocytosis. With time–lapse imaging of fluorescently conjugated antibodies (C, inset of A), the endocytosis of β_1-integrin from plasma membrane microdomains into vesicles destined for endosomes is evident.

into a large vesicle, but rather the entire ribbon of staining internalized slightly and then subsequently dissipated into several individual small vesicles that were visible at the 6-min mark (Fig. 2.2, arrows). The high spatiotemporal resolution of this technique can also be used to visualize the formation of new microdomains of β_1-integrin at the plasma membrane as well as trafficking of internalized β_1-integrin to endosomal compartments.

6.1.2. Immunocytochemistry

Cos-7 cells were rinsed in PBS to remove DMEM, fixed in 4% paraformaldehyde for 15 min at room temperature (RT), followed by quenching of excess PFA in $0.15\text{-}M$ glycine for 10 min. Cells were blocked in 1% donkey serum and 1% fatty acid–free BSA for 30 min after which primary antibody (1:100) was applied overnight at 4°. The antibody solution consisted of blocking solution supplemented with 0.01% Triton-X-100. Following washout of the primary antibody solution with PBS, the secondary antibody (Alexa Fluor555 goat anti-mouse; 1:200) was applied for 1 h at RT followed by several rinses with PBS. Cells were imaged the same day as secondary antibody application, using the epifluorescence microscope described above.

6.2. Endosome protection biotinylation assay

As a quantitative adjunct to the visualization of β_1-integrin endocytosis, we have developed an endosome protection biotinylation assay. The basic principle, which is applicable to any plasma membrane protein with an extracellular epitope, is to discern the partitioning of protein remaining on the cell surface from the fraction that has become internalized. Intact cells are biotinylated to label the pool of proteins remaining exposed to the surface during the test period, allowing the subsequent removal of cell-surface proteins on strepavidin beads. Thus the pool of protein that is internalized is protected from biotinylation and is quantified.

The experiment can be performed as follows: cells are grown in 100-mm Petri dishes for 24 h, and then incubated with anti-β_1-integrin antibody (4B7R, Santa Cruz) for 30 min at 37° to allow antibody to be internalized with β_1-integrin, and washed. Whole cells are subsequently biotinylated using Pierce EZ-Link Sulfo-NHS-LC-Biotin at >20-fold molar excess for 30 min on ice at 4°. Two control conditions should be performed: (1) one dish should not be biotinylated as a control for total β_1-integrin, and (2) one dish should be exposed to antibody and then immediately biotinylated as a negative control. Following washout of excess biotin, cells are lysed in lysis buffer (25 mM HEPES, 150 mM NaCl, 1% NP40, 0.5% deooxycholate, 0.1% SDS, 1mM EDTA, 1 μM NaF, and protease inhibitor cocktail) and biotinylated antibody was cleared using a 40-min incubation with strep-avidin beads and centrifugation in a microfuge at low speed for 20 sec. A total cell lysate sample should be saved, to later determine that equal amounts of initial protein were used. To recover the protected anti-β_1 integrin antibody, the cleared supernatant is incubated in gamma-bind sepharose beads (GE Healthcare BioSciences) for 40 min followed by three rounds of washing in 1× lysis buffer. Proteins are released from the beads using 1.5× Laemmli buffer containing β_1-mercaptoethanol at 95° for 3 min. Samples are run on an 8% SDS–PAGE gel, transferred to a PVDF membrane, probed for β_1-integrin antibody, which appears as the 50-kDa heavy chain on the western blot. Variations on this assay can be performed to determine the effect of various signaling intermediates or trafficking pathways, as cells can be transfected or exposed to pharmacological inhibitors prior to initiating the assay.

7. IMAGING CELL–MATRIX INTERACTIONS IN 3D COLLAGEN GELS

The use of MPLSM, which uses non-linear excitation to image a narrowly defined plane deep within tissue noninvasively, has the advantages of reduced phototoxicity (Squirrell et al., 1999) and the ability to image deeper into turbid samples (SHG)(Centonze and White, 1998) than other live cell imaging techniques. Moreover, its nonlinear excitation method is compatible with another nonlinear optical method—second harmonic generation (Campagnola *et al.*, 1999). Since collagen fibers are strong harmonophores, they can be readily imaged without additional contrast or staining approaches. This makes MPLSM an ideal approach to image cells and simultaneously image the 3D collagen matrix.

7.1. Cell culture

Mouse embryonic fibroblasts (MEFs) and the highly invasive MDA-MB-231 breast cancer cell line are maintained in T75 tissue culture flasks (Corning, Lowell, MA) containing DMEM supplemented with 10% (v/v) fetal bovine serum (FBS; Gibco, Carlsbad, CA) at 37° and 10% CO_2. Cells are passaged by trypsinization (0.5%; Cellgro, Herndon, CA) followed by 1:5 to 1:10 dilution into fresh media every 3 to 4 days.

7.2. Stable transfection

Due to the fact that the transfection efficiency in breast epithelial cells is relatively low, generation of stably expressing cells may by desirable. To generate stable MBD-MB-231 cells lines expressing GFP-vinculin, or GFP as a control, cells at ~85% confluency are transfected with Lipofectamine 2000 (Invitrogen, Carlsbad, CA). Prior to transfection, 20 μg of DNA (in 1.9 ml of serum-free DMEM) is combined with 75 μl of Lipofectamine 2000 (in 1.9 ml of serum-free DMEM) and DNA-Lipofectamine complexes are allowed to form for 20 min. The DNA-Lipofectamine complexes are then added to the culture plate with 12 ml of DMEM plus 10% FBS for 18 h. After 18 h, the cells are passed 1:20 into multiple T75 flasks containing culture medium supplemented with 1 mg/ml G418, where the cells grow and then die off over a 10- to 14-day period. Stable transfectants are then pooled and sorted under flow cytometry to obtain a population of cells with the desired level of expression.

7.3. Transient transfection

Since it is not always necessary or unproblematic to create a stably expressing cell line, transient transfection of the desired construct may be utilized. However, due to the low transfection efficiency for breast epithelial cells (typically 20 to 30%) a fluorescent marker, such as GFP, is necessary for cell biology experiments. For transient transfection, seed ~5 × 10^5 MEFs or MDA-MB-231 cells into 12-well culture plates 2 days prior to imaging. After 24 h, or after the cells are more than 80% confluent, cells are transfected with complexes resulting from the mixture of 1 μg of DNA (in 100 μl of serum-free media) and 4 μl of Lipofectamine 2000 (in 100 μl of

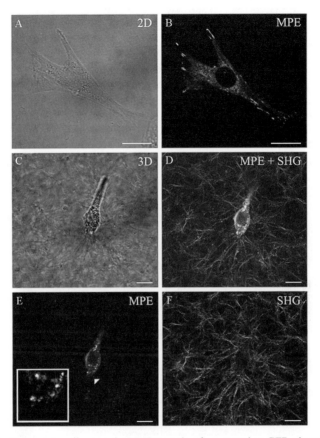

Figure 2.3 3D matrix adhesions in MEFs transiently expressing GFP-vinculin. Use of combined multiphoton excitation (MPE) and second harmonic generation (SHG) microscopy facilitates imaging of both the cell–matrix interaction and reorganization of the extracellular matrix (ECM). (A) Bright-field transmission image of a MEF on a collagen-coated coverslip for 24 h. (B) MPE image of GFP-vinculin–positive, 2D focal adhesions in the cell shown in (A) following two-photon excitation at 890 nm and filtering of the emission signal to isolate fluorescence from GFP (480-to-550–nm band-pass filter). (C) Bright-field transmission image of a MEF in a 3.0 mg/ml collagen gel after 24 h. (D) Concurrently acquired (nonfiltered) MPE/SHG image of the cell shown in (C) following two-photon excitation at 890 nm to simultaneously excite GFP and generate second harmonic signals from collagen. (E) MPE image of GFP-vinculin–positive, 3D matrix adhesions in the cell shown in (C) following two-photon excitation at 890 nm and filtering of the emission signal to isolate fluorescence from GFP (480-to-550–nm band-pass filter). The arrow in (E) indicates the region that is enlarged in the subpanel displayed in (E) showing punctuate 3D matrix adhesions at the leading edge of the cell where collagen has been pulled in toward the cells resulting in reorganization of the extracellular matrix. (F) SHG image of fibrillar collagen, following excitation at 890 nm and filtering of the emission signal to isolate SHG (445-nm, narrow band-pass filter), that has been reorganized by a contractile cell transmitting force through 3D matrix adhesions to the ECM. Note: See color pallet for pseudocolored image with merged MPE and SHG signals. Scale bars = 25 μm.

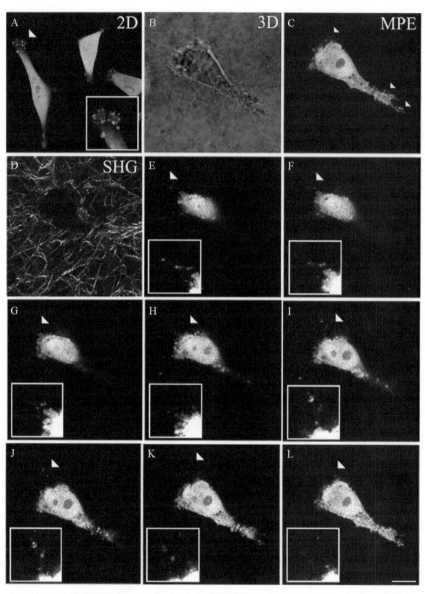

Figure 2.4 3D matrix adhesions in highly invasive and migratory MDA–MB–231 breast carcinoma cells stably expressing GFP-vinculin. Multiphoton excitation (MPE) and second harmonic generation (SHG) microscopy facilitate imaging of 3D matrix adhesions and extracellular matrix reorganization. (A) MPE image of GFP-vinculin–positive, 2D focal adhesions in MDA–MB–231 cells plated on collagen-coated glass coverslips for 24 h. GFP is excited at 890 nm and the emission signal filtered to isolate fluorescence from GFP (480-to-550–nm band-pass filter). (B) Bright-field transmission image of a

serum-free media). At 18 h post-transfection, pools of transfected cells are harvested by trypsinization and seeded into collagen gels.

7.4. Three-dimensional collagen gels

Following transfection, transient transfectants or stably expressing GFP-vinculin cells are detached and cells collected by centrifugation. Cells are resuspended in standard culture media and 1 to 5×10^5 cells seeded into a 3.0 mg/ml type I collagen gel. For 2D control experiments, cells are plated onto collagen coated (30 μg/ml) glass bottom culture dishes. Cell-seeded gels are allowed to polymerize at 37° and maintained under standard culture conditions for 24 h. After 24 h, the gels are moved to glass bottom dishes (MatTek Corp., Ashland, MA) and imaged with multiphoton laser-scanning microscopy.

7.5. Nonlinear optical imaging of collagen–3D matrix adhesion interaction

Multiphoton laser-scanning microscopy is used to generate multiphoton excitation (MPE) and SHG (Cox et al., 2003; Provenzano et al., 2006; Zipfel et al., 2003; Zoumi et al., 2002). In combination, these methodologies allow high-resolution live cell imaging as well as imaging of collagen structure. In MPE two or more low-energy (usually near-infrared) photons simultaneously excite a fluorophore, with the emission dependent on the square of the intensity, the probability of which is steeply dependent on the plane of focus. Thus, MPE produces optical sectioning that improves the axial resolution over standard confocal imaging (Centonze and White, 1998;

MDA-MB-231 cells in a 3.0-mg/ml collagen gel after 24 h. (C) MPE image of GFP-vinculin–positive, 3D matrix adhesions (examples of adhesions illustrated with arrows) in the cell shown in (B) following two-photon excitation at 890 nm and filtering of the emission signal to isolate fluorescence from GFP (480-to-550–nm band-pass filter). (D) SHG image of fibrillar collagen, following excitation at 890 nm and filtering of the emission signal to isolate SHG (445-nm, narrow band-pass filter), demonstrating cell-mediating collagen reorganization that correlates with 3D matrix adhesion location. (E to L) Progressive images taken at 1-μm intervals from an 8-μm z-stack showing GFP-vinculin–positive, 3D matrix adhesions (arrows indicate the region that is enlarged in the inset). On 2D substrates, matrix-engaged focal adhesions are located at the cell sub-strate interface (A), and are therefore primarily located within a narrow imaging window "below" the cell. In contrast, cells in 3D microenvironments (B to L) possess 3D matrix adhesions around the volume of the cell, and therefore are in multiple focal planes. Note: See color pallet for pseudocolored image with merged MPE and SHG signals. All images are the same magnification; scale bar = 25 μm. (Color pallet: Merged MPE and SHG images from Fig. 3(A) and Fig. 4(B). MPE of GFP-vinculin isolated with a 480-to-550–nm band-pass filter pseudocolored in green. SHG image of collagen isolated with a 445-nm, narrow band-pass filter pseudocolored in red.) (See color insert.)

Denk *et al.*, 1990). In contrast, SHG signals depend on nonlinear interactions of illumination with a noncentrosymmetric environment, such as the highly crystalline structure of fibrillar collagen, resulting in a coherent signal at exactly half the wavelength of the excitation (Mohler *et al.*, 2003).

MEF or MDA–MB-231 cell-seeded collagen gels are imaged with a MPLSM workstation assembled around a Nikon Eclipse TE300 (Wokosin *et al.*, 2003). Excitation is produced with a 5W mode-locked Ti:sapphire laser (Spectra-Physics-Millennium/Tsunami, Mountain View, CA) tuned to 890 nm. The beam is focused onto the collagen gels with a Nikon 60× Plan Apo water-immersion lens (N.A. = 1.2). To discriminate MPE and back-scattered SHG signals, a 445-nm, narrow band-pass filter is used to isolate SHG emission, while GFP signals are isolated with a 480-to-550–nm (band-pass) filter (all filters: TFI Technologies, Greenfield, MA). Without the use of these filters, a single image of GFP and collagen can be obtained; separation of the signals is not possible. The power of this approach is shown in Figs. 2.3 and 2.4, in which MPLSM imaging of GFP-vinculin expressing cells and their interaction with the collagenous matrix shows 3D matrix adhesions at the cell–collagen interface, as well as cell-mediated reorganization of the collagen matrix.

REFERENCES

Barcellos-Hoff, M. H., Aggeler, J., Ram, T. G., and Bissell, M. J. (1989). Functional differentiation and alveolar morphogenesis of primary mammary cultures on reconstituted basement membrane. *Development* **105,** 223–235.

Burridge, K., Turner, C. E., and Romer, L. H. (1992). Tyrosine phosphorylation of paxillin and pp125FAK accompanies cell adhesion to extracellular matrix: A role in cytoskeletal assembly. *J. Cell Biol.* **119,** 893–903.

Campagnola, P. J., Wei, M. D., Lewis, A., and Loew, L. M. (1999). High-resolution nonlinear optical imaging of live cells by second harmonic generation. *Biophys. J.* **77,** 3341–3349.

Centonze, V. E., and White, J. G. (1998). Multiphoton excitation provides optical sections from deeper within scattering specimens than confocal imaging. *Biophys. J.* **75,** 2015–2024.

Cox, G., Kable, E., Jones, A., Fraser, I., Manconi, F., and Gorrell, M. D. (2003). 3-dimensional imaging of collagen using second harmonic generation. *J. Struct. Biol.* **141,** 53–62.

Cukierman, E., Pankov, R., Stevens, D. R., and Yamada, K. M. (2001). Taking cell-matrix adhesions to the third dimension. *Science* **294,** 1708–1712.

Denk, W., Strickler, J. H., and Webb, W. W. (1990). Two-photon laser scanning fluorescence microscopy. *Science* **248,** 73–76.

Kleinman, H. K., McGarvey, M. L., Liotta, L. A., Robey, P. G., Tryggvason, K., and Martin, G. R. (1982). Isolation and characterization of type IV procollagen, laminin, and heparan sulfate proteoglycan from the EHS sarcoma. *Biochemistry* **21,** 6188–6193.

Mohler, W., Millard, A. C., and Campagnola, P. J. (2003). Second harmonic generation imaging of endogenous structural proteins. *Methods* **29,** 97–109.

Muthuswamy, S. K., Li, D., Lelievre, S., Bissell, M. J., and Brugge, J. S. (2001). ErbB2, but not ErbB1, reinitiates proliferation and induces luminal repopulation in epithelial acini. *Nat. Cell Biol.* **3,** 785–792.

Paszek, M. J., Zahir, N., Johnson, K. R., Lakins, J. N., Rozenberg, G. I., Gefen, A., Reinhart-King, C. A., Margulies, S. S., Dembo, M., Boettiger, D., Hammer, D. A., and Weaver, V. M. (2005). Tensional homeostasis and the malignant phenotype. *Cancer Cell* **8,** 241–254.

Provenzano, P. P., Eliceiri, K. W., Campbell, J. M., Inman, D. R., White, J. G., and Keely, P. J. (2006). Collagen reorganization at the tumor-stromal interface facilitates local invasion. *BMC Med.* **4,** 38.

Roeder, B. A., Kokini, K., Sturgis, J. E., Robinson, J. P., and Voytik-Harbin, S. L. (2002). Tensile mechanical properties of three-dimensional type I collagen extracellular matrices with varied microstructure. *J. Biomech. Eng.* **124,** 214–222.

Squirrell, J. M., Wokosin, D. L., White, J. G., and Bavister, B. D. (1999). Long-term two-photon fluorescence imaging of mammalian embryos without compromising viability. *Nat. Biotechnol.* **17,** 763–767.

Weaver, V. M., Lelievre, S., Lakins, J. N., Chrenek, M. A., Jones, J. C., Giancotti, F., Werb, Z., and Bissell, M. J. (2002). beta4 integrin-dependent formation of polarized three-dimensional architecture confers resistance to apoptosis in normal and malignant mammary epithelium. *Cancer Cell* **2,** 205–216.

Wokosin, D. L., Squirrell, J. M., Eliceiri, K. E., and White, J. G. (2003). An optical workstation with concurrent, independent multiphoton imaging and experimental laser microbeam capabilities. *Rev. Sci. Instru.* **74,** 193–201.

Wozniak, M. A., Desai, R., Solski, P. A., Der, C. J., and Keely, P. J. (2003). ROCK-generated contractility regulates breast epithelial cell differentiation in response to the physical properties of a three-dimensional collagen matrix. *J. Cell Biol.* **163,** 583–595.

Wozniak, M. A., and Keley, P. J. (2005). Use of three-dimensional collagen gels to study mechanotransduction in T47D breast epithelial cells. *Biol. Proc. Online* **7,** 144–161.

Zipfel, W. R., Williams, R. M., Christie, R., Nikitin, A. Y., Hyman, B. T., and Webb, W. W. (2003). Live tissue intrinsic emission microscopy using multiphoton-excited native fluorescence and second harmonic generation. *Proc. Natl. Acad. Sci. USA* **100,** 7075–7080.

Zoumi, A., Yeh, A., and Tromberg, B. J. (2002). Imaging cells and extracellular matrix *in vivo* by using second-harmonic generation and two-photon excited fluorescence. *Proc. Natl. Acad. Sci. USA* **99,** 11014–11019.

CHAPTER THREE

Integrins in Cell Migration

Keefe T. Chan,* Christa L. Cortesio,† and Anna Huttenlocher‡

Contents

Abstract

Integrins are cell-surface adhesion receptors that play a central role in regulating cell migration by mediating interactions between the extracellular matrix and the actin cytoskeleton. Substantial progress has been made in understanding the mechanisms by which the formation and breakdown of adhesions are regulated. Here we describe general methods used to study integrin-mediated cell migration. Furthermore, we outline detailed procedures to examine focal adhesion assembly and disassembly using time-lapse fluorescent microscopy. Finally, we provide methods for the analysis of podosomes, which are highly dynamic adhesive structures that form in immune cells and invasive cancer cells.

1. Introduction

Cell migration plays a central role in both normal and pathological processes, including embryonic development, wound healing, tumor metastasis, and inflammation. Integrins are cell-surface $\alpha\beta$ heterodimers that mediate interactions between the extracellular environment and the actin cytoskeleton and play a critical role in regulating cell migration, both by modulating adhesive interactions and by serving as cell signaling receptors. Many integrin receptors have been implicated in cell migration, including $\alpha 5\beta 1$ and $\alpha 4\beta 1$ integrins on fibronectin, $\alpha 2\beta 1$ integrins on collagen, and $\beta 2$ integrins on

* Department of Molecular and Cellular Pharmacology, University of Wisconsin, Madison, Wisconsin
† Department of Biomolecular Chemistry, University of Wisconsin, Madison, Wisconsin
‡ Departments of Pediatrics and Molecular and Cellular Pharmacology, University of Wisconsin, Madison, Wisconsin

Methods in Enzymology, Volume 426
ISSN 0076-6879, DOI: 10.1016/S0076-6879(07)26003-3

© 2007 Elsevier Inc.
All rights reserved.

47

intercellular cell adhesion molecule (ICAM) (reviewed by Humphries *et al.*, 2006). Furthermore, correlations between integrin expression and migratory capacity have been described in cancer cells where $\alpha v \beta 3$ integrin has been implicated in tumor cell metastasis. Thus, an understanding of how integrins regulate cell migration through the dynamic regulation of adhesions is fundamental to both normal physiology and human disease.

Cell migration requires a dynamic interaction between the extracellular matrix (ECM) and the actin cytoskeleton. To migrate, cells initially extend a directed protrusion at the leading edge through the localized polymerization of actin and subsequent integrin-mediated stabilization of adhesions. After stabilization of the protrusion or lamellipodium, cells generate traction and the contractile force required for cell movement. The final step in the migratory cycle involves release of adhesions at the rear of the cell to allow for forward progression. These classic steps are representative of the migratory cycle of many adherent cells such as fibroblasts, and although these principles generally apply to other cell types, leukocytes tend to display a more gliding movement with less force applied to the ECM as they migrate. General principles, however, apply to many cell types: adhesions at the front of the cell must be strong enough to withstand contractile forces generated by the cell front during migration, while adhesions at the rear must be weak enough to allow the cell to detach from its substrate. Therefore, asymmetry in the strength of adhesions at the cell front and rear is likely important for efficient cell migration. Several parameters may contribute to the strength of adhesion between a cell and its environment including ligand concentration, number of adhesion receptors, affinity between receptor and ligand, strength of receptor–cytoskeletal linkages, and organization of receptors on the cell surface (Gallant *et al.*, 2005; Gupton and Waterman-Storer, 2006; Huttenlocher *et al.*, 1996; Miyamoto *et al.*, 1995; Palecek *et al.*, 1998).

Cell migration requires the dynamic assembly and disassembly of integrin-mediated adhesions. These sites of adhesions contain integrin clusters that recruit multiple signaling and structural proteins that can form organized adhesion complexes known as focal adhesions (Lo, 2006; Sastry and Burridge, 2000; Zamir and Geiger, 2001). In general, cells that form less organized adhesion complexes, including leukocytes or tumor cells, tend to be more motile as compared to fibroblasts, which form more organized focal adhesions and are generally less migratory. However, some kind of linkage between integrins and the actin cytoskeleton is likely required for most forms of migration. Therefore, studies that address the dynamics of adhesions during migration are crucial for a better understanding of the mechanisms by which integrins regulate cell migration. In this article, we describe the general methods used to study integrin-mediated cell migration and for the live imaging of adhesion complex dynamics during cell migration. Specifically, we discuss the methods used to study the morphology and dynamics of two distinct types of integrin-mediated adhesions. First, we discuss imaging of focal adhesion dynamics using standard live imaging techniques of focal adhesion proteins that are tagged with fluorophores such as green fluorescent protein (GFP). We subsequently present methods used to study podosomes, the more dynamic adhesions formed in invasive cancer cells and leukocytes. We emphasize the practical considerations and general methodology used in the study of cell–substrate contact dynamics and some basic tools used for image analysis.

2. METHODS FOR ANALYSIS OF INTEGRIN-MEDIATED CELL MIGRATION

There are many methods available for the analysis of cell migration that have recently been reviewed, including the transwell assay (Shaw, 2005), wound healing assays (Rodriguez *et al.*, 2005), time-lapse microscopy, and Dunn chemotaxis chamber (Wells and Ridley, 2005). These types of assays can be used to measure distinct types of motility. For example, wound assays and chemotaxis assays using time-lapse microscopy or Dunn chemotaxis chambers are used to measure directed migration. Chemotaxis assays are used to measure migration in a gradient of soluble attractant, such as a growth factor or chemoattractant. In contrast, the wound assay measures directed migration in response to the clearing of cells in a monolayer to allow for directed migration into the denuded area. Other assays used to measure migration include random cell migration assays using time-lapse microscopy and cell tracking. This is commonly used to measure fibroblast migration and is generally used in a two-dimensional format with non–tissue culture plates coated with an ECM ligand, such as fibronectin. However, recent studies have also described methods to analyze migration within fibroblast-derived three-dimensional matrices (Cukierman, 2005). Random migration measures the ability of a substrate to support cell migration without a gradient of substrate (haptotaxis) or chemoattractant (chemotaxis). An important consideration in this context is the density of ligand used to analyze cell migration because most cell types display a biphasic dependence on ligand density with maximum migration observed at intermediate substrate concentrations (Cox and Huttenlocher, 1998). The optimal density of ECM ligand will need to be determined for every cell type and ligand.

The transwell assay can be used to measure random, haptotactic, and chemotactic cell migration. The transwell assay is generally used in a haptotactic configuration to detect migration toward the ligand by coating only the lower surface of the membrane with fibronectin or other substrates. Transwell assays can also be used to measure random migration by coating both the upper and lower surfaces of the membrane with substrate. Random migration shows an optimum cell migration speed at intermediate concentrations of substratum, while haptotactic migration requires higher concentrations of substrate for optimum migration with persistence of migration observed at higher coating concentrations. Finally, chemotactic migration can also be studied using the transwell assay configured in a random format with both lower and upper surfaces coated with substrate and the presence of chemoattractant only in the lower well (Lokuta *et al.*, 2003). An important consideration for all migration assays is the selection of media. Migration can be performed in the presence of serum-containing media, but it is important to consider that this will introduce other ECM components and may complicate studies that focus on a specific ligand–integrin interaction. In general, when studying $\alpha 5\beta 1$-mediated migration on fibronectin, we recommend using serum-free media that contains growth factors such as CCM1 (Hyclone, Logan, UT) or serum-free media with 0.5% bovine serum albumin (BSA) (Sigma Aldrich, St. Louis, MO). If one is interested in a specific integrin–ligand interaction, it is also important to select a cell type that requires the specific integrin for adhesion to the ligand. For example, Chinese hamster ovary (CHO)-K1 cell adhesion and migration on fibronectin require $\alpha 5\beta 1$ integrin (Huttenlocher *et al.*, 1996).

2.1. Imaging of adhesion dynamics

Advances in live fluorescence cell imaging have continued to improve our knowledge of the molecular architecture and dynamics of adhesive complexes (Ezratty *et al.*, 2005; Webb *et al.*, 2004; Zaidel-Bar *et al.*, 2004). Focal complexes are early adhesions that form at the cell periphery, and are more highly dynamic than focal adhesions. Focal adhesions, in contrast, are localized at the end of actin stress fibers at both central and peripheral regions of the cell, and contain some of the same components as focal complexes but also have components such as zyxin and α-actinin that are generally found in more mature adhesions. Figure 3.1A illustrates focal adhesion staining in HeLa cells. Other types of adhesions have also been described including fibrillar adhesions, which are more centrally localized, and contain distinct cytoskeletal components such as tensin (Zaidel-Bar *et al.*, 2003). In the following, we discuss the general methods used for the analysis of cell–matrix structure and dynamics.

2.1.1. Selection of cell type and ECM ligand

Many cell types are appropriate for studies of adhesion dynamics including HeLa cells (Franco *et al.*, 2006), NIH 3T3 cells (Franco *et al.*, 2004), CHO-K1 cells (Bhatt and Huttenlocher, 2003), and murine embryonic fibroblasts (MEF) (Webb *et al.*, 2004). For the purposes of this article, we will focus on the adhesion dynamics of two cell types: HeLa cells for focal adhesion dynamics and the rat breast carcinoma cell line MTLn3 for podosome dynamics. An additional consideration for the investigator is the selection of ECM used for the imaging studies. For general analysis of cell motility, frequently a mixed substrate is used such as Matrigel^{TM} (BD Biosciences), which includes basement membrane components such as

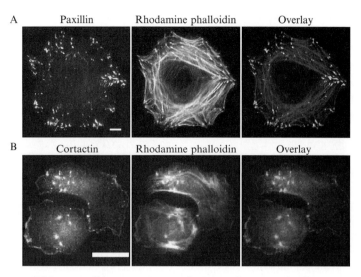

Figure 3.1 (A) Immunofluorescent image of a HeLa cell costained with an anti-paxillin antibody (green) and rhodamine phalloidin (red). Bar, 10 microns. (B) Immunofluorescent image of an MTLn3 cell co-stained with an anti-cortactin antibody (green) and rhodamine phalloidin (red). Bar, 10 microns. (See color insert.)

laminins and collagen IV. We generally use fibronectin to coat the coverslips for live cell imaging as described below. It is important to consider what integrins may be mediating the adhesion to the specific substrate that is selected (see Table 3.1 for commonly used ligands and associated integrins).

2.1.2. Transient transfection protocol

1. HeLa cells were purchased from the American Type Culture Collection (ATCC, Manassas, VA) and cultured in Dulbecco's modified Eagle's medium (DMEM) (Cellgro, Mediatech Inc., Herndon, VA) supplemented with nonessential amino acids, 10% fetal bovine serum (FBS) (HyClone, Logan, UT), 100 U/ml penicillin, and 100 μg/ml streptomycin (Cellgro, Mediatech Inc., Herndon, VA) (DMEM complete).

2. Plate HeLa cells 24 h before transfection at 8.5×10^5 cells on a 6-cm tissue culture dish in DMEM complete without penicillin/streptomycin so that the cells are ~70 to 80% confluent at the time of transfection. Avoid using penicillin/streptomycin in media prior to and during transfection to prevent cytotoxicity. The transfection mixture is prepared by first diluting 2.5 μg of DNA in 125 μl OPTI-MEM (Gibco, Invitrogen, Carlsbad, CA) and 5 μl of Lipofectamine 2000 (Invitrogen, Carlsbad, CA) in 125 μl OPTI-MEM, incubating each at room temperature for

Table 3.1 Integrin receptors and major ligands

Ligand	Integrin receptor
Laminin	$\alpha_1\beta_1$, $\alpha_2\beta_1$, $\alpha_3\beta_1$, $\alpha_6\beta_1$, $\alpha_7\beta_1$, $\alpha_{10}\beta_1$, $\alpha_6\beta_4$
Collagen	$\alpha_1\beta_1$, $\alpha_2\beta_1$, $\alpha_{10}\beta_1$, $\alpha_{11}\beta_1$, $\alpha_X\beta_2$
Fibronectin	$\alpha_4\beta_1$, $\alpha_5\beta_1$, $\alpha_8\beta_1$, $\alpha_4\beta_7$, $\alpha_V\beta_1$, $\alpha_V\beta_3$, $\alpha_V\beta_6$, $\alpha_{IIb}\beta_3$
Thrombospondin	$\alpha_2\beta_1$, $\alpha_3\beta_1$, $\alpha_4\beta_1$, $\alpha_V\beta_3$, $\alpha_{IIb}\beta_3$
Vitronectin	$\alpha_8\beta_1$, $\alpha_V\beta_3$, $\alpha_V\beta_5$, $\alpha_{IIb}\beta_3$
Osteopontin	$\alpha_4\beta_1$, $\alpha_5\beta_1$, $\alpha_8\beta_1$, $\alpha_9\beta_1$, $\alpha_4\beta_7$, $\alpha_V\beta_1$, $\alpha_V\beta_3$, $\alpha_V\beta_5$, $\alpha_V\beta_6$
Fibrinogen	$\alpha_V\beta_3$, $\alpha_{IIb}\beta_3$, $\alpha_X\beta_2$, $\alpha_M\beta_2$
von Willebrand factor	$\alpha_V\beta_3$, $\alpha_{IIb}\beta_3$
Matrigel^{TM} (laminin, collagen IV, entactin) (BD Biosciences)	$\alpha_1\beta_1$, $\alpha_2\beta_1$, $\alpha_3\beta_1$, $\alpha_6\beta_1$, $\alpha_7\beta_1$, $\alpha_{10}\beta_1$, $\alpha_6\beta_4$
Intercellular cell adhesion molecule (ICAM)	$\alpha_L\beta_2$, $\alpha_X\beta_2$, $\alpha_M\beta_2$, $\alpha_D\beta_2$
Vascular cell adhesion molecule 1 (VCAM-1)	$\alpha_4\beta_1$, $\alpha_9\beta_1$, $\alpha_4\beta_7$, $\alpha_D\beta_2$
Platelet endothelial cell adhesion molecule 1 (PECAM-1)	$\alpha_V\beta_3$
E-cadherin	$\alpha_E\beta_7$

Note: Reviewed in Humphries *et al.*, 2006; van der Flier *et al.*, 2001; Stupack and Cheresh, 2002; Hynes, 1987; and Buck and Horwitz, 1987.

5 min, mixing them together, and incubating the mixture at room temperature for another 20 min before adding to cells.

3. HeLa cells are highly sensitive to abrupt changes in serum concentration, so we typically add the transfection mixture directly to the cells in serum-containing medium and incubate at 37° and 5% CO_2 overnight. Aspirate the medium and wash once with Dulbecco's phosphate-buffered saline (PBS) (Cellgro, Mediatech Inc., Herndon, VA). Replace with 3 ml of DMEM complete without antibiotics, and incubate at 37° and 5% CO_2 24 to 48 h until the cells are ready to be used for experiments.

2.1.3. Fluorescent imaging of focal complexes in live cells

Live imaging of fluorescently tagged adhesion complex proteins is a powerful method to observe adhesion complex dynamics. A variety of focal adhesion markers fused to GFP have previously been described such as the α_5-integrin (Laukaitis *et al.*, 2001), α-actinin (Edlund *et al.*, 2001), and vinculin (Balaban *et al.*, 2001). In our studies, we have used dual live imaging to study the spatial and temporal dynamics of GFP-talin (Franco *et al.*, 2004) and paxillin-dsRed2 (generously provided by Rick Horwitz) (Fig. 3.2A). Below, we describe the general methodology to perform live fluorescent imaging of adhesion complexes in HeLa cells.

Acid washing coverslips

1. Acid wash 22-mm coverslips in a large flat-bottom beaker containing 20% (v/v) H_2SO_4. Gently mix the beaker periodically and incubate the coverslips in acid overnight. Pour off acid and neutralize coverslips with a solution of 0.1 M NaOH. Following neutralization of acid waste, rinse coverslips twice with distilled water. Remove coverslips and place them individually between two sheets of Whatman paper (Clifton, NJ) to dry. Coverslips can be stored in a Petri dish between two sheets of Whatman paper.
2. Prior to use, sterilize coverslips by rinsing them once with ethanol. Rinse coverslips twice with PBS to completely remove ethanol.

Preparation of glass bottom plates

1. Acid wash 22-mm coverslips as described above.
2. Bore 20-mm holes in the bottom of non–tissue culture, 35-mm Petri dishes. University machine shops generally can make these easily.

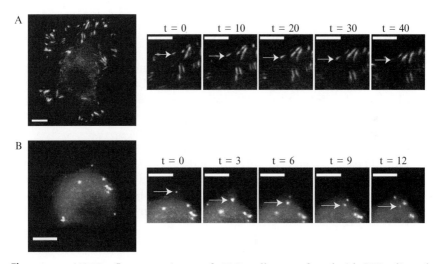

Figure 3.2 (A) Live fluorescent image of a HeLa cell cotransfected with GFP-talin and paxillin-dsRed2. Both talin and paxillin localize to focal complexes and show 100% colocalization. Time-lapse sequences of talin are shown at 10-min intervals. Arrow denotes a focal complex tracked over time. Bar, 10 microns. (B) Live fluorescent image of an MTLn3 cell transfected with GFP-cortactin. GFP-cortactin localizes to podosomes. Time-lapse sequences of cortactin are shown at 3-min intervals. Arrow denotes a podosome tracked over time. Bar, 10 microns.

3. Use optical adhesive to glue acid-washed coverslips to the bottom outer surface of the bored 35-mm Petri dishes. Cure the optical glue by exposure to long-wavelength UV light for 12 h. Using optical adhesive is critical as other adhesives may dissolve into the media during the experiment.

4. Coat the plate with ECM protein. For live fluorescence imaging of focal adhesion proteins, coat glass bottom dishes with 5 to 10 μg/ml of fibronectin diluted in PBS by incubating at 37° for 1 h or at 4° overnight. Wash the plate once with PBS, block with 2% BSA at 37° for 30 min, and wash twice with PBS. Coated plates containing 2 to 3 ml PBS may be stored at 4° for up to 1 week before use.

Calibrating stage before live fluorescent imaging   Calibration of the stage is necessary to maintain viability of the cells during the experiment. Two important requirements for successful live imaging are that the cells are kept at physiological temperature and that focus instabilities caused by fluctuations in the temperature of the microscope components are kept at a minimum. We use a temperature control system known as the *cube&box* (Life Imaging Services, Reinach, Switzerland). The *cube* is a recirculating heater that precisely compensates for external disturbances in temperature. The *box* is a transparent Makrolon^{TM} enclosure for the microscope that functions as an insulator. The following methods for calibration of the stage should be performed at least 1 h before imaging.

1. Turn on the *cube* and set the temperature to 37° the day before the experiment.

2. Turn on the microscope and select the appropriate oil–immersion objective (usually 60×) for visualizing focal contacts. Lay a drop of immersion oil on the objective and allow the oil temperature to equilibrate with the objective temperature. Before imaging it is also important to allow the plate to equilibrate with the objective (see below).

3. Turn on 5% CO_2 and equilibrate for 30 min. Check the pH of the medium using pH paper. A temperature of 37° and a pH of 7.2 should be maintained for a minimum of 15 min before live fluorescence imaging. For other cell types, pretreatment of the medium with 20-mM HEPES buffer at pH 7.2 (Cellgro, Mediatech Inc., Herndon, VA) may eliminate the need for CO_2 during the duration of the experiment.

Fluorescent imaging of focal complexes in live cells

1. Calibrate the stage as described above. Warm medium and reagents to 37° in a water bath prior to use. To minimize background fluorescence we recommend using a phenol red free medium, such as Ham's/F-12 medium (HyClone, Logan, UT). Pretreat Ham's/F-12 medium containing 2% FBS at 37° and 5% CO_2 for 1 h before use. We find that the

viability of HeLa cells is significantly reduced in serum-free conditions. For other cell types, such as NIH 3T3 fibroblasts or HEK 293, we generally use Ham's/F-12 with 0.2% BSA since FBS may increase background fluorescence.

2. Tissue culture dishes should not be used for resolution of focal contacts because of high background fluorescence and limited working distance. We use glass-bottom plates as described above.

3. Wash cells once with PBS, add a solution of 0.25% trypsin/0.02% EDTA in PBS (0.5 ml for 60-mm dish), and incubate cells at 37° until cells are detached ($\sim$5 min). Trypsin may cleave cell-surface receptors, so it is important to neutralize the trypsin as soon as the cells detach.

4. Resuspend the cells in 5 ml of DMEM containing 2% FBS, and centrifuge the cells at 1500g for 2 min. An additional wash with medium may be performed to ensure removal of trypsin/EDTA.

5. Add 2 to 3 ml of DMEM containing 2% FBS and 1×10^5 cells to the glass-bottom plate. The cell density should be low enough to allow the visualization of single cells that are not contacting neighboring cells for live imaging. Incubate the cells at 37° and 5% CO_2 for 1 h.

6. Remove medium from the plate and wash once with PBS. Replace with 2 to 3 ml pretreated Ham's/F-12 media containing 2% FBS as described above (Step 1). Set the plate on the heated stage.

7. Layer 1 to 2 ml of mineral oil over the medium to prevent evaporation. Focus on the cells using phase contrast or DIC (to minimize exposure to UV light) and allow the plate to equilibrate with the stage at least 30 min before imaging.

8. Scan the plate for a representative cell that expresses the fluorescently tagged protein at an optimum level for imaging. Low expression may require longer exposure times, which is likely to result in photobleaching. High expression may obscure adhesion complex localization. It is also important to make sure that the cell is not in contact with neighboring cells. Minimize the amount of UV exposure to the cells by using a neutral density filter while scanning. After selecting a cell for imaging, adjust optics so that the charge-coupled device (CCD) camera may capture images.

9. The software used for controlling the CCD camera may vary, but most will allow for the control of gain, exposure time, region imaged, and binning. It is critical to adjust settings so that exposure of the cells to UV is minimized without significant reduction in image quality. When visualizing adhesion complex dynamics, we use gain = 1, exposure time = 50 to 100 msec, region imaged = center 640×580, and bin = 2. The advantage of binning is that it allows you to reduce exposure times, but it also may compromise resolution. However, we find that setting the camera bin to 2 does not significantly reduce image resolution.

10. Select a timeframe and interval at which images will be captured. The interval will depend on the kinetics of the adhesion complex protein that is being studied. We find that the focal adhesion proteins talin, paxillin, and FAK, all exhibit similar kinetics. In general, we collect images every 2 min for 2 h. We use similar methods to perform dual imaging of fluorescent proteins in live cells. It is important to minimize exposure of the cells to UV by reducing exposure times or collecting images at less-frequent intervals.

11. Begin capturing images with the software package. Adjust focus of the cell during imaging since the cell may change shape and affect the focal plane of the adhesion complex protein. Fluctuations in ambient temperature may also affect microscope optics and result in changes in focus. Calibration of the stage and equilibration of the plate before imaging should help to reduce fluctuations in temperature.

Quantification of adhesion dynamics Recent developments in live imaging analysis have provided a framework for calculating the kinetics of adhesion complex formation and dissolution. There are at least four kinetic parameters that contribute to the duration of an adhesion complex. The assembly rate is the rate at which the fluorescently tagged protein is incorporated into an adhesion complex. The disassembly rate is the rate at which the fluorescently tagged protein is removed from an adhesion complex. The time elapsed during periods of assembly and disassembly contributes to the total duration of the adhesion complex. In addition, the time spent in the change from assembly to disassembly contributes to the duration of adhesions. In the following, we describe a recently published method developed by Webb and Horwitz and modified by our laboratory to quantify adhesion dynamics (Franco *et al.*, 2004; Webb *et al.*, 2004).

1. Using a high-pass filtration algorithm as previously described (Zamir *et al.*, 1999), remove diffuse fluorescence background from time-lapse sequences generated from live fluorescence imaging. Alternatively, most imaging software programs should have built-in capabilities to subtract background fluorescence.

2. Select a region encompassing an adhesion from its initial formation to its disappearance. Be careful to select an area in which another adhesion does not enter within the duration of the adhesion. Adhesions should be selected from various regions of the cell (protrusion, central, retraction) and logged.

3. Use appropriate imaging software to record the integrated fluorescence intensity at each frame of the adhesion. We use MetaMorph and Meta-Vue Imaging software (Universal Imaging Corporation, Downington, PA) for our studies. Log the fluorescence intensities of 10 to 25 adhesions per cell.

4. Transfer the fluorescence intensity values and time points to spreadsheet software such as Microsoft Excel. The time of initial appearance of the adhesion is defined as $t = 0$.

5. Generate a scatter plot of fluorescence intensity versus time. Two characteristic features of the plot should be a period of exponential growth (assembly) and a period of exponential decay (disassembly).

6. Generate semilogarithmic plots by taking the natural logarithm of the fluorescence intensities and plot against time. The assembly phase occurs between the initial appearance of the adhesion and the time of maximum fluorescence intensity. Likewise, the disassembly phase occurs between the time of maximum fluorescence intensity and the time at which the adhesion disappears. Assembly is defined as $Ln([I]/[I_0])$, and disassembly is defined as $Ln([I_0]/[I])$, where I_0 is the initial fluorescence intensity and I is the fluorescence intensity at subsequent time points.

7. Fit a trend line for each semilogarithmic plot and calculate the R^2 value to determine if this quantification model may be used for the selected adhesion. Typically, an R^2 value of at least 0.7 is acceptable. To calculate the rate constant for assembly or disassembly expressed as inverse time, determine the slope of the trendline. The duration is the time elapsed between the initial and final frames of an observed adhesion.

8. Measurements should be made for at least six cells for a total of 50 to 100 adhesions.

2.2. Imaging of podosomes

In contrast to focal adhesions, highly migratory and invasive cells form another type of integrin-mediated adhesion complex called the podosome. Podosomes have been observed in cells of monocytic lineage such as osteoclasts, macrophages and dendritic cells as well as transformed fibroblasts and carcinoma cells (reviewed by Linder and Aepfelbacher, 2003). More recently, podosomes have been discovered in smooth muscle cells (Hai et al., 2002) and endothelial cells (Tatin et al., 2006) in response to PKC-activating phorbol esters. Podosomes are formed on a variety of ligands/ ECM components, including fibronectin (Tarone et al., 1985), type I collagen (Shibutani et al., 2000), and VCAM-1 (Johansson et al., 2004).

Podosome architecture is defined by an actin-rich core, where the actin polymerizing machinery and actin regulatory proteins function to drive membrane protrusion and cell motility. This core is surrounded by a ring structure composed of signaling and adaptor proteins including cortactin and focal adhesion proteins such as talin and paxillin (Buccione et al., 2004). (See Table 3.2 for a list of components found at podosomes and focal adhesions.) Integrins, including $\beta 1$, $\beta 2$, and $\beta 3$ integrins, also localize to podosomes and are found within both the actin-rich core and the surrounding ring structure (Linder and Aepfelbacher, 2003). Podosomes are highly

Table 3.2 Components found at podosomes and focal adhesions

Component	Function	Reference
Unique to Podosomes		
Cortactin	Actin binding	Hiura *et al.*, 1995
Gelsolin	Actin binding	Chellaiah *et al.*, 2000
WASP	Actin binding	Linder *et al.*, 1999
Pyk2	Signaling	Duong and Rodan, 2000
PI3K	Signaling	Chellaiah *et al.*, 2001
N-WASP	Actin binding	Mizutani *et al.*, 2002
AFAP-110	Actin binding	Gatesman *et al.*, 2004
Arp2/3	Actin nucleation	Linder *et al.*, 2000
Actin	Cytoskeleton	Tarone *et al.*, 1985
$\alpha_M\beta_2$	Integrin	Duong and Rodan, 2000
$\alpha_X\beta_2$	Integrin	Gaidano *et al.*, 1990
Common to Focal Adhesions and Podosomes		
Talin	Cytoskeletal associated	Burridge and Connel, 1983; Marchisio *et al.*, 1988
Paxillin	Cytoskeletal associated	Turner *et al.*, 1990
Tensin	Actin binding	Bockholt *et al.*, 1992; Hiura *et al.*, 1995
Zyxin	Cytoskeletal associated	Crawford and Beckerle, 1991; Kaverina *et al.*, 2003
Vinculin	Actin binding	Bloch and Geiger, 1980; Tarone *et al.*, 1985
c-src	Signaling	Gavazzi *et al.*, 1989; Rohrschneider, 1980;
β_1	Integrin	Marcantonio and Hynes, 1988; Marchisio *et al.*, 1988; Nermut *et al.*, 1991
β_3	Integrin	Kramer *et al.*, 1990; Pfaff and Jurdic, 2001
Dynamin	Other	Ezratty *et al.*, 2005; Lee and De Camilli, 2002
VASP	Actin binding	Reinhard *et al.*, 1992; Spinardi *et al.*, 2004

dynamic structures that may mature into invadopodia that protrude into the ECM and are associated with ECM degradation (Baldassarre *et al.*, 2003). Few studies have addressed the mechanisms that regulate podosome dynamics in live cells. A challenge for future investigators will be to define the temporal and spatial regulation of podosome dynamics and the mechanisms that regulate the maturation of podosomes into invasive structures. Below, we discuss methods for the analysis of podosome structure and dynamics in both fixed and live cells.

2.2.1. Selection of cell type and ECM ligand

We describe methods in the following that use a metastatic rat mammary adenocarcinoma (MTLn3) cell line (Segall *et al.*, 1996) to visualize podosomes. MTLn3 cells are easily grown in Minimum Essential Medium (MEM) α medium with GlutaMAX (Invitrogen, Carlsbad, CA) supplemented with 5% FBS, 50 U/ml penicillin, and 50 μg/ml streptomycin (α-MEM complete), and are an excellent model system to study cell migration and invasion. MTLn3 cells have been reported to express the following integrins: α2β1, α3β1, and α5β1, which mediate interactions with the ECM (Beviglia and Kramer, 1999; Genersch *et al.*, 1996). Podosomes are rapidly formed (within 1 h) when MTLn3 cells are plated on fibronectin-coated plates. Furthermore, MTLn3 cells are easy to transfect with high efficiency using the methods outlined here.

Transient transfection protocol

1. MTLn3 cells were a generous gift from John Condeelis and Jeffrey Segall and cultured at 37° and 5% CO_2 as described above.
2. Plate MTLn3 cells 24 h prior to transfection at 1×10^5 cells on a 6-well tissue culture in α-MEM complete without penicillin/streptomycin. The cells should be ~70% confluent at the time of transfection. For a typical transfection, use a mixture of 2 μg of DNA and 4 μl of Lipofectamine 2000. We prepare transfection reagent as described by the manufacturer. The concentration of DNA in the transfection mixture may be varied (1 μg to 5 μg) to control the expression level. Allow the mixture to incubate for 30 min prior to treating cells.
3. Wash cells once with PBS and add 2 ml of OPTI-MEM to the plates. Add transfection reagent and incubate the cells at 37° and 5% CO_2 for 45 min. Remove transfection mixture and wash twice with 1 ml of α-MEM complete without antibiotics. Culture cells in this medium for 24 to 72 h prior to testing for transgene expression.

Transfection efficiency will vary depending on the protein being exogenously expressed. In efficient transfections 60 to 90% of cells will express the transgene; however, optimization of the ratio of DNA to transfection reagent may be necessary. Alternative approaches for expressing exogenous protein in MTLn3 cells are available. Retroviral infection is an alternative to transient transfection, which we have used to make stable cell lines (Perrin *et al.*, 2006).

2.2.2. Immunofluorescent imaging of podosomes in fixed cells

A variety of microscopic methods have been used to study podosome composition and morphology in fixed cells. Previous studies have identified proteins that serve as markers of podosomes when they colocalize with actin. These proteins include Arp 2/3 complex and N-WASP (Yamaguchi *et al.*, 2005), cortactin (Hiura *et al.*, 1995), and gelsolin (Chellaiah *et al.*, 2000). Figure 3.1B shows podosome morphology in MTLn3 breast cancer cells using staining for actin and cortactin.

Plating and staining MTLn3 cells for podosome markers

1. Acid wash 12-mm coverslips as described previously in the "Acid Washing Coverslips" section. Place coverslips in a 24-well plate. Coat coverslips with 10 μg/ml fibronectin diluted in PBS by incubating at 37° for 1 h or at 4° overnight. Wash the plate once with PBS, block with 2% BSA at 37° for 30 min, and wash twice with PBS. Coverslips may be stored in PBS at 4° for up to 1 week prior to use.

2. Rinse cells with PBS, aspirate and detach cells from tissue culture plate using a 0.25% trypsin/0.02% EDTA solution in PBS. MTLn3 cells should completely detach from plate within 5 min. Check plate under a microscope to ensure that cells have detached.

3. After cells have detached, add an equal volume of α-MEM complete medium. Spin cells down at 1500$\times$g for 2 min and resuspend in desired volume of medium. Plate 2×10^4 cells per coverslip. Incubate cells at 37° and 5% CO_2 for approximately 3 h.

4. Fix cells between 2 to 4 h after plating for optimum viewing of podosomes. For cell fixation, wash cells with 500 μl of PBS and add 500 μl of 3.7% formaldehyde (Polysciences, Inc., Warrington, PA) in PBS for 10 min.

5. Quench formaldehyde with 1 ml of a freshly made solution of 0.15 M glycine (Sigma Aldrich, St. Louis, MO) in PBS for 10 min and rinse twice with 500 μl of PBS.

6. Permeabilize cells with 500 μl of 0.2% Triton X-100 in PBS for 10 min and then rinse with 500 μl PBS twice, and block the coverslip with 500 μl of 5% goat serum (Biomeda, Foster City, CA) in PBS (filter sterilized through 0.22-μm filter) for 1 h at room temperature or overnight at 4°.

7. Transfer coverslips cell side up to a piece of Parafilm in a humidified chamber for staining. Dilute the primary antibody to the appropriate concentration in 5% goat serum and incubate for 30 min. For example, we generally stain cells with anti-cortactin antibody (Upstate Biotechnology, Inc., Lake Placid, NY) and the F-actin binding rhodamine phalloidin (Molecular Probes, Invitrogen, Carlsbad, CA) to label podosomes.

8. Rinse three times with PBS. Simultaneously aspirate and add PBS (allow 100 μl to remain on the coverslip after last wash). Let coverslip incubate for 5 min. Repeat twice.

9. Add 100 μl fluorescently conjugated secondary antibody diluted in 5% goat serum to coverslips and incubate for 30 min.

10. Rinse three times with PBS as in Step 8 and allow a fourth wash to remain on coverslip to prevent drying. Mount cells using a mounting media.

Mounting media is commercially available or can be prepared (6% [v/v] glycerol, 24 mM Tris-Cl, pH 8.5, 0.6 mg/ml phenylaminediamine, and 4.8% [w/v] polyvinyl alcohol) and stored at $-80°$. Mounting media should appear pale orange and should be discarded if it turns brown.

11. Mount coverslips on clean, labeled glass slides with 5 to 10 μl of mounting media. Carefully pick up coverslip with forceps and rinse in distilled water to remove PBS. This will prevent crystal formation during the drying process. Aspirate any excess liquid and mount cell side down on the glass slide.

2.2.3. Podosome degradation of ECM

A number of studies suggest that podosomes enhance the invasive properties of cells (reviewed by Gimona and Buccione, 2006). A variety of techniques are available to monitor ECM degradation such as invasion chambers (modified Boyden chambers) and fluorescently labeled ECM-coated coverslips, where podosome markers have been shown to localize at areas of degraded ECM. This matrix degradation occurs, in part, due to the enhanced expression and secretion of matrix metalloproteases (MMPs) (Linder and Aepfelbacher, 2003). Below, we describe a recently published method developed by Artym and Mueller and modified by our lab for coating coverslips with fluorescent gelatin, which we use to assess invadopodia formation in MTLn3 cells (Artym et al., 2006).

Preparation of Oregon-green conjugated gelatin coverslips

1. Dissolve Oregon-green 488-conjugated gelatin (Invitrogen, Carlsbad, CA) to 0.2% in PBS.
2. Coat pre-washed, acid-washed 12-mm coverslips with 150 μl of 50 μg/ml poly-L lysine (Sigma Aldrich, St. Louis, MO) in PBS for 20 min.
3. Rinse coverslips with 500 μl of PBS and add 500 μl of 0.5 % glutaraldehyde (Sigma Aldrich, St. Louis, MO). Incubate for 15 min. Rinse well with PBS (about 4 ml per coverslip).
4. Invert coverslips onto 80 μl of gelatin solution (from Step 1) on a piece of Parafilm. Incubate for 10 min.
5. Carefully place coverslips gelatin side up onto a new piece of Parafilm, and quench residual glutaraldehyde reactive groups with 150 μl of a freshly made 5 mg/ml sodium borohydrate (Sigma Aldrich, St. Louis, MO) solution in PBS. Rinse well with PBS (about 4 ml per coverslip).
6. Place coverslips in 24-well plates. Store in sterile PBS at 4° until use.

Plating cells on gelatin matrix and staining for invadopodia

1. Rinse cells with PBS, aspirate and detach cells from tissue culture plate using 0.25% trypsin/0.02% EDTA in PBS.

2. After cells have detached, add an equal volume of α-MEM medium. Spin cells down at $1500 \times g$ for 2 min and resuspend in medium. Plate 2×10^4 cells per coverslip. Incubate cells at 37° and 5% CO_2 for 6 h.

3. Fix cells at 6 h after plating for viewing ECM degradation. Fix and permeabilize cells as described above in Steps 4 to 6 in "Plating and Staining MTLn3 Cells for Podosome Markers" section.

4. Transfer coverslips to a piece of Parafilm in a humidified chamber for staining. Dilute primary antibody to the appropriate concentration in 5% goat serum and incubate for 30 min. To identify invadopodia, we generally stain cells with the anti-cortactin antibody described previously.

5. Rinse three times with PBS. Simultaneously aspirate and add PBS (allow 100 μl to remain on the coverslip after last wash). Let coverslip incubate for 5 min. Repeat twice.

6. Add 100 μl fluorescently conjugated secondary antibody diluted in 5% goat serum to coverslips and incubate for 30 min. We use a rhodamine red-conjugated secondary antibody (Invitrogen, Carlsbad, CA) to observe podosome markers that co-localize with degraded ECM on the green gelatin background.

7. Rinse three times with PBS as in Step 5 and allow a fourth wash to remain on coverslip to prevent drying. Mount cells on clean slides using mounting media as described above in "Plating and Staining MTLn3 Cells for Podosome Markers."

Invadopodia formation is assessed by counting the number of cells containing podosome structures (identified by staining for the podosome marker cortactin) that co-localize with areas of ECM degradation. In the assay described above, ECM degradation appears as black spots on a green gelatin background (Fig. 3.3). The area of degraded ECM will vary per cell, as will the percentage of cells actively degrading ECM in each experiment.

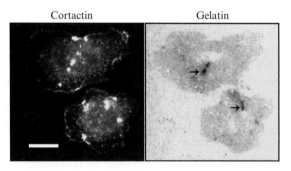

Cortactin   Gelatin

Figure 3.3 Immunofluorescent image of an MTLn3 cell stained with an anti-cortactin antibody on a coverslip coated with fluorescently labeled gelatin. Arrows depict areas of degradation. Bar, 10 microns.

We typically find between 50 and 75% of MTLn3 cells to contain invadopodia structures when assayed as described above.

2.2.4. Fluorescent imaging of podosomes in live cells

Live imaging of GFP-cortactin or fusion proteins to other podosome components provides a powerful tool to study podosome dynamics in real time. Time-lapse analysis of MTLn3 cells stably expressing GFP-actin has previously shown that podosomes are highly dynamic structures and that a subset of podosomes will form more stable structures that are associated with matrix degradation (Yamaguchi *et al.*, 2005). Below, we describe the live imaging techniques that we use to study podosome dynamics in MTLn3 cells that express the podosome marker, GFP-cortactin (Fig. 3.2B). An appropriate control is the analysis of cells that express GFP alone.

1. Coat a glass bottom dish with ECM protein as described above in the "Preparation of Glass Bottom Plates" section. We typically use 2 to 3 ml of 10 μg/ml fibronectin in PBS for coating the plate.
2. Prewarm Ham's/F-12 medium containing 5% FBS and 20 mM HEPES at pH 7.2.
3. Wash cells that express GFP-cortactin once with PBS. Lift cells with 0.25% trypsin/0.02% EDTA in PBS and incubate cells until detached.
4. After the cells have detached, add an equal volume of α-MEM complete medium. Spin cells down at 1500×g for 2 min in a 15-ml conical tube and wash cells twice with 2 ml of medium.
5. Plate 1 × 10^5 cells on a glass bottom plate containing 2 ml of prewarmed α-MEM complete medium. Cells are plated at low density in order to observe single cells that are not in contact. Allow the cells to adhere for 2 h at 37° and 5% CO$_2$.
6. Wash cells once with PBS to remove cells that have not adhered and add 2 ml of Ham's/F-12 medium (from Step 2). Layer 2 ml of mineral oil on top of the medium to prevent evaporation and place plate on heated stage.
7. Focus on the cells (we use a 60× objective) and allow the plate to adjust to the stage for approximately 15 min prior to imaging.
8. Scan the plate to identify representative cells that express GFP-cortactin for live imaging. GFP-cortactin should clearly localize at punctate dot structures (podosomes) (Fig. 3.1B) in many of the cells.
9. Determine the interval necessary to capture podosome dynamics. Because podosomes are highly dynamic, we capture images every 30 to 60 sec for 1 h.
10. Conduct fluorescent time-lapse video microscopy by capturing images using a CCD camera and MetaVue imaging software. Choose settings that minimize exposure time yet maintain image quality. We typically use gain = 1, exposure time = 100 to 200 msec, and camera bin = 2.

11. Podosome assembly and disassembly rates are determined using the methods discussed in the "Quantification of Adhesion Dynamics" section.

REFERENCES

Artym, V. V., Zhang, Y., Seillier-Moiseiwitsch, F., Yamada, K. M., and Mueller, S. C. (2006). Dynamic interactions of cortactin and membrane type 1 matrix metalloproteinase at invadopodia: Defining the stages of invadopodia formation and function. *Cancer Res.* **66,** 3034–3043.

Balaban, N. Q., Schwarz, U. S., Riveline, D., Goichberg, P., Tzur, G., Sabanay, I., Mahalu, D., Safran, S., Bershadsky, A., Addadi, L., and Geiger, B. (2001). Force and focal adhesion assembly: A close relationship studied using elastic micropatterned substrates. *Nat. Cell Biol.* **3,** 466–472.

Baldassarre, M., Pompeo, A., Beznoussenko, G., Castaldi, C., Cortellino, S., McNiven, M. A., Luini, A., and Buccione, R. (2003). Dynamin participates in focal extracellular matrix degradation by invasive cells. *Mol. Biol. Cell* **14,** 1074–1084.

Beviglia, L., and Kramer, R. H. (1999). HGF induces FAK activation and integrin-mediated adhesion in MTLn3 breast carcinoma cells. *Int. J. Cancer* **83,** 640–649.

Bhatt, A. K., and Huttenlocher, A. (2003). Dynamic imaging of cell–substrate contacts. *Methods Enzymol.* **361,** 337–352.

Bloch, R. J., and Geiger, B. (1980). The localization of acetylcholine receptor clusters in areas of cell-substrate contact in cultures of rat myotubes. *Cell* **21,** 25–35.

Bockholt, S. M., Otey, C. A., Glenney, J. R., Jr., and Burridge, K. (1992). Localization of a 215-kDa tyrosine-phosphorylated protein that cross-reacts with tensin antibodies. *Exp. Cell Res.* **203,** 39–46.

Buccione, R., Orth, J. D., and McNiven, M. A. (2004). Foot and mouth: Podosomes, invadopodia and circular dorsal ruffles. *Nat. Rev. Mol. Cell Biol.* **5,** 647–657.

Buck, C. A., and Horwitz, A. F. (1987). Cell surface receptors for extracellular matrix molecules. *Annu. Rev. Cell Biol.* **3,** 179–205.

Burridge, K., and Connell, L. (1983). Talin: A cytoskeletal component concentrated in adhesion plaques and other sites of actin-membrane interaction. *Cell Motil.* **3,** 405–417.

Chellaiah, M. A., Biswas, R. S., Yuen, D., Alvarez, U. M., and Hruska, K. A. (2001). Phosphatidylinositol 3,4,5–trisphosphate directs association of Src homology 2–containing signaling proteins with gelsolin. *J. Biol. Chem.* **276,** 47434–47444.

Chellaiah, M., Kizer, N., Silva, M., Alvarez, U., Kwiatkowski, D., and Hruska, K. A. (2000). Gelsolin deficiency blocks podosome assembly and produces increased bone mass and strength. *J. Cell Biol.* **148,** 665–678.

Cox, E. A., and Huttenlocher, A. (1998). Regulation of integrin-mediated adhesion during cell migration. *Microsc. Res. Tech.* **43,** 412–419.

Crawford, A. W., and Beckerle, M. C. (1991). Purification and characterization of zyxin, an 82,000–dalton component of adherens junctions. *J. Biol. Chem.* **266,** 5847–5853.

Cukierman, E. (2005). Cell migration analyses within fibroblast-derived 3-D matrices. *Methods Mol. Biol.* **294,** 79–93.

Duong, L. T., and Rodan, G. A. (2000). PYK2 is an adhesion kinase in macrophages, localized in podosomes and activated by beta(2)-integrin ligation. *Cell Motil. Cytoskeleton* **47,** 174–188.

Edlund, M., Lotano, M. A., and Otey, C. A. (2001). Dynamics of alpha-actinin in focal adhesions and stress fibers visualized with alpha-actinin-green fluorescent protein. *Cell Motil. Cytoskeleton* **48,** 190–200.

Ezratty, E. J., Partridge, M. A., and Gundersen, G. G. (2005). Microtubule-induced focal adhesion disassembly is mediated by dynamin and focal adhesion kinase. *Nat. Cell Biol.* **7,** 581–590.

Franco, S. J., Rodgers, M. A., Perrin, B. J., Han, J., Bennin, D. A., Critchley, D. R., and Huttenlocher, A. (2004). Calpain-mediated proteolysis of talin regulates adhesion dynamics. *Nat. Cell Biol.* **6,** 977–983.

Franco, S. J., Senetar, M. A., Simonson, W. T., Huttenlocher, A., and McCann, R. O. (2006). The conserved C-terminal I/LWEQ module targets Talin1 to focal adhesions. *Cell Motil. Cytoskeleton* **63,** 563–581.

Gaidano, G., Bergui, L., Schena, M., Gaboli, M., Cremona, O., Marchisio, P. C., and Caligaris-Cappio, F. (1990). Integrin distribution and cytoskeleton organization in normal and malignant monocytes. *Leukemia* **4,** 682–687.

Gallant, N. D., Michael, K. E., and Garcia, A. J. (2005). Cell adhesion strengthening: Contributions of adhesive area, integrin binding, and focal adhesion assembly. *Mol. Biol. Cell* **16,** 4329–4340.

Gatesman, A., Walker, V. G., Baisden, J. M., Weed, S. A., and Flynn, D. C. (2004). Protein kinase Calpha activates c-Src and induces podosome formation via AFAP-110. *Mol. Cell Biol.* **24,** 7578–7597.

Gavazzi, I., Nermut, M. V., and Marchisio, P. C. (1989). Ultrastructure and gold-immunolabelling of cell-substratum adhesions (podosomes) in RSV-transformed BHK cells. *J. Cell Sci.* **94**(Pt. 1), 85–99.

Generisch, E., Schuppán, D., and Lichtner, R. B. (1996). Signaling by epidermal growth factor differentially affects integrin-mediated adhesion of tumor cells to extracellular matrix proteins. *J. Mol. Med.* **74,** 609–616.

Gimona, M., and Buccione, R. (2006). Adhesions that mediate invasion. *Int. J. Biochem. Cell Biol.* **38,** 1875–1892.

Gupton, S. L., and Waterman-Storer, C. M. (2006). Spatiotemporal feedback between acto-myosin and focal-adhesion systems optimizes rapid cell migration. *Cell* **125,** 1361–1374.

Hai, C. M., Hahne, P., Harrington, E. O., and Gimona, M. (2002). Conventional protein kinase C mediates phorbol-dibutyrate-induced cytoskeletal remodeling in a7r5 smooth muscle cells. *Exp. Cell Res.* **280,** 64–74.

Hiura, K., Lim, S. S., Little, S. P., Lin, S., and Sato, M. (1995). Differentiation dependent expression of tensin and cortactin in chicken osteoclasts. *Cell Motil. Cytoskeleton* **30,** 272–284.

Humphries, J. D., Byron, A., and Humphries, M. J. (2006). Integrin ligands at a glance. *J. Cell Sci.* **119,** 3901–3903.

Huttenlocher, A., Ginsberg, M. H., and Horwitz, A. F. (1996). Modulation of cell migration by integrin-mediated cytoskeletal linkages and ligand-binding affinity. *J. Cell Biol.* **134,** 1551–1562.

Hynes, R. O. (1987). Integrins: A family of cell surface receptors. *Cell* **48,** 549–554.

Johansson, M. W., Lye, M. H., Barthel, S. R., Duffy, A. K., Annis, D. S., and Mosher, D. F. (2004). Eosinophils adhere to vascular cell adhesion molecule-1 via podosomes. *Am. J. Respir. Cell Mol. Biol.* **31,** 413–422.

Kaverina, I., Stradal, T. E., and Gimona, M. (2003). Podosome formation in cultured A7r5 vascular smooth muscle cells requires Arp2/3–dependent de-novo actin polymerization at discrete microdomains. *J. Cell Sci.* **116,** 4915–4924.

Kramer, R. H., Cheng, Y. F., and Clyman, R. (1990). Human microvascular endothelial cells use beta 1 and beta 3 integrin receptor complexes to attach to laminin. *J. Cell Biol.* **111,** 1233–1243.

Laukaitis, C. M., Webb, D. J., Donais, K., and Horwitz, A. F. (2001). Differential dynamics of alpha 5 integrin, paxillin, and alpha-actinin during formation and disassembly of adhesions in migrating cells. *J. Cell Biol.* **153,** 1427–1440.

Lee, E., and De Camilli, P. (2002). Dynamin at actin tails. *Proc. Natl. Acad. Sci. USA* **99**, 161–166.

Linder, S., and Aepfelbacher, M. (2003). Podosomes: Adhesion hot-spots of invasive cells. *Trends Cell Biol.* **13**, 376–385.

Linder, S., Higgs, H., Hufner, K., Schwarz, K., Pannicke, U., and Aepfelbacher, M. (2000). The polarization defect of Wiskott-Aldrich syndrome macrophages is linked to dislocalization of the Arp2/3 complex. *J. Immunol.* **165**, 221–225.

Linder, S., Nelson, D., Weiss, M., and Aepfelbacher, M. (1999). Wiskott-Aldrich syndrome protein regulates podosomes in primary human macrophages. *Proc. Natl. Acad. Sci. USA* **96**, 9648–9653.

Lo, S. H. (2006). Focal adhesions: What's new inside. *Dev. Biol.* **294**, 280–291.

Lokuta, M. A., Nuzzi, P. A., and Huttenlocher, A. (2003). Calpain regulates neutrophil chemotaxis. *Proc. Natl. Acad. Sci. USA* **100**, 4006–4011.

Marcantonio, E. E., and Hynes, R. O. (1988). Antibodies to the conserved cytoplasmic domain of the integrin beta 1 subunit react with proteins in vertebrates, invertebrates, and fungi. *J. Cell Biol.* **106**, 1765–1772.

Marchisio, P. C., Bergui, L., Corbascio, G. C., Cremona, O., D'Urso, N., Schena, M., Tesio, L., and Caligaris-Cappio, F. (1988). Vinculin, talin, and integrins are localized at specific adhesion sites of malignant B lymphocytes. *Blood* **72**, 830–833.

Miyamoto, S., Akiyama, S. K., and Yamada, K. M. (1995). Synergistic roles for receptor occupancy and aggregation in integrin transmembrane function. *Science* **267**, 883–885.

Mizutani, K., Miki, H., He, H., Maruta, H., and Takenawa, T. (2002). Essential role of neural Wiskott-Aldrich syndrome protein in podosome formation and degradation of extracellular matrix in src-transformed fibroblasts. *Cancer Res.* **62**, 669–674.

Nermut, M. V., Eason, P., Hirst, E. M., and Kellie, S. (1991). Cell/substratum adhesions in RSV-transformed rat fibroblasts. *Exp. Cell Res.* **193**, 382–397.

Palecek, S. P., Huttenlocher, A., Horwitz, A. F., and Lauffenburger, D. A. (1998). Physical and biochemical regulation of integrin release during rear detachment of migrating cells. *J. Cell Sci.* **111**, 929–940.

Perrin, B. J., Amann, K. J., and Huttenlocher, A. (2006). Proteolysis of cortactin by calpain regulates membrane protrusion during cell migration. *Mol. Biol. Cell* **17**, 239–250.

Pfaff, M., and Jurdic, P. (2001). Podosomes in osteoclast-like cells: Structural analysis and cooperative roles of paxillin, proline-rich tyrosine kinase 2 (Pyk2) and integrin alphaV-beta3. *J. Cell Sci.* **114**, 2775–2786.

Reinhard, M., Halbrugge, M., Scheer, U., Wiegand, C., Jockusch, B. M., and Walter, U. (1992). The 46/50 kDa phosphoprotein VASP purified from human platelets is a novel protein associated with actin filaments and focal contacts. *EMBO J.* **11**, 2063–2070.

Rodriguez, L. G., Wu, X., and Guan, J. L. (2005). Wound-healing assay. *Methods Mol. Biol.* **294**, 23–29.

Rohrschneider, L. R. (1980). Adhesion plaques of Rous sarcoma virus-transformed cells contain the src gene product. *Proc. Natl. Acad. Sci. USA* **77**, 3514–3518.

Sastry, S. K., and Burridge, K. (2000). Focal adhesions: A nexus for intracellular signaling and cytoskeletal dynamics. *Exp. Cell Res.* **261**, 25–36.

Segall, J. E., Tyerech, S., Boselli, L., Masseling, S., Helft, J., Chan, A., Jones, J., and Condeelis, J. (1996). EGF stimulates lamellipod extension in metastatic mammary adenocarcinoma cells by an actin-dependent mechanism. *Clin. Exp. Metastasis* **14**, 61–72.

Shaw, L. M. (2005). Tumor cell invasion assays. *Methods Mol. Biol.* **294**, 97–105.

Shibutani, T., Iwanaga, H., Imai, K., Kitago, M., Doi, Y., and Iwayama, Y. (2000). Use of glass slides coated with apatite-collagen complexes for measurement of osteoclastic resorption activity. *J. Biomed. Mater. Res.* **50**, 153–159.

Spinardi, L., Rietdorf, J., Nitsch, L., Bono, M., Tacchetti, C., Way, M., and Marchisio, P. C. (2004). A dynamic podosome-like structure of epithelial cells. *Exp. Cell Res.* **295,** 360–374.

Stupack, D. G., and Cheresh, D. A. (2002). Get a ligand, get a life: integrins, signaling and cell survival. *J. Cell Sci.* **115,** 3729–3738.

Tarone, G., Cirillo, D., Giancotti, F. G., Comoglio, P. M., and Marchisio, P. C. (1985). Rous sarcoma virus-transformed fibroblasts adhere primarily at discrete protrusions of the ventral membrane called podosomes. *Exp. Cell Res.* **159,** 141–157.

Tatin, F., Varon, C., Genot, E., and Moreau, V. (2006). A signalling cascade involving PKC, Src and Cdc42 regulates podosome assembly in cultured endothelial cells in response to phorbol ester. *J. Cell Sci.* **119,** 769–781.

Turner, C. E., Glenney, J. R., Jr., and Burridge, K. (1990). Paxillin: A new vinculin-binding protein present in focal adhesions. *J. Cell Biol.* **111,** 1059–1068.

van der Flier, A., and Sonnenberg, A. (2001). Function and interactions of integrins. *Cell Tissue Res.* **305,** 285–298.

Webb, D. J., Donais, K., Whitmore, L. A., Thomas, S. M., Turner, C. E., Parsons, J. T., and Horwitz, A. F. (2004). FAK-Src signalling through paxillin, ERK and MLCK regulates adhesion disassembly. *Nat. Cell Biol.* **6,** 154–161.

Wells, C. M., and Ridley, A. J. (2005). Analysis of cell migration using the Dunn chemotaxis chamber and time-lapse microscopy. *Methods Mol. Biol.* **294,** 31–41.

Yamaguchi, H., Lorenz, M., Kempiak, S., Sarmiento, C., Coniglio, S., Symons, M., Segall, J., Eddy, R., Miki, H., Takenawa, T., and Condeelis, J. (2005). Molecular mechanisms of invadopodium formation: The role of the N-WASP-Arp2/3 complex pathway and cofilin. *J. Cell Biol.* **168,** 441–452.

Zaidel-Bar, R., Ballestrem, C., Kam, Z., and Geiger, B. (2003). Early molecular events in the assembly of matrix adhesions at the leading edge of migrating cells. *J. Cell Sci.* **116,** 4605–4613.

Zaidel-Bar, R., Cohen, M., Addadi, L., and Geiger, B. (2004). Hierarchical assembly of cell–matrix adhesion complexes. *Biochem. Soc. Trans.* **32,** 416–420.

Zamir, E., and Geiger, B. (2001). Molecular complexity and dynamics of cell–matrix adhesions. *J. Cell Sci.* **114,** 3583–3590.

Zamir, E., Katz, B. Z., Aota, S., Yamada, K. M., Geiger, B., and Kam, Z. (1999). Molecular diversity of cell–matrix adhesions. *J. Cell Sci.* **112,** 1655–1669.

Integrin Cytoskeletal Interactions

Yatish Lad, David S. Harburger, *and* David A. Calderwood

Contents

Abstract

Integrin adhesion receptors mediate cell–cell and cell–substratum adhesion and provide a continuous link for the bidirectional transmission of mechanical force and biochemical signals across the plasma membrane. Integrin-dependent cellular activities such as adhesion, migration, proliferation, and survival rely upon the dynamic interaction of integrin cytoplasmic tails with intracellular integrin-binding proteins. In this review, we describe some of the methods that we have used to identify and characterize the interactions between integrin cytoplasmic tails and cytoskeletal proteins, as well as highlight methods to

Department of Pharmacology and Interdepartmental Program in Vascular Biology and Transplantation, Yale University School of Medicine, New Haven, Connecticut

Methods in Enzymology, Volume 426
ISSN 0076-6879, DOI: 10.1016/S0076-6879(07)26004-5

© 2007 Elsevier Inc.
All rights reserved.

decipher the regulation of integrin tail interactions with intracellular ligands. Specifically, we describe recombinant models of integrin cytoplasmic tails and their use in protein–protein interaction studies.

 ## 1. INTRODUCTION

Integrin adhesion receptors are heterodimeric glycoproteins essential for normal tissue development and homeostasis in multicellular animals. Cell–cell and cell–substratum adhesion is mediated by the binding of integrin extracellular domains to diverse protein ligands; however, cellular control of these adhesive interactions and their translation into dynamic cellular responses, such as cell spreading or migration, requires integrin cytoplasmic tails. These short tails bind to intracellular ligands that connect the receptors to signaling pathways and cytoskeletal networks (Brakebusch and Fassler, 2003; Calderwood *et al.*, 2000; Critchley, 2000; Geiger *et al.*, 2001; Giancotti and Ruoslahti, 1999; Liu *et al.*, 2000). Hence, by binding both extracellular and intracellular ligands, integrins provide a transmembrane link for the bidirectional transmission of mechanical force and biochemical signals across the plasma membrane regulating cell adhesion, migration, proliferation, and death.

Integrins are composed of noncovalently associated α and β transmembrane subunits. Each subunit is a type I transmembrane glycoprotein with a relatively large multidomain extracellular portion, a single membrane spanning helix and a short (20 to 70 amino acids) largely unstructured cytoplasmic tail (Hynes, 2002). The exception to this picture is the $\beta 4$ integrin subunit, which has a large (~1000 amino acid) cytoplasmic domain containing independently folding domains (de Pereda *et al.*, 1999; LaFlamme *et al.*, 1997). Humans contain 18 α and 8 β subunits that combine to produce at least 24 different heterodimers, each of which can bind to a specific repertoire of extracellular ligands such as fibronectin, collagen, laminin, or fibrinogen, or to cell surface counter-receptors like ICAM-1 or VCAM-1 (Hynes, 2002). Integrin binding to extracellular ligand is regulated by conformational changes in the integrin extracellular domains. These changes are induced in response to interactions of the short integrin cytoplasmic domains with intracellular cytoskeletal and signaling proteins (Liddington and Ginsberg, 2002). Furthermore, following the binding of extracellular ligand, complex multiprotein assemblies of cytoskeletal, scaffolding, and signaling proteins are recruited to the integrin cytoplasmic face that serve to link integrin to the actin cytoskeleton and convey signals into the cell (Miranti and Brugge, 2002). Cataloguing, characterizing, and elucidating the temporal and spatial regulation of integrin cytoplasmic tail interactions with intracellular proteins are therefore central to our understanding of integrin adhesion receptors and how they govern cellular activities.

A variety of methods have been employed to identify proteins that bind directly or indirectly to integrin cytoplasmic tails, including co-localization, co-purification, *in vitro* binding and yeast two-hybrid assays (Hemler, 1998; Liu *et al.*, 2000). Reported integrin tail-binding proteins now number more than 73, and exhibit varying integrin specificities. We have been most interested in integrin cytoskeletal interactions, and particularly in the actin-binding proteins that associate with integrin β subunit cytoplasmic tails. We have shown that talin binds integrin β tails and that this interaction is a key step in integrin activation (Calderwood *et al.*, 1999; Tadokoro *et al.*, 2003). Likewise we have seen that the binding of filamin to integrin β tails can modulate cell migration (Calderwood *et al.*, 2001), and that filamin and talin have overlapping binding sites on the β tail that results in competition for binding to integrins (Kiema *et al.*, 2006), and provides an additional avenue for regulation of integrin function. Our studies have generally relied on the ability to identify β tail binding proteins from cell lysates using pull-down or affinity chromatography assays. Following identification of integrin-binding proteins we have localized the integrin-binding site by performing binding assays with cell lysates containing proteolysed fragments of the target protein or transiently expressed epitope-tagged domains of the proteins of interest. We test whether the interaction is direct using smaller bacterially expressed and purified fragments of the protein, and also use purified protein–protein binding assays to quantify the affinity of the interaction. We screen for point mutations in either the integrin β tail or the binding protein that selectively perturb the interaction, using structural biology, where possible, to inform our choice of mutations. This facilitates investigation of the functional significance of the interaction in cells through expression of mutants and dominant negative constructs and by reconstitution of deficient cells. This general strategy has allowed us, in collaboration with others, to characterize integrin β tail interactions with talin, filamin, CD98, the kindler syndrome protein kindlin-1, and PTB-domain containing proteins (Calderwood *et al.*, 1999, 2001, 2003; Kloeker *et al.*, 2004; Zent *et al.*, 2000), while others have used it to investigate other proteins, most notably integrin β tail–Src family (Arias-Salgado *et al.*, 2003) and Syk family kinase interactions (Woodside *et al.*, 2001), and $\alpha4$ integrin binding to paxillin (Liu *et al.*, 1999). In some cases, these studies have now been further extended to *in vivo* experiments using knockin mouse models (Chen *et al.*, 2006; Feral *et al.*, 2005; Czuchra *et al.*, 2006).

Our strategy relies on a facile versatile integrin tail-binding assay. This has been made possible through the use of a recombinant, bacterially expressed, model of integrin cytoplasmic tails. In this chapter, we will describe the recombinant integrin tail model proteins; and provide methods for their generation, expression, purification, and immobilization onto affinity matrix, and use in standard pull-down assays and protein–protein interaction studies. We will also briefly highlight some more advanced uses to which the model proteins have been put.

2. MODELS OF INTEGRIN CYTOPLASMIC TAILS

Our models of integrin tails were first described and generated in Mark Ginsberg's lab; initially as chemically synthesized polypeptides (Muir *et al.*, 1994), and subsequently as recombinant bacterially produced proteins (Pfaff *et al.*, 1998). The following features of integrin transmembrane and cytoplasmic domains need to be considered in designing models:

1. Integrin subunits contain a single hydrophobic transmembrane domain that is predicted to adopt a helical conformation followed by a short, largely unstructured cytoplasmic tail.
2. Integrins are heterodimers of α and β subunits.
3. Integrin cytoplasmic tails emerge from the plasma membrane at defined points that specify the length of the tails and their stagger with respect to one another.
4. α and β tails can interact and this interaction serves to regulate integrin function and can be modulated by integrin tail-binding proteins.
5. Integrin transmembrane domains can interact, and the formation of homotypic (i.e., α-α or β-β) or heterotypic (α–β) interactions may modulate integrin function.

The model proteins we use (Fig. 4.1A) are designed to address these considerations, to promote production of soluble protein, and to facilitate purification of the model protein and their immobilization on affinity matrices. Each subunit consists of an N-terminal His-tag® sequence followed by a thrombin cleavage site, a cysteine-residue linker, a coiled-coil–forming helical sequence, a variable zero-to-three residue glycine spacer, and the integrin cytoplasmic domain. The N-terminal His-tag® allows purification of the recombinant model tails from bacteria by metal ion–affinity chromatography and also allows immobilization of the purified integrin tail on Ni^{+2}-NTA resin for use in binding assays. The thrombin cleavage site permits proteolytic removal of the His-tag®. The cysteine linker allows dimerization of integrin tails and ensures formation of a parallel coiled-coil with the correct vertical stagger to orient the integrin tails correctly with respect to one another. The coiled-coil region is designed to mimic the helical structure of integrin transmembrane domains and to mediate dimerization of subunits in aqueous solution. The original model proteins, and those in general use in our laboratory, contain a 28–amino acid amphiphilic sequence composed of four heptad repeats from the protein tropomyosin known to form a helical secondary structure and coiled-coil tertiary and quaternary structures in aqueous solutions (Lau *et al.*, 1984). These may be used to produce either heterotypic or homotypic dimers to mimic α–β dimers or the β-β or α–α interactions present in activated clustered integrins (Li *et al.*, 2003; Liu *et al.*, 1999; Pfaff *et al.*, 1998).

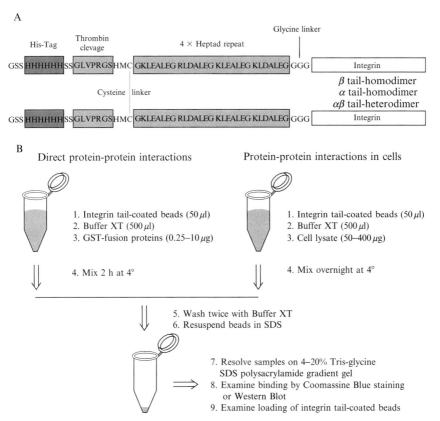

Figure 4.1 (A) Model of integrin cytoplasmic tails containing an N-terminal His-tag® sequence followed by a thrombin cleavage site, a cysteine-residue linker, a coiled-coil–forming helical sequence, a three-residue glycine spacer, and the integrin cytoplasmic domain. (B) Schemata of the integrin tail-binding assay with either cell lysates or purified proteins.

For applications requiring the use of heterodimers, Ginsberg's lab has now improved this system by replacing the tropomyosin repeats with two variants of four heptad repeats derived from the GCN4 transcription factor that preferentially heterodimerize (Ginsberg *et al.*, 2001). Initial experiments with the model proteins showed that propagation of helical structure from the coiled-coil–forming region into the integrin tail could impact integrin tail interactions with cytoskeletal proteins (Pfaff *et al.*, 1998). To limit this effect, a glycine spacer is inserted between the coiled-coil–forming repeats and the integrin tail sequence. Circular dichroism spectroscopy confirmed that introduction of three glycines prevented propagation of the helical structure (Pfaff *et al.*, 1998); therefore, we routinely use this form of the model tails in our assays.

These model proteins, and similar related models produced in other labs (Travis *et al.*, 2004), have now been extensively used to identify integrin tail-binding proteins from cell lysates and measure binding to other purified proteins (Calderwood *et al.*, 1999, 2001, 2002, 2003, 2004; Fenczik *et al.*, 2001; Kiema *et al.*, 2006; Ulmer *et al.*, 2003; Woodside *et al.*, 2002; Yan *et al.*, 2001; Zent *et al.*, 2000). They have been validated by fluorescence quenching, NMR spectroscopy, CD spectroscopy, size-exclusion chromatography and epitope mapping (Ginsberg *et al.*, 2001; Muir *et al.*, 1994; Pfaff *et al.*, 1998; Ulmer *et al.*, 2001), and represent a powerful versatile tool for investigating integrin tail interactions.

3. EXPRESSION CONSTRUCTS FOR INTEGRIN TAIL MODEL PROTEINS

Recombinant model proteins are produced using a modified version of the pET-15b bacterial expression vector (Novagen). The vector encodes an N-terminal 6-histidine His-tag® followed by a thrombin cleavage site. DNA coding for the four heptad repeats of the coiled-coil sequence and the glycine spacer is cloned in-frame after the thrombin cleavage site using *Nde*I and *Hind*III restriction sites, and the integrin tail coding sequence is inserted using *Hind*III and *Bam*HI restriction sites. This modular design makes exchanging the integrin tail coding sequences straightforward—DNA encoding different wild-type, mutant, or chimeric integrin tails with *Hind*III and *Bam*HI restriction sites added is simply generated by PCR and subcloned into the expression vector.

4. PURIFICATION OF INTEGRIN CYTOPLASMIC TAILS

Recombinant integrin cytoplasmic tail model proteins are produced and purified using the protocol detailed below, which is adapted from the handbook for His-tagged fusion proteins (Amersham). Modified pET15b vectors encoding the model proteins are transformed into BL21 *Escherichia coli* strains that permit expression from the T7 promoter, maximize protein expression, and reduce protein degradation. Ampicillin-resistant colonies are screened for recombinant protein production, and high-yielding clones are selected for large-scale purification experiments. The protocol described below is routinely used for production of wild-type and mutant β1A, β3, and β7 tails in our laboratory, but can also be applied to other α or β tails; however, individual results may vary and conditions should be optimized for each integrin or integrin mutant. The majority of integrin β tails are present in the bacterial inclusion bodies; however, we find that αIIb cytoplasmic tails express well in the soluble phase. Therefore, production of aIIb tails does not require preparation of the inclusion bodies as described below.

1. Seed 500 ml of Luria broth (LB) containing 50 μg/ml of ampicillin with a culture of BL21 E. *coli* containing the pET-15bM model tail expression construct, and culture overnight at 37° in a shaking bacterial incubator.

2. Harvest the bacteria from the overnight culture by centrifugation, resuspend in 30 ml of LB, and use to inoculate 6 liters of LB containing 50 μg/ml ampicillin. Grow the cultures at 37° with shaking until the OD_{600} reaches approximately 0.4. For the production of αIIb tails° cultures are grown to an approximate OD_{600} of 0.6.

3. Add isopropyl-beta-D-thiogalactopyranoside (IPTG) to a final concentration of 1 mM, to induce protein expression and culture for a further 2.5 h.

4. Harvest the bacteria by centrifugation and resuspend in 450 ml of lysis buffer (20 mM Tris-HCl pH 7.9, 0.5 M NaCl, 5 mM imidazole, 1% Triton X-100, 100 mM PMSF, and 1 μg/ml aprotinin), and lyse on ice by sonication. Sonicate in 10-sec bursts followed by 50 sec on ice to avoid overheating the bacterial extract, and repeat until the extract is no longer viscous.

5. Pellet the insoluble debris by centrifugation at 20,000×g at 4° for 20 min. The precipitate contains the inclusion bodies where most of the recombinant integrin tails are found. In the case of αIIb model tails, the supernatant contains most of the recombinant proteins and this should be used directly in Step 9.

6. Resuspend and solubilize the integrin containing inclusion bodies by sonication in 150 ml of binding buffer with urea (20 mM Tris-HCl, pH 7.9; 0.5 M NaCl, 5 mM imidazole, 0.2% Triton X-100, and 8 M urea).

7. Centrifuge at 30,000×g at 4° for 20 min to remove insoluble debris and collect the supernatant.

8. During the final centrifugation step equilibrate 10 ml of Ni^{2+}-NTA resin (Novagen). Wash with 15 volumes of dH_2O followed by 15 volumes of charge buffer (50 mM $NiCl_2$). Wash the beads with a final 15 volumes of binding buffer with urea and mix with the collected supernatant from Step 7.

9. Incubate the Ni^{2+}-NTA resin with the bacterial lysate overnight at 4° with slight agitation to avoid bead sedimentation.

10. Load the resin onto a column and wash with two column volumes of binding buffer with urea followed by 10 column volumes of wash buffer with urea (20 mM Tris-HCl pH 7.9, 0.5 M NaCl, 60 mM imidazole, 0.2% Triton X-100, and 8 M urea).

11. Elute the integrin tails with seven column volumes of elution buffer with urea (20 mM Tris-HCl pH 7.9, 0.5 M NaCl, 1 M imidazole, 0.2% Triton X-100, and 8 M urea), and collect the eluate in 10-ml fractions.

12. Analyze elution fractions by SDS-PAGE on a 15% Tris–glycine SDS polyacrylamide gel to identify fractions containing the purified integrins tails, which are approximately 10 to 15 kDa.

13. Extensively dialyze positive fractions against 0.1% trifluoroacetic acid (TFA) and 5 mM dithiothreitol using 3.5-kDa MW cutoff dialysis tubing (Spectra/Por RC).

14. Filter the dialyzed fractions through a 0.2-μm filter and load onto a reverse-phase C18 high-pressure liquid chromatography column. Elute the integrin tails from the column using a 20 to 80% gradient of acetonitrile/0.1% TFA.

15. At this stage, fractions containing purified integrin tails can be analyzed by electrospray-ionization mass spectroscopy to confirm that they have the predicted mass and then lyophilized and stored in a desiccator.

5. Preparation of the Affinity Matrix

In our studies, we have primarily investigated the binding of proteins to integrin β subunit cytoplasmic tails. For this purpose, we have used the homodimeric β-tail model proteins. Experiments using heterodimeric tail proteins require the formation and purification of heterodimers as described by Ginsberg *et al.* (2001). For use in binding assays, the model tails need to be immobilized on an affinity matrix. For this we take advantage of the N-terminal His-tag® and capture the tails on Ni^{2+}-charged NTA-resin beads. This orients the model protein with the integrin tail pointing out from the bead. Coating is performed in the presence of the denaturant urea to prevent precipitation of the tails, which might otherwise result in association of insoluble aggregates with the beads and increase nonspecific background binding to the coated beads. The density of tail coating is potentially important; the standard protocol described below results in high levels of coated tails, and for some applications such as competition assays, this is problematic. In these circumstances, the amount of tail added to the beads needs to be reduced to obtain conditions where a robust tail–ligand interaction is maintained with the minimum coating of integrin tails.

1. Dissolve 1 mg of purified integrin tails in 1 ml of binding buffer containing 8 M of urea by gently rocking at room temperature.

2. Mix dissolved tail with 50 μl Ni^{2+}-charged NTA resin with gentle agitation for 1 h at room temperature.

3. Wash the coated resin sequentially with binding buffer containing 8 M, 4 M, and 2 M of urea, and then in binding buffer without urea. Finally, wash the resin in buffer XT (50 mM NaCl, 10 mM Pipes, 150 mM sucrose, 50 mM NaF, 40 mM $Na_4P_2O_7$.10 H_2O, 1 mM Na_3VO_4, 0.05% Triton X-100, pH 6.8), and resuspend the resin in 500 μl of buffer XT to give 10% (v/v). The coated beads can then be stored at 4° until used.

6. Binding Assays Using Cell Lysates

Model integrin tail proteins immobilized on an affinity matrix can be used to identify binding proteins from cell lysates. Our general protocols for lysing cells and performing binding assays are described in Fig. 4.1B. When designing these assays it is imperative to include suitable negative controls. Since most of our studies involve identification of proteins that bind selectively to integrin β subunit tails, we usually include αIIb model proteins as a negative-control affinity matrix. In many cases, selective β-tail point mutations are also useful negative controls, and we generally use β tails containing a Tyr to Ala substitution in the first NP(I/L)Y motif, which impairs the binding of many β tail-binding proteins.

6.1. Cell lysis

1. Grow tissue culture cells (e.g., Chinese hamster ovary cells)—either untransfected or transfected with mammalian expression constructs—to confluency in 10-cm tissue culture plates.
2. Wash cells with cold PBS and lyse by scraping in 0.5-ml cell lysis buffer (50 mM NaCl, 10 mM Pipes, 150 mM sucrose, 50 mM NaF, 40 mM Na$_4$P$_2$O$_7$.10H$_2$O, 1 mM Na$_3$VO$_4$, pH 6.8, 0.5% Triton X-100, 0.1% sodium deoxycholate and EDTA-free protease inhibitor tablet [Roche]) on ice.
3. Incubate cell lysates on ice for 10 min before clarifying by centrifugation at 20,000×g for 10 min at 4°.
4. Determine the protein concentration of the clarified lysates using a BCA protein assay reagent (Pierce), and use lysates in binding assays immediately or store at −20° until required.

6.2. Binding assays

1. Add 50 to 400 μg of cell lysate to 500 μl of buffer XT containing 3 mM of MgCl$_2$. To this mixture, add 50 μl of the 10% (v/v) suspension of integrin-coated beads (described above), mix overnight at 4°.
2. Wash the integrin-coated resin twice with 1 ml of buffer XT to remove unbound proteins.
3. Elute bound proteins by resuspending the beads in SDS sample buffer and heating to 95° for 5 min.
4. Resolve samples on a 4 to 20% Tris-glycine SDS polyacrylamide gradient gel, and transfer to nitrocellulose membranes for immunoblot detection with appropriate antibodies. Loading of the integrin tails on the Ni^{2+}-coated resin can be assessed by Coomassie Blue staining of the 5- to 20-kDa portion of the polyacrylamide gel.

7. DIRECT PROTEIN–PROTEIN BINDING ASSAYS

The assays described above are useful for identifying integrin-binding proteins and for localization of the domains involved in binding, but they do not indicate whether the binding is direct or indirect. To test for direct protein–protein interactions, we use purified integrin tails and purified binding partners (Fig. 4.1B). It is often convenient to produce the integrin-binding proteins as recombinant GST–fusion proteins in bacteria. This allows purification of the recombinant protein, and in cases where we lack a specific antibody to the integrin-binding protein or domain of interest, the GST tag acts as a useful epitope for detection with anti-GST antibodies. If necessary, the GST can be removed from the fusion proteins by thrombin cleavage at a site between the GST tag and the fusion partner. The protocol described below for production of GST-fusion proteins is essentially that described in the handbook for GST-fusion protein purification (Amersham). We have scaled the purification to obtain sufficient protein for several experiments, but conditions vary for different expression constructs and need to be optimized in each case.

7.1. Production of GST-fusion proteins

1. Inoculate 100 ml of LB containing the appropriate antibiotics with BL21 *E. coli* culture containing the GST-tagged fusion protein expression construct. Grow overnight at $37°$ in a shaking incubator.
2. Use the overnight culture to inoculate 2 liters of LB with appropriate antibiotic selection. Grow the cultures to an approximate OD_{600} of 0.4.
3. Induce the bacteria with IPTG at a final concentration of 0.2 mM and grow for another 3 h.
4. Harvest the bacteria by centrifugation and resuspend in 60 ml of GST lysis buffer (phosphate buffered saline [PBS], pH 7.4, 1% Triton X-100, 1 mM dithiothreitol and protease inhibitor tablet [Roche]) on ice. Lyse by sonication, and repeat until the bacterial extract is no longer viscous.
5. Centrifuge at $30,000 \times g$ for 25 min at $4°$ and collect the supernatant.
6. During this final centrifugation step, prepare 0.5 ml glutathione-sepharose beads (Amersham). Wash with 10 volumes of PBS, and then mix with the supernatant from Step 5.
7. Incubate the glutathione-sepharose beads with the bacterial lysate overnight at $4°$ with slight agitation to avoid bead sedimentation.
8. Load the beads onto a column. Wash once with 20 volumes of PBS, 1% Triton X-100, and 1 mM dithiothreitol followed by two washes with 20 volumes of GST wash buffer (0.1 M Tris-HCl, pH 7.5, 0.1 M NaCl, 1% Triton X-100, 1 mM dithiothreitol and protease inhibitor tablet).
9. Elute GST fusion proteins with 20 volumes of 20-mM glutathione in GST wash buffer and collect the eluate in 1-ml fractions.

10. Analyze elution fractions on a 10% Tris–glycine SDS polyacrylamide gel to identify fractions containing the GST-fusion protein. Determine the concentration of fusion protein in positive fractions and store at −20° until required for use.

7.2. Binding assays with purified GST-fusion proteins

1. Add 0.25 to 10 μg of purified GST-fusion proteins to 500 μl of buffer XT. To this mixture, add 50 μl of the 10% (v/v) suspension of integrin-coated beads (described above) and mix at 4° for 2 h.
2. Wash the beads and elute bound proteins in SDS sample buffer as described above, and analyze bound proteins by SDS–PAGE followed by Coomassie Blue staining or immunoblotting of proteins after transfer to nitrocellulose membranes. Loading of integrin tails on the Ni^{2+}-coated resin can be assessed by Coomassie Blue staining of the 5- to 20-kDa portion of the polyacrylamide gel.

8. Advanced Applications

The assays described above have proved useful for identification of integrin tail-binding proteins, demonstration of direct integrin-binding interactions, localization of binding sites, and characterization of mutations that perturb binding. Furthermore, by performing dose–response experiments with increasing amounts of purified integrin tail-binding proteins, these assays allow estimation of apparent affinity constants. Nonetheless, the model integrin tails can be used in more advanced assays to provide additional information on the regulation of integrin tail interactions and to obtain structural information on integrin tails. Some of these applications are highlighted below.

8.1. Competition assays

By modifying the protein–protein assays to include three factors, we can obtain additional information on competitive interactions between proteins. In these experiments, it is important to optimize coating of the integrins on the Ni^{2+}-NTA resin to limit the available free tails and maximize competition. Using this system we have shown that filamin and talin can compete for binding to the $\beta7$ integrin tail (Kiema *et al.*, 2006). This was achieved by mixing two GST-tagged fusion proteins of different sizes, one containing the integrin-binding site of talin, and the other containing the integrin-binding site of filamin with integrin β tails. The binding of the talin fragment to integrin β tails in the presence of increasing concentrations of the filamin fragment, and of the filamin fragment in the presence of increasing concentrations of the talin fragment, was assessed by Coomassie staining of proteins eluted from the integrin-coated beads. Competition experiments can also be performed using inhibitory peptides; thus, we find that filamin binding to $\beta 7$ tails is inhibited by peptides corresponding to the filamin-binding site in $\beta 7$, but not by peptides containing phosphorylated residues that inhibit filamin binding.

8.2. Investigation of the role of integrin phosphorylation

As noted above competition experiments with phosphorylated peptides can reveal a role for phosphorylation in regulating integrin tail interactions with intracellular ligands. Phospho-blocking or phospho-mimicking mutations can also be introduced into the integrin model proteins to test the role of phosphorylation on ligand binding. Using such techniques we have shown that phosphorylation of the threonine residues in the $\beta 7$ tail can inhibit filamin binding (Kiema *et al.*, 2006). Model tails can also be used as substrates in kinase assays and proved invaluable in mapping phosphorylation sites in the $\alpha 4$ integrin tail and in demonstrating that this phosphorylation regulates paxillin binding (Han *et al.*, 2001).

8.3. ELISA using model tail proteins

The direct protein–protein assays have now been modified to permit analysis on 96 well plates in enzyme-linked immunosorbent assays (Arias-Salgado *et al.*, 2003, 2005). This minimizes reagent usage, makes sample handling easier, and allows higher throughput assays to be performed. Adaptation to the ELISA format involved modification of the model protein to introduce a 15-residue peptide between the thrombin cleavage site and the coiled-coil–forming region. This peptide incorporates biotin *in vivo* upon expression in *E. coli*, and the new biotinylated model protein can be purified using the standard protocol detailed earlier. For use in ELISA assays, the biotinylated model proteins are captured on neutravidin-coated micro-titer plates, and subsequent binding of purified GST-fusion proteins to the immobilized integrin tails is assessed using anti-GST antibody followed by detection with horseradish peroxidase–conjugated secondary antibody and a chromogenic substrate.

8.4. Surface plasmon resonance assays

While the model tail proteins can be used in standard protein–protein interaction assays or ELISA to estimate the affinity of interactions, these pull-down assays are not ideally suited to measuring affinity constants and kinetic parameters. Consequently, we adapted the model proteins for use in

BIAcore surface plasmon-resonance assays to obtain real-time data on the interactions between integrin β tails and talin or Syk kinases (Calderwood *et al.*, 2002; Woodside *et al.*, 2002; Yan *et al.*, 2001). Modification of the single cysteine in the model tails with biotinyl maleimide permits their immobilization in a unique orientation on the surface of streptavidin-coated sensor chips, and the binding of purified proteins flowed over the model tail-coated surface at concentrations of 10 to 200 nM allows calculation of on and off rates and apparent affinity constants.

8.5. Structural studies of model proteins

To obtain structural information on integrin cytoplasmic domains, the integrin tail model proteins have been investigated by NMR spectroscopy. This confirmed that the coiled-coil region has a folded helical structure (Ginsberg *et al.*, 2001; Ulmer *et al.*, 2001) and indicated that β3 tails are largely flexible and unstructured, although residues Arg724–Ala735 have a propensity to form a helical structure and residues Asn744–Tyr747 (NPLY) have a propensity to adopt reverse-turn conformations (Ulmer *et al.*, 2001). NMR analysis of model β3 tails has also facilitated identification of residues involved in interactions with talin (Garcia-Alvarez *et al.*, 2003; Ulmer *et al.*, 2003).

 9. CONCLUDING REMARKS

Understanding the complex interplay between the growing number of integrin tail-binding proteins that connect integrins with the cytoskeleton and cellular signaling networks remains a major challenge. The integrin tail model system and methods described here have allowed the identification of integrin tail-binding proteins, aided characterization of the functional signif-icance of these interactions, and is contributing to the deciphering of how these interactions are spatially and temporally regulated by signaling net-works and through competition with other integrin tail-binding proteins. The continued use of these systems, in association with a variety of other *in vitro* and *in vivo* methods promises further insights into how integrins control cell adhesion, migration, and survival, *in vivo*.

ACKNOWLEDGMENTS

The authors' work is supported by grants from the National Institutes of Health. DSH is supported by a National Science Foundation Graduate Research Fellowship.

REFERENCES

Arias-Salgado, E. G., Lizano, S., Sarkar, S., Brugge, J. S., Ginsberg, M. H., and Shattil, S. J. (2003). Src kinase activation by direct interaction with the integrin β cytoplasmic domain. *Proc. Natl. Acad. Sci. USA* **100**, 13298–13302.

Arias-Salgado, E. G., Lizano, S., Shattil, S. J., and Ginsberg, M. H. (2005). Specification of the direction of adhesive signaling by the integrin β cytoplasmic domain. *J. Biol. Chem* **280**, 29699–29707.

Brakebusch, C., and Fassler, R. (2003). The integrin-actin connection, an eternal love affair. *EMBO J.* **22**, 2324–2333.

Calderwood, D. A., Fujioka, Y., de Pereda, J. M., Garcia-Alvarez, B., Nakamoto, T., Margolis, B., McGlade, C. J., Liddington, R. C., and Ginsberg, M. H. (2003). Integrin β cytoplasmic domain interactions with phosphotyrosine-binding domains: A structural prototype for diversity in integrin signaling. *Proc. Natl. Acad. Sci. USA* **100**, 2272–2277.

Calderwood, D. A., Huttenlocher, A., Kiosses, W. B., Rose, D. M., Woodside, D. G., Schwartz, M. A., and Ginsberg, M. H. (2001). Increased filamin binding to β-integrin cytoplasmic domains inhibits cell migration. *Nat. Cell Biol.* **3**, 1060–1068.

Calderwood, D. A., Shattil, S. J., and Ginsberg, M. H. (2000). Integrins and actin filaments: reciprocal regulation of cell adhesion and signaling. *J. Biol. Chem.* **275**, 22607–22610.

Calderwood, D. A., Tai, V., Di Paolo, G., De Camilli, P., and Ginsberg, M. H. (2004). Competition for talin results in trans-dominant inhibition of integrin activation. *J. Biol. Chem.* **279**, 28889–28895.

Calderwood, D. A., Yan, B., de Pereda, J. M., Alvarez, B. G., Fujioka, Y., Liddington, R. C., and Ginsberg, M. H. (2002). The phosphotyrosine binding-like domain of talin activates integrins. *J. Biol. Chem.* **277**, 21749–21758.

Calderwood, D. A., Zent, R., Grant, R., Rees, D. J., Hynes, R. O., and Ginsberg, M. H. (1999). The talin head domain binds to integrin β subunit cytoplasmic tails and regulates integrin activation. *J. Biol. Chem.* **274**, 28071–28074.

Chen, H., Zou, Z., Sarratt, K. L., Zhou, D., Zhang, M., Sebzda, E., Hammer, D. A., and Kahn, M. L. (2006). *In vivo* $\beta1$ integrin function requires phosphorylation-independent regulation by cytoplasmic tyrosines. *Genes Dev.* **20**, 927–932.

Critchley, D. R. (2000). Focal adhesions—the cytoskeletal connection. *Curr. Opin. Cell Biol.* **12**, 133–139.

Czuchra, A., Meyer, H., Legate, K. R., Brakebusch, C., and Fassler, R. (2006). Genetic analysis of B1 integrin "activation motis" in mice. *J. cell. Biol.* **174**, 889–899.

de Pereda, J. M., Wiche, G., and Liddington, R. C. (1999). Crystal structure of a tandem pair of fibronectin type III domains from the cytoplasmic tail of integrin $\alpha6\beta4$. *EMBO J.* **18**, 4087–4095.

Fenczik, C. A., Zent, R., Dellos, M., Calderwood, D. A., Satriano, J., Kelly, C., and Ginsberg, M. H. (2001). Distinct domains of CD98hc regulate integrins and amino acid transport. *J. Biol. Chem.* **276**, 8746–8752.

Feral, C. C., Nishiya, N., Fenczik, C. A., Stuhlmann, H., Slepak, M., and Ginsberg, M. H. (2005). CD98hc (SLC3A2) mediates integrin signaling. *Proc. Natl. Acad. Sci. USA* **102**, 355–360.

Garcia-Alvarez, B., de Pereda, J. M., Calderwood, D. A., Ulmer, T. S., Critchley, D., Campbell, I. D., Ginsberg, M. H., and Liddington, R. C. (2003). Structural determinants of integrin recognition by talin. *Mol. Cell* **11**, 49–58.

Geiger, B., Bershadsky, A., Pankov, R., and Yamada, K. M. (2001). Transmembrane extracellular matrix–cytoskeleton crosstalk. *Nat. Rev. Mol. Cell Biol.* **2**, 793–805.

Giancotti, F. G., and Ruoslahti, E. (1999). Integrin signaling. *Science* **285**, 1028–1032.

Ginsberg, M. H., Yaspan, B., Forsyth, J., Ulmer, T. S., Campbell, I. D., and Slepak, M. (2001). A membrane-distal segment of the integrin α IIb cytoplasmic domain regulates integrin activation. *J. Biol. Chem.* **276,** 22514–22521.

Han, J., Liu, S., Rose, D. M., Schlaepfer, D. D., McDonald, H., and Ginsberg, M. H. (2001). Phosphorylation of the integrin α 4 cytoplasmic domain regulates paxillin binding. *J. Biol. Chem.* **276,** 40903–40909.

Hemler, M. E. (1998). Integrin associated proteins. *Curr. Opin. Cell Biol.* **10,** 578–585.

Hynes, R. O. (2002). Integrins: Bidirectional, allosteric signaling machines. *Cell* **110,** 673–687.

Kiema, T., Lad, Y., Jiang, P., Oxley, C. L., Baldassarre, M., Wegener, K. L., Campbell, I. D., Ylanne, J., and Calderwood, D. A. (2006). The molecular basis of filamin binding to integrins and competition with talin. *Mol. Cell* **21,** 337–347.

Kloeker, S., Major, M. B., Calderwood, D. A., Ginsberg, M. H., Jones, D. A., and Beckerle, M. C. (2004). The Kindler syndrome protein is regulated by transforming growth factor-β and involved in integrin-mediated adhesion. *J. Biol. Chem.* **279,** 6824–6833.

LaFlamme, S. E., Homan, S. M., Bodeau, A. L., and Mastrangelo, A. M. (1997). Integrin cytoplasmic domains as connectors to the cell's signal transduction apparatus. *Matrix. Biol.* **16,** 153–163.

Lau, S. Y., Taneja, A. K., and Hodges, R. S. (1984). Synthesis of a model protein of defined secondary and quaternary structure. Effect of chain length on the stabilization and formation of two-stranded α-helical coiled-coils. *J. Biol. Chem.* **259,** 13253–13261.

Li, R., Mitra, N., Gratkowski, H., Vilaire, G., Litvinov, R., Nagasami, C., Weisel, J. W., Lear, J. D., DeGrado, W. F., and Bennett, J. S. (2003). Activation of integrin αIIbβ3 by modulation of transmembrane helix associations. *Science* **300,** 795–798.

Liddington, R. C., and Ginsberg, M. H. (2002). Integrin activation takes shape. *J. Cell Biol.* **158,** 833–839.

Liu, S., Calderwood, D. A., and Ginsberg, M. H. (2000). Integrin cytoplasmic domain-binding proteins. *J. Cell Sci.* **113,** 3563–3571.

Liu, S., Thomas, S. M., Woodside, D. G., Rose, D. M., Kiosses, W. B., Pfaff, M., and Ginsberg, M. H. (1999). Binding of paxillin to α4 integrins modifies integrin-dependent biological responses. *Nature* **402,** 676–681.

Miranti, C. K., and Brugge, J. S. (2002). Sensing the environment: A historical perspective on integrin signal transduction. *Nat. Cell Biol.* **4,** E83–E90.

Muir, T. W., Williams, M. J., Ginsberg, M. H., and Kent, S. B. (1994). Design and chemical synthesis of a neoprotein structural model for the cytoplasmic domain of a multisubunit cell-surface receptor: Integrin α IIb β 3 (platelet GPIIb-IIIa). *Biochemistry* **33,** 7701–7708.

Pfaff, M., Liu, S., Erle, D. J., and Ginsberg, M. H. (1998). Integrin β cytoplasmic domains differentially bind to cytoskeletal proteins. *J. Biol. Chem.* **273,** 6104–6109.

Tadokoro, S., Shattil, S. J., Eto, K., Tai, V., Liddington, R. C., de Pereda, J. M., Ginsberg, M. H., and Calderwood, D. A. (2003). Talin binding to integrin β tails: A final common step in integrin activation. *Science* **302,** 103–106.

Travis, M. A., van der Flier, A., Kammerer, R. A., Mould, A. P., Sonnenberg, A., and Humphries, M. J. (2004). Interaction of filamin A with the integrin β 7 cytoplasmic domain: Role of alternative splicing and phosphorylation. *FEBS Lett.* **569,** 185–190.

Ulmer, T. S., Calderwood, D. A., Ginsberg, M. H., and Campbell, I. D. (2003). Domain-specific interactions of talin with the membrane-proximal region of the integrin β3 subunit. *Biochemistry* **42,** 8307–8312.

Ulmer, T. S., Yaspan, B., Ginsberg, M. H., and Campbell, I. D. (2001). NMR analysis of structure and dynamics of the cytosolic tails of integrin α IIb β 3 in aqueous solution. *Biochemistry* **40,** 7498–7508.

Woodside, D. G., Obergfell, A., Leng, L., Wilsbacher, J. L., Miranti, C. K., Brugge, J. S., Shattil, S. J., and Ginsberg, M. H. (2001). Activation of Syk protein tyrosine kinase through interaction with integrin β cytoplasmic domains. *Curr. Biol.* **11,** 1799–1804.

Woodside, D. G., Obergfell, A., Talapatra, A., Calderwood, D. A., Shattil, S. J., and Ginsberg, M. H. (2002). The N-terminal SH2 domains of Syk and ZAP-70 mediate phosphotyrosine-independent binding to integrin β cytoplasmic domains. *J. Biol. Chem.* **277,** 39401–39408.

Yan, B., Calderwood, D. A., Yaspan, B., and Ginsberg, M. H. (2001). Calpain cleavage promotes talin binding to the β 3 integrin cytoplasmic domain. *J. Biol. Chem.* **276,** 28164–28170.

Zent, R., Fenczik, C. A., Calderwood, D. A., Liu, S., Dellos, M., and Ginsberg, M. H. (2000). Class- and splice variant–specific association of CD98 with integrin β cytoplasmic domains. *J. Biol. Chem.* **275,** 5059–5064.

CHAPTER FIVE

CELL SURVIVAL IN A THREE-DIMENSIONAL MATRIX

Alireza Alavi *and* Dwayne G. Stupack

Contents

Abstract

Integrin-mediated adhesion acts as a pluripotent mediator of cell signaling, triggering many pathways that promote proliferation and permit them to resist exogenous proapototic insults. To date, most studies have focused on apoptosis among cells adherent to rigid tissue-culture plastic substrates that tends to maximize integrin survival signaling. The physiological interpretation of such studies remains unclear. Here we describe methods to study integrin-mediated cell survival using matched cell populations that differ only in integrin expression, using a three-dimensional (3D) extracellular matrix culture. The preparation of appropriate cell types as well as the use of 3D collagen and fibrin gels is described. Methods to assess apoptosis and their application in the model are detailed.

Department of Pathology, School of Medicine, Moores Cancer Center, University of California at San Diego, La Jolla, California

Methods in Enzymology, Volume 426

ISSN 0076-6879, DOI: 10.1016/S0076-6879(07)26005-7

© 2007 Elsevier Inc.

All rights reserved.

These techniques will offer an opportunity to study cell survival in the context of a non-rigid 3D extracellular matrix.

1. INTRODUCTION

The precise control of cell survival and cell death is crucial for the existence of metazoans (Chinnaiyan and Dixit, 1996). Programmed cell death is critical for both development and to maintain tissue homeostasis in mature organisms. In the latter respect, the selective culling of aberrant cells through programmed cell death has long been thought to prevent a number of pathologies ranging from autoimmune disease to cancer (Thompson, 1995).

Integrins are a family of 24 different heterodimeric glycoproteins that function as the principal adhesion receptors for the extracellular matrix (ECM) (Hynes, 2002), thus providing constant information to the cell regarding its immediate microenvironment. The constant interaction with the environment positions integrins well as a source of information to influence cell fate decisions. Although widely appreciated for their relatively mundane role in mediating cell anchorage, integrins are also responsible for initiating many chemical (Giancotti, 1997) and mechanical (Schwartz and Ingber, 1994) signaling events. Integrins appear to be obligate cofactors for signaling via the receptors for platelet-derived growth factor, basic fibroblast growth factor, epidermal growth factor, and vascular, endothelial cell growth factor, among others (Giancotti and Ruoslahti, 1999).

Integrin-mediated biochemical signaling events have been studied extensively and are triggered following recruitment of cytosolic signaling elements to substrate-ligated integrins. While signaling among different integrins will vary slightly, all involve the activation of downstream signaling pathways that include nonreceptor tyrosine kinases such as src family kinases and focal adhesion kinase (Kornberg et al., 1992; Thomas and Brugge, 1997), small GTPases of the Ras (Schlaepfer et al., 1994) and Rho (DeMali et al., 2003) families, and serine/threonine kinases, ranging from p21-activated kinases upstream to mitogen-activated protein kinases downstream (Schlaepfer et al., 1994). Integrins also activate lipid kinase signaling via phosphoinositide kinases (Hanks et al., 2003). Most of these signaling cascades regulate cell survival pathways, and have been directly implicated in cellular resistance to apoptosis.

Mechanotransduction also appears to play a critical role in regulating cell survival (Ingber, 2002). Although less well studied, it is clear that the capacity to generate tensile forces impacts cell survival, and mediates cell signaling as a factor of cell shape (Ingber, 1997). A cell with integrins ligated at distributed sites permits the establishment of isometric tension (Choquet et al., 1997). This may occur via several means, including scaffolding effects, the transfer of force between actomyosin and microtubular cytoskeletons, or by mechanically induced conformational changes in cell-signaling elements (Ingber, 2006).

Mechanotransduction is intimately tied to biochemical signaling, but has been studied to a limited degree. Our lack of understanding of the mechanisms of mechanotransduction and their integration with biochemical signaling (either direct or via scaffolding functions) has limited efforts to study cell survival. Most of our current understanding of the regulation of cell survival in adherent cells has been performed *in vitro* using tissue culture plastic as a substrate scaffold. The rigidity of the plastic substrate (~1 GPa) (Howe *et al.*, 1998) maximizes the production of high isometric tension in tissue culture cells, creating large areas of integrin contact with the ECM called focal adhesions. Yet studies of integrin contacts within mixed 3D matrices suggest that the large focal adhesion complexes that anchor stress fibers observed in two-dimensional culture do not occur in 3D culture (Cukierman *et al.*, 2001), and in fact the rigidity of tissue culture plastic is several orders of magnitude greater than a 3D matrix such as collagen. This makes it difficult to interpret cell survival studies performed on rigid 2D surfaces *in vitro* (where mechanotransduction is maximized) with respect to a physiological environment *in vivo*. Indeed, the phenomenon in which highly adherent cells resist apoptotic cues has been documented (Dalton, 2003; Ingber, 2002), but it remains unclear to what degree such mechanisms regulate cell survival *in vivo*.

Methods have been established to examine cells living and migrating within a 3D ECM. Our studies have focused on a matrix principally composed of a single defined ECM component—typically collagen or fibrin, although mixed gels, such as Matrigel, are appropriate and also work well. Aside from eliminating issues associated with inappropriate rigidity, 3D ECM gels are porous and do not offer a significant surface area to facilitate deposition of cell-secreted ECM. The use of defined ECM has permitted investigations of the role of integrins not only in the positive detection of matrix, but also in the role that specific integrins may play when their ligand is absent, or when they are antagonized.

For this purpose, it is useful to have cell lines that vary selectively in the expression of the integrin of interest. Such cell lines can be obtained from appropriate knockout mice, but can also be prepared by selection via serial cell sorting and *in vitro* transfection techniques. This chapter will focus on the preparation of defined 3D matrices and their use to study cell survival within a defined, non–rigid ECM.

 ## 2. Protocols

2.1. Isolation of matched cell lines expressing differing levels of integrins

To determine the role of integrins in mediating cell survival, it is useful to select cells expressing differing levels of the integrin of interest. For example, we previously sorted populations of T24 tumor cells that lacked expression of integrin $\beta 3$, and later reconstituted these cells with ectopic $\beta 3$ encoded on pcDNA-neo, a plasmid-bearing, neomycin resistance gene. While selection based on antibiotic incorporation is important to establish stable cell lines, the major basis for selection in all cases was expression of the integrin, as assessed by flow cytometry and cell sorting. This technique has been used to isolate populations of melanoma cells and neuroblastoma cells with differing integrin expression. The technique is not applicable to most primary cell populations, as repeated rounds of sorting and amplification are required.

2.1.1. Materials

cDNA construct encoding the appropriate integrin with selectable marker
   (e.g. pcDNA3.1-β3 integrin, encoding neomycin resistance)
Lipofectamine plus reagent (Invitrogen, or other gene delivery/transfection
   reagent)
Tissue-culture grade G418 (neomycin) stock solution (Invitrogen)
Primary antibody specific for integrin of interest (e.g., monoclonal antibody)
   LM609 specific for $\alpha v \beta 3$ integrin (serologicals)
Phycoerythrin-conjugated anti-mouse secondary antibody (Southern Biotech)
Sterile phosphate buffered saline
Fetal bovine serum (Cell Gro)
Complete tissue culture media suitable for the cell line

2.1.2. Equipment
Sorting flow cytometer

2.1.3. Procedure

1. 20×10^6 T24 cells are suspended in 5 ml of PBS, 3% FBS (PBSS), and incubated on ice with antibody to integrin (25 μg/ml) for 45 min. Cells are pelleted by centrifugation at $100 \times g$ for 5 min and resuspended in 15 ml of PBSS. The procedure is repeated once. Cells are next pelleted and resuspended in 1 ml of 1:40 dilution of stock secondary antibody in PBSS. Cells are incubated on ice for 30 min, with gentle mixing every 10 min. Cells are washed three times and resuspended in 3 to 4 ml of PBSS.
2. The relative expression of integrin is assessed by flow cytometry. Collection gating is set to isolate those cells expressing the lowest level of integrin. The lowest 2% or less are sorted into a 5-ml tube containing complete media, supplemented to 20% FBS. At least 25,000 cells need to be collected. It is important to gate the lowest 2% strictly, as cells tend to drift back toward initial expression levels with prolonged culture.
3. The cells are transferred to a single well on a 24-well plate, and cultured in complete medium, expanding the culture to 100-mm dishes when necessary. Cells are expanded until sufficient cells are available to permit a second round of flow cytometry and sorting. Repeat Steps 1 and 2 above. Repeat the process until populations that stably lack expression of integrin (designated with an "L" to indicate "lacking") are isolated. Typically this requires four or five iterations of Steps 1 to 3.

4. The integrin-lacking T24-L daughter cells are plated on a 100-mm tissue culture dish at 30% confluency, and cells are transfected with 6 μg of cDNA-encoding integrin β3, using lipofectamine according to the manufacturer's instructions. After 2 days, neomycin (500 μg/ml) is added to select for cells stably transfected with the plasmid. Stable cells are expanded until sufficient numbers for sorting, as described in Step 1, are obtained. Initially, approximately 25% of cells that survive selection will actually express integrin, while the remainder of the cells are neomycin resistant, but integrate plasmids in a manner not conducive to expression of the integrin.

5. The transfected cells are sorted according to the procedure using the Steps 1 to 3 outlined above: Cells are sorted to select for both those that express the highest (1 to 2%) level of integrin expression, as well as those cells with the lowest 1 to 2% (integrin negative). The low populations are cells that are resistant to neomycin and have integrated the transfected plasmid, but do not express the integrin. These cells can be used as matched controls in experiments using the integrin-reconstituted cells. After several rounds of sorting, the integrin-expressing cell population should be 100% positive for integrin, with relative expression levels as high or higher than the original parental cell line.

Isolation of these types of matched cell lines are useful for comparing responses based on integrin expression. Cells varying in integrin expression (such as the mother, daughter, and granddaughter cell lines described above) can then be directly compared in cell survival assays. In addition, while not dealt with specifically here, the option exists to express integrin mutants during reconstitution (Step 4). This permits the generation of cell lines with defined integrin deficiencies, and in turn permits analysis of structure–function relationships.

Alternatively, cell expression of integrins can be suppressed using a number of different mechanisms, including stable knock-down by expression of shRNA. In this case, transfection (see Step 4) is performed with selectable plasmids encoding shRNA for integrins (Open Biosystems). Cells are otherwise sorted and selected based on high and low expression, as described. Unlike serial sorting, the shRNA option is more limited in reconstitution experiments, and may require reconstitution with an orthologous gene.

2.2. Preparation of three-dimensional collagen culture

Our studies have focused largely on the role of integrins in modulating the capacity of cells to survive during cell invasion. For these studies, it is not necessary to challenge cells with exogenous apoptotic stimuli such as starvation, irradiation, chemotherapeutics, or death receptor agonists to induce cell death. Rather, the focus is on the capacity of the cells to interact with the local ECM, and live or die based on that interaction (Stupack *et al.*, 2001). However, the system is amenable to approaches that employ exogenous stimuli to induce programmed cell death.

To initiate the study, cells are suspended and interspersed within a 3D matrix, and individual cells are allowed to live and move within the ECM. Cells are monitored for viability as a function of time, using morphological criteria, annexin-V staining, and the incorporation of DNA-binding dyes as indicators of cell death.

2.2.1. Materials

Purified acid-soluble type I collagen (Millipore)
Tissue-culture grade 7.5% sodium bicarbonate solution (Sigma)
$10\times$ M199 solution (Invitrogen)
Complete media suitable for the cell line used
24-well tissue-culture plate
Cell tracker dye stock (5 mM in DMSO) (Invitrogen) (optional)

2.2.2. Procedure

1. Three dimensional collagen gels are prepared by mixing 800 μl of chilled, acid-soluble type I collagen (4 mg/ml) with 93 μl of $10\times$ M199 media (pH 7.4), and then adding 30 μl of 7.5% sodium bicarbonate. The ingredients are mixed thoroughly on ice, with care taken not to introduce bubbles into the gel. The pH of the collagen gel should be neutralized by the presence of the bicarbonate buffer, which can be tracked using pH sensitive dyes (such as phenol red) commonly present in tissue culture media.
2. Add the collagen mixture to 70 μl of cell suspension media containing $\sim$50,000 cells, and ensure that the suspension is well distributed within the mixture. Care should again be taken to prevent introducing air bubbles to the mixture. At this point, work relatively quickly. Once the collagen mixture is removed from the ice, it will begin to aggregate and form the collagen gel.
3. *Optional:* If fluorescent assessment of cell morphology is preferred over brightfield examination, cells may be prelabeled with cell tracker fluors (Invitrogen). One million cells are resuspended in 8 ml of serum-free media containing 1–2 μM celltracker. Cells are incubated at 37° for 10 min, and then pelleted by centrifugation to wash out unbound dye. Concentrations of cell tracker up to 5 μM are not thought to impact cell behavior.
4. After the cell suspension has been diluted into the collagen, use a micropipette to quickly "bead" 15 to 20 μl drops to a 24-well, sterile tissue culture plate. Work relatively quickly to avoid the cells in the gel settling and adhering to the rigid plastic surface rather than remaining within the

collagen matrix. Three drops can be dotted within each well. After all wells have been seeded, replace the cover and carefully invert the plate to create "hanging drops" of collagen and cells. Inverting the plate will minimize sedimentation of cells while the collagen sets, preventing cell accumulation and adhesion to the underlying plastic surface. Move the plate to a 37° humidified tissue culture incubator and allow the collagen gel to harden for 15 to 30 min. Formed gels appear slightly more opaque than the stock collagen solution, and will not move appreciably when the plate is tapped. When the gel has set, the plate is returned to the upright position and 500 μl of media is added to each well.

5. *Optional:* Reagents that impact cell survival can be added at this point. These include antibodies or receptor antagonists or known proapoptotic stimuli. Growth factors or other prosurvival reagents can similarly be added. For example, one might add an antibody to integrin β1 (10 μg/ml) (P4C10, Chemicon), which would block integrin-mediated cell adhesion to collagen, or integrin β3 (10 μg/ml), which is unligated in this ECM. Alternatively, agonistic antibodies to Fas (CH11, 100 ng/ml) or chemotherapeutics such as doxorubicin (500 nM) could be added to induce apoptosis and the impact of collagen culture compared to cells cultured on plastic.

2.3. Scoring apoptosis in collagen gels

Cells can be scored for death immediately, as an indication of the viability of the population of cells at the time of seeding, and may subsequently be followed for several days. There are several mechanisms to score cell survival, including the uptake of membrane impermeable dyes, morphological analysis, the binding of fluorescent probes to activated caspases and the specific staining for structures associated with cell death. In most cases, death occurs via apoptosis, which produces several hallmark features that can be detected, including caspase activation, the presence of negatively charged lipids on the cell surface, cleavage of DNA, and so-called plasma membrane "blebbing," in which cellular components are packaged into small vacuoles/liposomes to permit clearance by other cells (Majno and Joris, 1995). All of these mechanisms can be studied using 3D ECM systems. In fact, the characteristic "satellite" appearance of the blebs surrounding the apoptotic cell body is preserved within the collagen gel. The "entrapping" effect retains vesicles adjacent to the dying cell even after the vacuoles have physically dissociated from the surface. Over a period of hours and days the blebbed vacuoles will eventually decompose, leaving a shrunken nuclear body that can be easily recognized and enumerated. In contrast to this situation in 3D culture, the "blebbing" appearance is largely transient in 2D culture, where the shrunken cell and satellite system will dissociate from one another due to Brownian effects and are dispersed.

2.3.1. Materials

Fluor-conjugated Annexin V stock solution such as fluorescein isothiocyanate (FITC)-coupled-annexin V. Other fluors such as TRITC-coupled
annexin V, Alexa-488 annexin V, or EGFP-Annexin V can be substituted with the correct excitation and emission filter sets. In solution as
supplied by the manufacturer, typically use 20 μg/ml.
Propidium iodide stock solution (100 μg/ml in 20-mM HEPES, pH 7.4)
Hoechst 33342 stock solution (100 μg/ml in 20-mM HEPES, pH 7.4)

2.3.2. Equipment
Confocal microscope or fluorescence microscope

2.3.3. Procedure: "Noninvasive" assessment of cell survival
Counting cells based on morphology (Fig. 5.1) is the least-invasive mechanism to score apoptosis, as the only disturbance to the cells is dependent on
the time that the plate spends outside the incubator and the intensity of the
light source used to examine the cells. Thus, this assay minimizes alterations
in cell survival that occur due to handling, but requires the most care and
practice to score.

1. Five regions are selected in each collagen gel. Four fields at 90-degree
   intervals around the collagen gel are commonly counted, as well as a field
   directly in the center of the collagen gel.
2. Focus should be initially set adjacent to the surface of the gel, with the
   microscope focus traveling through appropriate focal planes in the gel to
   permit scoring of all cells. It is necessary to adjust focus, as cells will be
   distributed throughout the z dimension of the gel (Fig. 5.2). Cells are
   scored as dead or live. Do not count cells that are living at the interface of
   the plastic tissue culture dish and the collagen.

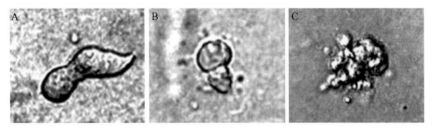

Figure 5.1 Brightfield images of individual neuroblastoma cells within a 3D collagen
culture. (A) Live cell is migrating. (B) Cells exhibit a very late apoptotic phenotype,
with atrophy of adjacent blebbed vacuoles and shrunken cell bodies. (C) Cell displays
the characteristic blebbing morphology of a cell undergoing apoptosis.

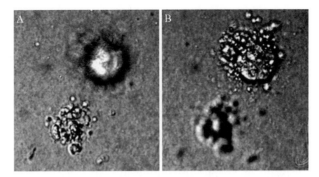

Figure 5.2 Focal depth adjustment is required when scoring apoptosis. Cells are distributed throughout the z axis, and positive identification of live or dead cells requires adjustment of focus. In the left panel, an apoptotic cell is shown, but the status of the second cell in the upper right corner is not clear. Adjustment of the focal plane permits scoring of this cell as apoptotic.

3. Within each well containing three collagen gels, 15 fields can be scored, and 200 to 300 cells counted. Cell survival is assessed as the ratio of living cells relative to the total number of cells observed.

2.4. Immunofluorescence techniques for assessing cell survival

Only slightly more invasive are surface-staining techniques using prelabeled fluorescent probes, or the introduction of dyes that are excluded from live cells. These methods introduce generally inert reagents, and transiently expose the cells to low levels of radiation required to permit excitation of the appropriate fluors. The cell morphology method described above can also be performed among cells prelabeled with one of the cell tracker dyes (Invitrogen), which may offer increased sensitivity in studying cell morphology depending on the microscope system used. The fluorescent method can be combined with the use of Hoechst 33342 dye to study the nuclear morphology and chromatin condensation in parallel. We commonly assess cell survival using staining with annexin V and propidium iodide, adapting methods originally used in FACS analysis (Bossy-Wetzel and Green, 2000), together with either brightfield analysis of cell morphology or the use of a cell tracker orange dye, which emits between the green and red channels.

2.4.1. Procedure

1. FITC-conjugated annexin V is introduced to the tissue culture medium surrounding the collagen gel at a final concentration of 100 ng/ml, together with propidium iodide (100 ng/ml). These agents will readily penetrate the collagen gels to label the cells in them.

2. Cells are incubated at 37° for 15 min, and the plate is then moved to a fluorescence microscope for analysis.

3. Cells are quantitated by first counting the number within a brightfield as described above, and then switching the filters to examine the cells by immunofluorescence. Dead cells will stain in the green channel (indicating annexin V staining and the presence of negatively charged lipids on the cell surface) or in the (far) red channel (propidium iodide nuclear staining, indicating membrane permeability).

4. For serial studies, the media should be aspirated and replaced to prevent nonspecific uptake and increased background staining. This prevents the eventual leakage of propidium iodide into live cells that occurs with extended exposure, and minimizes exposure to propidium iodide in general. The collagen matrix will retain shape and should remain anchored to the well bottom, which permits complete aspiration of media from a given well and replacing it.

5. The staining procedure is repeated at each time point, as required. For primary cells such as HUVEC, each time point will typically be taken after 4 to 6 h. For immortalized cells, time points may be every 12 h to every day.

Paired fluors such as FITC-annexin V and propidium iodide provide complementary measures of cell survival, and can be combined with bright-field assessments of cell morphology. We have found that while these methods alone (morphology, annexin staining, propidium iodide uptake) are comparable in the final indices of survival that they provide, they do not absolutely correspond within a given field. Thus, propidium iodide may not penetrate the lipid bilayer of an early apoptotic cell, although the cell will be clearly positive by annexin V staining and morphological criteria. Similarly, dead cells sometimes do not stain with annexin V despite apparent membrane permeability, as assessed by propidium iodide staining. Late-stage apoptotic cell bodies are often shrunken and sometimes weakly stain either agent, making brightfield examination very useful.

2.5. Determining cell survival by caspase activation

Although annexin V has been used widely as an indicator of apoptosis, we sometimes observe annexin V staining on the surface of viable cells following transition between different temperatures (moving from room temperature to 37°), as well as along the edges of migrating, spreading, or dividing cells that are not apoptotic. Therefore, it is useful to assess additional indicators of cell death to increase the sensitivity of the assay and the level of confidence in the results. In this respect, the activation of caspases is a hallmark of apoptosis. To determine whether cell death is proceeding via apoptosis, the activation of caspases can be monitored by covalent incorporation of a caspase-specific, membrane-permeable fluorescent substrate. In particular, fluorescent inhibitors of the caspases that covalently bind to the catalytic site have been used as indicators of caspase activity.

2.5.1. Materials
FITC-zVADfmk (Promega) stock solution
OptiMEM (Invitrogen)

2.5.2. Equipment
Confocal microscope or fluorescence microscope

2.5.3. Procedure

1. Complete media is aspirated from the tissue culture wells and replaced with serum-free media (OptiMEM) containing FITC-zVADfmk (20 μM). The label is incubated for 1 h with the cells at 37°.
2. The label-containing media is aspirated and replaced with complete media containing propidium iodide. The cells are scored for apoptosis based on morphology and immunofluorescence as described above.
3. This assay can be repeated at secondary time points by performing Steps 1 and 2. In this way, caspase activation can be monitored as a function of time.

2.6. Adaptation of three-dimensional culture to existing immunodetection methods

Depending on the nature of the study, it is sometimes desirable to assess additional measures of apoptosis within the collagen gel, such as the presence of cleaved caspase substrates or DNA fragmentation. The 3D ECM can be adapted for use in most cell-staining procedures as follows:

1. Media is aspirated from the wells. The cells and the collagen matrix are fixed with PBS/1% paraformaldehyde (diluted from a 16% stock, Fisher Scientific) for 5 min.
2. The paraformaldehyde solution is then aspirated, and washed twice with PBS/1% BSA.
3. To permeabilize the cells, 500 μl of 0.02% Triton/PBS is added. The collagen-embedded cells can then be stained for intracellular markers by incubation of the gel, with PBS/BSA containing an appropriate primary antibody (such as anti-active caspase 3 or anticleaved PARP, a caspase substrate).
4. Independently or in conjunction with these studies, the collagen gels can be subjected to TUNEL staining (Kaufmann et al., 2000) to detect cleaved DNA. The gel can be counterstained with DNA-binding dyes to indicate nuclei and permit quantitation of apoptotic cells.

5. Increased resolution of the cells within the gel is obtained by dislodging the gels from the tissue culture plastic with the side of a pipette tip or blunt forceps, and transferring the collagen gels to a coverslip to image and score on a confocal or fluorescence microscope.

2.7. Preparation of three-dimensional fibrin cell culture

Collagen is a ligand for integrins of the $\beta1$ family ($\alpha1\beta1$, $\alpha2\beta1$, $\alpha10\beta1$, $\alpha11\beta1$), but is not ligated by most other integrins. The provisional ECM component fibrin is bound by several integrins, including $\alpha5\beta1$, $\alpha v\beta3$, and $\alpha v\beta5$ integrins. It is sometimes helpful to use a second defined ECM to study cell survival *in vitro*; fibrinogen, a principal component of provisional ECM, is used to contrast with type I collagen, the most abundant component of the anatomical ECM.

2.7.1. Materials

Purified fibrinogen (fibronectin- and factor XIII–free) (American Diagnostica)
Thrombin (BD Biosciences)
OptiMEM (Invitrogen)

2.7.2. Protocol

1. Purified fibronectin-free fibrinogen (American Diagnostica) is suspended (8 mg/ml) in serum-free cell culture media.
2. Fibrinogen stock solution is mixed with an equal volume of serum-free cell culture media containing 100,000 cells/ml.
3. The fibrinogen/cell suspension is placed on ice and 5 U/ml of thrombin are added and quickly mixed to suspend throughout the gel. The thrombin will immediately catalyze the formation of the fibrin gel. Keeping the solution on ice will slow the reaction, but will not prevent it.
4. Working quickly, wells are dotted with 15 to 20 μl drops of fibrinogen and cells. The tissue culture plates are inverted to allow polymerization to occur in the tissue culture hood.
5. Fibrin gels will set much faster than collagen gels. If the assay being performed is comparing the action of fibrinogen and collagen directly, remember to also add 5 U/ml of thrombin to the collagen gels as a control. (Although thrombin will not catalyze the formation of collagen gels, it is necessary to include as a control.)
6. Perform any of the assays listed for evaluation of apoptosis or survival as described above. These assays can be performed using identical steps.

2.8. Alternative approaches and concerns

For a laminin and collagen IV–containing 3D ECM, Matrigel can be used. Growth factor–free Matrigel is thawed on ice overnight and then diluted in tissue culture medium to a final concentration of 4 mg/ml containing 50,000 cells/ml. For subsequent handling purposes, the Matrigel can be handled as described for the collagen gels.

It is worth noting that increasing cell density has two major effects on these survival assays. Increasing cell density appears to aid the capacity of cells to survive in 3D collagen gels. This is not the case among lone cells, but rather results from increasing incidence of cell–cell contact, as cells in aggregates exhibit a survival advantage. Even at lower cell densities, surviving cells will tend to "seek" each other out by day 3 or 4 and form aggregates (Fig. 5.3A). Interestingly, aggregates tend to persist and proliferate, while cells that venture into the local ECM will often form a corona consisting of apoptotic bodies. A second problem presented is that aggregates comprising more than a few cells will be difficult to assess for cell number and survival with any precision. For these reasons, cell numbers are typically maintained at levels that permit dispersed cells to be quantified and assessed with respect to apoptosis.

Note: It is possible to assess the capacity of individual cells to form colonies and survival in the collagen gel as a surrogate marker of survival. The preparation of collagen and the concentration of cells is the same as for the apoptosis-scoring assays, but larger gel volumes are used (500 μl vs. 20 μl) and the gel is polymerized in a six-well dish. Cells are incubated for 10 days in complete medium, and then colonies are visualized and counted via

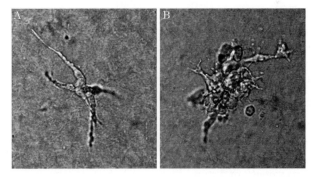

Figure 5.3 Prolonged assay lengths of several days, or high cell-culture seeding densities, result in aggregate formation. In this study, NB8 neuroblastoma cells have migrated toward each other (right panel) and formed aggregates in the collagen gel (right panel). The cell–cell contact has the effect of increasing overall cell survival within the aggregate while making it difficult to quantitate cell number.

microscopy, or by staining the entire gel with a filtered 0.1% crystal violet (Sigma) in 20% methanol and ddH$_2$O solution for 20 min. The dye solution is aspirated and unbound dye removed by filling the well with water and allowing time for the dye to exit the collagen gel. Water is exchanged several times until the gel is clear. Colonies can then be observed and imaged macroscopically (Stupack *et al.*, 2006).

2.9. Studies on cells cultured on the surface of three-dimensional gels

In addition to studying cell survival as they migrate within a 3D matrix, survival can be evaluated in populations of cells adherent to the surface of 3D gels. Placing cells in a 2D setting can have a general effect of increasing cell survival. Depending on the cell type studied, culture on the surface of a non-rigid 3D ECM may be more appropriate than culture on a plastic surface, or within a 3D matrix. This may better mimic the presence of a basement membrane for culturing epithelial cells. The preparation of these matrices proceeds very similarly to that described above. However, cells are added late in the process, and additional manipulations can be performed to the gels without impacting cells directly. For example, additional matrix components could be added to the gels, allowed to bind (or chemically cross-link), and then washed out prior to seeding cells on the gels.

2.9.1. Procedure: Preparation of "two-dimensional" collagen gels

1. Two-dimensional collagen gels are prepared by mixing 800 μl of chilled, acid-soluble type I collagen (4 mg/ml) with 100 μl of 10× M199 pH 7.4 media, and 30 μl of bicarbonate and 70 μl of ddH$_2$O. As with preparing a 3D gel culture, the ingredients are mixed thoroughly on ice, with care taken to avoid introducing bubbles into the gel.
2. The collagen can then be transferred to a 35-mm dish modified with a coverslip bottom (Electron Microscopy Sciences 7670-02). While regular tissue culture dishes can also be used, imaging using an inverted microscope will have to penetrate the culture dish and the collagen layer to image the cells growing on the surface, and results are improved when coverslip bottoms are used. Pipette enough collagen gel to cover the surface of the coverslip to a depth of 2 mm.
3. The dish is then placed in a 37° incubator and the mixture allowed to set for 30 min.
4. Cells are prelabeled with Cell Tracker dye (as described above) and resuspended in complete media (100,000 cells/ml) and 2 ml are added to the plate. Cells are allowed to adhere to the surface for 4 h.

5. Sterile low-melting–temperature agarose (1.2% in PBS) (SeaKem) is heated until dissolved and cooled to 40°.

6. Nonadherent cells are gently removed by pipetting serum free media over the plate. Excess liquid is aspirated.

7. The agarose mixture is quickly mixed with an equal volume of tissue culture media prewarmed to 37°, then layered over the surface of the cells and allowed to attach. The agarose will solidify and create a barrier that will prevent dislodgement of apoptotic cells, permitting more accurate assessments of cell survival.

8. After the agarose has hardened, the cells can be overlaid with complete media containing reagents such as antibodies, receptor agonists/antagonists, or other small molecule inhibitors, as outlined in the 3D gel system. In situations where the impact of a single growth factor is of interest, it may be necessary to grow cells in a minimal media (i.e., 1% serum) to minimize exposure to other growth factors.

9. Apoptosis can then be assessed using the same techniques as outlined for cell culture within 3D gels above. For example, annexin V staining can be used to assess apoptotic cells (Fig. 5.4).

Note: Similar to 3D culture systems, there is a prosurvival effect as cell density and therefore cell:cell contact increases. The tissue culture dish will contain cells adherent to the tissue culture plastic adjacent to the collagen gel. These cells can be assessed for apoptosis as well, and the relative impact of culture on a non-rigid surface can be compared to standard tissue culture conditions. The impact of both factors is shown for human umbilical vein endothelial cells in Fig. 5.5.

FITC-annexin V Cell tracker red Merge

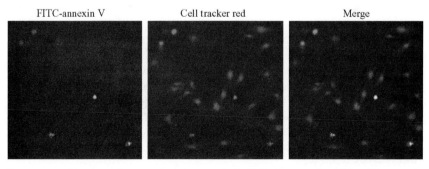

Figure 5.4 Scoring of apoptosis among endothelial cells spread on 3D collagen gels. Endothelial cells labeled with Cell Tracker Orange (red channel) were plated on 35-mm dishes on top of a 3D collagen matrix, and 0.6% agarose was layered on top of the cells. Apoptosis was detected by incubating cells in FITC-annexin V (green channel) and confocal imaging. Apoptotic cells appear yellow in the merge. (See color insert.)

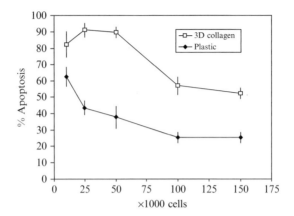

Figure 5.5 The survival of endothelial cells on collagen gels is dependent on seeding density. The indicated number of cells were plated on 35-mm tissue-culture dishes containing regions of 3D collagen. Cells were cultured in minimal media (1% FBS) overnight and apoptosis was scored by annexinV staining.

 ## 3. CONCLUSIONS AND PERSPECTIVE

Adhesion to rigid tissue culture plastic increases cellular resistance to apoptosis, and in the field of therapeutics "cell adhesion–mediated drug resistance" (CAM-DR) (Damiano, 2002) remains a significant factor (Li and Dalton, 2006). The methods presented here permit an investigation of the role of specific integrin-mediated adhesion events on non-rigid surfaces. Moreover, these systems permit cells to be placed in environments where integrins can remain unligated or antagonized. It has become evident that while substrate-ligated integrins convey positive signals that maintain cell survival, unligated or antagonized integrins appear to be capable of transmitting negative signals. Some of these result in the induction or acceleration of programmed cell death—events that occur in response to integrin antagonism among adherent cells *in vivo*. The use of these methods to study cell survival may be useful to reconcile differences observed in the biology of cells *in vivo* relative to those cultured on plastic.

REFERENCES

Bossy-Wetzel, E., and Green, D. R. (2000). Detection of apoptosis by annexin V labeling. *Methods Enzymol.* **322,** 15–18.
Chinnaiyan, A. M., and Dixit, V. M. (1996). The cell-death machine. *Curr. Biol.* **6,** 555–562.

Choquet, D., Felsenfeld, D. P., and Sheetz, M. P. (1997). Extracellular matrix rigidity causes strengthening of integrin-cytoskeleton linkages. *Cell* **88**, 39–48.

Cukierman, E., Pankov, R., Stevens, D. R., and Yamada, K. M. (2001). Taking cell-matrix adhesions to the third dimension. *Science* **294**, 1708–1712.

Dalton, W. S. (2003). The tumor microenvironment: Focus on myeloma. *Cancer Treat. Rev.* **29**(Suppl. 1), 11–19.

Damiano, J. S. (2002). Integrins as novel drug targets for overcoming innate drug resistance. *Curr. Cancer Drug Targets* **2**, 37–43.

DeMali, K. A., Wennerberg, K., and Burridge, K. (2003). Integrin signaling to the actin cytoskeleton. *Curr. Opin. Cell Biol.* **15**, 572–582.

Giancotti, F. G. (1997). Integrin signaling: Specificity and control of cell survival and cell cycle progression. *Curr. Opin. Cell Biol.* **9**, 691–700.

Giancotti, F. G., and Ruoslahti, E. (1999). Integrin signaling. *Science* **285**, 1028–1032.

Hanks, S. K., Ryzhova, L., Shin, N. Y., and Brabek, J. (2003). Focal adhesion kinase signaling activities and their implications in the control of cell survival and motility. *Front Biosci.* **8**, d982–d996.

Howe, A., Aplin, A. E., Alahari, S. K., and Juliano, R. L. (1998). Integrin signaling and cell growth control. *Curr. Opin. Cell Biol.* **10**, 220–231.

Hynes, R. O. (2002). Integrins: Bidirectional, allosteric signaling machines. *Cell* **110**, 673–687.

Ingber, D. E. (1997). Tensegrity: The architectural basis of cellular mechanotransduction. *Annu. Rev. Physiol.* **59**, 575–599.

Ingber, D. E. (2002). Mechanical signaling and the cellular response to extracellular matrix in angiogenesis and cardiovascular physiology. *Circ. Res.* **91**, 877–887.

Ingber, D. E. (2006). Cellular mechanotransduction: Putting all the pieces together again. *FASEB J.* **20**, 811–827.

Kaufmann, S. H., Mesner, P. W., Jr., Samejima, K., Tone, S., and Earnshaw, W. C. (2000). Detection of DNA cleavage in apoptotic cells. *Methods Enzymol.* **322**, 3–15.

Kornberg, L., Earp, H. S., Parsons, J. T., Schaller, M., and Juliano, R. L. (1992). Cell adhesion or integrin clustering increases phosphorylation of a focal adhesion-associated tyrosine kinase. *J. Biol. Chem.* **267**, 23439–23442.

Li, Z. W., and Dalton, W. S. (2006). Tumor microenvironment and drug resistance in hematologic malignancies. *Blood Rev.* **20**, 333–342.

Majno, G., and Joris, I. (1995). Apoptosis, oncosis, and necrosis: An overview of cell death. *Am. J. Pathol.* **146**, 3–15.

Schlaepfer, D. D., Hanks, S. K., Hunter, T., and van der Geer, P. (1994). Integrin-mediated signal transduction linked to Ras pathway by GRB2 binding to focal adhesion kinase. *Nature* **372**, 786–791.

Schwartz, M. A., and Ingber, D. E. (1994). Integrating with integrins. *Mol. Biol. Cell* **5**, 389–393.

Stupack, D. G., Puente, X. S., Boutsaboualoy, S., Storgard, C. M., and Cheresh, D. A. (2001). Apoptosis of adherent cells by recruitment of caspase-8 to unligated integrins. *J. Cell Biol.* **155**, 459–470.

Stupack, D. G., Teitz, T., Potter, M. D., Mikolon, D., Houghton, P. J., Kidd, V. J., Lahti, J. M., and Cheresh, D. A. (2006). Potentiation of neuroblastoma metastasis by loss of caspase-8. *Nature* **439**, 95–99.

Thomas, S. M., and Brugge, J. S. (1997). Cellular functions regulated by Src family kinases. *Annu. Rev. Cell Dev. Biol.* **13**, 513–609.

Thompson, C. B. (1995). Apoptosis in the pathogenesis and treatment of disease. *Science* **267**, 1456–1462.

CHAPTER SIX

Platelet Integrin Adhesive Functions and Signaling

Nicolas Prévost, Hisashi Kato, Laurent Bodin, *and*
Sanford J. Shattil

Contents

Abstract

Integrin-mediated cellular events affect all cell types and functions, in physiological as well as pathological settings. Blood platelets, because of their unique nature, have proven to be a powerful cell model with which to study the adhesive and signaling properties of integrins. The characterization of the structural and molecular mechanisms regulating the main platelet integrin, $\alpha IIb\beta 3$, has provided some essential clues as to how integrins are regulated in general. The present chapter details the various protocols and reagents currently in use in our laboratory to study $\alpha IIb\beta 3$ adhesive responses and signaling in both human and murine cell models.

Division of Hematology-Oncology, Department of Medicine, University of California, San Diego, La Jolla, California

Methods in Enzymology, Volume 426
ISSN 0076-6879, DOI: 10.1016/S0076-6879(07)26006-9

© 2007 Elsevier Inc.
All rights reserved.

1. Introduction

The preservation of the vascular system integrity in case of hemorrhage is primarily mediated by the formation of a platelet aggregate, or thrombus, at the site of injury. At the molecular level, platelet aggregation is mediated by integrin $\alpha IIb\beta 3$, a receptor for adhesive macromolecules such as fibrinogen and von Willebrand factor. $\alpha IIb\beta 3$ undergoes highly regulated structural changes during platelet activation, leading to its transition from a low- to a high-affinity state (Adair *et al.*, 2002; Takagi *et al.*, 2002; Vinogradova *et al.*, 2002). The molecular mechanisms regulating integrin activation have recently become a topic of growing interest in fields such as cardiovascular and cancer biology (Pauhle *et al.*, 2005). Platelets have proven a good model in which to study integrin-mediated adhesion and signaling. Platelets are readily available, easily separated from other blood cells, and contain a signaling apparatus similar to other cells. Using human and murine platelets and $\alpha IIb\beta 3$-expressing cell lines as models, it has been established that biochemical signaling pathways interacting with $\alpha IIb\beta 3$ can be grouped into two categories: "inside-out" signals leading to the activation of the integrin and "outside-in" signals leading to rearrangements of the platelet actin cytoskeleton and changes in platelet morphology. The aim of the present chapter is to introduce the techniques commonly used in our laboratory for the study of integrin $\alpha IIb\beta 3$ adhesive and signaling functions.

2. Cell Models

2.1. Platelets

2.1.1. Preparation of washed platelets from human blood

Using a 19-G butterfly needle, venous blood is taken from healthy donors who have not taken any medication for at least 10 days and anticoagulated 5:1 with acid-citrate-dextrose (ACD) (65 mM trisodium citrate, 70 mM citric acid, 100 mM dextrose, pH 4.4). Red and white blood cells are removed by centrifugation at $100 \times g$ for 20 min at room temperature. After collection, the platelet-rich plasma (PRP) is centrifuged at $220 \times g$ for 10 min at room temperature. Platelets are then resuspended in wash buffer (150 mM NaCl, 20 mM HEPES, pH 6.5) and centrifuged at $220 \times g$ for 10 min at room temperature in the presence of 1 U/ml apyrase and 1 μM prostaglandin E1 (PGE1). Finally, the platelet pellet is resuspended in Walsh's buffer (137 mM NaCl, 20 mM PIPES, 5.6 mM dextrose, 1 g/liter BSA, 1 mM MgCl$_2$, 2.7 mM KCl, 3.3 mM NaH$_2$PO$_4$, pH 7.4). Prior to any experimental procedure, platelets are typically left at room temperature for 30 min; after 30 min, most of the PGE1 has become inactive.

2.1.2. Preparation of washed platelets from mouse blood

Two methods are routinely being used for the isolation of blood from mice: cardiac puncture and venepuncture. The main appeal to cardiac punctures is that they are not time consuming and do not require the use of anesthetics, because animals are euthanized through CO_2 asphyxiation. On the other hand, because of the loss of vasculature tonus in dead animals, blood has to be collected swiftly before platelets become activated. Cardiac punctures also present the disadvantage of yielding smaller blood volumes than venepunctures do. Blood collection through cardiac puncture is performed by the insertion of a 25-G needle through the diaphragm of the animal, immediately beneath its sternum. For venepunctures, mice are anesthetized with 1 mg of pentobarbital per 10 g of weight. After dissection of the abdomen, blood is drawn from the hepatic portal vein in heparin (15 U/ml of blood) using a 22-G needle. The blood is then diluted with one volume of wash buffer (150 mM NaCl, 20 mM PIPES, pH 6.5) and centrifuged at $60 \times g$ for 7 min. After collection, the PRP is centrifuged at $240 \times g$ for 10 min at room temperature. Platelets are resuspended in Walsh's buffer as described in Section 2.1.1. Alternative anticoagulants, such as sodium citrate, are used when platelets will subsequently be exposed to thrombin (see Sections 4.3 and 4.4) or for aggregation studies. For mouse platelet aggregation studies, blood is collected into 1/9 volume 147 mM sodium citrate, pH 6.5. One volume of Walsh's buffer is then added to the sample before centrifugation at $60 \times g$ for 7 min at room temperature. After PRP recovery, the platelet concentration is adjusted to 2 to 4 $\times$ 10^8 platelets per milliliter with Walsh buffer. The platelet suspension is supplemented with 1 mM $CaCl_2$ before aggregations are performed.

2.1.3. Preparation of gel-filtered platelets from human blood

An alternative to the washed platelet protocol is gel filtration of the PRP through a sepharose column. Gel-filtered platelets are typically more responsive to adenosine diphosphate stimulation than washed platelets, especially in the context of aggregation studies. PRP is prepared as described in paragraph 2.1.1. The sepharose 2B column is prepared as follows: sepharose 2B (GE Healthcare, Cat# 17-0130-01) is poured into a 60-ml syringe containing a nylon mesh disc with a pore size less than 50 μm (seven to eight volumes of gel for each volume of PRP). The column is then carefully degassed and washed with four gel volumes of Walsh's buffer. PRP is deposited drop by drop on the column without disturbing the gel–PRP interface. The PRP is then allowed to enter the column by gravity and the platelets are eluted in Walsh's buffer. The platelet eluate is collected in 1-ml fractions. After use, the column is washed with four volumes of Walsh's buffer and one volume of ddH$_2$O, and stored at 4° in 0.05% sodium azide. Columns can be re-used up to 10 times.

2.2. Cell lines

2.2.1. Integrin-expressing cell lines

Because of the limitations anucleate platelets present as a cell model for studies of protein overexpression or knockdown, alternate model systems have been developed for studies of platelet integrin function. These include Chinese hamster ovary (CHO) cells, "SYF" murine fibroblasts deficient for the Src family kinases Src, Yes, and Fyn (Klinghoffer *et al.*, 1999), and murine megakaryocytes. Integrin αIIbβ3–expressing CHO and SYF cells were established as described by Diaz-Gonzalez *et al.* (1996) and O'Toole *et al.* (1990). As discussed in detail elsewhere, megakaryocytes are produced *ex vivo* through the differentiation of murine bone marrow (Shiraga *et al.*, 1999) or embryonic stem cells (Eto *et al.*, 2003).

2.2.2. Cell culture and transfection

Cell culture   CHO and SYF cells stably expressing integrin αIIbβ3 are maintained in Dulbecco's Modified Eagle's Medium (DMEM) supplemented with 10% fetal calf serum (FCS), 100 U/ml of penicillin, 0.1 mg/ml of streptomycin, 2 mM of L-glutamine, and 1% nonessential amino acids at 37° in 6% CO_2. Cells are never grown beyond 80 to 90% confluency. Cell reseeding is performed as follows: cells are first washed in phosphate–buffered saline (PBS) and then detached by 0.5-mM EDTA–0.05% trypsin treatment for 2 to 3 min at 37°. Trypsin is immediately neutralized by the addition of at least three volumes of culture medium. Cells are collected by centrifugation at 150×g for 3 min, resuspended in medium, and replated at a confluency of approximately 40 to 50%. Typically, cells are split every 2 to 4 days, and experiments are always carried out with cells that have been passaged no more than five times.

Transfection   Transfection is usually carried out to test the effects of overexpression of specific recombinant proteins on αIIbβ3 function. Twelve to 16 h before transfection, cells are seeded at 1×10^6 cells per 100-mm cell culture dish. The next day, upon reaching 50 to 80% confluency, cells are washed twice with unsupplemented DMEM. A transfection mix is prepared by mixing 300 μl of DMEM, 1 to 5 μg of DNA, and 15 μl of lipofectamine (Invitrogen, Cat# 18324-020) per 100-mm culture dish, followed by a 20-min incubation at room temperature. Four milliliters of DMEM are then added back to the mixture before incubation with the cells. After 6 h at 37°, the transfection medium is removed and replaced with complete DMEM. Transfected cells are typically ready for assaying integrin functions after 24 to 48 h.

3. Microscopy

3.1. Cell spreading assay

On the eve of the assay, coverslips are flamed in ethanol and then coated overnight at 4° with 100 µg/ml of fibrinogen (Enzyme Research Laboratories, Cat# FIB-1) in PBS, pH 8.0. The next morning, coverslips are washed three times in PBS and blocked with 5 mg/ml of heat-denatured bovine serum albumin (BSA) for 2 h at room temperature and then washed three times in PBS.

Platelets are prepared (described in Section 2.1) and resuspended at 10^7 platelets/ml in Walsh's buffer supplemented with 1 mM $CaCl_2$. When working with cells in culture, the final concentration is adjusted to 10^6 cells/ml in Walsh's buffer or Tyrode's buffer (137 mM NaCl, 2.68 mM KCl, 11.9 mM $NaHCO_3$, 3.3 nM NaH_2PO_4, 2 mM $CaCl_2$, 1 mM $MgCl_2$, 5.5 mM glucose, 5 mM HEPES, 0.35% BSA, pH 7.4).

Cells are incubated on fibrinogen coated coverslips at 37° for 15 to 60 min in the absence or presence of agonist. After removal of unbound cells by three PBS washes, adherent cells are fixed with 3.7% of methanol-free formaldehyde (Polysciences, Inc. Cat# 04018) for 10 min at room temperature. Adherent cells are then washed three times with PBS, permeabilized with 0.2% Triton X-100 for 5 min at room temperature, and washed another three times in PBS. Cells are then ready for immunostaining.

3.2. Immunostaining

Coverslips are blocked in 10% serum/PBS for 30 min at room temperature. The serum is preferentially from the same animal species as the one that the primary detecting antibody was raised in. Coverslips are then incubated with the primary antibody (at 1/100 to 1/1000 in 10% serum-PBS) for 45 min at 37°, washed three times with PBS, and incubated for 45 min at 37° with fluorophore-labeled secondary antibody (Invitrogen and Jackson Immunochemicals) at 1/500 to 1/1000 in 10% serum-PBS. For staining of intracellular actin, coverslips are incubated with 10 U/ml rhodamine-phalloidin (Invitrogen, Cat# R415) in 5% BSA/TBST (50 mM Tris/HCl, 150 mM NaCl, 0.1 % [v/v] Tween 20, pH 7.5) for 30 min at room temperature, and then washed three times with PBS. Negative controls include cells stained with an irrelevant primary immunoglobulin and cells stained with secondary antibody. After mounting in Citifluor (Ted Pella, Inc. Cat# 19470), coverslips are stored at 4° in the dark.

4. Assaying Integrin Adhesive Responses

4.1. Flow cytometry analysis of integrin function

4.1.1. Introduction to ligands and antibodies used

The affinity of integrin $\alpha IIb\beta3$ for fibrinogen and other ligands is modulated through conformational changes of the integrin on the surface of platelets. The unmasking of specific epitopes in the extracellular domain of the integrin is one of the defining structural changes that allow its transition from a resting state to a high-affinity state. Therefore, it is possible to monitor the binding of fluorophore-labeled soluble fibrinogen as a function of integrin activation on the cell surface.

Because fibrinogen is a polyvalent ligand, more discriminating tools have also been developed in order to monitor the changes in integrin affinity at the single molecule level. One such tool is PAC-1 Fab, an RGD-containing antibody Fab fragment specific for high-affinity human $\alpha IIb\beta3$ (Abrams *et al.*, 1994), derived from the pentameric IgM PAC-1 (Shattil *et al.*, 1985). Because PAC-1 is specific for human $\alpha IIb\beta3$, POW-2, a re-engineered version of PAC-1 capable of recognizing both human and mouse high-affinity $\alpha IIb\beta3$ (and $\alpha V\beta3$), was subsequently produced (Bertoni *et al.*, 2002). Other antibodies routinely used in the laboratory include A2A9 and 1B5, blocking antibodies that prevent fibrinogen binding to human and mouse $\alpha IIb\beta3$, respectively (Bennett *et al.*, 1983; Lengweiler *et al.*, 1999). SSA6 and D57 are function-independent antibodies used for the detection of $\beta3$ integrins and $\alpha IIb\beta3$ on the surface of cells, respectively (O'Toole *et al.*, 1994; Silver *et al.*, 1987). General information regarding the aforementioned antibodies has been summarized in Table 6.1.

4.1.2. Platelets

Platelets are prepared (described in Section 2.1) and adjusted to a final concentration of 10^8 platelets/milliliter in Walsh's buffer. Fifty to 100 μl of the platelet suspension are used for each reaction carried out in a polystyrene tube suitable for flow cytometry (Becton Dickinson, Cat# 352008). Platelets are incubated with the detection reagent with or without a platelet agonist or inhibitor for 30 min at room temperature, away from light. For a soluble fibrinogen-binding assay, platelets are incubated with either FITC- or Alexa-Fluor488–conjugated fibrinogen (Invitrogen, Cat# F13191) at a concentration of 200 μg/ml of fibrinogen in the presence of 1 mM CaCl$_2$. For the detection of Alexa Fluor488–conjugated PAC-1 binding to the platelet surface, an antibody concentration of 150 μg/ml is used. A secondary antibody is used to detect binding of POW-2 Fab (Bertoni *et al.*, 2002). Four hundred microliters of PBS are added to the

Table 6.1 Antibodies commonly used to study integrin $\alpha IIb\beta 3$

Antigen	Antibody name	Species/isotype	Application
Human $\alpha_{IIb}\beta_3$	PAC-1	Mouse IgMκ	Ligand mimetic; FACS
Human/murine $\alpha_{IIb}\beta_3$	POW-2 Fab	Re-engineered from mouse IgG1κ	Ligand mimetic; FACS
Human $\alpha_{IIb}\beta_3$	A2A9	Mouse IgG2aκ	Blocking; FACS
Murine $\alpha_{IIb}\beta_3$	1B5	Hamster IgG3κ	Blocking; FACS
Human $\alpha_{IIb}\beta_3$	D57	Mouse	FACS, IF
Human β_3	SSA6	Mouse IgG1	IP, FACS, IF

FACS, fluorescence assisted cell sorting; IF, immunofluorescence; IP, immunoprecipitation.

platelets and samples are analyzed by flow cytometry (Shattil *et al.*, 1987). Changes in the F-actin content of platelets can be assessed by flow cytometry using BODIPY FL-phallacidin (Invitrogen, Cat# B607) as described in Shattil *et al.* (1994).

4.1.3. Cell lines

PAC-1/POW-2 binding   Each reaction is performed on 0.5 to 7 $\times$ 10^6 cells in Walsh's buffer. For each sample, 50 μl of cell suspension are incubated with or without 5 μl of an agonist and/or inhibitor for 10 min at room temperature. Control samples are treated with 5 μl of 100 mM EDTA, pH 7.35, or a specific $\alpha IIb\beta 3$ antagonist, such as 10 of μM integrilin or 2 mM of RGDS (to estimate nonspecific binding), or with 5 μl of 10 mM MnCl$_2$ to stimulate $\alpha IIb\beta 3$ extrinsically. PAC-1 or POW-2 is added to each sample at a final concentration of 150 $\mu g/ml$, followed by a 20-min incubation at room temperature. Cells are then washed with 1 ml of Walsh's buffer, collected by centrifugation at 115$\times g$ for 5 min, and resuspended in 50 μl of secondary antibody solution (FITC-labeled anti-IgM 1/400 in Walsh's buffer for PAC-1, AlexaFluor488-labeled antimurine IgG1 for POW-2) (Biosource, Cat# AMI3608; Invitrogen, Cat# A11068). Cells are then incubated for 20 min on ice in the dark, after which time 1 ml of ice-cold Walsh's buffer is added to the cells. Finally cells are centrifuged at 115$\times g$ for 5 min in the dark, at 4° and resuspended in 400 μl of 1-$\mu g/ml$ propidium iodide (PI) in Walsh's buffer. Ligand binding is assessed on single, living cells by flow cytometry, as described by O'Toole *et al.* (1990).

Soluble fibrinogen binding   Fibrinogen-binding assays are per- formed with FITC-, Alexa-Fluor488-, or biotin-labeled soluble fibrinogen. Cells are resuspended in Walsh's or Tyrode's buffer in the presence of 1 mM of CaCl$_2$ at a final concentration of 0.5 to 3 × 10^6 cells/ml. Twenty-five microliters of cell suspension are added to a FACS tube containing 200 μg/ml fibrinogen in the presence or absence of an agonist or inhibitor. Samples incubated with 1 mM of MnCl$_2$ serve as a positive control and/or 10 mM of EDTA or a specific αIIbβ3 antagonist serves as a control for nonspecific binding. For biotin-labeled fibrinogen, cells are incubated for another 25 min with 100 μg/ml of fluorophore-labeled streptavidin at room temperature. Cells are then washed with 500 μl of ice-cold Walsh's or Tyrode's buffer, collected by centrifugation at 150×g for 3 min at 4°, and resuspended in 400 μl of ice-cold PBS containing 2 μg/ml of PI for flow cytometry analysis.

4.2. Adhesion assay

The number of platelets adhering to immobilized fibrinogen can be quantified in a colorimetric assay by measuring the activity of the platelet lysosomal enzyme acid phosphatase through the release of a p-nitrophenyl phosphate enzymatic product. One day before the experiment, a 96-well plate (Fisher #14-245-61) is coated with fibrinogen (150 μl per well of a 200 μg/ml PBS solution), or, as a negative control, heat-inactivated BSA (at 4 mg/ml in PBS) overnight at 4°. The following day the plate is blocked with 5 mg/ml of heat-inactivated BSA for 1 h at room temperature and washed twice with PBS before adding platelets.

Platelets are prepared as described in Section 2.1 and adjusted to a final concentration of 10^8 platelets/ml in Walsh's buffer. After adding 50 μl of platelet suspension per well, the plate is pre-incubated for 5 min at 37°. Twenty-five microliters of Walsh's buffer (containing 3 mM of CaCl$_2$, with or without agonist) are subsequently added to each well. The plate is incubated at room temperature for 20 to 60 min, after which it is washed four times with PBS. The substrate solution (5 mM p-nitrophenyl phosphate in 0.1 M citrate/0.1% Triton, pH 5.4) is then added at 150 μl per well. After a 1-h incubation at room temperature in the dark, the enzymatic reaction is stopped by the addition of 100 μl of a 2 N NaOH solution to each well. The plate is read immediately at 405 nm. Values for platelet adhesion are typically expressed as a percentage of platelets added to the well. The latter is estimated by subjecting all the platelets (adherent and nonadherent) in a control well to the enzyme reaction.

4.3. Platelet aggregation

Human platelets are prepared by gel filtration. Platelets are resuspended at a final concentration of 1 to 4 × 10^8 platelets/milliliter in the presence of 1 mM of CaCl$_2$ and 300 μg/ml of fibrinogen. Mouse platelets are isolated as described in Section 2.1.2. Aggregation studies are performed at 37° in a Chrono-Log optical aggregometer using siliconized glass cuvettes, under constant stirring (900 rpm) after the platelets are allowed to sit at 37° for 1 to 5 min.

4.4. Clot retraction

Blood is drawn into 1/9 volume of 147 mM sodium citrate, pH 6.5. To each milliliter of blood, 500 μl of Walsh's buffer are added. After centrifugation at 60×g for 7 min, the PRP is supplemented with 1 mM of CaCl$_2$ and adjusted to 2 × 10^8 platelets/ml with Walsh's buffer containing 1 mM CaCl$_2$ and 300 μg/ml fibrinogen. (For slower clot retraction kinetics, lower platelet concentrations may be used. The final platelet concentration shall not be lower than 6 × 10^5 platelets/ml.) In the event of platelet count disparities between mice of different genetic backgrounds, blood collections are to be performed in ACD. Each blood sample is then diluted with one volume of wash buffer (150 mM NaCl, 20 mM PIPES, pH6.5) and centrifuged at 60×g for 7 min. The platelets are then isolated by centrifugation at 240×g for 10 min at room temperature and resuspended to matching platelet concentrations in the sodium citrate-anticoagulated platelet-free plasma of a normal mouse sacrificed simultaneously. The platelet suspension is then supplemented with 1 mM of CaCl$_2$.

Clot retraction assays are performed at 37° in siliconized glass cuvettes on 200-μl platelet suspension samples in the presence of 2 U of thrombin. Clots are photographed at 5, 15, 30, 45, and 60 min. At each time point, the clot exudate is collected and weighed for quantification purposes. The greater the volume of exudate, the more the clot has retracted.

5. BIOCHEMICAL ANALYSIS OF INTEGRIN-BASED SIGNALING

5.1. Initiation of integrin αIIbβ3 "outside-in" signaling

Ligand binding to integrin αIIbβ3 initiates a variety of signaling events immediately downstream of the integrin. These responses are commonly referred to as "outside-in signaling" and result in the recruitment of signaling molecules and cytoskeletal proteins directly or indirectly to the cytoplasmic tails of the integrin (Arias-Salgado et al., 2005a,b; Han et al., 2006. Tadokoro et al., 2003).

It is possible to artificially trigger these responses independently of agonist-induced integrin activation. Cell adhesion to immobilized fibrinogen under static conditions offers one of two experimental settings in which to study outside-in signaling downstream of αIIbβ3. Alternatively, cells can be treated by a combination of soluble fibrinogen and MnCl$_2$ while in suspension.

After fibrinogen ligation to $\alpha IIb\beta 3$ has been achieved by either means, cells are lysed in ice-cold NP40 buffer (1% NP40, 150 mM NaCl, 50 mM Tris-HCl pH 7.4), or ice-cold RIPA buffer (1% Triton X100, 1% deoxycholate, 0.1% sodium dodecyl sulfate, 5 mM Na$_2$EDTA, 158 mM NaCl, 10 mM Tris-HCl pH 7.2), in the presence of 1 mM of Na$_3$VO$_4$, 0.1 mM of NaF, and a protease inhibitor cocktail (Roche, Cat# 11873580001). For western blotting studies, cells will preferentially be lysed in RIPA buffer.

Cell adhesion is carried out in 100 mm of non-tissue culture–treated Petri dishes after overnight coating with 100 μg/ml fibrinogen in PBS, pH 8.0, at 4° and 1 h blocking with 5 mg/ml BSA in PBS, pH 7.4, at room temperature. Control plates that do not support adhesion are coated overnight with 5 mg/ml BSA in PBS, pH 7.4, at 4°. For each 100-mm dish, 2 ml of either 2 × 10^8 platelets/ml or a 10^6 CHO or SYF cells/ml cell suspension are incubated for 5 to 30 min at room temperature. After incubation, control suspension cells not adhering to BSA plates are collected by centrifugation at 11,000×g for 1 sec (platelets) or 300×g for 3 min (cell lines) and lysed. For fibrinogen-coated dishes, nonadherent cells are removed with a single gentle washing in PBS, and adherent cells are lysed on ice by adding 200 μl of lysis buffer. Typically, lysates from three to five individual plates are pooled for each immunoprecipitation for a final lysate volume of 0.6 to 1 ml.

The activation of integrin $\alpha IIb\beta 3$ in a cell suspension is performed on a 0.5- to 1-ml aliquot of 4 × 10^8 platelets/ml or a 3-ml aliquot of 2 × 10^6 nucleated cells/ml. All incubations are performed in Walsh's buffer in the presence of 0.5 to 1 mM of MnCl$_2$ and 250 μg/ml of fibrinogen, at 37° for 2 to 30 min. Cells are collected by centrifugation at 11,000×g for 1 sec (platelets) or 300×g for 3 min (cell lines) and lysed immediately in 0.6 to 1 ml of lysis buffer.

5.2. Immunoprecipitation and co-immunoprecipitation

Typically, RIPA buffer is used for cell lysis in immunoprecipitation studies and NP40 buffer in co-immunoprecipitation studies (Table 6.2). Cell lysates are incubated on ice for 10 min with repeated vortexing. Insoluble material is pelleted by centrifugation at 21,000×g for 10 min at 4°; the protein content of the recovered soluble fraction is determined by colorimetric quantification, through the use of a bicinchoninic acid–based kit (Pierce, Cat# 23225). Each immunoprecipitation is performed on 0.3 to 1 mg of total protein, typically in a final volume of 0.5 ml (see Table 6.2 for details). The lysates are precleared for 2 to 3 h at 4° with 50 μl of protein A- or G-sepharose slurry (50% beads, 50% lysis buffer). The immunoprecipitating antibody is then added to the precleared lysate, and immunoprecipitations are carried out overnight at 4°, unless otherwise stated (Table 6.2). The next morning, the immunoprecipitate is isolated from the lysate with 50 μl of

Table 6.2 Examples of antibodies used to detect co-immunoprecipitation of proteins with integrin $\alpha IIb\beta 3$

IP antibody	IB antibody (antigen)	HRP-conjugate	Lysis buffer	Lysate amount		IP time
				Platelets	Cell line	
SSA6 (human $\beta 3$)	Santa Cruz sc-6627 ($\beta 3$)	Anti-Goat-HRP	RIPA/NP40	300 μg	500 μg	$\geq$8 h
	Santa Cruz sc-19 (Src)	Protein A-HRP	NP40	600 μg	1 mg	$\geq$8 h
	Cell signaling #2107S (pY529 c-Src)	Protein A-HRP	NP40	900 μg	1 mg	$\geq$8 h
	Biosource #44–660G (pY418 c-Src)	Protein A-HRP	NP40	900 μg	1 mg	$\geq$8 h
Santa Cruz sc-6627 (mouse $\beta 3$)	Santa Cruz sc-14021 (human PTP-1B)	Protein A-HRP	NP40	900 μg	1 mg	$\geq$8 h
	Santa Cruz sc-209 (PKC-β)	Protein A-HRP	NP40	900 μg	1 mg	4 h
	Santa Cruz sc-286 (Csk)	Protein A-HRP	NP40	600 μg	1 mg	$\geq$8 h

a 50%-protein A- or G-sepharose slurry (GE Healthcare, Cat# 17-0780-01 and Cat#17-0618-01), for 3 h at 4°. The beads are then washed three times in PBS containing 1 mM of NaVO$_4$, 0.1 mM of NaF and protease inhibitor cocktail (Roche, Cat# 11873580001) After the washing step, the beads are boiled for 5 to 10 min at 100° in 80 μl of 2× Laemmli buffer (62.5 mM Tris-HCl, pH 6.8, 2% sodium dodecyl sulfate [SDS], 25% glycerol, 0.01% bromophenol blue) plus 5% β-mercaptoethanol (v/v). Samples are then separated by SDS polyacrylamide gel electrophoresis (SDS-PAGE) and transferred to nitrocellulose or polyvinylidene difluoride (PVDF) membranes (Biorad, Cat# 162-0112 and Cat# 162-0181). For the detection of phosphorylated tyrosine residues, PVDF membranes and BSA (MP Biomedicals, Cat# 103700) are used; in the absence of such residues, nitrocellulose membranes and non-fat dry milk are used. Membranes are first blocked for 30 min in 5% blocking agent–TBST, and then incubated for 1 h at room temperature with the primary antibody in 1% blocking agent–TBST. After three 15-min washes in TBST, membranes are incubated with the secondary reagent (Protein A-HRP, Upstate, Cat# 18-160; HRP-conjugated antibodies are from Jackson Immunochemicals) in 1% blocking agent-TBST for 1 h at room temperature. Finally, membranes are washed three times for 15 min in TBST, and incubated for 1 min with a peroxidase chemiluminescent substrate (Pierce, Cat# 34080).

REFERENCES

Abrams, C., Deng, Y. J., Steiner, B., O'Toole, T., and Shattil, S. J. (1994). Determinants of specificity of a baculovirus-expressed antibody Fab fragment that binds selectively to the activated form of integrin alpha IIb beta 3. *J. Biol. Chem.* **269**, 18781–18788.

Adair, B. D., and Yeager, M. (2002). Three-dimensional model of the human platelet integrin alpha IIbbeta 3 based on electron cryomicroscopy and x-ray crystallography. *Proc. Natl. Acad. Sci. USA* **99**, 14059–14064.

Arias-Salgado, E. G., Haj, F., Dubois, C., Moran, B., Kasirer-Friede, A., Furie, B. C., Furie, B., Neel, B. G., and Shattil, S. J. (2005a). PTP-1B is an essential positive regulator of platelet integrin signaling. *J. Cell Biol.* **170**, 837–845.

Arias-Salgado, E. G., Lizano, S., Shattil, S. J., and Ginsberg, M. H. (2005b). Specification of the direction of adhesive signaling by the integrin beta cytoplasmic domain. *J. Biol. Chem.* **280**, 29699–29707.

Bennett, J. S., Hoxie, J. A., Leitman, S. F., Vilaire, G., and Cines, D. B. (1983). Inhibition of fibrinogen binding to stimulated human platelets by a monoclonal antibody. *Proc. Natl. Acad. Sci. USA* **80**, 2417–2421.

Bertoni, A., Tadokoro, S., Eto, K., Pampori, N., Parise, L. V., White, G. C., and Shattil, S. J. (2002). Relationships between Rap1b, affinity modulation of integrin alpha IIbbeta 3, and the actin cytoskeleton. *J. Biol. Chem.* **277**, 25715–25721.

Diaz-Gonzalez, F., Forsyth, J., Steiner, B., and Ginsberg, M. H. (1996). Trans-dominant inhibition of integrin function. *Mol. Biol. Cell* **7**, 1939–1951.

Eto, K., Leavitt, A. L., Nakano, T., and Shattil, S. J. (2003). Development and analysis of megakaryocytes from murine embryonic stem cells. *Methods Enzymol.* **365**, 142–158.

Han, J., Lim, C. J., Watanabe, N., Soriani, A., Ratnikov, B., Calderwood, D. A., Puzon-McLaughlin, W., Lafuente, E. M., Boussiotis, V. A., Shattil, S. J., and Ginsberg, M. H. (2006). Reconstructing and deconstructing agonist-induced activation of integrin alphaIIbbeta3. *Curr. Biol.* **16,** 1796–1806.

Klinghoffer, R. A., Sachsenmaier, C., Cooper, J. A., and Soriano, P. (1999). Src family kinases are required for integrin but not PDGFR signal transduction. *EMBO J.* **18,** 2459–2471.

Lengweiler, S., Smyth, S. S., Jirouskova, M., Scudder, L. E., Park, H., Moran, T., and Coller, B. S. (1999). Preparation of monoclonal antibodies to murine platelet glycoprotein IIb/IIIa (alphaIIbbeta3) and other proteins from hamster–mouse interspecies hybridomas. *Biochem. Biophys. Res. Commun.* **262,** 167–173.

O'Toole, T. E., Loftus, J. C., Du, X. P., Glass, A. A., Ruggeri, Z. M., Shattil, S. J., Plow, E. F., and Ginsberg, M. H. (1990). Affinity modulation of the alpha IIb beta 3 integrin (platelet GPIIb-IIIa) is an intrinsic property of the receptor. *Cell Regul.* **1,** 883–893.

O'Toole, T. E., Katagiri, Y., Faull, R. J., Peter, K., Tamura, R., Quaranta, V., Loftus, J. C., Shattil, S. J., and Ginsberg, M. H. (1994). Integrin cytoplasmic domains mediate inside-out signal transduction. *J. Cell Biol.* **124,** 1047–1059.

Paulhe, F., Manenti, S., Ysebaert, L., Betous, R., Sultan, P., and Racaud-Sultan, C. (2005). Integrin function and signaling as pharmacological targets in cardiovascular diseases and in cancer. *Curr. Pharm. Des.* **11,** 2119–2134.

Shattil, S. J., Hoxie, J. A., Cunningham, M., and Brass, L. F. (1985). Changes in the platelet membrane glycoprotein IIb.IIIa complex during platelet activation. *J. Biol. Chem.* **260,** 11107–11114.

Shattil, S. J., Cunningham, M., and Hoxie, J. A. (1987). Detection of activated platelets in whole blood using activation-dependent monoclonal antibodies and flow cytometry. *Blood* **70,** 307–315.

Shattil, S. J., Haimovich, B., Cunningham, M., Lipfert, L., Parsons, J. T., Ginsberg, M. H., and Brugge, J. S. (1994). Tyrosine phosphorylation of pp125FAK in platelets requires coordinated signaling through integrin and agonist receptors. *J. Biol. Chem.* **269,** 14738–14745.

Shiraga, M., Ritchie, A., Aidoudi, S., Baron, V., Wilcox, D., White, G., Ybarrondo, B., Murphy, G., Leavitt, A., and Shattil, S. (1999). Primary megakaryocytes reveal a role for transcription factor NF-E2 in integrin alpha IIb beta 3 signaling. *J. Cell Biol.* **147,** 1419–1430.

Silver, S. M., McDonough, M. M., Vilaire, G., and Bennett, J. S. (1987). The *in vitro* synthesis of polypeptides for the platelet membrane glycoproteins IIb and IIIa. *Blood* **69,** 1031–1037.

Tadokoro, S., Shattil, S. J., Eto, K., Tai, V., Liddington, R. C., de Pereda, J. M., Ginsberg, M. H., and Calderwood, D. A. (2003). Talin binding to integrin beta tails: A final common step in integrin activation. *Science* **302,** 103–106.

Takagi, J., Petre, B. M., Walz, T., and Springer, T. A. (2002). Global conformational rearrangements in integrin extracellular domains in outside-in and inside-out signaling. *Cell* **110,** 599–611.

Vinogradova, O., Velyvis, A., Velyviene, A., Hu, B., Haas, T., Plow, E., and Qin, J. (2002). A structural mechanism of integrin alpha(IIb)beta(3) "inside-out" activation as regulated by its cytoplasmic face. *Cell* **110,** 587–597.

CHAPTER SEVEN

DEVELOPMENT OF MONOCLONAL ANTIBODIES TO INTEGRIN RECEPTORS

E. A. Wayner* *and* B. G. Hoffstrom[†]

Contents

* Antibody Development Laboratory, Fred Hutchinson Cancer Research Center, Seattle, Washington
† Department of Biological Sciences, Columbia University, New York, New York

Methods in Enzymology, Volume 426
ISSN 0076-6879, DOI: 10.1016/S0076-6879(07)26007-0

© 2007 Elsevier Inc.
All rights reserved.

Abstract

Integrins are heterodimeric cell surface receptors composed of an α and a β subunit. They are involved in homotopic and heterotopic cell adhesion and also function as receptors for extracellular matrix molecules such as collagen, fibronectin and laminin. The family to which an integrin belongs is defined by the presence of a particular β subunit paired with a unique α subunit. In this chapter we describe methods to produce monoclonal antibodies to the family of integrin subunits characterized by $\beta 1$ and provide detailed instructions for the development of a monoclonal antibody to the $\alpha 6$ integrin receptor expressed by human prostate carcinoma cells (PC3 cells). Data are presented that correlate the functional capabilities of an antibody with its biochemical characterization.

1. INTRODUCTION

Integrin receptors are integral membrane proteins composed of an alpha subunit complexed with a beta subunit. These receptors traverse the plasma membrane and provide a link between the extracellular environment and the intracellular cytoskeleton via both "inside-out" and "outside-in" signaling pathways. This family of receptors is divided into subfamilies based on the molecular make up of the integrin heterodimer, the composition of which determines a receptor's ligand-binding specificity. For example, the $\beta 1$ family comprises 12 different heterodimers in mammals, each of which shares a common beta subunit ($\beta 1$) paired with a unique α subunit. A second major subfamily, the αV family, shares a common α subunit (αV) paired with one of five different β subunits (from a total of eight, reviewed by Hynes, 2002). At least two other alpha subunits ($\alpha 4$, $\alpha 6$) can associate with more than one β subunit ($\beta 7$ or $\beta 4$, respectively). Despite this apparent diversity in alpha and beta subunits, only 24 $\alpha\beta$ heterodimers have been described (Hynes, 2002).

Basic rules exist regarding integrin receptor composition and function (reviewed by Albeda and Buck, 1990; Hemler, 1990; Hynes, 1987, 2002). In general, the $\beta 2$ integrins interact with members of the complement (C3bi) and Ig superfamily (ICAM-1) of adhesion molecules, and are involved primarily in mediating cell–cell adhesion as opposed to cell–extracellular matrix (ECM) adhesion. The rest of the integrins ($\beta 1$, αV subfamilies) are the major receptors used by cells for attachment to ECM proteins fibronectin, collagen, laminin, and vitronectin, and can be categorized into those that interact with the arg-gly-asp (RGD) or leu-asp-val (LDV) sequences in fibronectin (Pierschbacher and Ruoslahti, 1984; Pytela *et al.*, 1985a; Wayner and Kovach, 1992), and vitronectin (RGD) (Cheresh *et al.*, 1989; Pytela *et al.*, 1985b); those that interact with collagen (Takada *et al.*, 1988; Wayner and Carter, 1987); and those that interact with laminin (Carter *et al.*, 1991). Integrins are expressed on all cells, although certain subfamilies of receptors are restricted to cells belonging to certain lineages. This information can be exploited when making antibodies to integrin receptors. For example, the $\beta 2$ integrins (LFAs) and $\alpha 4$ with $\beta 7$ are restricted to hematopoietic cells. Mesenchymal cells express the fibronectin receptors $\alpha 5\beta 1$ and $\alpha v\beta 3$; epithelial cells express the laminin receptors $\alpha 3\beta 1$ and $\alpha 6$ (complexed either with $\beta 1$ or $\beta 4$) and the vitronectin receptor $\alpha v\beta 5$ (not $\alpha v\beta 3$); and melanoma cells express $\alpha 4\beta 1$, and the vitronectin receptor, $\alpha v\beta 3$, and lower levels of $\alpha v\beta 5$. Finally, in some cell populations, many integrin receptors are maintained in an inactive or "low avidity" state. Such is the case with platelet and leukocyte integrins. In such cell populations, the avidity or activation state of the receptor can be modulated by various stimuli such as cytokines, growth factors, mitogens, or even other antibodies (Hynes, 2002; Wayner and Kovach, 1992). Therefore, when designing strategies to obtain antibodies to integrin receptors, the potential activation state of the target receptor should be considered as well.

1.1. Monoclonal antibodies to integrin receptors

The two principal approaches to the study of receptors for the ECM are (1) affinity chromatography with extracellular matrix (ECM) ligands (Mould *et al.*, 1990; Pytela *et al.*, 1985a,b), and (2) the production of monoclonal antibodies that block adhesion of cells to purified ECM ligands (Kunicki *et al.*, 1988; Lam *et al.*, 1989; Leavesley *et al.*, 1992; Wayner and Carter, 1987; Wayner *et al.*, 1988, 1989, reviewed by Hynes, 2004). Many of these functionally defined antibodies to integrin receptors are well documented and their binding sites have been mapped to certain domains in each receptor (Kamata *et al.*, 1994, 1995; Schiffer *et al.*, 1995; Takada and Puzon, 1993). In addition to antibodies that inhibit integrin function, some antibodies are known to stimulate integrin-binding activities. For example, 8A2 is a well-known activating anti-$\beta 1$ monoclonal antibody (Kovach *et al.*, 1992) and monoclonal antibodies to $\alpha 4$ can stimulate intercellular aggregation (Pulido *et al.*, 1991; Schiffer *et al.*, 1995). Still other monoclonal antibodies to $\beta 1$, such as P4G11, can stimulate adhesive ligand binding, but only in the presence of divalent cations (Takada and Puzon, 1993). Interestingly, some of the function-inhibiting and function-stimulating monoclonal antibodies map to the same region within the $\beta 1$ receptor (Takada and Puzon, 1993), further confusing the characterization of antibodies to these complex receptors.

   Not all antibodies to integrin receptors are equivalent, nor can all antibodies to a particular receptor be used for all purposes. Since there are several functional domains within each integrin receptor, multiple antibodies to the same receptor can exhibit vastly different binding or functional character- istics depending on the epitope to which it binds (Kamata *et al.*, 1994; Schiffer *et al.*, 1995; Wayner, 1990). For instance, some antibodies to the integrin β1 subunit do not inhibit function but are particularly good at identifying the denatured β1 subunit by western blot (e.g., A1A5, P5D2). Likewise, many well-characterized, functionally inhibitory monoclonal antibodies to β1 do not western blot, but bind only when the receptor is in a "native" state (e.g., P4C10). Still others will not even bind to the receptor unless divalent cations are present or unless the receptor is in an "active" conformation. For example, P1H5 binds to the resting α2 receptor expressed by T lymphocytes, but P4B4, P1E6, and P1H6 recognize epitopes that require activation of the α2β1 complex (Wayner, unpublished; Kamata *et al.*, 1994). Other monoclonal antibodies can modulate an integrin receptor differently even when they can be shown to bind to the same epitope. For example, P4G9 stimulates aggregation of T cells via α4, but P3E3 does not, even though they bind to the same epitope in α4 (Kamata, *et al.*, 1994; Schiffer *et al.*, 1995; Wayner, 1990). Another factor to consider when characterizing monoclonal antibodies to integrin receptors is that several of these receptors (such as α3β1 and α6β1) (Berditchevski *et al.*, 1995) can exist in a complex with still other cell surface receptors. This adds another level of complexity when attempting to develop antibodies to integrin receptors or to define the specificity of an antibody by immune precipitation.

Many antibodies to integrin receptors are now commercially available. The Linscott Directory (www.linscottdirectory.com) offers a complete listing of many of these antibodies including their function-blocking cap- abilities and commercial sources. However, many of the commercially available antibodies are not function modulating, certain species-specific reagents may not be available, and other problems can exist such as quality control among batches. In this paper, we will describe some of the methods that we have used to generate function-modulating antibodies to the β1 and αV integrins (Kunicki *et al.*, 1988; Leavesley *et al.*, 1992; Wayner and Carter, 1987; Wayner *et al.*, 1988, 1989). Additionally, we will describe in detail a protocol that was used to generate an inhibitory mouse mono- clonal antibody to the human α6 laminin-1 receptor expressed on PC3 prostate carcinoma cells.

1.2. Production of monoclonal antibodies with hybridoma technology

An antigen is defined as any substance that, if foreign to the host, will stimulate the production of an antibody directed to each antigenic determinant or epitope. If such a foreign substance or antigen is injected into a vertebrate host, such as a mouse, some of the immune system's B lymphocytes will differentiate into plasma cells and start to produce antibodies that recognize the antigen. Each antigen-specific B cell produces a unique antibody characterized by specific Ig gene rearrangements and affinity maturation (clonal selection). Thus, during an immune response, many structurally different antibodies are produced that bind to various epitopes of the antigen. This natural mixture of antibodies found in serum is known as a polyclonal antibody. In the 1970s, it was known that some B-cell cancers, myelomas, could be propagated *in vitro* and that some of these could produce a single type of antibody or "paraprotein." However, it was not possible to produce a single type or "monoclonal antibody" of predetermined specificity. The process of producing monoclonal antibodies secreted by specifically immunized and immortalized B cells was invented by Kohler and Milstein (1975). Their Nobel Prize–winning experiment was to use a myeloma cell line that had lost its ability to produce paraproteins and to fuse these cells to healthy Ig-producing B cells and select for the successfully fused and immortalized "hybridomas."

To produce monoclonal antibodies one removes B cells from the spleen of an immunized animal or one that has been challenged several times with the antigen of interest. These B cells are then fused with non-Ig secreting immortal myeloma cells. The fusion is accomplished by making the cell membranes more permeable by the addition of polyethylene glycol, electroporation, or infection with a virus. The fused cells (hybridomas) multiply rapidly and indefinitely. The antibody of interest can therefore be produced indefinitely and in unlimited quantities. The hybridomas can be diluted sufficiently to ensure clonality and grown. The antibodies from different clones are then tested for their ability to bind to the antigen in a test such as an ELISA, immuno-dot blot, western blot, or immune precipitation reaction. The best antibody with the highest affinity can then be selected and propagated. The general scheme for the production of any monoclonal antibody is shown in Fig. 7.1.

2. METHODS

2.1. Production of monoclonal antibodies to integrin receptors using hybridoma technology

Our primary protocol for the production of an antibody to an integrin receptor is summarized in the flow chart in Fig. 7.1. However, first the antigen, the test system, and the immunization and fusion strategy must be chosen. We decided for the purposes of this paper to make a function-blocking mouse monoclonal antibody to the human α6 integrin receptor. The decisions we made in order to accomplish this and the reasons for each decision will be described in detail. Data have been recorded over the course of developing this reagent and the results are compared to GoH3

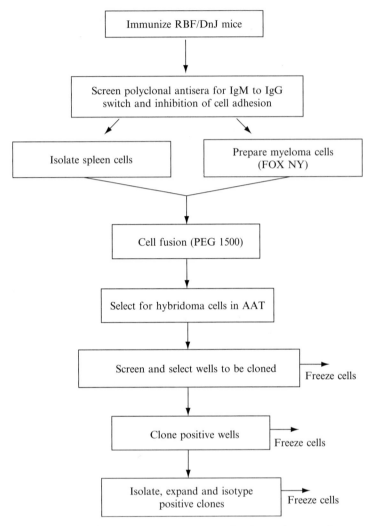

Figure 7.1 Flow chart showing the general scheme for the production of any monoclonal antibody via the PEG fusion of immune mouse splenocytes and mouse myeloma cells (FOX-NY) and selection in AAT media.

(Sonnenberg *et al.*, 1987), a well known rat anti-α6 monoclonal antibody. All antibodies except GoH3 are available from E. Wayner.

2.2. Choice of the antigen (native versus non-native receptor)

When initiating the development of an antibody to an integrin receptor, it is useful to consider several points. The most important of these is to decide what characteristics are required from the antibody; in other words, identify what you want the antibody to do. For example, do you want the antibody to block the binding of a receptor to its ligand (a function-blocking antibody), or to identify the receptor in various reactions such as immunohistochemistry (IHC), flow cytometry, western blot, or immune precipitation? Generally, depending on the desired application of the antibody, this will determine the immunization protocol, the antigen used to prepare the immune spleen, and the system used to screen the resulting hybridomas. For example, if a western blotting antibody is desired, it is best to use denatured protein as the immunogen. Such immunogens could be recombinant proteins expressed in bacteria (no post-translational modifications), gel-purified and eluted material (completely denatured), or receptor purified and eluted from an affinity column. Receptors can also be purified via immune precipitation with specific monoclonal antibodies. The ligand-receptor or receptor-immuno beads can then be used as immunogens. Such protocols were used to identify and develop antibodies to the fibronectin and vitronectin receptors. For the purposes of this paper, we will assume that function-modifying antibodies are required. In our experience, the best way to generate function-modulating monoclonal antibodies is to immunize with the cells that naturally express the receptor (see below). However, it is useful to briefly discuss other types of antigen preparations.

2.2.1. Peptides, fusion proteins, and denatured antigens

Historically, peptides conjugated to carrier molecules have been shown to produce good immune responses to integrin receptors, particularly when interested in reagents specific for the cytoplasmic tail. Such reagents have been useful for immunoblotting and other studies. However, peptides as immunogens are less useful when antibodies are required that react with "native" receptors expressed on the surface of cells or when antibodies are required that inhibit ligand binding. When peptides or bacterial fusion proteins are used to immunize we recommend that two (or more) preparations of each be made. For example, a glutathione-S-transferase (GST) fusion protein can be used to immunize, and a maltose-binding protein (MBP) or his-tagged protein can be used to screen the resulting hybridoma culture supernatants. Alternatively, a KLH-peptide conjugate can be used to immunize, and then the resulting hybridomas can be screened by ELISA on a BSA-peptide conjugate. Sometimes we recommend the use of three peptide conjugates: KLH for immunization, ovalbumin (OVA) for boosting, and BSA for screening. In this way, B-cell clones secreting antibody specific for the peptide can be positively selected *in vivo* before the fusion. The use of one antigen preparation for immunization and another for screening we refer to as a "differential" immunization and screening protocol. This concept can also be applied to screening positive and negative cell pairs by cell-based ELISA or flow cytometry. Another technique we have found to be particularly useful is to use cells that express the receptor to immunize, and then use a purified fusion protein-containing sequences of interest to screen hybridoma supernatants by ELISA. This is particularly helpful when antibodies to native epitopes are required and the cell population cannot be used for a cell-based ELISA (e.g., nonadherent cell population) to screen for fusion positives. We have found that antibodies initiated with the "native" receptor expressed on cells and then identified by ELISA on fusion protein results in a higher proportion of antibodies that recognize the "native" form of the receptor than if the fusion protein is used throughout. This technique is also particularly useful when antibodies are required to specific epitopes (immunizing with cells and screening on overlapping fusion protein segments or peptides). We refer to this as differential epitope ELISA screening.

2.2.2. Antibodies to denatured epitopes
In this case, SDS-PAGE gel-purified material would be appropriate. The desired protein can be immune precipitated with a commercial antibody, run under nonreducing conditions, and then the area of the gel containing the integrin (based on its molecular weight) can be excised and emulsified with an adjuvant such as incomplete Freund's (IFA). Alternatively, affinity-purified receptor can be run on an SDS-PAGE gel; the specific area of the gel corresponding to the protein of interest can be excised, emulsified, and injected. Proteins transferred to nitrocellulose can also be injected. In this case, the piece of nitrocellulose containing the protein of interest is frozen in LN2, pulverized with a mortar and pestle, and then emulsified with IFA and injected. A large-gauge emulsifying needle is required (16 gauge, Popper #7969). It is important to be sure to inject such antigens intraperitoneally (i.p.), and it is necessary to monitor the health of animals injected with acrylamide. A final strategy is to electro-elute the desired receptor from an unfixed gel, and then emulsify and inject it.

2.2.3. Antibodies to "native" epitopes
When considering how to generate the desired monoclonal antibody to an integrin, it is very important to clearly define the properties required in the resulting antibody. For example, in order for a monoclonal antibody to western blot it must be able to detect denatured proteins. If an antibody is desired that reacts with the surface of unfixed or "native" cells by FACS or IHC on frozen sections or for modulating the function of a receptor (inhibition or activation), one essential characteristic is that it must react with the "native" receptor. Therefore, in order to obtain such antibodies, SDS-PAGE purified or denatured proteins would not be used to immunize. This conundrum (native vs. denatured) can be problematic, especially if the only source of receptor is purified recombinant proteins made in bacteria, which is essentially one step above denatured protein. Therefore, we recommend immunizing with cells that express a functional receptor on the cell surface and screening with the recombinant protein. This approach can

produce antibodies that represent the best of both worlds; since native epitopes were used for immunization, one can end up with a "do-it-all" antibody useful for reacting with the native receptor as well as the denatured receptor. However, as a general rule, antibodies that modulate function do not react with denatured receptors via western blot, and immune precipitation reactions are necessary for their biochemical characterization. Sometimes, it is not possible to use cells to immunize. If this is the case, antibodies that recognize the native receptor can still be obtained, but a larger number of hybridomas may need to be screened.

The way to increase the number of positives in any fusion is to hypersensitize the spleen to the antigen in the final immunization phase. The way to achieve this is to use a promiscuous injection protocol with 50 μg of protein emulsified in RIBI adjuvant injected on Days −3, −2, and −1 before sacrifice (Day 0). A watchful eye needs to be kept on the mice injected with such a protocol since they can become ill or die of an anaphylactic shock reaction. With this protocol, it is possible to end up with 100% of the fusion wells positive for the immunizing antigen. In the case of developing an "activating" anti-integrin antibody, it is logical to assume that if one injects cells expressing the receptor in an active conformation, it is more likely that antibodies to so-called "activation-dependent epitopes" will be obtained. Therefore, it might be reasonable to inject a promiscuous cell line such as PC3 cells or HT1080 cells (or lymphokine-activated killer cells) and then screen for the ability to activate resting lymphocyte binding in a cell adhesion assay. Antibodies that induce activation of lymphocyte binding to ECM proteins and other cells have been described (Kovach et al., 1992).

2.2.4. Use of cells as the immunogen

For functionally defined antibodies to integrin receptors, it is usually best to use native antigen as the immunogen. The cells that naturally express the receptor are the best source of such antigens. Whenever cells are used as the immunogen, especially cultured cells, it is essential to wash them free of any residual fetal bovine serum or human plasma since serum and plasma contain many highly immunogenic proteins. Table 7.1 summarizes some general information concerning expression of the major integrin receptors by different cell types and a proposed method for obtaining a particular antibody using each cell type. For example, for an antibody to $\alpha v \beta 3$, it would be best to use a melanoma cell population that expresses large quantities of this receptor, and if making antibodies to another RGD-dependent αv-containing integrin—for example, $\alpha v \beta 5$—it is best to use a population of cells that does not express $\alpha v \beta 3$, such as UCLA P3 cells (Leavesley et al., 1992).

Another technique to be taken advantage of is to exploit the exquisite ligand specificity of certain receptors. For example, $\alpha 5 \beta 1$ and $\alpha 4 \beta 1$ differentially recognize the central cell-binding domain on an 80-kDa fragment or the carboxy terminal cell-binding domain on a 38-kDa fragment of

Table 7.1 Expression of integrin receptors by selected cells and methods to make inhibitory monoclonal antibodies

Integrin receptor	Cell population to use	Method to use
α1β1 (180-kDa α subunit)	LAK cells (IL2-activated T cells)	Look for 180-kDa subunit in IP matrix (see Figures 4 and 6)
α2β1	HT1080 fibrosarcoma cells	Specific inhibition of adhesion to COL I or III but not FN
α3β1	Normal human keratinocytes or carcinoma cells	Inhibition of adhesion to matrix secreted by HFKs (laminin 5)
α4β1	Melanoma cells (Mel 13) or T cells (Jurkat)	Specific inhibition of adhesion to 38-kDa FN fragment and LDV containing peptides or activation of T-cell aggregation
α5β1	HT1080 fibrosarcoma cells or T cells (Jurkat)	Specific inhibition of adhesion to 80-kDa FN fragment and RGD containing peptides
α6	PC3 carcinoma cells	Specific inhibition of adhesion to laminin 1
β1 inhibiting	HT1080 fibrosarcoma cells	Concurrent inhibition of cell adhesion to LAM, FN, and COL I or III
β1 activating	HT1080 or LAK cells	Concurrent activation of lymphocyte adhesion to LAM, FN, and COL I or III
β2	Leukocytes, PBL	Inhibition of spontaneous aggregation of LAK or B cells
αvβ3	Melanoma cells (high αvβ3, no αvβ5)	Inhibition of adhesion to VN or RGD-containing peptides
	KCA (B cell line, no αvβ5)	Inhibition of adhesion to VN
αvβ5	UCLA P3 carcinoma cells (no αvβ3)	Inhibition of adhesion to VN, RGD-containing peptides

COL, collagen; FN, fibronectin; HFKs, human foreskin keratinocytes; IP, immune precipitation; LAK, lymphokine activated killer cells; LAM, laminin; LDV, leu-asp-val; RGD, arg-gly-asp; VN, vitronectin.

fibronectin, respectively (Wayner et al., 1989). Often it is beyond the scope of a particular laboratory to generate large quantities of purified ECM proteins (fibronectin, collagen, or laminin) or purified fragments. In this case, peptides conjugated to carrier molecules can be used to generate differential adhesion assay screens. For example, activated lymphocytes or melanoma cells will use $\alpha v\beta 3$ (Cheresh et al., 1989) or $\alpha 5\beta 1$ to bind to RGD-containing peptides (Pierschbacher and Ruoslahti, 1984) and $\alpha 4\beta 1$ to bind to LDV-containing peptides, such as CS-1 (Wayner and Kovach, 1992). Therefore, melanoma cells or activated T cells such as Jurkat could be used to immunize mice and the resulting hybridomas could be screened on LDV or RGD-containing peptides in order to obtain antibodies to $\alpha v\beta 3$, $\alpha 5\beta 1$, and $\alpha 4\beta 1$ (Table 7.1).

In our specific example, if an antibody to the $\alpha 6$ receptor is required, the first step is to find a cell line that expresses a large amount of this receptor. We performed a number of immune precipitation screens on several cell lines, and found one, a prostate carcinoma cell line (PC3) that expressed large amounts of $\alpha 6$ when compared to other $\beta 1$-containing integrins (Fig. 7.2A). Thus, if we could find a test system dependent on $\alpha 6$, it would be reasonable to use PC3 as the immunogen or starting material.

2.3. Choice of a test system

It is always essential when creating monoclonal antibodies that inhibit the function of an integrin receptor to choose a test system that is absolutely dependent on the receptor–ligand interaction. This is not always the case as we learned upon making antibodies to the lymphocyte fibronectin receptors $\alpha 4$ and $\alpha 5$ (Wayner et al., 1989). However, it is easier to interpret the results of an adhesion assay if the data are unequivocal (see below).

Since $\alpha 6$ has been reported to be a receptor for laminin-1, we decided to determine whether PC3 cell adhesion to laminin-1–coated surfaces could be used as a test system to obtain a functionally inhibitory monoclonal antibody to $\alpha 6$. PC3 cells did in fact adhere to purified laminin-1 (LAM-1), plasma fibronectin (Figs. 7.2B and 7.3C) and collagen I (not shown). All of these ligands were purified according to published protocols (Carter et al., 1990; Wayner and Carter, 1987; Wayner et al., 1989). We could inhibit adhesion of PC3 cells to all ligands including laminin-1 (Fig. 7.2B) with an inhibitory antibody to $\beta 1$, P5D2 (Wayner et al., 1993), confirming that a $\beta 1$ integrin was involved in PC3 adhesion to laminin-1 (not shown for FN and collagen). Our functionally defined inhibitory monoclonal antibody P1D6 to the FN receptor, $\alpha 5\beta 1$, inhibited PC3 cell adhesion to FN (Fig. 7.3D) but not LAM-1 (Fig. 7.2B). Importantly, our well-defined anti-$\alpha 3$, P1B5, also did not significantly inhibit adhesion of PC3 cells to laminin-1 (Figs. 7.2B and 7.3B). Since $\alpha 3\beta 1$ has been reported to be a receptor for some forms of laminin, in particular laminin 5 (Carter et al., 1991), this strongly suggested that PC3 cells use $\alpha 6$ as a receptor for laminin-1. Therefore, we decided to use

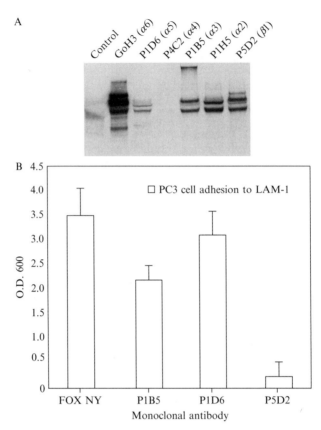

Figure 7.2 (A) Immune precipitation of α6 and other β1-containing integrins from PC3 cells. Cells were biotinylated and extracted with 1% Triton-X-100 according to the protocol described here. The integrin receptors were immune precipitated with monoclonal antibodies specific for α6 (GoH3) (GoH3 was a generous gift from A. Sonnenberg), α5 (P1D6, Wayner *et al.*, 1988), α4 (P4C2, Wayner *et al.*, 1989), α3 (P1B5, Wayner and Carter, 1987), α2 (P1H5, Wayner and Carter, 1987), and P5D2 (Wayner *et al.*, 1993). Immune precipitates were run on 4 to 12% Bis-Tris/MES Nu-PAGE gels under nonreducing conditions and transferred to nitrocellulose (5% MeOH). Biotinylated and immune precipitated proteins were detected with HRP-conjugated streptavidin (1/100,000) and ECL (GE Healthcare #1059243/1059250). (B) Adhesion of PC3 cells to laminin-1 (LAM-1) in the presence of functionally inhibitory antibodies to α3 (P1B5), α5 (P1D6), and β1 (P5D2). Laminin-1 was purified according to published protocols (Carter *et al.*, 1990; Wayner and Carter, 1987), and coated on 48-well plates at 5 µg/ml. Culture supernatant derived from the hybridomas (200 µl per well) was used to inhibit LAM-1 adhesion. The adhesion assay was performed exactly as described here.

a differential adhesion screen with FN as a negative control to search for inhibitory monoclonal antibodies that could inhibit PC3 adhesion to laminin-1. Use of PC3 cells as the immunogen and this test system should result in an antibody to α6.

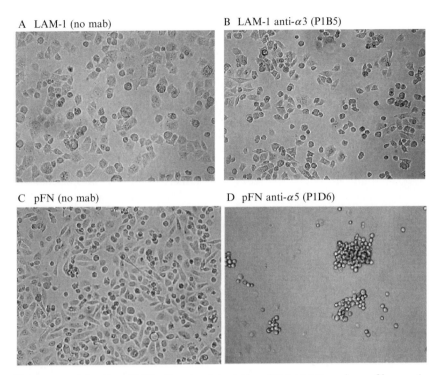

A LAM-1 (no mab)

B LAM-1 anti-α3 (P1B5)

C pFN (no mab)

D pFN anti-α5 (P1D6)

Figure 7.3   Adhesion of PC3 cells to laminin-1 (A and B, LAM-1) or plasma fibronectin (C and D, pFN) at 5 μg/ml in the presence (B and D) or absence (A and C) of functionally inhibitory antibodies to the laminin-5 receptor, α3 (P1B5, B) or the fibronectin receptor, α5 (P1D6, D).

2.4. Immunization strategy

2.4.1. Choice of host species

For the purposes of this paper we assume that protocols for obtaining mouse anti-human integrin receptors are required. However, it is possible to create rat anti-mouse (or human) and mouse anti-rat integrin-receptor antibodies. In addition, it is also possible to create rabbit monoclonal antibodies thanks to the development of a rabbit myeloma, 240E, by Katherine Knight and co-workers (Spieker-Polet et al., 1995). This myeloma has been successfully used in our laboratory to create rabbit antipeptide monoclonal antibodies directed to human laminin 5. We have also used a rat myeloma (Y3-D10, Lydia Sorkin) to create rat monoclonal antibodies directed to the mouse α3β1 integrin receptor (Sigle et al., 2004). The basic protocol for the production of both rat and rabbit monoclonal antibodies is the same as the one described here. The Y3-D10 cell line is available from E. A. Wayner, and the 240E cell line is available from Katherine Knight (Loyola University) via MTA.

2.4.2. Mouse strain and myeloma

There are many protocols for the production of monoclonal antibodies (Harlow and Lane, 1988; Howard and Bethell, 2001; Kohler and Milstein, 1975; Oi and Herzenberg, 1980; Peters and Baumgarten, 1992). The method that we have used for the past 15 years is based on the protocol described by Oi and Herzenberg (1980), and makes use of the Robertsonian (Rb (8.12) 5 Bnr) mouse strain (RBF/DnJ), and the FOX-NY myeloma cell line described by Taggart and Samloff (1983). FOX-NY was isolated as a mutant of the NS-1 cell line and possesses a double-enzyme deficiency (APRT as well as HPRT). Thus, this cell line can be eliminated with media containing either HAT (hypoxanthine/aminopterin/thymidine) or AAT (adenine/aminopterin/thymidine). In the Robertsonian 8.12 mouse, the heavy-chain Ig locus on chromosome 12 and the selectable enzyme marker APRT on chromosome 8 are linked. Thus, exposure of a cell fusion mixture to medium requiring APRT activity eliminates both unfused myelomas and APRT-deficient hybridomas. Theoretically, this protocol would provide a selective advantage for cells that retain the heavy-chain Ig locus and thus increase the numbers of hybridomas secreting Ig. Use of AAT provides another advantage in that AAT is an inherently nontoxic growth additive, in contrast to hypoxanthine, which can be toxic. In fact, we have shown that some hybridomas selected with AAT cannot be propagated in HAT. As can be seen in Fig. 7.4C, the RBF/FOX-NY fusion system with AAT selection is much more successful at producing antibodies useful in immune-precipitating receptors from surface-labeled HT1080 cells when directly compared to two other fusion systems involving BALB/c mice and the SP2/0 myeloma (Fig. 7.4A and B). In fact, this fusion (Fig. 7.4C) was particularly rich and yielded the antibodies P1H5, P1B5, P1D6, and P4C10 (Carter et al., 1990; Wayner and Carter, 1987; Wayner et al., 1988). This system produces good antibodies no matter what the screening criteria, including inhibition of cell adhesion (see below). Therefore, the system we describe here will be for immunization of RBF/DnJ mice (Jackson Labs, stock number 00726) with human PC3 cells and fusion of the immune splenocytes with FOX-NY myeloma cells. One important point to make is that any myeloma cell line used should be tested for sensitivity to AAT (or HAT) every time the cell line is thawed out and used for a fusion.

2.5. Choice of adjuvant and immunization protocol

2.5.1. Adjuvants

Various adjuvants have been reported to provide strong stimulation of antibody responses in mice. We have used several of these to promote antibody responses to integrin receptors. These include RIBI (now available from Sigma, M-6536), complete (CFA) and incomplete (IFA) Freund's (Pierce, ##77140 and #77145, respectively), and adjuvant peptide from the *Mycobacterium* cell wall (muramyl dipeptide, Sigma A-9519). More recently,

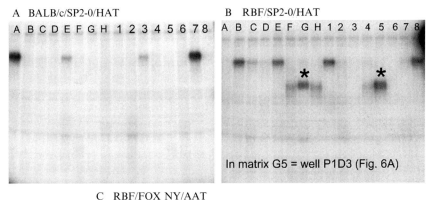

A BALB/c/SP2-0/HAT

B RBF/SP2-0/HAT

In matrix G5 = well P1D3 (Fig. 6A)

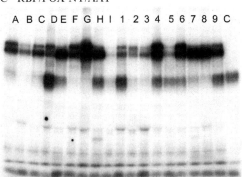

C RBF/FOX NY/AAT

Figure 7.4 Efficiency of RBF/FOX NY fusion system (C) with AAT selection versus two other systems that make use of BALB/c splenocytes, another myeloma (SP2–0), and HAT selection (A and B). Data represent immune-precipitated proteins from surface-labeled HT1080 cells run on 10% SDS-PAGE gels as described (Wayner and Carter, 1987). Tubes A to H and 1 to 10 represent mixtures of culture supernatants derived from the fusions indicated. As can be seen from the data, according to these criteria, immune precipitation of surface-labeled proteins via the RBF/FOX-NY/AAT fusion system is much more efficient at producing useful antibodies. The matrix described in Fig. 7.6A was used here in B; the asterisks indicate the location of a specific antibody specificity duplicated in tubes G and 5. G5 represents the antibody secreted in well P1D3 (go to matrix in Fig. 7.6A and find G5).

bacterial DNA sequences that contain unmethylated CpG motifs (CpG ODN) have been reported to promote Th1 immune responses characterized by the secretion of IFN-gamma, TNF alpha, and IL12 cytokines. This complement of cytokines promotes IgG subtype switching that results in a predominantly IgG2a response (Chu et al., 1997). In our experience, long-term immunization with IFA adjuvant (more than five injections) results in a Th2-type response characterized by a predominant IgG1 response. This is particularly true when purified fusion proteins or peptides are used as the immunogen. We have used all of these adjuvants with varying degrees of success. Generally, when we require an IgG2a response, we now use CpG repeats, with or without IFA.

Our primary protocol when injecting cells into RBF mice is to use an adjuvant when immunizing (first injection), but not for subsequent injections (boosts) unless we want to induce Ig2a subtype switching. In general, the more complex the antigen (cells vs. peptides), the less likely it will be to require the use of an adjuvant to promote a desirable immune response or subtype switching. If using recombinant receptors produced in bacteria, receptors denatured by elution from affinity columns, or peptides or proteins eluted from SDS–PAGE gels, a T–dependent adjuvant such as RIBI or bacterial CpG should be considered. Again, we should stress that if antibodies are required to native epitopes, it is best to use native immunogens, such as intact cells or renatured protein.

2.5.2. Immunization schedule and route of antigen administration

Microemulsifying needles can be purchased from Fisher; they are made by Popper and come in various gauges. If using cells, we usually do not emulsify them in CFA or IFA. We usually emulsify only purified proteins or peptides (25 gauge, #7974). However, cells can be emulsified in Freund's if a large-gauge emulsification needle is used (16 gauge, Popper #7969). Immunogens such as cells can also be emulsified in RIBI adjuvant using a probe sonicator (three 10-sec pulses). If using adjuvants, our first injection involves injecting one mouse with antigen plus CFA (40% aqueous/60% oil) and one mouse with antigen plus RIBI. Thus, we routinely use two mice per treatment group. We use two mice per treatment group, even when no adjuvant is used, in case one of the animals dies before the spleen is harvested. We then wait for 3 weeks, and then boost twice without adjuvant with each injection 7 to 10 days apart. Ten days after the last boost, test bleeds are taken from the orbital sinus or tail vein, and the antiserum is tested via ELISA or in a cell adhesion assay. This is absolutely critical. If the polyclonal repertoire (see below) does not contain the desired antibody, it is very unlikely that it can be obtained by fusion. If, after the second boost the required antibody response is not elicited, then use of an axillary adjuvant should be considered. We have found the combination of CFA and RIBI or CFA and CpG to be successful in eliciting antibodies to difficult antigens or epitopes. However, this immunization protocol is very hard on the mice, and they tend to develop an immune malaise syndrome characterized by weight loss, lethargy, and the presence of ascites fluid due to the inflammation caused by injecting the antigen into the peritoneal cavity. Thus, it could be desirable to switch to subcutaneous injections when using a more promiscuous immunization protocol (CFA plus RIBI or CpG). It is also necessary to be aware that bleeding the mice either from the tail or from the orbital sinus is also very hard on them, and care must be taken so that they do not develop secondary infections from these procedures.

2.6. Polyclonal antisera screening

2.6.1. IgM to IgG switch

The polyclonal antisera must be screened via each protocol that will be used to obtain the monoclonals. It is also very important to be sure that there has been an adequate IgM to IgG switch so that IgGs can be obtained in the fusion. It has been our experience that if there is a large amount of antigen-specific IgM in the polyclonal serum, then the fusion will contain predominantly IgMs. We usually continue with the immunizations until there has been a complete IgM to IgG switch with literally no IgM directed to the antigen observable in the immune sera. As strange as it sounds, this happens even with complex antigens such as cells (Fig. 7.5A).

As seen in Fig. 7.5A, after three injections with 1×10^8 PC3 cells, the serum from the RBF 1L mouse (1L indicates one ear punch on the left ear) exhibited an almost complete IgM to IgG switch when tested by a cell-based ELISA using PC3 cells. Interestingly, this mouse was injected with PC3 cells emulsified in RIBI adjuvant. Polyclonal sera against cells and the presence of IgGs or IgMs can also be assessed by flow cytometry on the immunizing cells. However, as long as the cells adhere to gelatin–coated plates and can be fixed to the plates using 2% formaldehyde, a cell-based ELISA (see below) is a quick and simple way of evaluating the efficiency of the immunization protocol. When performing such tests, it is absolutely critical that the secondaries do not cross-react. We have tested such reporter molecules from many companies, and have found that the reagents from Southern Biochemical (Birmingham, AL) are the best. This includes their fluorescent as well as their HRP-conjugated secondaries. Additionally, their secondaries to unique mouse isotypes (IgG1, IgG2a, IgG2b, IgG3) are high quality and also do not cross-react. Regardless of which company is chosen for the supply of such reagents, it is essential to test them for cross-reactivity on known standards by ELISA. When polyclonal mouse antisera are tested for their utility in predicting the success of a fusion, it is best to run titrations from 1/100 to 1/100,000. If the serum does not show a good dose–response curve to the antigen with the IgG activity titrating out in the 1/10,000 range, it is best to boost at least two more times at 7- to 10-day intervals.

2.6.2. Inhibition of cell adhesion by polyclonal antiserum

An example of an adhesion assay testing the ability of our 1L antiserum generated to PC3 cells for its ability to inhibit PC3 cell adhesion to laminin-1 and plasma fibronectin is shown in Fig. 7.5B. As seen in these data, after one immunization and two boosts, this animal's antibody repertoire contains inhibitory antibodies that block adhesion to laminin-1, and to fibronectin to a lesser extent. Therefore, this repertoire potentially contains antibodies to $\alpha 5$-, $\alpha 6$-, and $\beta 1$-containing integrins. The decision was made to prepare the

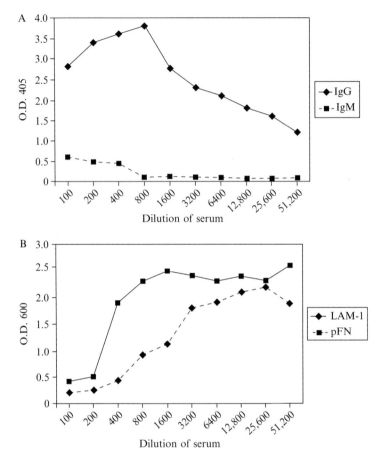

Figure 7.5 (A) IgG (triangles) versus IgM (squares with broken line)-dependent assay of RBF 1L antiserum on PC3 cell-based ELISA. PC3 cells were grown on gelatin-coated plates, fixed in 2% formaldehyde, and then reacted with the 1L antiserum at the indicated dilutions. HRP-conjugated goat anti-mouse antibodies specific for either IgG or IgM were added at 1/2000 per well. Bound anti-IgG or anti-IgM was reacted with ABTS and the plates were read at O.D. 405. (B) Effect of RBF 1L antiserum on the ability of PC3 cells to adhere to laminin-1 (LAM-1) or plasma fibronectin (pFN) coated surfaces. Adhesion proteins were coated at 5 μg/ml and the cell adhesion assay was carried out exactly as described here except that instead of culture supernatant the 1L antiserum was added in the adhesion buffer at the indicated dilutions. PC3 cells were added, reacted with the antiserum for 15 min, and then allowed to adhere to the surfaces for 30 min at 37°. At the end of this time, the plates were washed, and then the cells were fixed and stained with crystal violet and read at O.D. 600 (after transfer of stain to 96-well ELISA plate).

1L mouse for fusion. At this point in the cell adhesion assay, it is impossible to know for sure that the antibody or antibodies that inhibit are IgGs, but we make this assumption based on results of the IgM to IgG switch ELISA.

2.7. Preparation of the spleen for fusion

The final injection is prepared exactly 3 days (Day −3) before fusion. This final injection could contain an adjuvant to increase the specific polyclonal B-cell response and to further push the B cells into the appropriate differentiation state for successful fusion. At this point, however, it is not recommended to use a highly promiscuous adjuvant since usually it would be disastrous for the animal to die of immune malaise or anaphylactic shock. We usually choose to combine the final injection of cells with 50 μl of adjuvant peptide (stock at 1 mg/ml). On Day 0, a final bleed is obtained for use as a positive control in any ELISAs or cell adhesion assays. The animal is sacrificed via an IRB/IACUC-sanctioned protocol and the spleen is removed. In our specific example, on Day −3 we boosted the 1L mouse with 1×10^8 PC3 cells in 0.5 ml i.p.

2.8. Design the strategy for screening the fusion

Large numbers of hybridoma wells can be screened for the presence of antibodies to integrin receptors in various ways. They can be screened for function (see below) using a cell adhesion assay. They can also be screened for the presence of specific antibodies to integrin receptors via the use of a matrix protocol (Fig. 7.6A) and immune precipitation (IP). This can be used directly, or in conjunction with western blot, to locate antibodies to certain family members such as the αV or β1 family members (Fig. 7.6B). In this technique, a worksheet is used to place each supernatant in each of two tubes (either A to H or 1 to 12) with well A1 being placed into tubes A and 1, A2 placed into tubes B and 1, and so on for the duration of the particular matrix being designed (see Fig. 7.6A). Then, for a direct test, the IPs are performed with surface-labeled cell extracts, and each IP is run on an SDS gel (Fig. 7.2A,B, and C). For antibodies to the αv- or β1 integrin-containing receptors, a matrix is designed with supernatants placed in tubes A to H and 1 to 10, unlabeled cell extracts are immune precipitated, run on SDS-PAGE gels under nonreducing conditions, transferred to nitrocellulose, and then purified and biotinylated antibodies to the β1 (approximate molecular weight 116 kDa) or αV (approximate molecular weight 150 kDa) subunits are used to western blot the specific receptors immune-precipitated by monoclonal antibodies located in the matrix. Antibodies for detection of integrin receptors by western blot are commercially available (Linscott Directory), and can be biotinylated with LC-biotin from Pierce. This technique has been used successfully to identify αV and β1 integrins expressed in *Xenopus laevis* (Cohen *et al.*, 2000), and to identify monoclonal antibodies to human integrin receptors (Wayner and Carter, 1987).

A

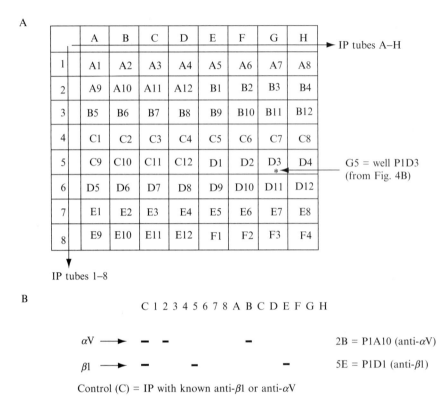

B

C 1 2 3 4 5 6 7 8 A B C D E F G H

αV ⟶   −   −   −   2B = P1A10 (anti-αV)

$\beta 1$ ⟶   −   −   −   5E = P1D1 (anti-$\beta 1$)

Control (C) = IP with known anti-$\beta 1$ or anti-αV

Figure 7.6 (A) Description of matrix and worksheet used to screen large numbers of hybridoma culture supernatants by immune precipitation or immune precipitation followed by western blot (B). In this design, each hybridoma supernatant is represented twice in each of two tubes (A to H or 1 to 10). Cells are either labeled with biotin and extracted (for direct IP) or extracted without labeling for IP followed by western blot with integrin-specific monoclonal antibodies to the αv or $\beta 1$ receptors (B). Conceivably, it could be possible that both αv and $\beta 1$ would light up in the event that a particular cell line expressed the $\alpha v \beta 1$ receptor. As an example of a direct IP matrix assay, the protein identified in lane G and again in lane 5 (shown in Fig. 7.4B) was immune precipitated by the antibody in well P1D3.

2.9. Screen fusions via a cell-adhesion assay

Antibodies that inhibit the function of integrin receptors are very easy to spot in a standard cell adhesion assay. The data with a true function-blocking antibody should be unequivocal (see Fig. 7.7). In our differential screen (data shown for Plate 5 only, Fig. 7.7) several antibodies were obtained that could inhibit adhesion to laminin (triangles) or both laminin and fibronectin (squares). For example, P5G10 was very good at inhibiting adhesion of PC3 cells to laminin–1 but not to fibronectin (Fig. 7.7). In Figure 7.8A and B, the difference between a positive control well

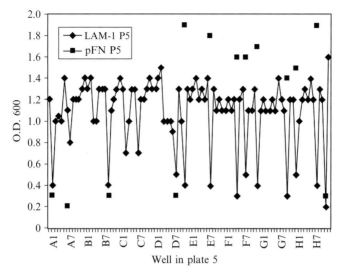

Figure 7.7 Inhibition of PC3 cell adhesion to laminin-1 (LAM-1, triangles and solid line) or plasma fibronectin (pFN, squares) by the hybridoma supernatants in plate 5 from the RBF/1L spleen fusion (selected in AAT).

(no antibody, Fig. 7.8A) and P5G10 (Fig. 7.8B) can clearly be seen at the light microscopic level. This fusion (1L spleen/FOX-NY myeloma cells) resulted in several inhibitory monoclonal antibodies per plate. Some inhibited adhesion to both fibronectin and laminin (see Fig. 7.7, well P5B9) while others were similar to P5G10. In fact, there were several functional antibodies on each plate. When we compared the results with the 1L RBF spleen to a spleen from a BALB/c mouse immunized with PC3 cells and fused and screened simultaneously, we obtained only one functional antibody in the entire BALB/c-FOX-NY/AAT fusion. Therefore, we strongly recommend the protocol we describe here.

2.10. Biochemical characterization of anti-integrin antibodies by immune precipitation

Cell surface receptors involved in cell adhesion can be identified by immune precipitation analysis. PC3 cell surfaces are labeled with biotin; the receptors are then extracted from the membranes with nonionic detergents in the presence of protease inhibitors and immunoprecipitated with the function-blocking antibodies identified by the cell adhesion assay. Functionally defined antibodies are bound to protein A agarose beads coated with rabbit anti-mouse IgG and added to the biotinylated cell extract. Bound receptors are run on an SDS-PAGE, blotted onto nitrocellulose, and detected by the addition of streptavidin-HRP conjugate followed by substrate (ECL) and

A LAM-1 (no mab)   B LAM-1 anti-α6 (P5G10)

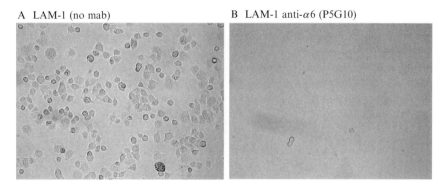

C Identification of P5G10 as an α6 specific mab by IP from
   PC3 extracts

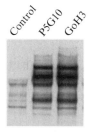

Figure 7.8 Identification of P5G10 as an α6–specific monoclonal antibody. A and B show adhesion of PC3 cells in the presence (B) or absence (A) of culture supernatant derived from well G10 on plate 5 identified in the fusion screen shown in Fig. 7.7 C shows immune precipitation by P5G10 culture supernatant as compared to immune precipitation of the α6 complex from PC3 cells by GoH3 or control (FOX-NY supernatant).

compared to the proteins immune precipitated by known anti-integrin monoclonal antibodies. Some issues to be considered are (1) whether to include divalent cations in the biotinylation and/or extraction buffers, and (2) whether to trypsinize cells before initiating the protocol. Association of α4 with β1, for instance, has been shown to require divalent cations (Wayner et al., 1989; reviewed by Hemler, 1990) and some integrin subunits, such as a4, also contain proteolytic cleavage sites. Thus, we recommend that adherent cells be scraped off culture dishes with a rubber policeman or that they be released via the use of EDTA and that immune precipitations be performed in the presence of Ca and Mg (1 to 2 mM).

When we performed immune precipitations with Triton X-100 extracts from biotinylated PC3 cells (Fig. 7.8C) with the inhibitory monoclonal antibodies identified in the cell adhesion assay (Fig. 7.7), one of these (P5G10) reacted with an integrin complex identical to that immune precipitated with GoH3 a-well known rat anti-α6 monoclonal antibody (Fig. 7.8C) (GoH3 was a generous gift from Dr. A. Sonnenberg). In fact, sequential immune precipitation analysis (Wayner et al., 1989) with GoH3 revealed that once the GoH3-reactive proteins were removed, P5G10 could no longer immune precipitate specific bands (not shown), which strongly suggests that these monoclonal antibodies react with the same receptor complex in PC3 cells. Several of the other antibodies reacted with $\beta1$, such as P5B9 (see Fig. 7.7), and several could not be used in IP reactions at all.

2.11. Identification of integrin receptors by immune precipitation/Western blot

While no specific examples are shown here, one method of screening for anti-integrin antibodies is to take advantage of the fact that the αv, $\beta1$, and $\beta2$ subunits define entire families of receptors and that all have different molecular weights (150 kDa, 116 kDa or 90 kDa, respectively). Thus, it is possible to screen for antibodies that will immune precipitate any αv-, $\beta1$-, or $\beta2$-containing heterodimer via western blot detection of immunoprecipitated integrin complexes with purified and biotinylated antibodies to αv, $\beta1$, or $\beta2$ (Fig. 7.6B). Once the immune precipitates are run on gels and transferred to nitrocellulose, the western blot can be probed simultaneously for all three integrin-containing heterodimers since these subunits migrate at significantly different molecular weights. This type of screen is accomplished by performing immunoprecipitations on unlabeled cell extracts in a pooled matrix format (Fig. 7.6A) followed by western blot detection using any number of commercially available antibodies to any or all three of these receptors. The keys to success here are to use purified and biotinylated antibodies for western blot detections (otherwise, the primary antibody [for the IP step] gets in the way) and nonreducing conditions should be used when the SDS-PAGE gels are run. This protocol has been used successfully to isolate antibodies to the *Xenopus laevis* $\alpha v\beta3$, $\alpha5\beta1$, and $\alpha3\beta1$ receptors (Cohen *et al.*, 2000; Hoffstrom, unpublished).

3. PROTOCOLS

3.1. Cell-based ELISA

3.1.1. Materials

PC3 cells passage 6 or 7 (ATCC #CRL-1435) grown in RPMI 1640 supplemented with 10% FBS
96-well TC plates (Falcon #3072)
0.02% gelatin (2% stock from Sigma, G-1393), dilute 1/100 in PBS for coating plates
PBS as wash buffer (Do not add detergents to this buffer unless you want to extract or permeabilize the cells.)
2% formaldehyde (dilute 37% stock with PBS)

Trypsin 0.25%/EDTA solution (Gibco #25200)

Blocking solution: 5% nonfat dry milk, NFDM (Carnation) 2% normal goat serum, GS (Gibco #16210–064)

HRP conjugated goat anti-mouse IgG (gamma-chain specific, Southern Biotechnologies, Birmingham, AL #1030–05)

HRP-conjugated goat anti-mouse IgM (mu-chain specific, Southern Biotechnologies, #1020–05)

Substrate for HRP (ABTS turns blue green in the presence of HRP, Kirkgaard and Perry #50–66–18)

SBTI (Sigma T-9777) stock solution at 1 mg/ml

Plate reader with filter set for O.D. 405

3.1.2. Procedure

1. Coat 96-well plates with gelatin for 1 to 2 h at room temperature or overnight at 4°.
2. Trypsinize PC3 cells (0.025% trypsin/EDTA) for 5 to 10 min at 37°.
3. Wash two times in RPMI supplemented with 10% FBS or 0.5 mg/ml SBTI.
4. Plate PC3 cells at 5×10^3 cells per well in RPMI 1640 supplemented with 10% FBS, and incubate at 37° for 1 to 2 days until the wells are confluent. Do not allow the cells to overgrow or they will not stay on the plates for the duration of the ELISA.
5. Wash plates three times with PBS and fix for 15 to 30 min at room temperature with 2% formaldehyde.
6. Wash plates three times with PBS and block for 30 min at room temperature with 5% NFDM/2% GS.
7. Wash plates once and dilute the antisera into the first well (A1) at 1:100 and perform doubling dilutions across the plate (1:100, 1:200, 1:400, etc.). Each sera should be run in duplicate (two rows) so that one row can receive HRP-conjugated anti-IgG and one row can receive HRP-conjugated anti-IgM.
8. Incubate plates with primary for 1 h, and wash three times with PBS.
9. Add HRP-conjugated anti-IgG or anti-IgM at 1/2000 (Southern Biotech) per well in block and incubate at RT for 30 min with rocking.
10. Wash plates three times with PBS and add ABTS.
11. Incubate plates for 10 to 15 min at RT and read on a plate reader at O.D. 405.

3.2. Cell adhesion assay

Cells can be labeled before the assay with radioactive isotopes (Cr 51) or fluorescent dyes such as calcein AM (Molecular Probes) or after (crystal violet staining) the completion of the assay. There is an excellent protocol for using calcein labeled cells in an adhesion assay on the Invitrogen website (Handbook Section 15.6, "Probes for Cell Adhesion"). The objective of the protocol described here is to stain the adherent cells with crystal violet for use with a standard plate reader (O.D. 600 filter set).

3.2.1. Materials

PC3 cells grown in 15-cm tissue culture dishes. You will need four 15-cm plates to screen 6 × 96-well plates in a single fusion.

Blocking solution: PBS supplemented with 10 g/l heat denatured BSA (dissolve BSA in PBS and heat to 80° for 3 min and plunge into 4° water bath until completely cooled; do not overheat)

Adhesion buffer: RPMI 1640-HEPES supplemented with 10 mg/ml HBSA (heat-denatured BSA)

48-well plates (TC or non-TC)

Protein solutions containing purified ECM components (Chemican, Linscott Directorylisting) made up at 2 to 5 μg/ml in PBS or Voller's Buffer (0.05 M of bicarbonate at pH 9.6). A higher pH will help solubilize the ECM proteins when diluted. It is advisable to run the purified proteins on a gel under reducing conditions to control for purity and protein concentration.

37% formaldehyde (10× stock)

For the crystal violet stain, 0.5% crystal violet (Sigma C-3886) in MeOH (5× stock)

3N NaCl (20× stock)

Methanol

3.2.2. Method
Prepare the plates

1. Decide on the number of plates you will need to screen the fusion. You will need two 48-well plates (Falcon #3078) per fusion plate. Do not use 96-well plates because the background adhesion will complicate interpretation of the results.

2. Dilute protein to 5 μg/ml with PBS or Voller's buffer (0.05 M bicarbonate, pH 9.6). Add 200 μl of protein solution per well. You will need 10 ml of protein solution per 48-well plate. Incubate the plates at 4° for 24 h.

3. Before you initiate the cell adhesion assay, wash the protein-coated plates three times with PBS and block with 0.5 ml PBS/HBSA per well for 30 min at room temperature. Wash once and add 50 μl of adhesion buffer per well.

4. While the plates are blocking, release cells from tissue culture dishes (trypsin, EDTA, or scraping), and wash two times with adhesion buffer supplemented with 20 μg/ml SBTI if trypsin was used.

5. Pellet cells after final wash and resuspend at 1×10^6 per milliliter in adhesion buffer. Keep at room temperature until used.

6. Add 200 μl of hybridoma supernatants per well. In the initial stages, duplicates cannot be run since there is not enough supernatant. We do not usually run duplicates until we are testing clones. Be sure to include both a positive (anti-β1) and a negative (no antibody) control well.

7. Add 50 μl of cell suspension to each well containing 5×10^4 cells per well and incubate with hybridoma supernatants on 48-well plates with rocking for 15 min at room temperature to pre-bind antibodies to the cell surface.

8. After 15 min, put plates in the incubator and incubate for 30 min.

9. It is not convenient to process too many plates at once so we do them in groups of two (two 48-well plates at a time).

10. After 30 min, check to be sure that adhesion was adequate and spread in the assay in the control wells. If the cells in the negative control wells (see Fig. 7.3) are not adhered and/or spread, let the assay run incubate for another 15 to 30 min.

11. Once the wells have been washed with PBS, fix the adherent cells with 3.7% formaldehyde for at least 3 h or overnight.

12. Wash off the formaldehyde with PBS at least four times and make the crystal violet stain (0.15 mM NaCl/0.1% crystal violet, 20% MeOH) and add 200 μl of stain to each well for 10 min at room temperature.

13. Dilute stain in each well by adding 1.0 ml of tap water or by filling the wells to the brim. Aspirate the stain and refill the wells with tap water two more times. It is very important to remove any residual stain. The stained cells are very easy to see. At this point the plates can be read in the microscope and the presence of functional antibodies can be determined.

14. Dump and blot the plates and let air dry. Release the crystal violet in the adherent cells with 200 μl of MeOH per well and transfer the contents of each well to a 96-well ELISA plate and read on a standard ELISA plate reader at O.D. 600.

If a functionally inhibitory antibody is obtained, the data are usually unequivocal and can be evaluated by simply looking at the wells in the microscope. Sometimes the wells with a function-blocking antibody are blank (Fig. 7.8B). Sometimes you will see aggregated cells in sheets (Fig. 7.3D), which often results from disruption of cell substrate adhesion triggering cell–cell adhesion (cell aggregation). Additionally, you might see partial inhibition where about 50% of the cells remain. Often this can indicate the presence of a functional antibody, and you should proceed to cloning. Once the antibody is cloned, its functional capabilities can be accurately characterized. Some functional antibodies are better than others. The reasons for this are unknown but could be due to overlapping epitope recognition, with one antibody actually binding to the receptor docking site while another simply overlaps, or it could result from concentration, antibody affinity, or avidity. In our experience, there is no correlation with the isotype of an antibody and its functional capabilities.

3.3. Production of monoclonal antibodies via cell fusion

The first step before performing a fusion with an immune spleen is to be sure that the polyclonal antiserum contains the appropriate antibody repertoire (see above). Once this has been established, the spleen is prepared for fusion with myeloma cells by performing the final injection (Day −3) or series of injections (Days −3, −2, −1). On Day 0, the mouse is sacrificed and the spleen is removed. The spleen should be quite large and could show evidence of germinal centers (white areas). If IFA had been used more than once, the spleen could be embedded in connective tissue in the abdomen. It should be blunt dissected, with care being taken not to puncture the stomach or intestines. If this happens and another mouse is not available, do not panic. Try to get the spleen out and dip it in 70% ethanol before smashing on a screen (see below).

3.3.1. Materials

Immunized RBF/DnJ mouse (Jackson Laboratories, stock #000726)
3- to 4-week BALB/cByJ mice for thymocyte feeder cells (Jackson Laboratories, stock #001026)
Sterile instruments
FOX-NY myeloma cells (available from this laboratory or the ATCC [CRL-1732])
FBS (fetal bovine serum, Hyclone Laboratories, Logan UT)
Serum-free RPMI-1640 (Gibco #11875) for washing the cells. Heat 30 ml to 37° for the fusion
AAT medium:
    RPMI-1640 media (Gibco #11875), 500 ml
    $100\times$ l-glutamine solution (Gibco #25030), 5 ml
    $100\times$ penicillin-streptomycin solution (Gibco #15140), 5 ml
    $100\times$ sodium pyruvate (Gibco #11360)
    $1000\times$ gentamicin solution (Sigma G-1522), 0.5 ml
    $50\times$ adenine/aminopterin/thymidine (AAT Sigma A-5539), 1 vial
50% polyethylene glycol 1500 (PEG 1500, Roche #783 641)
100-mesh screens (Sigma S3895–5EA)
CO_2 incubator set at 5% CO_2 in air
Glass beakers for preparing makeshift water baths for use in the culture hood ($\times 3$)
50-ml centrifuge tube

Centrifuge set at 400×g

Plastic pipettes (2- and 10-ml)

4 to 8 × 96-well plates

50× vial adenine/thymidine (AT, Sigma A-7422) to wean the resulting hybridomas once they have been cloned (In general, however, we maintain the selection pressure by growing the hybridomas in AAT media until they have been through several rounds of cloning.)

3.3.2. Method
Prepare the FOX-NY myeloma cells

1. Thaw FOX-NY cells and plate one-half into RPMI supplemented with 10% FBS and one-half into media supplemented with 1× AAT (see below). The purpose of this AAT test is to ensure that the myeloma cells have not reverted to an aminopterin-insensitive state. This will definitely happen if the FOX-NY cells are maintained for long periods of time in tissue culture. Thaw the FOX-NY cells only when they are needed. Store them in liquid nitrogen at all other times. We keep hundreds of vials in our LN2 bank that we have quality controlled for their ability to support fusion and antibody development. The unfused FOX-NY cells should all be dead in 3 to 4 days. If not, then you need to return to another earlier frozen seed or to your original stock.

2. Culture FOX-NY cells to 5×10^5 to 10^6 per milliliter in RPMI 1640 plus 10% FBS. You will need 1×10^8 FOX-NY cells in exponential growth for the fusion. We usually culture three T75 flasks with 50 to 75 ml of media for 3 days before fusion. Do not grow the FOX-NY in the presence of antibiotics. You want to know if these cells become contaminated.

3. On Day 0, pour the FOX-NY cells into 50-ml tubes, centrifuge at 400×g, and pool into one tube and wash three times in RPMI-1640 (no FBS). Resuspend in 20 ml and count in a hemocytometer. You will need 1×10^8 cells.

Preparation of the spleen for fusion

1. On Day −3, scrape PC3 cells with a rubber policeman from three 15-cm tissue culture plates. Wash the cells three times and resuspend them into a final volume of 0.5 ml and suck into a 1-ml syringe with an 18-gauge needle. Switch to a 23-gauge needle and inject the cells intraperitoneally into the RBF mouse chosen for fusion—in this case, the 1L mouse. Do not use an adjuvant. If using a purified protein, you may want to inject again on Days −2 and −1. If using a protein or peptide preparation, we often select RIBI for the final injection(s).

2. At Day 0, remove the spleen and place on a sterile screen in a 6-cm Petri dish. The screen can be purchased from Sigma and sterilized by autoclaving. We recommend a 100-mesh screen (Sigma #S389505EA). Keep the spleen dry on wet ice until processed to remove the cells.

3. To make a single cell suspension of splenocytes, add 2 ml of RMPI 1640 (no serum) to the spleen in the plate. Smash the spleen with the plunger of a sterile 10-ml syringe until there are no more red clumps visible. Some connective tissue will remain in the screen; discard it. Wash the splenocytes off the screen and Petri dish and into a 50-ml tube with serum-free RPMI-1640. Do not add serum, as this can stabilize the cell membranes. The whole purpose of the fusion protocol is to destabilize the cell membranes. Centrifuge at $400 \times g$ for 5 min at room temperature, and wash once with RPMI 1640 (no FBS). Resuspend spleen in 10 ml of RPMI (no FBS).

Fusion

1. Warm 5 ml of PEG 1500 and 30 ml of serum-free RPMI 1640 (no FBS) to 37°. When ready, move these into the hood in makeshift water baths. The preparation of the spleen should be synchronized with the preparation of the FOX-NY cells.
2. Wash FOX-NY cells (see above) and resuspend in 20 ml of RPMI (no FBS).
3. Count spleen cells and FOX-NY cells. For the fusion, 5×10^7 to 10^8 spleen cells are fused with 1×10^8 FOX-NY cells so that the fusion ratio is at least 0.5:1.0 spleen cells to FOX-NY. For 5×10^7 spleen cells, use four 96-well plates, and for 10×10^8 spleen cells, use eight 96-well plates. We do not set out more than 8 plates per fusion. You will need at least 5×10^7 spleen cells for a good fusion.
4. Once the fusion ratio is set and the PEG and RPMI are warmed to 37°, the protocol described in Oi and Herzenberg (1980) is followed exactly as described with a mini-water bath setup in the sterile hood to maintain the temperature as close to 37° as possible during fusion.
5. Mix together the spleen cells and the FOX-NY myeloma cells and centrifuge in a 50-ml tube at $400 \times g$ for 10 min at room temperature.
6. Remove all supernatant from the pellet by aspirating and warm the pellet to 37° in the makeshift water bath and keep at 37° for all further manipulations.
7. Using a 2-ml plastic pipette, add 1 ml of warm PEG 1500 (50% v/v in media) to the cell pellet with gentle stirring over a 1-min period. Use the tip of the pipette to stir the pellet and keep immersed in the water bath.
8. Continue to stir the pellet for 1 min.
9. With the same pipette, gradually add 1 ml of warm serum-free RPMI over a 1-min period with gentle stirring.
10. With the same pipette, add another 1 ml of warm media with gentle stirring.
11. Then over the next 2 to 3 min, with the same pipette add another 7 ml of warm media. This should bring the volume in the 50-ml tube up to 10 ml.
12. Centrifuge at $400 \times g$ for 10 min.

13. Use a 10-ml plastic pipette to break up the cell pellet and plate in AAT media (see above) supplemented with 20% FBS. This is a departure from the published Oi and Herzenberg (1980) protocol. Re-feed the fusion at Days 3 and 7 and screen at Day 10.

14. Thymocytes can be used as feeder cells during the fusion or later for cloning. If thymocytes are used, 3- to 4-week BALB/c mice are sacrificed and one thymus provides enough feeder cells for 4×96-well plates. Thymocytes can also be frozen at a ratio of 1 thymus per two freezer vials (freezing mixture is 90% FBS and 10% DMSO). Then one freezer vial is thawed per two 96-well plates.

By Day 3 there should be colonies visible in the bottoms of the wells (10 to 20 cells). If the plate ratio given above is followed for the number of spleen cells, three to four colonies per well should be visible by Day 7. If each well is screened by a capture ELISA (see below), in a good fusion every single well should be positive for a mouse IgG. On Day 10 or 11, the media should be quite yellow and the colonies should be visible to the naked eye when the plates are held up to the light. Screen the fusion for whatever criteria are decided upon, and clone directly from the master wells (96-well plates). These plates can be frozen or the cells can be transferred to multicluster tubes (Costar 4411) in freezing mixture (90% FBS and 10% DMSO) and frozen. Multicluster plates can be kept at $-80°$ or frozen in LN2 vapor for longer-term storage.

3.4. Cloning

Positive hybridoma culture wells are cloned via the following protocol. Two dilutions are made. The first is made so that the cells are diluted to 1×10^4/ml. Next, 100 μl of this cell suspension are further diluted to 10 ml in RPMI 1X AAT supplemented with 20% FBS and plated at 100 μl/well in a 96-well plate at 1 cell/well. Thymocyte feeder cells can be added at 1×10^5 per well (or, as we have done on occasion, use Hybridoma Cloning Factor [Bioveris Corporation #210001] at 10%). However, it is never advisable to allow your cells to get hooked on a growth factor. Other dilutions can be used (limit dilutions), but we have found that cloning at one cell per well falls within the range of most cell lines. It is advisable to clone as soon as possible, and if possible, directly from the original master well (96-well plate). Fusion plates can be frozen by aspirating the supernatant and the addition of 100 to 200 μl of freezing mixture directly to each well (90%FBS and 10% DMSO). The plates are then placed into a $-80°$ freezer and can be kept up to 1 year.

3.5. Isotyping

Hybridoma culture supernatants can be isotyped in one of two ways: (1) capture ELISA and reaction with HRP-conjugated specific secondaries (Southern Biotech), or (2) using a Mouse Monoclonal Antibody Isotyping

Kit (Roche #1–493–027). We routinely use both protocols, although the isostrips are rather expensive ($20 per strip). Therefore, we recommend the use of a capture ELISA. The protocol is identical to the one that we described for cells, except that instead of cells on the bottom of the wells, a purified goat anti-mouse reagent is coated at 1 to 2 μg/ml. Again, we recommend the use of Southern Biotech for these reagents, since we know that they exhibit low cross-reactivity and good batch-to-batch variation. When the capture reagent is coated at 1 to 2 μg/ml or at 1/1000, the wells are blocked and the hybridoma culture supernatants are added at 100 μl/well. The specific HRP-conjugated secondaries are then added at 1/2000 (Southern Biotech) per well in block for 30 min at RT. Then the plates are washed and the HRP-substrate ABTS is added at 100 μl/well. When the Roche isostrips are used, we have often seen prozone effects (Hoffstrom and Wayner, 1994), and recommend diluting the culture supernatant samples to at least 1:100 or even 1:1000.

3.6. Immunoprecipitation for integrin identification

3.6.1. Cell preparation

1. Culture cells and harvest before confluency either by trypsinization, scraping, or by treating with 2 mM EDTA for no more than 10 min.
2. Wash cells two times with PBS (see recipes below). If trypsin was used, the first wash should contain soybean trypsin inhibitor at 0.5 mg/ml.
3. Adjust cell density to 5 $\times$ 10^6/ml in PBS (cations).

3.6.2. Biotinylation and lysis of cells

1. Make a stock solution of biotin at 10 mg/ml in dimethyl sulfoxide (DMSO, nonsulfonated form, Pierce #21343) or water (sulfonated form Pierce #21217) immediately before use.
2. Add stock biotin to the cell suspension to a final concentration of 1:1000 (100 μg/ml) and mix.
3. Incubate at room temperature for 60 min with rocking to prevent cells from settling.
4. Wash cells three times with PBS plus 100 mM of glycine to block free amines, and once with PBS alone.
5. Resuspend cell pellet in 100 μl of formaldehyde-fixed *Staphylococcus aureus* bacteria 10% suspension (Sigma #P7155) or Pansorbin (Calbiochem #507858) directly from vial. Pipette to resuspend cell suspension with *S. aureus* bacteria.
6. Add 1.0 ml of extraction (lysis) buffer per 1 $\times$ 10^7 cells. The lysis buffer should contain 1% Triton X-100 (or NP-40) 1 mM of PMSF (200 mM stock in absolute ethanol), 1 mM of N-ethylmalemide (200 mM stock in absolute ethanol), and 10 μg/ml of SBTI as protease inhibitors. Lyse cells

for 30 to 45 min at 4° with occasional vortexing. Lysis buffer may contain divalent cations at 1 mM, especially when working with the α4 receptor.

7. Clarify cell lysate by centrifugation at 10,000×g for 30 min. This step is absolutely essential to remove nuclei and cytoskeletal components.

8. At this point lysates can be frozen at −80° or precleared for immune precipitation.

3.7. Preclearing of lysates

1. Add 500 μl of packed fetuin–agarose (Sigma Cat# F-3256) per milliliter of lysate in 1.7-ml microcentrifuge tubes. Incubate 4° for 1 h with rocking, and centrifuge at 10,000×g for 15 min at 4°.

2. Decant supernatant and add to fresh fetuin–agarose as in Step 1 and preclear lysate again.

3. Preclear lysates in this fashion a total of three times. The last preclear can go overnight if necessary.

3.8. Preparation and storage of protein A agarose

1. Swell 1 g of lyophilized protein A agarose (Sigma Cat# P-1406) in 40 ml of PBS pH 7.4 with 0.02% azide overnight on a mixer at 4° (1 g swells to approximately 4 ml).

2. Pellet gel by centrifugation, decant supernatant, and wash gel by resuspending bed to 50 ml with PBS/azide. Wash gel three times.

3. After final wash, resuspend gel in PBS/azide in a 10% (v/v) solution and store in the dark at 4°.

3.9. Coupling of rabbit anti-mouse IgG to protein A-agarose

The reason for using rabbit anti-mouse IgG to capture the mouse monoclonal antibodies on protein A beads is that rabbit IgGs have a very high affinity for protein A (higher than mouse IgGs). Rabbit IgGs also out-compete bovine IgG for sites on protein A. Hybridoma culture supernatants routinely contain up to 10% FBS, which can in turn contain up to 200 μg/ml bovine IgG. Therefore, if protein A is used for direct IPs without the benefit of prebinding the anti-mouse then there may not be sufficient binding of the mouse IgGs to the protein A beads to bind the protein to the beads. Furthermore, we do not recommend using protein G agarose since unless employing purified mouse antibody, culture supernatant is used, which most likely contains 10% FBS. Unless low-Ig FBS is used, most FBS batches contain 100 to 200 μg/ml of bovine IgG, which will outcompete the mouse IgG at a ratio of 10:1 to 50:1

(since most mouse hybridoma culture supernatant contains 2 to 10 μg/ml of specific antibody). Thus, if protein G is used, most likely there will not be enough mouse IgG on the beads to IP the protein of interest. For the best results with hybridoma culture supernatant, use rabbit anti-mouse serum or purified rabbit IgG to capture mouse IgGs to protein A agarose.

1. To calculate the amount of rabbit anti-mouse serum to use, figure on 10 μl of serum or 2 μg of purified IgG to 10 μl of packed protein A beads. Then figure on 10 μl of packed rabbit anti-mouse beads per IP reaction.
2. Solubilize rabbit anti-mouse antibody (Zymed Laboratories Cat# 61–6500) in 2 ml of deionized water. IgG not used can be aliquoted and stored at −80°.
3. Mix protein A agarose and rabbit anti-mouse at 4° for 1 to 2 h. Use a ratio of 0.1 ml rabbit anti-mouse to 1 ml of 10% protein A beads. With purified rabbit anti-mouse serum, use 2 μg per IP reaction, but keep the volumes the same (0.1 ml rabbit anti-mouse to 1 ml of 10% protein A beads). Coupling of rabbit anti-mouse to protein A beads can be done while preclearing the cell lysates.
4. Pellet rabbit anti-mouse–coupled protein A-agarose by centrifugation, decant supernatant and wash three times by bringing the total volume up to 10 bead volumes with 1× IP buffer (see below).
5. After the final wash, resuspend pellet in 1× IP buffer such that a 50-μl aliquot will contain 10 μl of packed gel (i.e., a 20% suspension).
6. Rabbit anti-mouse protein A beads can also be used to preclear lysates (see 3.7).

3.10. Coupling of test mouse antibody-containing supernatants to rabbit anti-mouse IgG protein A-agarose

1. To each 1.7-ml microcentrifuge tube, add 0.45 ml of test supernatant, 0.05 ml of 10× IP buffer (see below), and 0.05 ml of 20% rabbit anti-mouse IgG–coupled protein A-agarose.
2. Mix tubes with rocking for 2 h (can be overnight) at 4°.
3. Pellet the test mouse IgG coupled rabbit anti-mouse protein A-agarose beads by centrifugation (use microcentrifuge at highest setting for 3 to 5 min). Aspirate supernatant leaving a final volume of 50 μl and wash with 1.0 of 1× IP buffer once.

3.11. Immunoprecipitation of cell lysate

1. To each tube containing washed and pelleted IP beads, add 100 μl of precleared lysate. Make a final volume of 0.5 ml by adding 1× IP buffer per IP tube.

2. Incubate lysate with beads on a rocker for at least 2 h at 4°.
3. Pellet the beads by centrifugation (use a microfuge set a maximum for 3 to 5 min), aspirate supernatant leaving a final volume of 0.05 ml, and wash IPs with wash buffer (see below) four times with vortexing to resuspend pellet after each wash. The original supernatants can be saved and rotated to recycle the lysates for IP with another and distinct antibody, or they can be used for sequential preclear analysis to determine the identity of the test antibody. Generally, to prove that a particular antibody can preclear another antibody (i.e., they react with the same protein), it takes three rounds of preclear to complete.
4. After the final wash and pelleting, aspirate to approximately 0.025 ml and resuspend in 0.025 ml of 2× sample buffer (reduced or nonreduced) for SDS-PAGE gel analysis.

3.12. Releasing immunoprecipitated material from beads

1. Boil tubes from previous step for 5 min and then vortex and pellet beads as described above.
2. Transfer supernatant to a stacking gel for SDS-PAGE. For IPs it is best to use 1.5-mm gels so that the entire IP will fit in the well created by the comb. If a few beads are also transferred, they will not affect how the gel runs. We use Invitrogen SeeBlue Plus 2 molecular weight makers. They run well and transfer well.

3.13. Western blot to detect biotinylated and immune-precipitated integrin receptors

1. Run immune precipitates on SDS-PAGE gel and transfer to nitrocellulose.
2. Take off western and block nitrocellulose in PBS 3% BSA plus 2% normal goat serum (not heat denatured). Keep block sterile. Do not use blocking solutions made up using NFDM because milk contains endogenous biotin, which will create background. Incubate with rocking at RT for 30 min.
3. Rinse with PBS 0.5% Tween and add HRP conjugated streptavidin at 1/100,000 (BD 554066). Use no more than 10 ml per blot (0.5 μl/50 ml) in PBS 3% BSA.
4. Incubate at room temperature for *no longer* than 30 min.
5. Rinse extensively in PBS Tween 0.5% (around 20 times). Then wash at least four times with rocking for 5 to 10 min with extensive rinsing in between.
6. Apply ECL at 1 to 2 ml per blot and develop film.

3.14. IP buffers

3.14.1. Extraction (lysis) buffer

PBS pH 7.4	10 ml
1% Triton X-100	1 ml(10% stock in PBS 0.02% azide)
10 μg/ml SBTI	100 μl (1 mg/ml stock in PBS keep aliquoted at -20oC)
1 mM PMSF	50 μl (200 mM stock in absolute ethanol)
1 mM NEM	50 μl (200 mM stock in absolute ethanol)

Note: 10 mM of EDTA can be added or 1 mM of $CaCl_2$ and/or 1 mM of $MgCl_2$ depending on desired involvement of divalent cations.

3.14.2. 10× IP buffer

5% BSA	5 gm
500 mM Tris, pH 7.5	10 ml
10% Triton X-100	10 ml
0.2% Na azide	10 ml
H_2O (deionized)	to 100 ml

3.14.3. Wash buffer

50 mM Tris, pH 7.5
400 mM NaCl
1% Triton X-100
0.02% Na azide

REFERENCES

Albeda, S. M., and Buck, C. A. (1990). Integrins and other cell adhesion molecules. *FASEB J.* **4**, 2868–2880.

Berditchevski, F., Bazzoni, G., and Hemler, M. E. (1995). Specific association of CD63 with the VLA-3 and VLA-6 integrins. *J. Biol. Chem.* **270**, 17784–17790.

Carter, W. G., Kaur, P., Gil, S. G., Gahr, P., and Wayner, E. A. (1990). Distinct functions for integrins a3b1 in focal adhesion and a6b4/bullous pemphigoid antigen in a new stable anchoring contact (SAC) of keratinocytes: Relation to hemedesmosomes. *J. Cell Biol.* **111**, 3141–3154.

Carter, W. G., Ryan, M. C., and Gahr, P. G. (1991). Epiligrin, a new cell adhesion ligand for integrins in epithelial basement membranes. *Cell* **65**, 599–610.

Carter, W. G., Wayner, E. A., Bouchard, T. S., and Kaur, P. (1990). The role of integrins a2b1 and a3b1 in cell–cell and cell–substrate adhesion of human epidermal cells. *J. Cell Biol.* **110**, 1387–1404.

Cheresh, D., Smith, J., Cooper, H., and Quaranta, V. (1989). A novel vitronectin receptor integrin (avbx) is responsible for distinct adhesive properties of carcinoma cells. *Cell* **57**, 59–69.

Chu, R. S., Targoni, O. S., Krieg, A. M., Lehmann, P. V., and Harding, C. V. (1997). CpG oligodeoxynucleotides act as adjuvants that switch on T helper 1 (Th1) immunity. *J. Exp. Med.* **186,** 1623–1631.

Cohen, M. W., Hoffstrom, B. G., and Desimone, D. W. (2000). Active zones on motor nerve terminal contain alpha 3 beta 1 integrin. *J. Neurosci.* **20,** 4912–4921.

Harlow, E., and Lane, D. (1988). "Antibodies: A Laboratory Manual." Laboratory Press, Cold Spring Harbor, NY.

Hemler, M. E. (1990). VLA proteins in the integrin family: Structures, functions, and their role on leukocytes. *Ann. Rev. Immunol.* **8,** 365–400.

Hoffstrom, B. G., and Wayner, E. A. (1994). Immunohistochemical techniques to study the extracellular matrix and its receptors. *Methods Enzymol.* **245,** 316–347.

Howard, G. C., and Bethel, D. R. (2001). "Basic Methods in Antibody Production and Characterization." CRC Press, New York.

Hynes, R. O. (1987). Integrins: A family of cell surface receptors. *Cell* **48,** 549–554.

Hynes, R. O. (2002). Integrins: Bidirectional allosteric signaling machines. *Cell* **110,** 673–687.

Hynes, R. O. (2004). The emergence of integrins: A personal and historical perspective. *Matrix Biol.* **23,** 333–340.

Kamata, T., Puzon, W., and Takada, Y. (1994). Identification of putative ligand binding sites within I domain of integrin a2b1 (VLA-2, CD49b, CD29). *J. Biol. Chem.* **269,** 9659–9663.

Kamata, T., Puzon, W., and Takada, Y. (1995). Identification of putative ligand-binding sites of the integrin a4b1 (VLA-4, CD49d/CD29). *Biochem. J.* **305,** 945–951.

Kohler, G., and Milstein, C. (1975). Continuous cultures of fused cells secreting antibody of predefined specificity. *Nature* **256,** 495–497.

Kovach, N. L., Carlos, T. M., Yee, E., and Harlan, J. M. (1992). A monoclonal antibody to beta 1 integrin (CD29) stimulates VLA-dependent adherence of leukocytes to human umbilical vein endothelial cells and matrix components. *J. Cell Biol.* **116,** 499–509.

Kunicki, T. J., Nugent, D., Wayner, E. A., and Carter, G. (1988). The human fibroblast class II extracellular matrix receptor (ECMR II) mediates platelet adhesion to collagen and is identical to platelet glycoproteins Ia/IIa. *J. Biol. Chem.* **263,** 4516–4519.

Lam, S. C., Plow, E. F., D'Souza, S. E., Cheresh, D. A., Frelinger, A. L., and Ginsberg, M. H. (1989). Isolation and characterization of a platelet membrane proteins related to the vitronectin receptor. *J. Biol. Chem.* **264,** 3742–3749.

Leavesley, D. I., Ferguson, G., Wayner, E. A., and Cheresh, D. A. (1992). Requirement of the integrin b3 subunit for carcinoma cell spreading or migration on vitronectin and fibrinogen. *J. Cell Biol.* **117,** 1101–1107.

Mould, P., Wheldon, L. A., Komoriya, A., Wayner, E. A., Yamada, K. M., and Humphries, M. J. (1990). Affinity chromatographic isolation of the melanoma adhesion receptor ofr the IIICS region of fibronectin and its identification as the integrin a4b1. *J. Biol. Chem.* **256,** 4020–4024.

Oi, V. T., and Herzenberg, L. A. (1980). Immunoglobulin-producing hybrid cell lines. *In* "Selected Methods in Cellular Immunology" (B. B. Mishell and S. M. Shiigi, eds.), pp. 351–372. Academic Press, New York.

Peters, J. H., and Baumgarten, H. (1992). "Monoclonal Antibodies." Springer-Verlag, New York.

Pierschbacher, M. D., and Ruoslahti, E. (1984). Variants of the cell recognition site of fibronectin that retain attachment promoting activity. *Proc. Natl. Acd. Sci USA* **81,** 5989–5988.

Pulido, R., Elices, M. J., Campanero, M. R., Osborn, L., Schiffer, S., García-Pardo, A., Lobb, R., Hemler, M. E., and Sánchez-Madrid, F. (1991). Functional evidence for three distinct and independently inhibitable adhesion activities mediated by the human integrin VLA-4. *J. Biol. Chem.* **266,** 10241–10245.

Pytela, R., Pierschbacher, M. D., and Ruoslahti, E. (1985a). Identification and isolation of a 140 kDa cell surface glycoprotein with properties expected of a fibronectin receptor. *Cell* **40**, 191–198.

Pytela, R., Pierschbacher, M. D., and Ruoslahti, E. (1985b). A 125/115 kDa cell surface receptor specific for vitronectin interacts with the aginin-glycine-aspartic acid adhesion sequence derived from fibronectin.. *Proc. Natl. Acad. Sci. USA* **82**, 5766–5770.

Schiffer, S. G., Hemler, M. E., Lobb, R. R., Tizard, R., and Osborn, L. (1995). Molecular mapping of functional antibody binding sites of a4 integrin. *J. Biol. Chem.* **270**, 14270–14273.

Sigle, R. O., Gil, S. G., Bhattacharya, M., Ryan, M. C., Yang, T.-M., Brown, T. A., Boutaud, A., Miyashita, Y., Olerud, J., and Carter, W. G. (2004). Globular domains 4/5 of the laminin a3chain mediate deposition of precursor laminin 5. *J. Cell Sci.* **117**, 4481–4494.

Sonnenberg, A., Janssen, H., Hogervorst, F., Calafat, J., and Hilgers, J. (1987). A complex of platelet glycoproteins Ic and IIa identified by a rat monoclonal antibody. *J. Biol. Chem.* **262**, 10376–10383.

Spieker-Polet, H., Sethupathi, P., Yam, P. C., and Knight, K. (1995). Rabbit monoclonal antibodies: Generating a fusion partner to produce rabbit-rabbit hybridomas. *Proc. Natl. Acad. Sci. USA* **92**, 9348–9352.

Takada, Y., Wayner, E. A., Carter, W. G., and Hemler, M. (1988). Extracellular matrix receptors, ECMR II and ECMR I for collagen and fibronectin correspond to VLA 2 and VLA 3 in the VLA family of heterodimers. *J. Cell Biochem.* **37**, 385–393.

Takada, Y., and Puzon, W. (1993). Identification of a regulatory region of integrin b1 subunit using activating and inhibiting antibodies. *J. Biol. Chem.* **268**, 17597–17601.

Taggart, T. R., and Samloff, I. M. (1983). Stable antibody-producing murine hybridomas. *Science* **219**, 1228–1230.

Wayner, E. A., Orlando, R. A., and Cheresh, D. A. (1991). Integrins avb3 and avb5 contribute to cell attachment to vitronectin but differentially distribute on the cell surface. *J. Cell. Biol.* **113**, 919–929.

Wayner, E. A., and Carter, W. G. (1987). Identification of multiple cell adhesion receptors for collagen and fibronectin in human fibrosarcoma cells possessing unique alpha and common beta subunits. *J. Cell Biol.* **105**, 1873–1884.

Wayner, E. A., Carter, W. G., Piotrowicz, R. S., and Kunicki, T. J. (1988). The function of multiple extracellular matrix receptors in mediating cell adhesion to the extracellular matrix: Preparation of monoclonal antibodies that specifically inhibit cell adhesion to fibronectin and react with platelet glycoproteins Ic-IIa. *J. Cell Biol.* **107**, 1881–1891.

Wayner, E. A., Garcia-Pardo, A., Humphries, M. J., McDonald, J. A., and Carter, W. G. (1989). Identification and characterization of the lymphocyte adhesion receptor for an alternative cell attachment domain in plasma fibronectin. *J. Cell Biol.* **109**, 1321–1330.

Wayner, E. A. (1990). Characterization of the lymphocyte receptor for a new adhesion sequence, EILDVPST, located in the CS-1 domain of plasma fibronectin: Function in mediating heterotypic lymphocyte adhesion. *J. Cell Biochem.* (Suppl. 14A), 139–140.

Wayner, E. A., and Kovach, N. L. (1992). Activation-dependent recognition by hematopoietic cells of the LDV sequence in the V region of fibronectin. *J. Cell Biol.* **116**, 489–497.

Wayner, E. A., Gill, S., Murphy, G. F., Wilke, M. S., and Carter, W. G. (1993). Epiligrin, a component of epithelial basement membranes, is an adhesive ligand for a3b1 positive T lymphocytes. *J. Cell Biol.* **121**, 1141–1152.

CHAPTER EIGHT

CELL ADHESION, CELLULAR TENSION, AND CELL CYCLE CONTROL

Eric A. Klein, Yuval Yung, Paola Castagnino, Devashish Kothapalli, *and* Richard K. Assoian

Contents

Abstract

Cooperative signaling between growth factor receptor tyrosine kinases, integrins, and the actin cytoskeleton is required for activation of the G1-phase cyclin-dependent kinases and progression through G1-phase. Increasing evidence suggests that there is cell type specificity in these cooperative interactions and that the compliance of the underlying substratum can strongly affect adhesion-dependent signaling to the cell cycle. This chapter reviews our current methods for studying how cell type specificity and changes in substratum compliance can contribute to G1-phase cell cycle control. We also describe several of our current analytical procedures.

Department of Pharmacology, University of Pennsylvania School of Medicine, Philadelphia, Pennsylvania

Methods in Enzymology, Volume 426
ISSN 0076-6879, DOI: 10.1016/S0076-6879(07)26008-2

© 2007 Elsevier Inc.
All rights reserved.

1. INTRODUCTION

Cell adhesion to substratum plays a critical role in cell proliferation because integrin signaling cooperates with growth-factor-receptor tyrosine kinase signaling to activate the G1-phase cyclin-dependent kinases (cdks). Several studies by us and others, using immortalized fibroblast or epithelial cell lines, have shown that the ligation of integrins by cell adhesion to the extracellular matrix (ECM) is important for ERK activation, cyclin D1 expression, and regulation of the cip/kip family of cdk inhibitors (reviewed in Assoian and Schwartz, 2001). As a consequence, pocket protein phosphorylation and E2F activity are regulated by cell adhesion to the ECM. Some of the current challenges in this field include understanding whether there is cell type specificity in the relationship between integrin-mediated adhesion and cell cycle control and how ECM-dependent tensional properties may affect cell type–specific control of the cell cycle by cellular adhesion.

While experiments performed in our laboratory still exploit the strengths of the fibroblast model system (facile cell cycle synchronization, reasonable transfection efficiencies, and relative availability of knockout MEFs), we have put increasing emphasis on the MCF10A mammary epithelial cell line: this line undergoes an epithelial-mesenchymal transition (EMT) and is therefore well suited to address the issue of cell type–specific responses. Moreover, the redistribution of f-actin from cortical bundles to stress fibers that accompanies the EMT (Bhowmick *et al.*, 2001; Zhong *et al.*, 1997) provides an excellent model to address the potential role of cellular tension as a regulator of integrin signaling and of the cell cycle.

Many of our methods for fibroblast cell culture, synchronization, induction of cell cycle progression, as well as analysis of collected cells by western blotting and *in vitro* kinase assays have been previously described and are still current (Zhu *et al.*, 1999). The methodology associated with our current research focuses on cellular tension and cell type–specific integrin signaling, as well as new techniques for transfection and adenoviral infection, real-time quantitative PCR (QPCR), immunofluorescence, Rho family pull-down assays, and flow cytometry are described here.

2. PREPARATIVE METHODS

2.1. Cell culture

2.1.1. Coating dishes with purified matrix proteins

To examine the effect of specific matrix proteins on signal transduction cascades or activation of cdks, culture dishes are coated with purified matrix proteins as described (Zhu *et al.*, 1999). For fibroblasts, we currently use

1.27 μg/cm^2 of collagen, 0.635 μg/cm^2 fibronectin, and 0.635 μg/cm^2 vitronectin. For MCF10A cells, we have used 1.27 μg/cm^2 collagen or laminin-1, 0.318 μg/cm^2 fibronectin, and 0.635 μg/cm^2 vitronectin. The dishes are blocked with Dulbecco's Modified Eagle's Medium containing 1–2 mg/ml heat-inactivated fatty acid-free BSA (DMEM-BSA) prior to use, and poly-L-lysine or BSA-coated dishes are used as negative controls (Zhu *et al.*, 1999). Note that we reduce the amount of matrix protein used to coat dishes (typically to one-quarter of the standard amount) to see efficient inhibition of FAK autophosphorylation at Y397 by ectopically expressed FRNK.

2.1.2. Culturing fibroblasts

Our conditions for maintenance, serum starvation, and cell cycle stimulation of immortalized embryonic fibroblasts and early-passage human fibroblasts have been described (Zhu *et al.*, 1999). Early-passage mouse embryo fibroblasts (derived from day 12.5 to 14.5 mouse embryos) are cultured similarly to immortalized MEFs, except that a longer stimulation (typically 15 to 36 h for early-passage cells vs. 15 to 24 h for immortalized cells) is needed for efficient S-phase progression.

2.1.3. Cell cycle synchronization

In addition to the synchronization of cells at G0 by serum starvation, it can be helpful to arrest cells in different phases of cell cycle. We arrest immortalized mouse fibroblasts at the G1/S boundary or in mitosis (metaphase) with hydroxyurea and nocodazole, respectively. To induce the S-phase block, immortalized MEFs suspended in DMEM-10% FBS are seeded (1 to 1.5 $\times$ 10^6 cells per 100-mm dish) in the presence of 2 mM of hydroxyurea and incubated overnight. For metaphase arrest, the cells are seeded (1.5 to 2 $\times$ 10^6 cells per 100-mm dish) in the presence of 0.5 μg/ml of nocodazole and incubated overnight. In order to release cells back into the cell cycle, the medium is removed and the cell monolayers are washed three times with serum-free DMEM. We typically prepare stock solutions of hydroxyurea (500 mM in sterile filtered DMEM) and nocodazole (10 mg/ml in DMSO) and store them at −20°. Note that cells arrested in metaphase round up and are only loosely attached to the culture dish. A portion of the cells are analyzed by flow cytometry (see below) to assess the efficiency of cell cycle synchronization. The remainder is used for experimentation.

2.1.4. Culturing MCF10A mammary epithelial cells

MCF10A cells are maintained at or below 80% confluence in 1:1 low-glucose DMEM:Ham's F12 nutrient media supplemented with 5% horse serum, 10 mM HEPES pH 7.4, and a growth factor cocktail including 20 ng/ml EGF, 10 μg/ml insulin, 0.5 μg/ml hydrocortisone, and 100 ng/ml cholera toxin. To synchronize MCF10A cells in G0, cells are seeded at 2 to 3 $\times$ 10^6 per 100-mm dish or 1 $\times$ 10^6 per 60-mm dish and allowed to spread and attach overnight in maintenance medium prior to synchronization for 2 days in serum-free media (1:1 low-glucose DMEM:Ham's F12 nutrient media supplemented with 1 mg/ml heat-inactivated fatty acid–free BSA).

To stimulate cell cycle reentry, G0-synchronized MCF10A cells are trypsinized, collected by centrifugation, resuspended in 1:1 low-glucose DMEM:Ham's F12 nutrient media supplemented with growth-factor cocktail and 10% fetal bovine serum, and finally reseeded at 1.5×10^6 per 100-mm dish on collagen-coated dishes. The dishes are precoated with collagen as described above.

2.1.5. The EMT in MCF10A cells

MCF10A cells undergo a phenotypic transition to a fibroblast-like cell upon treatment with TGF-β or ectopic expression of Snail family transcription factors (Bhowmick *et al.*, 2001; Bolos *et al.*, 2003; Cano *et al.*, 2000; Thiery, 2003). We have used TGF-β to regulate the EMT in MCF10A cells. Cells are seeded in 100-mm dishes at low confluence (3×10^5 and 6×10^5 cells for control and TGF-β treated cells, respectively) in maintenance medium. To induce the EMT, cells are incubated with 3 ng/ml human recombinant TGF-β1 for 3 days prior to G0-synchronization in serum-free media (in the continued presence of TGF-β). Induction of cell cycle reentry is performed as described above, except that 3 ng/ml TGF-β is maintained in the medium of transitioned cells.

2.2. Transfection, adenoviral infection, and siRNA

Spontaneously immortalized mouse embryo fibroblasts (MEFs) or NIH-3T3 cells are seeded the day before transfection at 3×10^5 cells/100-mm dish in DMEM containing 10% FBS without antibiotics. The morning of transfection, 5 μg of DNA are added to 750 μl DMEM for each transfection. Twenty microliters of Lipofectamine Plus reagent (Invitrogen) is added to the DNA-DMEM solution and allowed to form a complex at room temperature for 15 min. Thirty microliters of Lipofectamine reagent (Invitrogen) is diluted into 750 μl of DMEM and then added dropwise to the DNA solution, mixed, and allowed to form a complex for an additional 15 min (total volume $=$ 1.5 ml). While the complexes are forming, the cells are washed twice with DMEM (without serum). After the final wash, the medium is aspirated and 5 ml of fresh DMEM is added to the cells. The DNA-Lipofectamine complexes are then added dropwise to the cells while swirling the dish. The cells are incubated for 3 h at 37° to allow the DNA complexes to enter the cells. The MEFs are allowed to recover by adding 6.5 ml of DMEM containing 20% FBS to the dishes and incubating overnight. The following day, the cells are starved by incubation for 48 h in DMEM, with 1 to 2 mg/ml of heat-inactivated, fatty acid–free BSA (DMEM-BSA). Expression of the expressed cDNA can be assessed 48 to

72 h post-transfection. Transfection efficiency can be determined by immunofluorescence.

2.2.1. Adenoviral infection

While spontaneously immortalized MEFs and 3T3 cells transfect at relatively high efficiency, we find that early-passage MEFs, human fibroblasts, and MCF10A cells are refractory to transient transfection. When using these cells, adenoviral infection is our method of choice for ectopic cDNA expression. We generally incorporate the infection into the serum-starvation step of our experiments. For adenoviral infection of fibroblasts, near confluent cells are incubated in DMEM-BSA for 8 to 12 h. Adenoviruses are added to the culture medium and incubated overnight. The medium is then replaced with fresh DMEM-BSA and the cells are incubated for an additional day to complete the 2-day serum-starvation procedure. MCF10A are infected similarly except that (1) we seed 2×10^6 cells per 100-mm dish and allow them to attach and spread the day before infection, and (2) we use serum-free DMEM:Ham's F12, 1 mg/ml BSA for the serum-free incubations.

2.2.2. Knock-down with siRNA

We typically use siRNAs for knocking down expression of cell cycle genes in fibroblasts and MCF10A cells; siRNA sequences and conditions are listed in Table 8.1. MEFs are trypsinized, suspended in DMEM-10% FBS without antibiotic, and reseeded in 35-mm (0.5×10^6 cells) or 100-mm (2×10^6 cells) dishes overnight. The next day, Lipofectamine 2000 (Invitrogen) (1 μl per 25,000 cells) and siRNA oligonucleotide (20 to 300 nM final concentration) solutions are each diluted into 0.5 (35- or 60-mm dishes) or 2 ml (100-mm dishes) OPTI-MEM (Invitrogen) and incubated for 5 min before being mixed together and incubated for an additional 20 min to allow for complex formation. During this time, the cultures are washed three times with serum-free OPTI-MEM. After the 20-min incubation is completed, the last wash is removed and the transfection mixture is applied (1 ml per 35- or 60-mm dish and 4 ml per 100-mm dish). The cells are washed with DMEM-BSA 4 to 6 h after transfection. If serum starvation is desired, the incubation in DMEM-BSA continues for an additional day. Otherwise, the cells are switched to serum-containing medium. The efficiency of knock-down is assessed by western blotting and QPCR $\sim$48 h after transfection.

For knock-down in MCF10A cells, we seed 1×10^6 cells in six-well dishes in maintenance medium without antibiotics the day before transfection. The cells should be fully confluent by the next day. The transfection procedure is very similar to that used for fibroblasts except that we use 10 μl of Lipofectamine 2000 per 10^6 seeded cells. Replacement of the transfection medium is not necessary, so the cells can be serum starved by 2-day incubation in the transfection medium.

Table 8.1 siRNA sequences

siRNA	Sequence	siRNA ID
Human β-catenin	sense: 5'-GGAAGAGGAUGUGGAUAACCtt-3' antisense: 5'-GGUAUCCACAUCCUCU-UCCtc-3'	2908
Human β-catenin	sense: 5'-GAUUGGUGUCUGCUAUUCUAtt-3' antisense: 5'-UACAAUAGCAGACAC-CAUCtg-3'	203060
Human E-cadherin	sense: 5'-GAGUGAAUUUUGAAGAUUGtt-3' antisense: 5'-CAAUCUUCAAAAUUCA-CUCtg-3'	44988
Human E-cadherin	sense: 5'-GCACGUACACAGCCCUAAUtt-3' antisense: 5'-AUUAGGGCUGUGUACG-UGCtg-3'	146381
Human Cdc42	sense: 5'-CCAUAUACUCUUGGACUUUtt-3' antisense: 5'-AAAGUCCAAGAGUAUA-UGGtt-3'	145994
Human Cdc42	sense: 5'-CCGCUGAGUUAUCCACAAAtt-3' antisense: 5'-UUUGUGGAUAACUCAG-CGGtc-3'	145995
Human Rac1	sense: 5'-GGAGAUUGGUGCUGUAAAAtt-3' antisense: 5'-UUUUACAGCACCAAUC-UCCtt-3'	27408
Human Rac1	sense: 5'-GAAUAUAUCCCUACUGUCUtt-3' antisense: 5'-AGACAGUAGGGAUAUA-UUCtc-3'	45358

Human cyclin D1	sense: 5'-GGAGGGUUGUGCUACAGAUtt-3' antisense: 5'-AUCUGUAGCACAACCCUCctc-3'	118853
Human cyclin D1	sense: 5'-GGGUUAUCUUAGAUGUUUCtt-3' antisense: 5'-GAAACAUCUAAGAUAACCCtt-3'	118854
Mouse Rac1	sense: 5'-CCGUCUUUGACAACUAUUCtt-3' antisense: 5'-GAAUAGUUGUCAAAGACGGtg-3'	214457
Mouse Rac1	sense: 5'-CCAGUGAAUCUGGGCCUAUtt-3' antisense: 5'-AUAGGCCCAGAUUCACUGGtt-3'	214461
Mouse cyclin D1	sense: 5'-CGAUUUCAUCGAACACUUCtt-3' antisense: 5'-GAAGUGUUCGAUGAAAUCGtg-3'	160053
Mouse cyclin D1	sense: 5'-GCGGUAGGGGAUGAAAUAGUtt-3' antisense: 5'-ACUAUUUCAUCCCUACCGCG-3'	160054
Mouse Skp2	sense: 5'-GCUUAGUCGGGAGAACUUUtt-3' antisense: 5'-AAAGUUCUCCCGACUAAGCtt-3'	188567
Mouse Skp2	sense: 5'-CCGAAGUGUUACUUUUUCAAtt-3' antisense: 5'-UUGAAAAAGUAACACUUCGGtg-3'	188568

Notes: These sequences of the siRNAs are the most commonly used in our lab. Silencer Pre-Designed siRNAs are ordered from Ambion using the indicated siRNA ID numbers. Each of the two siRNAs shown efficiently knocks down target protein expression, albeit with somewhat different efficiencies depending on conditions and cells used.

2.3. Preparing substrata of near physiological compliance using matrix proteins linked to polyacrylamide gels

Although cell culture dishes are widely used to characterize cellular behavior *in vitro*, a major difference between these dishes and the physiological state is the degree of substratum compliance: culture plastic is noncompliant while physiological tissue is quite compliant (Yeung *et al.*, 2005). To more closely mimic the compliance of physiological substrata, we and others have turned to polyacrylamide gels of known compliance, which are covalently attached to "reactive" glass coverslips and used as a support for the subsequent covalent attachment of purified matrix proteins such as fibronectin, collagen, and vitronectin. A second set of coverslips is siliconized and used as the top component of a "sandwich" to generate a uniformly spread, flat polyacrylamide gel suitable for coating with matrix proteins as described below. We prepare these coverslips using a modification of the protocol described by Yeung *et al.* (2005). Figure 8.1 summarizes the procedure, and Fig. 8.2 shows the distinct morphologies of fibroblasts attached to fibronectin-coated substrata of low and high compliance.

2.3.1. Preparation of the reactive and siliconized glass coverslips

Preparation of the "reactive" coverslips begins with several 25-mm glass coverslips (Fisher) placed in a Parafilm-lined 150-mm dish. Each coverslip is covered with a 1-ml droplet of 0.1-M NaOH for 3 min to etch the glass surface. The NaOH solution is removed by aspiration, and the coverslips are covered for 3 min with a 0.5-ml droplet of 97% 3-APTMS (3-aminopropyl-trimethoxysilane) (Sigma). After aspirating the 3-APTMS, the coverslips are washed well (three times for 10 min each) with deionized water. The coverslips are then dried by aspirating *all* remaining water, and they are placed into a new Parafilm-lined dish. (If the coverslips are not sufficiently washed, an orange precipitate will form in the next step and they will be unusable.) Each coverslip is then coated with a 0.5-ml droplet of 0.5% glutaraldehyde for 30 min, and then washed three times for 10 min each with deionized water. While these coverslips are being washed, the other set of coverslips are siliconized (several 25-mm coverslips are added to a 50-ml Falcon tube containing 10% surfasil siliconizing reagent [Pierce] in chloroform and rocked at room temperature for at least 10 min).

2.3.2. Preparation of reagents for the polyacrylamide gels

While the coverslips are being washed and siliconized, the reagents necessary for making the gels are prepared. These reagents include a saturated solution of acrylic acid N-hydroxy succinimide ester (A-NHS) in toluene, 10% ammonium persulfate (APS), 40% acrylamide, and 1% bis-acrylamide.

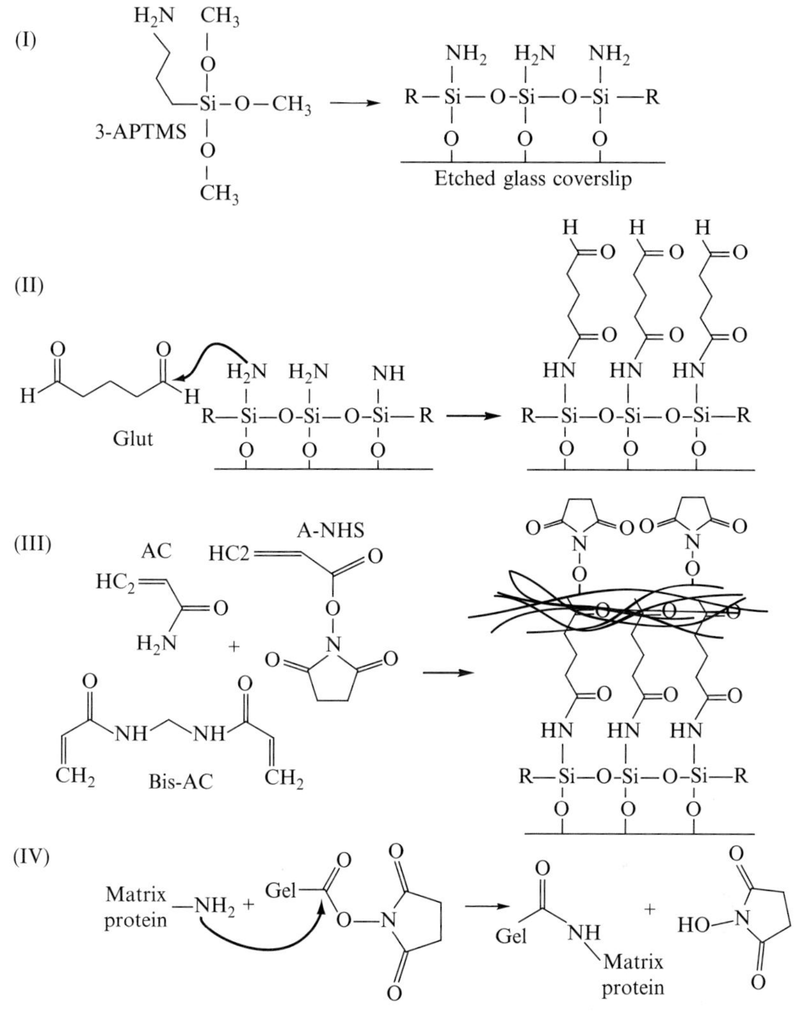

Figure 8.1 Reaction scheme for generating substrata of near physiological compliance. We generate substrata of near physiological compliance by covalently attaching a purified matrix protein to polyacrylamide gels that had been covalently bound to glass coverslips. (I) 3-APTMS reacts with the etched glass surface to form Si-O-Si bonds. The amine group from 3-APTMS remains free for further reaction. (II) Glutaraldehyde (glut) reacts with the exposed amines and will act as a cross-linker between the polyacrylamide gel and the 3-APTMS. (III) A mixture of acrylamide (AC), bis-acrylamide (Bis-AC), and acrylic acid N-hydroxy succinimide ester (A-NHS) is allowed to polymerize on the coverslip surface. The acrylic acid moiety on NHS allows the ester to integrate into the forming gel via free radical polymerization, and the free aldehyde groups from glutaraldehyde react with the polymerizing gel to covalently link the gel to the coverslip. (IV) The activated NHS ester reacts with amine groups on the ECM protein of choice to link it to the gel-coated coverslip.

Low compliance   High compliance
(9000 Pa)   (700 Pa)

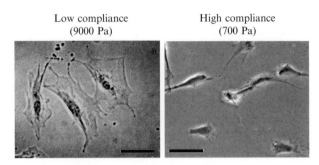

Figure 8.2   Morphology of cells on fibronectin-polyacrylamide gels of different compliance. MEFs were plated on fibronectin-polyacrylamide substrata of high or low compliance. After 9 h, the cells were imaged by phase-contrast microscopy. MEFs on the low compliance fibronectin subtratum are well spread as compared to those on the high-compliance substratum. Scale bar = 85 μm.

2.3.3. Assembly of polyacrylamide gel-coated coverslips

Working in the tissue culture hood, equal numbers of "reactive" and siliconized coverslips are dried by aspiration and placed reactive side up on strips of Parafilm. The coverslips are allowed to air dry completely. Six-well culture dishes are filled with 2 ml of PBS to serve as receptacles for the polymerized gels. An acrylamide solution is made in 0.8-ml aliquots (see below). The compliance of these gels can easily be varied by changing the acrylamide and bis-acrylamide concentrations according to the dose–response curves reported by Yeung *et al.* (2005). We currently keep the acrylamide concentration constant at 7.5% and vary the bis-acrylamide concentration between 0.03 and 0.3%, corresponding to shear moduli between 700 to 9000 Pa.

The reagents are mixed in the following order: water (variable in amount but resulting in a final volume of 0.8 ml), 150 μl 40% acrylamide, 24 to 240 μl of 1% bis-acrylamide (depending on the final compliance desired), 1 μl of TEMED, and 8 μl of APS. A-NHS (228 μl) is then added, the microcentrifuge tube is briefly vortexed and 125 μl drops are placed on each "reactive" coverslip. The acrylamide solution polymerizes very quickly, so the droplets must be placed on the coverslips *immediately* and a maximum of five coverslips can be plated at one time. As soon as the droplets are in place, a siliconized coverslip is carefully placed on top; the resulting capillary action should result in a uniformly spread polyacrylamide gel. After 2 to 3 min, the gels should be fully polymerized, the top (siliconized) coverslips are carefully removed, and the polymerized gels are placed in the prepared six-well dishes containing PBS. The gels are washed three times on a rocker (5 min per wash) in PBS to remove remaining toluene, and then coated with the appropriate ECM protein overnight at 4° as outlined above. The final amount for any particular experiment may be dependent on the cell type being studied and is determined by monitoring the rate of attachment and spreading in the low-compliance gels. The matrix protein-coated acrylamide gels are "blocked" for 30 min (37°) in DMEM-BSA prior to use. We typically seed 1×10^5 cells per well (for fibroblasts) or 3.25×10^5 cells per well (for MCF10A cells) on these acrylamide-matrix protein substrata for experimentation.

3. ANALYTICAL METHODS

3.1. Cell cycle analysis by propidium iodide staining and flow cytometry

Our laboratory uses flow cytometry of propidium iodide (PI)–stained cells to assess quiescence and monitor cell cycle progression. The PI staining solution is 10 mM Tris-HCl, 5 mM MgCl$_2$, pH 7, with 50 μg/ml PI and 0.1% NaN$_3$. The staining solution is stirred until the dye is completely dissolved, sterile filtered, and stored at 4° protected from light.

Cell monolayers of at least 5×10^5 cells are washed once with 0.05% Trypsin-0.5 mM EDTA and then incubated at 37° with enough trypsin-EDTA to cover the cell monolayer. As soon as cells have rounded, they are resuspended in 10 ml of ice-cold PBS, clumps are disrupted by gentle pipetting with a 1-ml pipette tip, and the suspension is transferred into a pre-chilled 15-ml polypropylene conical tube. From this point on, all cell manipulations are performed at 0 to 4°. After centrifugation (1000×g for 5 min) and removal of the supernatant, the cell pellet is resuspended in 250 μl of cold PBS, and clumps are again disrupted by pipetting. The suspended cells are fixed by dropwise addition of 1.75 ml of cold 80% ethanol in PBS for at least 30 min on ice. Cells can be stored in the fixing solution for up to a month at 4°.

After fixation, the cells are collected by centrifugation at 1000×g for 5 min and then washed once with cold PBS. The washed cell pellet is resuspended in 280 μl of PI staining and 20 μl of DNase-free RNase (0.5 mg/ml stock from Roche, Cat.# 1119915). The cells are again resuspended repeatedly with the 1-ml pipette tip and then incubated in a water bath at 37° for 2 h, protected from light. We transfer the stained cells to polystyrene tubes and analyze them for DNA content using an EPICS XL flow cytometer (Beckman-Coulter, Inc.).

3.2. Crystal violet staining to monitor cell proliferation

As a surrogate to direct cell counting, we use crystal violet staining to monitor the proliferation of cells in culture. Cells are trypsinized, and duplicate cultures are seeded in growth medium at a low density so that contact inhibition does not reduce the rate of cell proliferation (yet at a high enough density to allow for a detectable signal at time zero). We use 4×10^4 and 1×10^5 as starting values for immortalized MEFs and MCF10A cells, respectively, in six-well dishes. A set of duplicate wells lacking cells is included for every time point and used to assess background staining. The first set of samples is collected as soon as the cells are fully attached (typically 2 to 4 h) and well before cell division occurs. The crystal violet staining intensity of these samples corresponds to the starting (time 0) cell number. At collection time, the cells are washed three times with ice-cold PBS and then stained with a 0.5% crystal violet (Sigma, C3886) for 3 h at room temperature with gentle rocking. The crystal violet stock solution is prepared in 20% methanol, stirred overnight, and filtered through Whatman paper before use. The staining solution is removed, and the cell monolayer is washed five times with distilled H_2O to remove all excess crystal violet. The final wash is removed by aspiration, and a damp Q-tip is used to remove deposited crystal violet from the sides of the wells. The fully washed wells are stored at $4°$ until all samples have been collected, stained, and washed. After the last sample has been stained and washed, the stain is eluted with 0.5 ml of 0.1-M sodium citrate, pH 4.2, in 50% ethanol for 30 min at room temperature. We determine sample absorbance at 595 nm from duplicate 0.2-ml aliquots in a Bio–Rad Microplate Reader (model 550). Mean background absorbance is subtracted from the mean absorbance value of each sample.

3.3. Immunofluorescence

Our procedures for preparing coverslips of adherent, suspended, and cytochalasin D-treated cells have been described (Zhu *et al.*, 1999), but our current staining procedures are described here. Immunofluorescence is performed in a humidified chamber (with coverslips placed on a dampened piece of Whatman 3 paper or paper towel within a covered plastic dish). Collected coverslips are gently washed once in PBS before incubation with fixative (3.7% formaldehyde in PBS) for 10 to 15 min. Residual aldehydes are quenched by treating the fixed samples with 50-mM NH_4Cl in PBS for 10 min before rinsing twice with PBS. This procedure can reduce background fluorescence. (The fixed cells can be stored in PBS at $4°$ for 2 to 4 weeks.) During the staining procedure, antibodies, washes, and stains are carefully added to the coverslips as 50 to 100 μl droplets. Washes are typically 2 to 5 min. All procedures are performed at room temperature.

3.3.1. BrdU incorporation

To monitor entry into S-phase, the incubation of cells with FBS is performed in the presence of BrdU (3 μg/ml) (1000-fold dilution of Cell Proliferation Labeling Reagent; Amersham, Piscataway, NJ). Formaldehyde-fixed and NH_4Cl-treated fibroblasts are permeabilized with 0.2%

Triton X-100 in PBS for 10 min. The cells are washed three times with PBS and then incubated for 1 to 3 h with a 1000-fold dilution of sheep anti-BrdU (Biodesign, Saco, ME), and a 100-fold dilution of DNase stock solution in PBS containing 2% BSA. The stock solution is 500 units DNase (Sigma D4527)/ml in 20 mM Tris-Hcl, pH 7.5, 50 mM Nacl, 0.1 mg/ml BSA, and 50% glycerol. The cells are washed three times in PBS and then incubated for 1 h with anti-sheep secondary antibody and 2 μg/ml DAPI (Sigma) in PBS. Currently, we use a 1000-fold dilution Alexa Fluor® 594 donkey anti-sheep (Molecular Probes, Eugene, OR) as the secondary antibody, but have also used 200-fold dilutions of TRITC- or FITC-conjugated anti-sheep antibodies successfully. The stained coverslips are washed in PBS two to three times, dipped in water, gently dried, and then inverted and mounted in antifade reagent (Molecular Probes, #57461). After the coverslips dry (~10 to 30 min), they are sealed with clear nail polish. The percentage of BrdU-positive cells is assessed by epifluorescence microscopy, counting the number of BrdU positive and DAPI-positive cells, and typically counting ~150 nuclei in several separate fields of view per sample. The incubations and washes can also be performed in PBS with 0.1% Tween-20 and 1 mg/ml of BSA to reduce background.

BrdU incorporation in MCF10A cells is determined similarly, except that the permeabilization is performed with 0.5% Triton X-100 for 15 min, the washes are carried out in PBS, 0.1% Tween-20, and the secondary Alexa Fluor 594 antibody is used at a 1:500 dilution in PBS, 1 to 2% BSA, 0.1% Tween-20.

3.3.2. Immunostaining for individual proteins

The conditions for immunofluorescence analysis of individual proteins should be optimized for each primary antibody being used. For example, when detecting cyclin A, we incubate fixed and permeabilized cells with a 100-fold dilution of ammonium sulfate-fractionated rabbit polyclonal antibody against cyclin A in PBS-2% BSA for 1 h, wash three times in PBS, and then incubate the cells for 1 h with a 1000-fold dilution of an Alexa Fluor® 488 chicken anti-rabbit secondary antibody in PBS (Molecular Probes). Vinculin is immunostained at an elevated temperature (37°) using a 100-fold dilution of anti-vinculin (Sigma; V4505) for 1 h at 37°, and a 1000-fold dilution of Alexa Fluor 488 secondary antibody at room temperature. For E-cadherin staining, MCF10A cells are simultaneously fixed and blocked in PBS, 0.5% Triton X-100, 0.2% BSA, and 2% goat serum for 30 to 60 min. Anti–E-cadherin monoclonal antibody is used at a 500-fold dilution in fresh permeabilization/blocking buffer without Triton X-100.

As needed, we try to reduce background staining in immunofluorescence by increasing the stringency of the permeabilization, blocking, antibody incubations, or washes by using PBS or Tris-HCl buffers containing 2 to 5% BSA, goat serum, 150 to 300 mM NaCl, or 0.1 to 0.5% Tween-20 or Triton X-100. Although we prefer Alexa Fluor in our secondary antibodies, we have also used 100- to 200-fold dilutions of TRITC- or FITC-conjugated antibodies successfully.

3.4. Rho family GTPase pulldown assays

3.4.1. Rac and Cdc42 activity assays

Cells are grown to near confluence in 100–mm dishes and starved for 2 days. The cells are stimulated directly in the culture dish with media containing 10% FBS or the purified growth factors under investigation for 0 to 120 min. To collect samples, the dishes are chilled on ice and washed once with ice-cold PBS. The cells are scraped into 1 ml of PBS and collected by centrifugation at $8000 \times g$ for 2 min. The pellets are immediately lysed in 300 μl of buffer consisting of 50 mM of Tris-HCl, pH 7.4, 250 mM NaCl, 10 mM MgCl$_2$, 10% glycerol, 0.5% NP-40, 1 μg/ml aprotinin, 1 μg/ml leupeptin, and 20 μg of GST-PBD. The GST-PBD is prepared as a recombinant protein in bacteria as described by Welsh *et al.* (2001). A portion (10%, 30 μl) of the total lysate is removed and used as a loading control in the subsequent analysis by SDS gel electrophoresis. To pull down the GST-PBD-Rac complexes, 100 μl of washed glutathione-agarose beads are added to the remaining lysate and incubated at 4° for 30 min with rocking. The complexes are collected by centrifugation for 30 sec at $1000 \times g$. Ten percent of the supernatant is saved to measure the levels of inactive (GDP-bound) Rac. The beads are gently washed twice in 1 ml of ice-cold lysis buffer, and the collected complexes are eluted from the beads with 100 μl 1$\times$ SDS-sample buffer with reductant, boiling the samples for 5 min followed by brief centrifugation to pellet the spent beads. The resulting supernatant containing the GTP-bound Rac and Cdc42 is fractionated on a 12% reducing SDS-polyacrylamide gel, as is the total lysate loading control and the "depleted" supernatant. Rac is detected by western blotting with anti-Rac (UBI, 1:5000 dilution). Since GST-PBD pulls down GTP-bound Cdc42 as well as GTP-bound Rac, blotting the filters with anti-Cdc42 will allow an assessment of Cdc42 activation. Activated (V12 or Q61L) or dominant negative (N17) alleles of Rac or Cdc42 can be transfected into cells as positive and negative controls, respectively.

3.4.2. Rho activity assay

Although Rho pull-downs have traditionally been performed in solution using GST-Rhotekin, we currently use an ELISA-based kit (G-LISA; Cytoskeleton) to measure Rho activation. Cells are seeded in six-well dishes (one well/sample) so that after starvation they are $\sim$50% confluent ($\sim$10 to 20 $\times$ 10^4 cells). Measurement of Rho activity is done according to the G-LISA protocol with the supplied reagents. Briefly, cells are collected and lysed.

A portion (10%) of the lysate is removed to measure protein concentration, and equal amounts of protein are incubated for 1 h in the G-LISA ELISA plate. Each well is coated with the Rho-binding domain of Rhotekin, which selectively binds to GTP-loaded Rho. The bound Rho is then detected by adding an anti-Rho primary antibody and an HRP-labeled secondary antibody to the wells. Peroxidase substrates are added, and the resulting colorimetric reaction is quantified by measuring A_{490} in a microplate reader.

3.5. Assays for transcriptional regulation

3.5.1. Quantitative real-time RT-PCR

To measure mRNA levels of different genes, cells in 60- or 100-mm dishes are washed twice with PBS and collected in 0.5 to 1 ml of Trizol (Invitrogen). Alternatively, a cell pellet can be extracted in 0.5 ml of Trizol. Total RNA is isolated according to the manufacturer's protocol, and the final RNA pellet is dissolved in 10 μl of DEPC-treated water. A small aliquot of the RNA is quantified using a spectrophotometer, and the ratio of absorbance at 260 and 280 nm is measured to ensure RNA quality. The RNA is diluted to 10 to 50 ng/ml with DEPC-treated water, and 35 to 180 ng of RNA is used for reverse transcription with Applied Biosystems reagents (1× buffer, 5.5 mM MgCl$_2$, 500 μM of each dNTP, 2.5 μM random hexamers, 0.4 U/ml RNase inhibitor, and 1.25 U/ml Multiscribe Reverse Transcriptase) in a total volume of 10 to 20 μl. The samples are incubated at 25° for 10 min, followed by reverse transcription at 48° for 30 min, and reverse transcriptase inactivation at 95° for 5 min.

The subsequent real-time PCR reactions contain 2 μl of the reverse transcription mix, 1× TaqMan Universal PCR mix (Applied Biosystems), forward and reverse primers, and a probe that provides additional levels of specificity. The concentration of the primers and probes that we currently use to study G1-phase cell cycle genes are listed in Table 8.2. The solution is brought to a final volume of 22 μl with water. The PCR reaction starts with an incubation for 2 min at 50° followed by 40 cycles of denaturation (15 sec at 95°) and annealing/extension (1 min at 60°) using an Applied Biosystems Prism 7000 sequence detection system.

For each primer-probe set, we generate a standard curve from an RNA sample with good expression of the gene of interest, using ∼200 ng of RNA for the reverse transcription. The real-time PCR standard curve is then generated using four 5-fold serial dilutions of the reverse transcription reaction product. Relative changes in mRNA levels are quantified as changes in Threshold Cycle (C_T) relative to the standard curve using ABI Prism 7000 sequence detection system software. An $R^2 > 0.99$ is typically obtained for the standard curve, and the C_T values of the unknowns fall within the standard curve. Duplicate PCR reactions are run for each

Table 8.2 QPCR primer-probe sets

mRNA target	Sequence or assay ID# (Applied Biosystems)	Concentration
Mouse p21cip1	F: TCCACAGGCGATATCCAGACATT	900 nM
	R: CGGACATCACCAGGATTGG	900 nM
	P: 6FAM–AGAGCCACAGGCACC–MGBNFQ	250 nM
18S (mouse/human)	F: CCTGGTTGATCCTGCCAGTAG	150 nM
	R: CCGTGCGTACTTAGACATGCA	150 nM
	P: VIC–TGCTTGTCTCAAAGATTA–MGBNFQ	125 nM
Mouse Skp2	F: AGCGCGTGGGTGAAAGC	200 nM
	R: ATCACTGAGTTCGACAGGTCCAT	200 nM
	P: 6FAM–TCAGCTCTTTCCGGGTAC–MGBNFQ	150 nM
Mouse cdk4	F: GGCCTTTGAACATCCCAAT	900 nM
	R: TCAGTTCGGGAAGTAGCACAGA	900 nM
	P: VIC–ATCCATCAGCCGTACA–MGBNFQ	250 nM
Mouse cyclin A	Mm00438064_m1	
Mouse cyclin E	Mm00432367_m1	
Mouse p27kip1	Mm00438167_m1	
Mouse cyclin D1	F: TGCCATCCATGCGGAAA	900 nM
	R: AGCGGGAAGAACTCCTCTTC	900 nM
	P: 6FAM–CTCACAGACCTCCAGCAT–MGBNFQ	250 nM
Human cyclin D1	F: TGTTCGTGGCCTCTAAGATGAAG	300 nM
	R: AGGTTCCACTTGAGCTTGTTCAC	300 nM
	P: 6FAM–AGCAGCTCCATTTGCAGCAGCTCCT–TAMRA	250 nM

Human Skp2	Hs00261857_m1	
Human p27kip1	Hs00153277_m1	
Human cyclin A	Hs00171105_m1	
Human p21cip1	Hs00355782_m1	
Human cdk4	F: ACAAGTGGTGGAACAGTCAAGCT	150 nM
	R: GCATATGTGGACTGCAGAAGAACT	200 nM
	P: VIC–ATGGCACTTACACCCGTGGTTGTTACACTCT–TAMRA	150 nM
ChIP QPCR primers		
Mouse Skp2	F: TGGTGATGGAACGTTGCTAGT	900 nM
	R: GGTGTCCACTGATTCAGGA	900 nM
Mouse Cyclin A	F: CAGACAATCCTGGCGTGGA	900 nM
	R: ACTCAGAACCAAACAATGGCTGA	900 nM
Mouse p107	F: GTCCGAGGTCCATCTTCTTATCC	900 nM
	R: TTCATCGTCTGCCTCCGC	900 nM

Notes: We regularly use these sequences and working concentrations of the primer-probe sets for QPCR. We use Applied Biosystems Primer Express version 2.0 software to develop custom primer-probe sets for QPCR. These custom pimer-probe sets are purchased from Applied Biosystems. When we use Applied Biosystems Assay-On-Demand primer-probe sets, they are used under the conditions provided by the manufacturer. We use Primer3 software to select primers for ChIP QPCR performed with SYBR GREEN.

sample, and the reproducibility of duplicates should be excellent with a standard deviation of less than 5% of the mean C_T value.

In certain situations, such as when performing a high-throughput screening of many genes, it is impractical to design primer-probe sets for each transcript. In these cases, we use SYBR GREEN instead of a fluorescently labeled probe. We use 1× SYBR GREEN TaqMan buffer (Applied Biosystems) with 900-nM primers. As with primer-probe sets, we generate a standard curve to quantify changes in transcript abundance. Additionally, to verify the specificity of the primers, we follow each PCR reaction with a dissociation curve analysis. If the primers are specific, there will only be one peak in the dissociation curve that corresponds to the melting temperature of the desired amplicon. We typically design primers using the default parameters in Primer3 software (http://frodo.wi.mit.edu/cgi-bin/primer3/primer3_www.cgi).

3.5.2. Chromatin immunoprecipitation

We use chromatin immunoprecipitation (ChIP) to measure transcription factor binding to promoters in intact cells. The ChIPs are performed using the ChIP Assay Kit from Upstate Biotechnology (Lake Placid, NY) and follow the manufacturer's protocol with few modifications. To cross-link transcription factors to DNA, formaldehyde (1% final concentration) is added to the medium of cell culture monolayers (10 min at room temperature). Glycine is then added at a final concentration of 125 mM (5 min at room temperature) to quench unreacted formaldehyde. The cells are washed twice with cold PBS, scraped into 0.2 ml of mild lysis buffer (5 mM Pipes, pH 8.0, 85 mM KCl, 0.5% NP-40, 1 μg/ml aprotinin, 1 μg/ml leupeptin, and 1 mM freshly prepared phenylmethylsulfonyl fluoride), and incubated on ice for 10 min to allow for lysis of the plasma membrane. The nuclei are collected by centrifugation (2300×g, 5 min, 4°), lysed in 0.2 ml of the supplied SDS-lysis buffer (50 mM Tris-HCl, pH 8.1, 10 mM EDTA, 1% SDS), and then sonicated on ice (three 7-sec pulses separated by 30 sec on ice) using a Virsonic 475 (VirTis, Gardiner, NY) microprobe set at power level 3. The goal is to obtain sheared DNA of 200 to 1000 bp.

After centrifugation (16,000×g, 10 min, 4°), half of the supernatant is diluted 10-fold with the supplied ChIP dilution buffer (16.7 mM Tris-HCl, pH 8.1, 167 mM NaCl, 1.2 mM EDTA, 1.1% Triton X-100, 0.01% SDS). An aliquot (1%) of the diluted supernatant is saved and used as the loading control (input) for the PCR; the remainder is precleared by incubation with 60 μl protein A-agarose (Invitrogen) that had been washed several times in ChIP dilution buffer. After a 1-h incubation at 4° with rocking, the agarose beads are removed by centrifugation, and the precleared supernatants are incubated with 5 μg of the desired antibody or control antibody (ideally, preimmune IgG for polyclonal antibodies and isotype-matched controls for monoclonal antibodies) at 4° overnight with rocking. The immune complexes are collected by addition of supplied protein A-agarose beads (60 μl) that had been blocked with salmon sperm DNA. After a 1-h incubation at 4° with rocking, the immunoprecipitated complexes are washed with the supplied wash buffers as follows: once with 1 ml of low-salt wash buffer (20 mM Tris-HCl, pH 8.1, 150 mM NaCl, 2 mM EDTA, 0.1% SDS, 1% Triton X-100), once with 1 ml of high-salt wash buffer (20 mM Tris-HCl, pH 8.1, 500 mM NaCl, 2 mM EDTA, 0.1% SDS, 1% Triton X-100), once with 1 ml of LiCl wash buffer (10 mM Tris-HCl, pH 8, 1 mM EDTA, 0.25 M LiCl, 1% IGEPAL-CA630, 1% deoxycholic acid [sodium salt]), and twice with 1 ml of TE buffer (10 mM Tris-HCl, pH 8.0, 1 mM EDTA), with each wash lasting 5 min at 4°. The DNA-antibody complex is eluted twice (15 min per elution) into 0.1 ml of freshly prepared elution buffer (0.1 M NaHCO$_3$, 1% SDS). The cross-linking is reversed by heating the samples at 65° for 4 h, and the DNA is recovered with a MiniElute PCR Purification Kit (Qiagen, Valencia, CA). One-tenth of the resuspended DNA (5 μl) is analyzed by QPCR as described above, but using SYBR GREEN as the fluorescent indicator.

We have also performed ChIP after transient transfection of promoter-luciferase constructs in MEFs. MEFs (10^6 cells per 100-mm dish) were transfected as described above but using 10 ng of the desired promoter-luciferase plasmid. After a 24-h recovery in 10% FBS, ChIP is performed as described above. To prevent amplification of the endogenous promoter during the PCR, one of the primers should encode plasmid backbone sequence within the promoter-luciferase construct. The amplified PCR product ($\sim$300 bp) can be detected by electrophoresis on a 1.5% agarose gel.

3.5.3. Determining changes in transcription by assessing changes in pre-mRNA

We use QPCR to measure the abundance of newly transcribed, unspliced pre-mRNA. Total RNA is purified using Trizol reagent and reverse transcribed using random hexamer primers according to the protocol described earlier. The cDNA is amplified and quantified by real-time PCR using SYBR GREEN and primers that correspond to an intron and exon sequence in the RNA (to avoid amplifying spliced mRNA). To ensure that the results do not reflect amplification of genomic DNA contaminants we either (1) do a mock reverse transcription without the reverse transcriptase enzyme so no cDNA is made, or (2) treat the purified RNA with RNase-free DNase (Invitrogen) (1U DNase, 25 min) to digest contaminating genomic DNA prior to reverse transcription. When RNase-free DNase is used, the DNase must be inactivated by heating (65°, 10 min) in the presence of EDTA before the sample can be used in the reverse transcription reaction.

3.6. Special analytical procedures needed for experimentation on acrylamide–extracellular matrix substrata

3.6.1. Collection of protein and RNA

A strip of Parafilm is taped to the bench top and 100-μl droplets of the appropriate lysis buffer (reducing SDS-gel sample buffer for extraction of cellular protein or Trizol for isolation of total RNA) are spaced out on the Parafilm. The coverslips are washed once in PBS, and the gel is placed face-down on the drops of lysis buffer. After 1 to 2 min, the coverslips are rinsed, and the lysis buffer is pipetted into microcentrifuge tubes. Protein samples are boiled for 5 min to denature proteins.

3.6.2. Immunostaining of polyacrylamide/matrix protein-coated coverslips

The staining procedure for polyacrylamide/matrix protein-coated cover-slips requires several modifications of our standard procedure (see above) to accommodate the reduced diffusion within the polyacrylamide. All procedures are performed with submerged coverslips in six-well culture dishes. Cells should be fixed in 3.7% formaldehyde for at least 16 h at 4°, and if BrdU incorporation is being measured, the fixation time should be extended to 2 to 3 days, with daily changes of the formaldehyde solution to wash out the unincorporated BrdU. The cells are permeabilized with 1 ml of 0.4% Triton X-100 PBS with rocking for 15 min. The gels are then "blocked" with 2% BSA in PBS for 1 h at room temperature with rocking. Primary and secondary antibody incubations are performed with 0.5 ml so that the sample is completely covered. We have also found that all washes should be extended (to three to four times with 10 min per wash) to keep background staining low. Finally, the stained samples are washed in water and mounted on glass slides in a drop of antifade reagent. The images must be captured within 24 h because we do not seal these gels, and the images are distorted if the samples dry.

ACKNOWLEDGMENTS

We thank the present and former lab colleagues who helped in the development of the methodologies described here. We also thank Paul Janmey and Valerie Weaver for guidance in the development of polyacrylamide gel–matrix protein substrata technology in our lab. Work in our lab is supported by grants from the National Institutes of Health.

REFERENCES

Assoian, R. K., and Schwartz, M. A. (2001). Coordinate signaling by integrins and receptor tyrosine kinases in the regulation of G1 phase cell-cycle progression. *Curr. Opin. Genet. Dev.* **11,** 48–53.

Bhowmick, N. A., Ghiassi, M., Bakin, A., Aakre, M., Lundquist, C. A., Engel, M. E., Arteaga, C. L., and Moses, H. L. (2001). Transforming growth factor-beta1 mediates epithelial to mesenchymal transdifferentiation through a RhoA-dependent mechanism. *Mol. Biol. Cell* **12,** 27–36.

Bolos, V., Peinado, H., Perez-Moreno, M. A., Fraga, M. F., Esteller, M., and Cano, A. (2003). The transcription factor Slug represses E-cadherin expression and induces epithelial to mesenchymal transitions: A comparison with Snail and E47 repressors. *J. Cell Sci.* **116,** 499–511.

Cano, A., Perez-Moreno, M. A., Rodrigo, I., Locascio, A., Blanco, M. J., del Barrio, M. G., Portillo, F., and Nieto, M. A. (2000). The transcription factor snail controls epithelial-mesenchymal transitions by repressing E-cadherin expression. *Nat. Cell Biol.* **2,** 76–83.

Thiery, J. P. (2003). Epithelial-mesenchymal transitions in development and pathologies. *Curr. Opin. Cell Biol.* **15,** 740–746.

Welsh, C. F., Roovers, K., Villanueva, J., Liu, Y., Schwartz, M. A., and Assoian, R. K. (2001). Timing of cyclin D1 expression within G1 phase is controlled by Rho. *Nat. Cell Biol.* **3,** 950–957.

Yeung, T., Georges, P. C., Flanagan, L. A., Marg, B., Ortiz, M., Funaki, M., Zahir, N., Ming, W., Weaver, V., and Janmey, P. A. (2005). Effects of substrate stiffness on cell morphology, cytoskeletal structure, and adhesion. *Cell Motil. Cytoskeleton* **60,** 24–34.

Zhong, C., Kinch, M. S., and Burridge, K. (1997). Rho-stimulated contractility contributes to the fibroblastic phenotype of Ras-transformed epithelial cells. *Mol. Biol. Cell* **8,** 2329–2344.

Zhu, X., Roovers, K., Davey, G., and Assoian, R. K. (1999). Methods for analysis of adhesion-dependent cell cycle progression. *In* "Signaling through Cell Adhesion Molecules" (J.-L. Guan, ed.). CRC Press, Boca Raton, FL.

CHAPTER NINE

Analysis of Integrin Signaling by Fluorescence Resonance Energy Transfer

Yingxiao Wang* *and* Shu Chien[†]

Contents

Abstract

Fluorescence resonance energy transfer (FRET) has been proven to be a powerful tool to visualize and quantify the signaling cascades in live cells with high spatiotemporal resolutions. Here we describe the development of the genetically encoded and FRET-based biosensors for imaging of integrin-related signaling cascades. The construction of a FRET biosensor for Src kinase, an important tyrosine kinase involved in integrin-related signaling pathways, is used as an example to illustrate the construction procedure and the pitfalls involved. The design strategies and considerations on improvements of sensitivity and

* Department of Bioengineering and Molecular & Integrative Physiology, Neuroscience Program, Center for Biophysics and Computational Biology, Beckman Institute for Advanced Science and Technology, University of Illinois, Urbana-Champaign, Urbana, Illinios
† Departments of Bioengineering and Medicine, and Whitaker Institute of Biomedical Engineering, University of California at San Diego, La Jolla, California

Methods in Enzymology, Volume 426
ISSN 0076-6879, DOI: 10.1016/S0076-6879(07)26009-4

© 2007 Elsevier Inc.
All rights reserved.

specificity are also discussed. The FRET-based biosensors provide a complementary approach to traditional biochemical assays for the analysis of the
functions of integrins and their associated signaling molecules. The dynamic
and subcellular visualization enabled by FRET can shed new light on the molecular mechanisms regulating integrin signaling and advance our knowledge in the
understanding of integrin-related pathophysiological processes.

1. INTRODUCTION

The physiological functions of cells are largely dependent on the local
chemical and physical microenvironment in which they live. Cell adhesion
mediates the interactions between a cell with the extracellular matrix (ECM)
and neighboring cells. There is compelling evidence that cell adhesion on
ECM plays crucial roles in regulating cellular functions, including proliferation, migration, and chemotaxis (Hynes, 1999). Integrins, a group of
plasma membrane receptors, are the main mediators for cell–ECM adhesion
(Hynes, 1987). Research by many groups on integrins has led to tremendous
progress in this field since the mid-1990s. However, the detailed molecular
mechanisms by which integrins are activated remain unresolved.

The development of new fluorescence probes and the advancement
of fluorescence microscopy systems have made possible the visualization of
signaling molecules in live cells. In particular, with the development of
green fluorescence protein (GFP) and other fluorescence proteins (FPs)
with different colors, it is not a difficult task now to visualize the positions
of multiple signaling molecules simultaneously in the same live cell. The
technology of fluorescence resonance energy transfer (FRET) further allows
the dynamical imaging of molecular activities in live cells, particularly those
activities due to molecule–molecule interactions, molecular conformational
changes, or enzymatic activities. Genetically encoded fluorescence biosensors based on FRET have attracted a lot of attention because they can be
easily introduced into live cells and targeted to specific subcellular compartments. A variety of genetically encoded FRET biosensors have been developed to visualize the activation of integrins and their associated signaling
molecules. These biosensors have helped greatly in our understanding of
integrins and will continue to contribute significantly to integrin research in
the future. In this article, we will introduce the concept of FRET and the
development of FRET biosensors for integrins signaling. In particular, we
will provide detailed information on the development of a FRET biosensor
for an integrin-related signaling molecule, Src kinase, as an illustrative
example.

2. FLUORESCENCE PROTEINS AND FRET

2.1. Fluorescence proteins

Since the discovery of GFP (Shimomura *et al.*, 1962) and its subsequent cloning and sequencing by Prasher *et al.* (1992), GFP and its derivatives have provided powerful tools for cell biology. The chromophore of GFP is buried deep inside and protected by strands of β-barrels forming a can shape; hence GFP is relatively stable under various environments (Tsien, 1998). Because of their inertial nature, GFP and its derivatives have been widely applied and genetically fused to host proteins to monitor their positions in live cells. The limitation of GFP and its derivatives is that the emission peaks of these FPs are shorter than 529 nm (Tsien, 1998). Lukyanov's group cloned DsRed in 1999, a long-awaited red fluorescence protein (RFP) with excitation and emission peaks at 558 and 583 nm, respectively (Baird *et al.*, 2000; Matz *et al.*, 1999). However, DsRed tends to form tetramers, which hinders its application as a fusion tag for host proteins (Baird *et al.*, 2000). Through a directed DNA evolution strategy, a true monomer RFP (mRFP) has been developed (Campbell *et al.*, 2002). Although mRFP matures faster than DsRed, it has a relatively weak extinction coefficient, low quantum yield, and poor photo-stability (Campbell *et al.*, 2002). A series of monomeric FPs with different red colors have been further developed from mRFP with improved characteristics (Shaner *et al.*, 2004). With this palette of red-shifted FPs, together with GFP and its derivatives, multiple signaling molecules can be fused with FPs with different colors and monitored simultaneously in the same cells (Giepmans *et al.*, 2006). These FPs with a variety of colors also provide a high degree of flexibility in choosing various donor/acceptor pairs for FRET studies.

2.2. FRET

The development of functional mutants of GFP and RFP, fluorescent small-molecule probes, and advanced fluorescent microscope have made possible the visualization, quantification, and manipulation of signaling transduction in live cells with high temporal and spatial resolutions (Tour *et al.*, 2003; Tsien, 1998; Zhang *et al.*, 2002). In particular, FRET has been extensively employed to monitor active cellular signaling cascades in live cells. FRET is a phenomenon of quantum mechanics. When two chromophores are in proximity and the emission spectrum of one chromophore (donor) overlaps the excitation spectrum of the other (acceptor), the excitation of the donor will cause a sufficient energy transfer to the acceptor and result in the emission from the acceptor. The efficiency of energy transfer (FRET) is dependent on the relative orientations and the distance between these two chromophores (Tsien, 1998). Hence, any change of the orientation/distance between the two chromophores will alter the FRET efficiency and change the acceptor/donor emission ratio. A wide range of FRET-based biosensors have been developed to visualize calcium concentration, protein–protein interactions, and kinase activities (Miyawaki *et al.*, 1999, 1997; Mochizuki *et al.*, 2001; Persechini and Cronk, 1999; Sato *et al.*, 2002; Ting *et al.*, 2001; Truong *et al.*, 2001; Violin *et al.*, 2003; Zhang *et al.*, 2001). Genetically encoded biosensors consisting of FPs as donors and acceptors are popular because of their convenience for delivery into cells and targeting to different subcellular compartments. Among these FPs, cyan fluorescence protein (CFP) and yellow fluorescence protein (YFP) are the most popular donor/acceptor pair for FRET at the current stage because of their excellent extinction coefficients, quantum yield, and photo-stability (Tsien, 1998). Many biosensors based on CFP and YFP have been successfully developed to monitor various kinase activities in live cells with high spatial and temporal resolutions (Sato *et al.*, 2002; Ting *et al.*, 2001; Zhang *et al.*, 2001).

3. INTEGRIN SIGNALING

3.1. Integrin activation

Cell adhesion to the ECM is mediated by integrins, which consist of a large family (>20) of heterodimeric glycoprotein receptors, with each composed of an α subunit associated with a β subunit (Hynes and Lander, 1992). Each subunit contains an extracellular domain, a single transmembrane domain, and a short cytoplasmic domain. The large extracellular domain consists of an N-terminal globular headpiece and a long stalk proximal to the membrane. Integrins can transmit signals bidirectionally—"inside-out" and "outside-in." Inside-out signaling refers to the cytoplasmic tail–triggered regulation of integrin affinity for ligands, whereas outside-in signaling refers to the signaling cascades induced by extracellular ligand coupling (Ginsberg *et al.*, 2005). It has become clear that talin is an important molecule for inside-out signaling. Integrins are proposed to have two allosteric conformations (Carman and Springer, 2003). When integrins are in an inactive state, their long extracellular stalks are severely bent to form a V-shaped topology, such that the headpieces of α- and β-subunits are coupled together and are positioned proximal to the transmembrane domains, which are also connected through salt bridges. Upon activation, the phosphotyrosine binding (PTB) domain-like portion of talin can interact with the cytoplasmic tails of the integrin β-subunits.

This interaction induces a conformational change of cytoplasmic and transmembrane domain of the β subunit, which results in the disruption of salt bridges connecting α- and β-subunits. Integrins then undergo a "switch-blade–fashion" conformation change: the two extracellular stalks straighten up and cause the further separation of the transmembrane helices and cytoplasmic domains. This global conformational change leads to the activation of integrins with high affinity for ligands (Carman and Springer, 2003). This hypothesis is supported by the evidence that the suppression of talin expression by siRNA or the mutation of interaction domains on talin or integrin β subunit blocked the integrin activation (Tadokoro et al., 2003). For outside-in signaling, ligand ligation of integrin extracellular domains causes conformational changes that are propagated to transmembrane and cytoplasmic domains. The subsequent homotypical oligomerization of the transmembrane domains, together with integrin lateral diffusion, amplifies the signals and causes integrin clustering (Li et al., 2001, 2003; Wiseman et al., 2004). This results in the exposure of docking sites on the cytoplasmic domains and the subsequent recruitment of structural and signaling molecules, including talin, vinculin, α-actinin, and Src (Arias-Salgado et al., 2003). As such, integrins not only mechanically couple the cytoskeleton to the extracellular matrix, but also transmit molecular signaling cascades to regulate cellular functions.

3.2. Integrin-related signaling cascades

Integrin activation induces a wide range of signaling events (DeMali et al., 2003). Focal adhesion kinase (FAK) is a key molecule activated by integrin-mediated adhesion (Schlaepfer et al., 1999). Tyrosine 397 on FAK can be phosphorylated by integrin ligation and subsequently recruit Src family kinases (SFKs), which further induce a cascade of tyrosine phosphorylation and their related events. For example, SFKs will further phosphorylate tyrosine 576/577/925 on FAK to enhance FAK activity and create the binding site for Grb2 (Schlaepfer and Mitra, 2004). Paxillin and p130cas are also known substrate molecules phosphorylated by the FAK-SFKs complex (Abbi and Guan, 2002).

Rho small GTPase family members are another group of signaling molecules regulated by integrins. Small GTPase Rac and Cdc42 can be activated by integrins, in which Rac activation is dependent on the β-subunit of integrins (Berrier et al., 2002; Price et al., 1998). Another GTPase RhoA can be transiently inhibited by integrin engagement at the early phase and then activated at a later stage (Ren et al., 1999). Besides tyrosine kinases and Rho small GTPases, integrin engagement also regulates cytoskeleton and a variety of other signaling molecules, including actin (DeMali et al., 2002), microtubules (Bershadsky et al., 1996), tyrosine phosphatases (Maher, 1993; Sastry et al., 2002), serine/throenine protein kinase A (O'Connor and Mercurio, 2001), and protein kinase C (Disatnik et al., 2002).

4. FRET Analysis of Integrin Signaling

4.1. FRET visualization of integrin activation

Several studies have been conducted to visualize integrin activation by FRET in live cells. In general, CFP and YFP are fused to the subunits of integrins, and FRET has been utilized to measure the homo- or hetero-oligomerization between the subunits to assess the activation status of integrins. By fusing CFP to the $\beta2$ subunit and YFP to the αM subunit, Fu and colleagues (2006b) have demonstrated the hetero–oligomerization between $\beta2$ and αM subunits of integrin $\alpha M\beta2$. The same group has used FRET to further demonstrate homomeric interactions between $\beta2$ subunits (Fu et al., 2006a). By measuring the FRET between a rhodamine–containing lipid and a fluorescein small molecule, which binds to the tip of the extracellular domain of integrins, integrins $\alpha L\beta2$ and $\alpha4$ have been shown to have a bent extracellular domain when inactive and adopt extended forms upon activation (Chigaev et al., 2003; Larson et al., 2005). Springer's group fused CFP and YFP to the cytoplasmic tails of αL and $\beta2$ subunits of integrin $\alpha L\beta2$ (LFA-1), respectively. FRET between CFP and YFP was successfully applied to monitor the separation of cytoplasmic tails of αL and $\beta2$ subunits upon LFA-1 activation in live cells (Kim et al., 2003).

4.2. FRET visualization of the activation of integrin-related signaling molecules

Many FRET biosensors have been developed to monitor and analyze the temporal and spatial activation patterns of signaling molecules related to integrin signaling, including Src (Ting et al., 2001; Wang et al., 2005), small GTPases RhoA (Pertz et al., 2006; Yoshizaki et al., 2003), Rac1 (Itoh et al., 2002; Kraynov et al., 2000), Cdc42 (Itoh et al., 2002; Nalbant et al., 2004), PKA (Zhang et al., 2001, 2005), and PKC (Violin et al., 2003). In general, the biosensors are engendered by fusing CFP and YFP with either (1) the targeting molecules themselves, or (2) the pairing of a substrate peptide with a binding partner. The conformational changes of these biosensors are due to either (1) the structural alteration of the targeting molecules themselves, or (2) the binding adjustment between the substrate peptide and its partner upon the activation of targeting molecules. The changes of the emission ratio between CFP and YFP of these biosensors due to FRET are recorded and monitored to visualize and analyze the activation of these targeting molecules. In this article, we will specifically focus on a FRET-based biosensor for Src kinase and use it as an example to discuss the detailed designation strategies and experimental approaches for developing FRET biosensors.

4.2.1. The activation and functions of Src kinase

Src kinase belongs to the non–receptor tyrosine kinases family that consists of, sequentially, an N-terminal SH4 domain, a unique region, a SH3 domain, an SH2 domain, a catalytic domain, and a C-terminal regulatory sequence (Thomas and Brugge, 1997). In the inactive state, the SH3 and SH2 domains of Src kinase are coupled together by intramolecular interactions, and the catalytic kinase domain of Src is masked by its interaction with the C-terminal tail, thus preventing its action on substrate molecules. Several putative mechanisms by which the ligand-engagement of integrins leads to Src activation have been proposed (Arias-Salgado et al., 2003; Inoue et al., 2003; Kaplan et al., 1995):

1. The cytoplasmic tail of integrin $\beta 3$ may directly recruit the SH3 domain of Src for binding, facilitated by the myristoylation at the Src N-terminus. Such an interaction between integrin and Src causes a global conformational change of Src to turn on Src activity (Arias-Salgado et al., 2003).
2. Integrin engagement causes the myristoylation-mediated translocation of Src to focal adhesion sites (Thomas and Brugge, 1997). FAK Y397, in a high-affinity $_p$YAEI context, competes with Src $_p$Y527 for its intramolecular interaction and thus activates Src (Eide et al., 1995).
3. Recent evidence has shown that integrin-engagement leads to the association of integrin $\alpha v \beta 3$ with RPTPα, a well-characterized activator of Src family kinases (von Wichert et al., 2003). It is possible that integrins recruit RPTPα to dephosphorylate the Y527 on the C-terminal tail of Src and release it from the kinase domain, thus activating Src.

The activation of Src impacts on a variety of cellular functions, such as cell polarity, adhesion, focal adhesion assembly/disassembly, lamellipodia formation, and migration (Thomas and Brugge, 1997). For example, inhibition of Src causes impaired polarization toward migratory stimuli (Timpson et al., 2001). As the Src SH2 and SH3 domains are important in focal complex/adhesion assembly, the catalytic activity of Src induces adhesion disruption and turnover (Fincham et al., 2000; Kaplan et al., 1994). Src can also phosphorylate cortactin, an F-actin cross-linking protein enriched in cortical membrane ruffles (Wu and Parsons, 1993). The phosphorylated cortactin can associate and activate Arp2/3, a nucleator of actin filament polymerization (Weaver et al., 2001), to induce the growth of cortical actin network. Src, possibly through MAPK signaling pathway (Glading et al., 2001), also activates the calpain–calpastatin proteolytic system to cleave FAK and disrupt focal adhesion complex (Carragher et al., 2001, 1999, 2002). As a result, cell adhesion to extracellular matrix is reduced and cell motility can be enhanced. Furthermore, p190RhoGAP is a substrate for Src and an inhibitory molecule for RhoA. Src can phosphorylate p190RhoGAP and induce its binding to p120RasGAP (Belsches et al., 1997; Fincham et al., 1999). This action results in the inhibition of RhoA and subsequent dissolution of actin filaments, which ultimately lead to the inhibition of local contractile forces and lamellipodia formation (Arthur and Burridge, 2001; Arthur *et al.*, 2000; Settleman *et al.*, 1992). Recent evidence indicates that p130cas, a major substrate molecule for Src, is phosphorylated by Src. The phosphorylated p130cas can recruit Crk to form a complex with ELMO and DOCK180, a Rac GEF. The activated DOCK180 then causes Rac activation and the extension of lamellipodia (Brugnera *et al.*, 2002; Klemke *et al.*, 1998; Vuori *et al.*, 1996). Since Src has a multitude of functions, the ultimate result of Src activation is largely dependent on its dynamic nature and subcellular localization. Hence, it is crucial to monitor the Src activation with high spatiotemporal resolutions.

4.2.2. The development of FRET biosensors for Src kinase

Design strategies   The FRET technology allows the visualization of the Src activity in live cells with high spatiotemporal resolutions. A FRET biosensor for reporting Src kinase activity was initially developed in Roger Tsien's lab by Alice Ting (Ting *et al.*, 2001). This original Src reporter consists of an N-terminal–enhanced CFP with a truncated C-terminal tail (tECFP, 1–227) (Miyawaki *et al.*, 1997), an SH2 domain derived from Src kinase, a flexible linker (GSTSGSGKPGSGEGS) (Ting *et al.*, 2001), a substrate sequence (EIYGEF) that is selected through an *in vitro* peptide library screening method to be specifically phosphorylated by Src kinase (Songyang *et al.*, 1995), and a C-terminal citrine (a version of enhanced YFP [EYFP]) (Fig. 9.1). The truncated CFP (tECFP) and the SH2 domain are tightly locked in space and have very limited degrees of freedom relative to each other. Hence the relative distance/orientation among tECFP and citrine and the subsequent FRET signal is mainly determined by the flexible linker and the substrate region. It is therefore convenient to modify the linker and substrate peptide to alter and manipulate the FRET signals. The molecular activation mechanism of this Src reporter is hypothesized as follows: When the Src kinase is at its rest state, the tECFP and citrine are positioned in proximity as a consequence of the flexible linker and the juxtaposition of N- and C-terminals of the SH2 domain. Therefore, a strong FRET can occur and the excitation of tECFP at 433 nm leads to the emission of citrine at 527 nm. When Src kinase is activated, it phosphorylates the designed substrate peptide, which then displays a high affinity for and binds to the docking pocket of the SH2 domain. This action will lead to the separation of the citrine from the tECFP and decrease the FRET efficiency between these two FPs. The excitation of tECFP at

Figure 9.1   A schematic drawing of the Src reporter composition. (See color insert.)

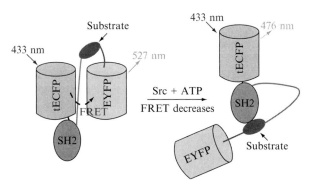

Figure 9.2 A schematic cartoon of the activation mechanism of the Src reporter. (See color insert.)

433 nm results in the emission from citrine at 476 nm, as shown in Fig. 9.2. Hence, the emission ratio of tECFP/citrine should serve as a good indicator of the status of Src activation.

The FRET level of this original Src reporter was found to be reduced *in vitro* by Src kinase and in cells by EGF which is known to activate Src kinase (Ting *et al.*, 2001). However, an inhibitor of Src kinase, PP1, and the mutation of the consensus tyrosine site in the substrate EIYGEF did not block the EGF-induced FRET change of this reporter. These results suggest that the FRET change of this Src reporter is not specifically due to Src activation, and the activation mechanism of this reporter is not clear. We suspected that the tyrosine sites in the SH2 domain (adjacent to the designed substrate peptide) may be phosphorylated by Src kinase and would thus cause the false-positive FRET signals. There are mainly four tyrosine sites (Y149, Y184, Y202, and Y229) positioned on the outer surface of the SH2 domain that are easily accessible to upstream kinases. We have mutated each of these four tyrosine sites to phenylalanine individually (#2 to 5 in Table 9.1, for Y149F, Y184F, Y202F, and Y229F, respectively). The further mutation (Y→F) of the tyrosine site in the substrate peptide of the Y202F mutant reporter blocked the EGF-induced FRET response, suggesting that the EGF-induced FRET response of the Y202F mutant reporter is due to the tyrosine phosphorylation of its substrate peptide. Further experiments revealed that 10 μM PP1 also blocked the FRET response of this Y202F mutant reporter. Hence, this mutant Y202F reporter appears to be able to report Src activity with a clear activation mechanism and high specificity. However, the FRET change of this modified reporter (Y202F) decreased to around 10% upon stimulation, in comparison to 20% of the reporter with the wildtype SH2 domain. We hence started a series of experiments to improve the sensitivity of this mutant reporter. We have made further mutations on the SH2 domain besides Y202F mutation and altered the reporter linker length by inserting or deleting amino acids

Table 9.1 Various Src biosensors with tECFP positioned at the N-termini and citrine positioned at the C-termini

	tECFP	SH2	Substrate	Linker	Citrine	FRET[a]
1	WT	WT	EIYGEF	GSTSGSGKPGSGEGS	WT	++
2	WT	Y149F	EIYGEF	GSTSGSGKPGSGEGS	WT	++
3	WT	Y184F	EIYGEF	GSTSGSGKPGSGEGS	WT	++
4	WT	Y202F	EIYGEF	GSTSGSGKPGSGEGS	WT	+
5	WT	Y229F	EIYGEF	GSTSGSGKPGSGEGS	WT	++
6	WT	Y202F Y213F	EIYGEF	GSTSGSGKPGSGEGS	WT	—
7	WT	Y202F Y149F Y184F Y229F	EIYGEF	GSTSGSGKPGSGEGS	WT	+
8	WT	Y202F	EIYGEF	GSTSGSGKPGSGEGS-GGSGGT	WT	+
9	A206K	Y202F	EIYGEF	GSTSGSGKPGSGEGS-GGSGGT	A206K	-
10	WT	Y202F	EIYGEF	SGSGKPGSGEGS	WT	-
11	WT	Y202F	EIYGEF	GSTSGSGKPGSGEGS-TKGGGTGGS	WT	-

#	WT	Y202F	GGSGGT GSTSGSGKPGSGEGS-TKGGGTGGS	EIYGEF	WT	
12	WT	WT			WT	–
13[a]	WT	WT	GSTSGSGKPGSGEGS	W/MEDYDYVHLQG (p130cas)	WT	+++
14	WT	WT	GSTSGSGKPGSGEGS	EETPYSYPTGN (paxillin)	WT	+
15	WT	WT	GSTSGSGKPGSGEGS	DEGIYDVPLLGP (sin)	WT	+
16	WT	WT	GSTSGSGKPGSGEGS	AQDIYQVPPSAG (p130cas)	WT	+
17[a]	A206K	WT	GSTSGSGKPGSGEGS	W/MEDYDYVHLQG	A206K	++++
18	A206K	WT	GSTSGSGKPGSGEGS	DEGIYDVPLLGP	A206K	+
19	A206K	WT	GSTSGSGKPGSGEGS	AQDIYQVPPSAG	A206K	+
20	A206K	WT	GSTSGSGKPGSGEGS	PQDIYDVPPVRG (p130cas)	A206K	+
21	A206K	WT	GSTSGSGKPGSGEGS	LLEVYDVPPSEV (p130cas)	A206K	+

[a] The dynamic ranges of FRET responses of various Src biosensors in live HeLa cells are graded with ++++ as the best (~40%) and — as the worst (~0%), with ++++ > +++ > ++ > + > - - - > —.

(Table 9.1, #6 to 8, #10 to 12). We also mutated the Alanine 206 to lysine in tECFP and citrine to eliminate the weak dimerization tendency of these FPs (Zacharias *et al.*, 2002) (Table 9.1, #9). None of these further mutations or modifications resulted in an improved sensitivity of this Y202F mutant reporter. We then changed the construction strategy by fusing the SH2 domain in the N-terminus, followed by tECFP, linker, citrine, and substrate peptide (Fig. 9.3) (Table 9.2, #1). It was expected that when this new Src reporter is not activated, FRET would be weak and the excitation of tECFP at 433 nm would result in its emission at 476 nm. When Src kinase is activated to phosphorylate the substrate peptide of the reporter, the phosphorylated substrate peptide would bind to SH2 domain and position citrine next to tECFP to enhance FRET, that is, the excitation of tECFP at 433 nm would then lead to the emission from citrine at 527 nm, as shown in Fig. 9.4. We also swapped the positions of citrine and tECFP in this reporter, and

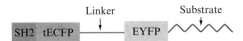

Figure 9.3 A schematic drawing of the Src reporter with tECFP and EYFP positioned between the SH2 domain and the substrate peptide and flanking the linker. (See color insert.)

Table 9.2 Various Src biosensors with tECFP and citrine positioned in the center

	SH2	FP1	Linker	FP2	Substrate	FRET[a]
1	WT	tECFP	GSTSGSGKP-GSGEGS	Citrine	EIYGEF	—
2	WT	Citrine	GSTSGSGKP-GSGEGS	tECFP GIT	EIYGEF	—
3	WT	Citrine	GSTSGSGKP-GSGEGS	tECFP GITLGMD	EIYGEF	—

[a] The dynamic ranges of FRET responses of various Src biosensors in live HeLa cells are graded with ++++ as the best (∼40%) and — as the worst (∼0%), with ++++ > +++ > ++ > + > - - - > —.

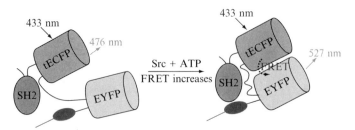

Figure 9.4 A schematic cartoon of the activation mechanism of the Src reporter with tECFP and EYFP positioned between the SH2 domain and the substrate peptide and flanking the linker. (See color insert.)

added different amino acids at the C-terminal tail of tECFP to provide more spatial flexibility for the substrate peptide to bind to the docking pocket of SH2 domain (Table 9.2, #2 to 3). Although all three new reporters displayed increased FRET changes upon Src addition during *in vitro* kinase assay, these reporters did not display the expected FRET changes upon EGF stimulation when they were expressed in HeLa cells.

We accidentally observed that the original Src biosensor (Table 9.1, #1) can undergo a large (~40%) FRET change upon the application of pervanadate, a general inhibitor of phosphatases. This result suggests that the design strategy of this original Src reporter can allow a large dynamic range of FRET change when the substrate peptide is phosphorylated by an appropriate kinase. This prompted us to suspect that the substrate peptide selected through the *in vitro* library screening method may not be an efficient substrate for Src kinase in cells. We therefore sought a better substrate peptide for the Src reporter. A series of peptides derived from Src substrate molecules, for example, p130cas, paxillin, and Sin, were examined (Table 9.1, #13 to 16). When the consensus substrate peptide WMEDYDYVHLQG is derived from p130cas (Table 9.1, #13), the reporter underwent a significant change of FRET (>30%) upon EGF stimulation for HeLa cells. We therefore decided to choose this biosensor as our first generation of the new Src biosensor (SR 1.0). In the latter part of this article, the Src biosensor only refers to this SR 1.0 unless otherwise specified.

Characterization of the Src biosensor in vitro

We first examined the FRET change of the Src biosensor (SR 1.0) in response to Src kinase *in vitro*. The gene sequence encoding the Src biosensor was subcloned into pRSETB vector using *Bam*HI and *Eco*RI sites for bacterial expression (Invitrogen, San Diego, CA). The chimeric Src biosensor proteins were expressed as N-terminal His$_6$ tag fusions in *Escherichia coli* JM109 (DE3), which was grown on LB plates containing ampicillin (0.1 mg/ml). Two milliliters of overnight culture were used to inoculate 100 ml of medium. The bacteria were grown at 37° to an optical density of 0.2 to 0.4 at 600 nm as measured by a photospectrometer. The bacteria were then diluted into 500 ml with a medium containing 0.1 mg/ml of ampicillin and 0.4 mM of isopropylthiogalactoside, and incubated overnight at room temperature. Cells were harvested by centrifugation and resuspended in 10 ml of B-PER II protein extraction reagent (Pierce, Rockford, IL) containing protease inhibitors (leupeptin, pepstatin A, and phenylmethylsulfonyl fluoride). The lysate was incubated and shaken for 10 min at room temperature and then subjected to centrifugation at 15,000 g at 4° for 10 min. The supernatant was then passed through 0.45-μm filter before being incubated and mixed with 1 ml of 50% (v/v) Ni-NTA slurry (Qiagen, Valencia, CA) at 4° for 1 h. The protein-Ni-NTA solution was loaded on a column and washed three times with 10-ml binding buffer (300 mM NaCl, 50 mM Tris-HCl, pH 7.4). The Src biosensor proteins were eluted with an elution buffer (300 mM NaCl, 100 mM imidazole, 50 mM Tris-HCl, pH 7.4), after washing for three times with 10 ml of the binding buffer (300 mM NaCl, 10 mM imidazole, 50 mM Tris-HCl, pH 7.4). Imidazole was removed by subjecting the eluted protein solution to dialysis in 1 liter of a kinase buffer (50 mM Tris, pH 8.0, 100 mM NaCl, 10 mM MgCl$_2$, 2 mM DTT) at 4° overnight. The protein concentration of the solution was obtained by determining the absorbance spectrum in a spectrophotometer (Fig. 9.5). Two biosensor protein solutions (10 μl each) were diluted in the SDS loading buffer (2% (w/v) SDS, 10% (w/v) glycerol, 0.1% (w/v) bromphenol blue, 300 mM 2-mercaptoethanol, 62.5 mM Tris-HCl, pH 6.8), with one of them further boiled. These two biosensor protein solutions were then subjected to SDS gel electrophoresis. As shown in Fig. 9.6A, boiling linearized the biosensor protein and caused an up-shift of the gel band. When this gel was excited at 360 nm, a strong fluorescence signal was emitted from the nonboiled biosensor band, but there was no fluorescence signal from the boiled band (Fig. 9.6B). Hence, boiling linearized the biosensor and destroyed the chromophores of its FPs. The majority of the purified proteins were biosensors in their native form with the correct fluorescence. Some minor bands remained after the purification process, possibly representing the isoforms of nonspecific protein EF–Tu (the most abundant protein in *E. coli*) (Fig. 9.6A). The purified biosensor protein was then subjected to an *in vitro* Src kinase assay at 37°, by measuring the fluorescence spectra of the biosensor in the kinase buffer with a fluorescence plate reader (TECAN Safire, with emission spectra scan from 470 nm to 550 nm at excitation wavelength 433 nm) for the detection of changes in FRET before and after the addition of kinases. The addition of Src kinase (1 μM) and ATP (1 mM) to the Src biosensor solution led to a decrease in citrine emission and an increase in tECFP emission (Fig. 9.7). This resulted in a

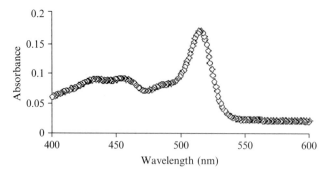

Figure 9.5 The absorbance spectrum of the eluted biosensor protein solution. The absorbance peaks at 440 nm and 510 nm indicate the abundant presence of tECFP and citrine, respectively.

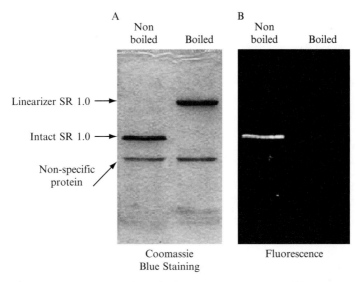

Figure 9.6 (A) The gel bands of purified protein solutions stained by Coomassie blue to illustrate the main types of proteins in the solution. Note that the biosensor protein is up-shifted in size after boiling. (B) The gel from (A) was excited at 360 nm and a fluorescence band can be clearly observed from the nonboiled native biosensor protein, but not the boiled group.

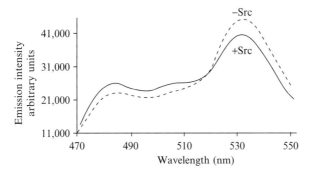

Figure 9.7 Emission spectra of the Src biosensor before (dotted line) and after (solid line) Src kinase addition in an *in vitro* kinase assay.

25% of FRET loss of the biosensor, whereas the addition of other kinases (including Yes, FAK, EGFR, Abl, Jak2, or ERK1) only caused insignificant FRET change (<2%). Fyn, another member of Src family kinases, caused a moderate 10% FRET change of the biosensor (Fig. 9.8). All these results indicate that the Src biosensor is capable of reporting Src activity with specificity *in vitro*.

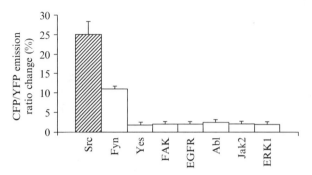

Figure 9.8 The emission ratio change of tECFP/citrine of the Src biosensor when treated with various kinases as indicated.

Characterization of the Src biosensor in live cells   For expression in mammalian cells, such as HeLa cells, the chimeric Src biosensor was subcloned in frame into pcDNA3 (Invitrogen) behind a Kozak sequence using the *Bam*HI and *Eco*RI sites. The biosensor DNAs were introduced into HeLa cells for expression with the lipofectamine method. Cells were passed the night before the transfection so that the cell density can reach about 80% confluency when transfected. One microgram of DNA was gently mixed with 2 μl of lipofectamine and incubated for 20 min before adding to cells on an area of 10 cm^2. After 6.5 h of incubation, the cells were starved in a 0.5% FBS medium for 36 h before being subjected to imaging and stimulation. HeLa cells transfected with the Src biosensor were treated with 50 ng/ml EGF, which is known to activate Src kinase (Belsches *et al.*, 1997). EGF induced a significant change in the emission ratio of the biosensor, which was propagated from the plasma membrane to the nucleus (Fig. 9.9). A series of mutants were constructed to test whether the FRET response of the Src biosensor emanates from the designed intramolecular interaction between the SH2 domain and substrate peptide. Mutations of the putative Src phosphorylation sites at tyrosine 662 and 664 in the substrate peptide, either individually or in combination, blocked the EGF-induced FRET response in HeLa cells. A charge-silent mutation of Arginine 175, a residue in the SH2 domain forming an ion pair with phosphate group in the bound phosphorylated peptide, to eliminate its binding capability to phosphorylated peptides (Waksman *et al.*, 1993) resulted in the loss of response to EGF. These results validate the phosphorylation–induced intramolecular interaction as the mechanism for the observed FRET response. Immunoblotting revealed that EGF-induced tyrosine phosphorylation of the Src biosensor was abolished by the double mutations of both tyrosine 662 and 664, confirming that Src-directed tyrosine phosphorylation occurred at the designed substrate peptide. The disruption of SH2 domain by R175V mutation also blocked the EGF-induced tyrosine phosphorylation of the Src

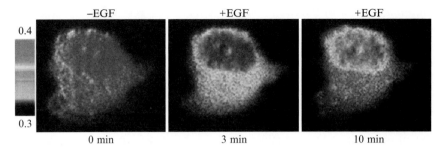

Figure 9.9 EGF induced a FRET change of the Src biosensor. HeLa cells were transfected with the Src biosensor and stimulated with EGF (50 ng/ml) for various time periods as indicated. The color scale bar on the left represents the tECFP/citrine emission ratio, with cold colors indicating low Src activity and hot colors indicating high levels of Src activation. The color images represent the emission ratio maps of the Src biosensor in the cell. (See color insert.)

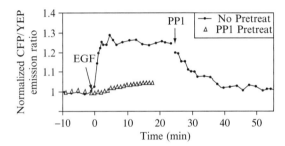

Figure 9.10 Time courses of the normalized tECFP/citrine emission ratio of the Src biosensor in response to EGF in HeLa cells with PP1 pretreatment (PP1 Pretreated) or without PP1 (No Pretreat).

biosensor. Hence the SH2 domain of the Src biosensor may help the binding to the activated Src and facilitate the phosphorylation process.

Experiments were performed to test the specificity of the Src biosensor in live cells. The EGF-induced FRET response can be reversed by the addition of PP1, a specific inhibitor of Src family tyrosine kinases. When HeLa cells were treated with PP1, EGF only induced a weak change (~5%) in emission ratio (Fig. 9.10), suggesting that the Src biosensor specifically reports the Src family activity in live cells. To further differentiate the different members in Src family, the Src biosensor was transfected into wildtype or Src/Yes/Fyn triple-knockout mouse embryonic fibroblasts (SYF −/−). The normal PDGF-induced FRET response of the Src biosensor was ablated in SYF −/− cells. The introduction of c-Src, but not c-Fyn or a kinase-dead mutant of c-Src, into these SYF −/− cells fully restored the FRET response. These results indicate that the Src biosensor has a high specificity toward c-Src in a mammalian cell line.

Monomerization of the Src biosensor effect   It was recently reported that CFPs and YFPs have weak affinity toward each other. To eliminate the potential intermolecular FRET elicited from unintended dimerization, A206K mutations have been introduced in CFP and YFP to generate monomeric mCFP and mYFP (Zacharias *et al.*, 2002). We introduced A206K into both tECFP and citrine by using a Quickchange kit (Stratagene, La Jolla, CA) to examine whether this monomerization of FPs can improve the FRET dynamic range of the Src biosensor. *In vitro* kinase assay showed that the monomerized Src biosensor did have a higher sensitivity toward Src kinase, with an increase in the change in emission ratio from ~30 to 40%. We reasoned that the knockdown of the sticky interface between tECFP and citrine may release the tension between them, thus resulting in a position more favorable for FRET changes. EGF induced a rapid FRET response in HeLa cells, which is reversible by the washout of EGF. A second EGF application at a later time induced a FRET response with a lower magnitude, possibly due to the nonrecovered EGF receptor level on the plasma membrane following endocytosis triggered by the first EGF application (Fig. 9.11). We have also introduced A206K into other versions of Src biosensors with various substrate peptide sequences to examine whether this monomerization process can enhance the sensitivity of biosensors in general (Table 9.1) (#18 to #21). The monomerization did not have any observable effect on the other biosensors, suggesting that the FP monomerization did not affect the specificity of the Src biosensors.

Targeting of the Src biosensor to the plasma membrane   The translocation of Src to the plasma membrane is a prerequisite for its activation (Thomas and Brugge, 1997). To increase the local concentration of the biosensor and to position it close to the active Src, we have genetically modified the Src biosensor by fusing a small peptide derived from Lyn kinase to the N-terminus of Src biosensor so that the biosensor can be tethered to the lipid raft domains in the plasma membrane (Fig. 9.12A). The membrane-targeted tECFP was constructed by PCR amplification of the monomeric tECFP with a sense primer containing the codes for 16 N-terminal amino acids from Lyn kinase[18] to produce a membrane-targeted monomeric tECFP. This membrane-targeted tECFP was then cloned to replace the tECFP in the monomerized Src biosensor to generate the membrane-targeted Src biosensor. The FRET response of this membrane-targeted Src biosensor can be induced by the Src activator EGF and suppressed by the Src-specific inhibitor PP1 (Fig. 9.12B). These results indicate that the membrane-targeted Src biosensor is a sensitive and specific indicator, capable of reporting Src activity at the plasma membrane.

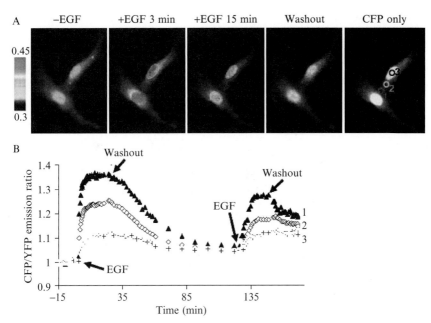

Figure 9.11 The FRET response of the monomeric Src biosensor is reversible. HeLa cells transfected with the monomerized Src biosensor were stimulated with EGF (50 ng/ml), washed with serum-free medium (washout), and subsequently subjected to EGF and washout for a second time. (A) The tECFP/citrine emission ratio images of the monomeric Src biosensor. The color scale bar on the left represents the tECFP/citrine emission ratio, with cold colors indicating low Src activity and hot colors indicating high levels of Src activation. The representative emission ratio images (control, 3 and 15 min after EGF, and after washout) are shown in color on the left, and the CFP-only image is shown in black and white on the far right. (B) Time courses of the average tECFP/citrine emission ratios of the monomeric Src biosensor from different regions of cells in (A). (See color insert.)

Retrovirus construction and primary cell infection The traditional transfection method to deliver DNA into mammalian cells by lipofectamine resulted in a low efficiency in primary endothelial cells, particularly human umbilical vein endothelial cells (HUVECs). To enhance transfection efficiency, the RetroMax retroviral expression system (Imgenex, San Diego, CA) was used to introduce genes into HUVECs. To generate retrovirus expression plasmids, the Src biosensor and its mutants were inserted into the *Hind*III/*Xho*I restriction sites of the PCLNCX retrovirus expression vector (Imgenex, San Diego, CA). Various Src biosensor constructs incorporated in the PCLNCX vector were co-transfected with PCL-Ampho packaging vector into *293* cells by using lipofectamine. The medium of *293* cells was collected after 2 days of

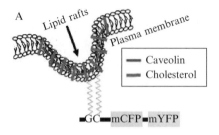

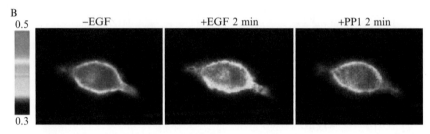

Figure 9.12 (A) A schematic drawing of the monomeric Src biosensor targeted to the plasma membrane. (B) The tECFP/citrine emission ratio images. HeLa cells transfected with the membrane-targeted Src biosensor were stimulated with EGF (50 ng/ml) and subsequently incubated with PP1 for the time periods as indicated. The color scale bar on the left represents the tECFP/citrine emission ratio, with cold colors indicating low Src activity and hot colors indicating high levels of Src activation. (See color insert.)

transfection and subjected to centrifugation for 10 min at $3000 \times g$. The retrovirus supernatant was filtered through 0.45-μm pore filters and immediately frozen at $-80°$ in aliquots. HUVECs were incubated and infected with 1:1 mixture of fresh endothelial growth medium, and the collected retrovirus supernatant in the presence of 8 μg/ml polybrene for 48 h before subjected to imaging. With this retrovirus method, 5 to 10% of HUVECs were consistently transfected in each experiment.

 ## 5. FUTURE DIRECTIONS

At the present time, the construction and optimization of ratiometric FRET-based biosensors is rather semi-rational. In particular, the improvement of biosensor specificity in cells is a challenging task. Successes in developing a biosensor with specificity have often been achieved after numerous trials and errors. With the rapid improvement and development of technologies such as fluorescence-activated cell sorter and directed DNA evolution strategies, a high-throughput method to screen and select an optimized biosensor out of a huge library has become possible. It is envisioned that the development of large-scale computational simulation of molecular dynamics can also help to predict the 3D orientations of biosensors and facilitate the optimization process. Recent developments of nanotechnology in single-cell sorting and identification may also lead to convenient and efficient screening assays for biosensor optimization.

ACKNOWLEDGMENTS

This work is supported by the National Institutes of Health (grants HL-064382, HL-080518, and HL-085159) (SC) and the Wallace H. Coulter Foundation (YW).

REFERENCES

Abbi, S., and Guan, J. L. (2002). Focal adhesion kinase: Protein interactions and cellular functions. *Histol. Histopathol.* **17,** 1163–1171.

Arias-Salgado, E. G., Lizano, S., Sarkar, S., Brugge, J. S., Ginsberg, M. H., and Shattil, S. J. (2003). Src kinase activation by direct interaction with the integrin beta cytoplasmic domain. *Proc. Natl. Acad. Sci. USA* **100,** 13298–13302.

Arthur, W. T., and Burridge, K. (2001). RhoA inactivation by p190RhoGAP regulates cell spreading and migration by promoting membrane protrusion and polarity. *Mol. Biol. Cell* **12,** 2711–2720.

Arthur, W. T., Petch, L. A., and Burridge, K. (2000). Integrin engagement suppresses RhoA activity via a c-Src-dependent mechanism. *Curr. Biol.* **10,** 719–722.

Baird, G. S., Zacharias, D. A., and Tsien, R. Y. (2000). Biochemistry, mutagenesis, and oligomerization of DsRed, a red fluorescent protein from coral. *Proc. Natl. Acad. Sci. USA* **97,** 11984–11989.

Belsches, A. P., Haskell, M. D., and Parsons, S. J. (1997). Role of c-Src tyrosine kinase in EGF-induced mitogenesis. *Front. Biosci.* **2,** d501–d518.

Berrier, A. L., Martinez, R., Bokoch, G. M., and LaFlamme, S. E. (2002). The integrin beta tail is required and sufficient to regulate adhesion signaling to Rac1. *J. Cell Sci.* **115,** 4285–4291.

Bershadsky, A., Chausovsky, A., Becker, E., Lyubimova, A., and Geiger, B. (1996). Involvement of microtubules in the control of adhesion-dependent signal transduction. *Curr. Biol.* **6,** 1279–1289.

Brugnera, E., Haney, L., Grimsley, C., Lu, M., Walk, S. F., Tosello-Trampont, A. C., Macara, I. G., Madhani, H., Fink, G. R., and Ravichandran, K. S. (2002). Unconventional Rac-GEF activity is mediated through the Dock180–ELMO complex. *Nat. Cell Biol.* **4,** 574–582.

Campbell, R. E., Tour, O., Palmer, A. E., Steinbach, P. A., Baird, G. S., Zacharias, D. A., and Tsien, R. Y. (2002). A monomeric red fluorescent protein. *Proc. Natl. Acad. Sci. USA* **99,** 7877–7882.

Carman, C. V., and Springer, T. A. (2003). Integrin avidity regulation: Are changes in affinity and conformation underemphasized? *Curr. Opin. Cell Biol.* **15,** 547–556.

Carragher, N. O., Fincham, V. J., Riley, D., and Frame, M. C. (2001). Cleavage of focal adhesion kinase by different proteases during SRC-regulated transformation and apoptosis. Distinct roles for calpain and caspases. *J. Biol. Chem.* **276,** 4270–4275.

Carragher, N. O., Levkau, B., Ross, R., and Raines, E. W. (1999). Degraded collagen fragments promote rapid disassembly of smooth muscle focal adhesions that correlates with cleavage of pp125(FAK), paxillin, and talin. *J. Cell Biol.* **147,** 619–630.

Carragher, N. O., Westhoff, M. A., Riley, D., Potter, D. A., Dutt, P., Elce, J. S., Greer, P. A., and Frame, M. C. (2002). v-Src–induced modulation of the calpain–calpastatin proteolytic system regulates transformation. *Mol. Cell Biol.* **22,** 257–269.

Chigaev, A., Buranda, T., Dwyer, D. C., Prossnitz, E. R., and Sklar, L. A. (2003). FRET detection of cellular alpha4-integrin conformational activation. *Biophys. J.* **85,** 3951–3962.

DeMali, K. A., Barlow, C. A., and Burridge, K. (2002). Recruitment of the Arp2/3 complex to vinculin: Coupling membrane protrusion to matrix adhesion. *J. Cell Biol.* **159,** 881–891.

DeMali, K. A., Wennerberg, K., and Burridge, K. (2003). Integrin signaling to the actin cytoskeleton. *Curr. Opin. Cell Biol.* **15,** 572–582.

Disatnik, M. H., Boutet, S. C., Lee, C. H., Mochly-Rosen, D., and Rando, T. A. (2002). Sequential activation of individual PKC isozymes in integrin-mediated muscle cell spreading: A role for MARCKS in an integrin signaling pathway. *J. Cell Sci.* **115,** 2151–2163.

Eide, B. L., Turck, C. W., and Escobedo, J. A. (1995). Identification of Tyr-397 as the primary site of tyrosine phosphorylation and pp60src association in the focal adhesion kinase, pp125FAK. *Mol. Cell Biol.* **15,** 2819–2827.

Fincham, V. J., Brunton, V. G., and Frame, M. C. (2000). The SH3 domain directs acto-myosin-dependent targeting of v-Src to focal adhesions via phosphatidylinositol 3-kinase. *Mol. Cell Biol.* **20,** 6518–6536.

Fincham, V. J., Chudleigh, A., and Frame, M. C. (1999). Regulation of p190 Rho-GAP by v-Src is linked to cytoskeletal disruption during transformation. *J. Cell Sci.* **112,** 947–956.

Fu, G., Wang, C., Wang, G. Y., Chen, Y. Z., He, C., and Xu, Z. Z. (2006a). Detection of constitutive homomeric associations of the integrins Mac-1 subunits by fluorescence resonance energy transfer in living cells. *Biochem. Biophys. Res. Commun.* **351,** 847–852.

Fu, G., Yang, H. Y., Wang, C., Zhang, F., You, Z. D., Wang, G. Y., He, C., Chen, Y. Z., and Xu, Z. Z. (2006b). Detection of constitutive heterodimerization of the integrin Mac-1 subunits by fluorescence resonance energy transfer in living cells. *Biochem. Biophys. Res. Commun.* **346,** 986–991.

Giepmans, B. N., Adams, S. R., Ellisman, M. H., and Tsien, R. Y. (2006). The fluorescent toolbox for assessing protein location and function. *Science* **312,** 217–224.

Ginsberg, M. H., Partridge, A., and Shattil, S. J. (2005). Integrin regulation. *Curr. Opin. Cell Biol.* **17,** 509–516.

Glading, A., Uberall, F., Keyse, S. M., Lauffenburger, D. A., and Wells, A. (2001). Membrane proximal ERK signaling is required for M-calpain activation downstream of epidermal growth factor receptor signaling. *J. Biol. Chem.* **276,** 23341–23348.

Hynes, R. O. (1987). Integrins: A family of cell surface receptors. *Cell* **48,** 549–554.

Hynes, R. O. (1999). Cell adhesion: Old and new questions. *Trends Cell Biol.* **9,** M33–M37.

Hynes, R. O., and Lander, A. D. (1992). Contact and adhesive specificities in the associations, migrations, and targeting of cells and axons. *Cell* **68,** 303–322.

Inoue, O., Suzuki-Inoue, K., Dean, W. L., Frampton, J., and Watson, S. P. (2003). Integrin alpha2beta1 mediates outside-in regulation of platelet spreading on collagen through activation of Src kinases and PLCgamma2. *J. Cell Biol.* **160,** 769–780.

Itoh, R. E., Kurokawa, K., Ohba, Y., Yoshizaki, H., Mochizuki, N., and Matsuda, M. (2002). Activation of rac and cdc42 video imaged by fluorescent resonance energy transfer-based single-molecule probes in the membrane of living cells. *Mol. Cell Biol.* **22,** 6582–6591.

Kaplan, K. B., Bibbins, K. B., Swedlow, J. R., Arnaud, M., Morgan, D. O., and Varmus, H. E. (1994). Association of the amino-terminal half of c-Src with focal adhesions alters their properties and is regulated by phosphorylation of tyrosine 527. *EMBO J.* **13,** 4745–4756.

Kaplan, K. B., Swedlow, J. R., Morgan, D. O., and Varmus, H. E. (1995). c-Src enhances the spreading of src-/- fibroblasts on fibronectin by a kinase-independent mechanism. *Genes Dev.* **9**, 1505–1517.

Kim, M., Carman, C. V., and Springer, T. A. (2003). Bidirectional transmembrane signaling by cytoplasmic domain separation in integrins. *Science* **301**, 1720–1725.

Klemke, R. L., Leng, J., Molander, R., Brooks, P. C., Vuori, K., and Cheresh, D. A. (1998). CAS/Crk coupling serves as a "molecular switch" for induction of cell migration. *J. Cell Biol.* **140**, 961–972.

Kraynov, V. S., Chamberlain, C., Bokoch, G. M., Schwartz, M. A., Slabaugh, S., and Hahn, K. M. (2000). Localized Rac activation dynamics visualized in living cells. *Science* **290**, 333–337.

Larson, R. S., Davis, T., Bologa, C., Semenuk, G., Vijayan, S., Li, Y., Oprea, T., Chigaev, A., Buranda, T., Wagner, C. R., and Sklar, L. A. (2005). Dissociation of I domain and global conformational changes in LFA-1: Refinement of small molecule-I domain structure-activity relationships. *Biochemistry* **44**, 4322–4331.

Li, R., Babu, C. R., Lear, J. D., Wand, A. J., Bennett, J. S., and DeGrado, W. F. (2001). Oligomerization of the integrin alphaIIbbeta3: Roles of the transmembrane and cytoplasmic domains. *Proc. Natl. Acad. Sci. USA* **98**, 12462–12467.

Li, R., Mitra, N., Gratkowski, H., Vilaire, G., Litvinov, R., Nagasami, C., Weisel, J. W., Lear, J. D., DeGrado, W. F., and Bennett, J. S. (2003). Activation of integrin alphaIIb-beta3 by modulation of transmembrane helix associations. *Science* **300**, 795–798.

Maher, P. A. (1993). Inhibition of the tyrosine kinase activity of the fibroblast growth factor receptor by the methyltransferase inhibitor 5′-methylthioadenosine. *J. Biol. Chem.* **268**, 4244–4249.

Matz, M. V., Fradkov, A. F., Labas, Y. A., Savitsky, A. P., Zaraisky, A. G., Markelov, M. L., and Lukyanov, S. A. (1999). Fluorescent proteins from nonbioluminescent *Anthozoa* species. *Nat. Biotechnol.* **17**, 969–973.

Miyawaki, A., Griesbeck, O., Heim, R., and Tsien, R. Y. (1999). Dynamic and quantitative Ca^{2+} measurements using improved cameleons. *Proc. Natl. Acad. Sci. USA* **96**, 2135–2140.

Miyawaki, A., Llopis, J., Heim, R., McCaffery, J. M., Adams, J. A., Ikura, M., and Tsien, R. Y. (1997). Fluorescent indicators for Ca2+ based on green fluorescent proteins and calmodulin. *Nature* **388**, 882–887.

Mochizuki, N., Yamashita, S., Kurokawa, K., Ohba, Y., Nagai, T., Miyawaki, A., and Matsuda, M. (2001). Spatio-temporal images of growth-factor-induced activation of Ras and Rap1. *Nature* **411**, 1065–1068.

Nalbant, P., Hodgson, L., Kraynov, V., Toutchkine, A., and Hahn, K. M. (2004). Activation of endogenous Cdc42 visualized in living cells. *Science* **305**, 1615–1619.

O'Connor, K. L., and Mercurio, A. M. (2001). Protein kinase A regulates Rac and is required for the growth factor-stimulated migration of carcinoma cells. *J. Biol. Chem.* **276**, 47895–47900.

Persechini, A., and Cronk, B. (1999). The relationship between the free concentrations of Ca2+ and Ca2+-calmodulin in intact cells. *J. Biol. Chem.* **274**, 6827–6830.

Pertz, O., Hodgson, L., Klemke, R. L., and Hahn, K. M. (2006). Spatiotemporal dynamics of RhoA activity in migrating cells. *Nature* **440**, 1069–1072.

Prasher, D. C., Eckenrode, V. K., Ward, W. W., Prendergast, F. G., and Cormier, M. J. (1992). Primary structure of the Aequorea victoria green-fluorescent protein. *Gene* **111**, 229–233.

Price, L. S., Leng, J., Schwartz, M. A., and Bokoch, G. M. (1998). Activation of Rac and Cdc42 by integrins mediates cell spreading. *Mol. Biol. Cell* **9**, 1863–1871.

Ren, X. D., Kiosses, W. B., and Schwartz, M. A. (1999). Regulation of the small GTP-binding protein Rho by cell adhesion and the cytoskeleton. *EMBO J.* **18**, 578–585.

Sastry, S. K., Lyons, P. D., Schaller, M. D., and Burridge, K. (2002). PTP-PEST controls motility through regulation of Rac1. *J. Cell Sci.* **115,** 4305–4316.

Sato, M., Ozawa, T., Inukai, K., Asano, T., and Umezawa, Y. (2002). Fluorescent indicators for imaging protein phosphorylation in single living cells. *Nat. Biotechnol.* **20,** 287–294.

Schlaepfer, D. D., Hauck, C. R., and Sieg, D. J. (1999). Signaling through focal adhesion kinase. *Prog. Biophys. Mol. Biol.* **71,** 435–478.

Schlaepfer, D. D., and Mitra, S. K. (2004). Multiple connections link FAK to cell motility and invasion. *Curr. Opin. Genet. Dev.* **14,** 92–101.

Settleman, J., Albright, C. F., Foster, L. C., and Weinberg, R. A. (1992). Association between GTPase activators for Rho and Ras families. *Nature* **359,** 153–154.

Shaner, N. C., Campbell, R. E., Steinbach, P. A., Giepmans, B. N., Palmer, A. E., and Tsien, R. Y. (2004). Improved monomeric red, orange and yellow fluorescent proteins derived from *Discosoma* sp. red fluorescent protein. *Nat. Biotechnol.* **22,** 1567–1572.

Shimomura, O., Johnson, F. H., and Saiga, Y. (1962). Extraction, purification and properties of aequorin, a bioluminescent protein from the luminous hydromedusan, *Aequorea*. *J. Cell Comp. Physiol.* **59,** 223–239.

Songyang, Z., Carraway, K. L., 3rd, Eck, M. J., Harrison, S. C., Feldman, R. A., Mohammadi, M., Schlessinger, J., Hubbard, S. R., Smith, D. P., Eng, C., Lorenzo, M. J., Ponder., B. A. J., Mayer,B. J., and Cantley, L. C. (1995). Catalytic specificity of protein-tyrosine kinases is critical for selective signalling. *Nature* **373,** 536–539.

Tadokoro, S., Shattil, S. J., Eto, K., Tai, V., Liddington, R. C., de Pereda, J. M., Ginsberg, M. H., and Calderwood, D. A. (2003). Talin binding to integrin beta tails: A final common step in integrin activation. *Science* **302,** 103–106.

Thomas, S. M., and Brugge, J. S. (1997). Cellular functions regulated by Src family kinases. *Annu. Rev. Cell Dev. Biol.* **13,** 513–609.

Timpson, P., Jones, G. E., Frame, M. C., and Brunton, V. G. (2001). Coordination of cell polarization and migration by the Rho family GTPases requires Src tyrosine kinase activity. *Curr. Biol.* **11,** 1836–1846.

Ting, A. Y., Kain, K. H., Klemke, R. L., and Tsien, R. Y. (2001). Genetically encoded fluorescent reporters of protein tyrosine kinase activities in living cells. *Proc. Natl. Acad. Sci. USA* **98,** 15003–15008.

Tour, O., Meijer, R. M., Zacharias, D. A., Adams, S. R., and Tsien, R. Y. (2003). Genetically targeted chromophore-assisted light inactivation. *Nat. Biotechnol.* **21,** 1505–1508.

Truong, K., Sawano, A., Mizuno, H., Hama, H., Tong, K. I., Mal, T. K., Miyawaki, A., and Ikura, M. (2001). FRET-based *in vivo* Ca^{2+} imaging by a new calmodulin-GFP fusion molecule. *Nat. Struct. Biol.* **8,** 1069–1073.

Tsien, R. Y. (1998). The green fluorescent protein. *Annu. Rev. Biochem.* **67,** 509–544.

Violin, J. D., Zhang, J., Tsien, R. Y., and Newton, A. C. (2003). A genetically encoded fluorescent reporter reveals oscillatory phosphorylation by protein kinase C. *J. Cell Biol.* **161,** 899–909.

von Wichert, G., Jiang, G., Kostic, A., De Vos, K., Sap, J., and Sheetz, M. P. (2003). RPTP-alpha acts as a transducer of mechanical force on alphav/beta3-integrin-cytoskeleton linkages. *J. Cell Biol.* **161,** 143–153.

Vuori, K., Hirai, H., Aizawa, S., and Ruoslahti, E. (1996). Introduction of p130cas signaling complex formation upon integrin-mediated cell adhesion: A role for Src family kinases. *Mol. Cell Biol.* **16,** 2606–2613.

Waksman, G., Shoelson, S. E., Pant, N., Cowburn, D., and Kuriyan, J. (1993). Binding of a high affinity phosphotyrosyl peptide to the Src SH2 domain: Crystal structures of the complexed and peptide-free forms. *Cell* **72,** 779–790.

Wang, Y., Botvinick, E. L., Zhao, Y., Berns, M. W., Usami, S., Tsien, R. Y., and Chien, S. (2005). Visualizing the mechanical activation of Src. *Nature* **434,** 1040–1045.

Weaver, A. M., Karginov, A. V., Kinley, A. W., Weed, S. A., Li, Y., Parsons, J. T., and Cooper, J. A. (2001). Cortactin promotes and stabilizes Arp2/3–induced actin filament network formation. *Curr. Biol.* **11,** 370–374.

Wiseman, P. W., Brown, C. M., Webb, D. J., Hebert, B., Johnson, N. L., Squier, J. A., Ellisman, M. H., and Horwitz, A. F. (2004). Spatial mapping of integrin interactions and dynamics during cell migration by image correlation microscopy. *J. Cell Sci.* **117,** 5521–5534.

Wu, H., and Parsons, J. T. (1993). Cortactin, an 80/85-kilodalton pp60src substrate, is a filamentous actin-binding protein enriched in the cell cortex. *J. Cell Biol.* **120,** 1417–1426.

Yoshizaki, H., Ohba, Y., Kurokawa, K., Itoh, R. E., Nakamura, T., Mochizuki, N., Nagashima, K., and Matsuda, M. (2003). Activity of Rho-family GTPases during cell division as visualized with FRET-based probes. *J. Cell Biol.* **162,** 223–232.

Zacharias, D. A., Violin, J. D., Newton, A. C., and Tsien, R. Y. (2002). Partitioning of lipid-modified monomeric GFPs into membrane microdomains of live cells. *Science* **296,** 913–916.

Zhang, J., Campbell, R. E., Ting, A. Y., and Tsien, R. Y. (2002). Creating new fluorescent probes for cell biology. *Nat. Rev. Mol. Cell Biol.* **3,** 906–918.

Zhang, J., Hupfeld, C. J., Taylor, S. S., Olefsky, J. M., and Tsien, R. Y. (2005). Insulin disrupts beta-adrenergic signalling to protein kinase A in adipocytes. *Nature* **437,** 569–573.

Zhang, J., Ma, Y., Taylor, S. S., and Tsien, R. Y. (2001). Genetically encoded reporters of protein kinase A activity reveal impact of substrate tethering. *Proc. Natl. Acad. Sci. USA* **98,** 14997–15002.

CHAPTER TEN

STUDIES ON INTEGRINS IN THE NERVOUS SYSTEM

Sumiko Denda* *and* Louis F. Reichardt[†]

Contents

Abstract

Integrins are of interest to neuroscientists because they and many of their ligands are widely expressed in the nervous system and have been shown to have diverse roles in neural development and function (Clegg *et al.*, 2003; Li and Pleasure, 2005; Pinkstaff *et al.*, 1998, 1999; Reichardt and Tomaselli, 1991;

* Shiseido Research Center 2, Kanazawa-ku, Yokohama, Japan
† Department of Physiology, Neuroscience Program, Howard Hughes Medical Institute, University of California, San Francisco, San Francisco, California

Methods in Enzymology, Volume 426
ISSN 0076-6879, DOI: 10.1016/S0076-6879(07)26010-0

© 2007 Elsevier Inc.
All rights reserved.

Schmid *et al.*, 2005). Integrins have also been implicated in control of pathogenesis in several neurodegenerative diseases, brain tumor pathogenesis, and the aftermath of brain and peripheral nervous system injury (Condic, 2001; Ekstrom *et al.*, 2003; Kloss *et al.*, 1999; Verdier and Penke, 2004; Wallquist *et al.*, 2004). Using integrin antagonists as therapeutic agents in a variety of neurological diseases is of great interest at present (Blackmore and Letourneau, 2006; Mattern *et al.*, 2005; Polman *et al.*, 2006; Wang *et al.*, 2006). In this chapter, we describe methods used in our laboratory to characterize neuronal responses to extracellular matrix proteins, and procedures for assessing integrin roles in neuronal cell attachment and differentiation.

1. INTRODUCTION

In this section, we provide a brief summary of the current state of our knowledge on the roles of integrin receptors in the nervous system, beginning with their roles in neural development. Since the 1980s, the many roles for this fascinating family of adhesion molecules have been documented, which has increased our understanding of neural development, synaptic function, and several neurological diseases.

Much of the initial interest in integrin roles in the nervous system was focused on their roles in the neural crest, a migratory population of cells derived from the dorsal neural tube that populate the sensory, autonomic, and enteric nervous systems as well as contributing to formation of many specialized sense organs, glands, the heart, cranial mesenchyme and bone, and supporting cells in peripheral nerves, including Schwann cells and endoneural fibroblasts (Le Douarin and Dupin, 2003). Neural crest cells express many integrins and migrate through an extracellular matrix (ECM)-rich environment (Bronner-Fraser, 1994; Kil *et al.*, 1998). Acute inhibition experiments in avian embryos have documented important roles for integrins in migration of the neural crest (Tucker, 2004). In mice, genetic ablation of $\beta 1$ integrins results in severe perturbations of the peripheral nervous system, including failure of normal nerve arborization, delay in Schwann cell migration, and defective neuromuscular junction differentiation (Pietri *et al.*, 2004). In addition to direct effects on migration, it has been shown that absence of specific integrin heterodimers compromises Schwann cell precursor survival, proliferation, and differentiation (Feltri *et al.*, 2002; Haack and Hynes, 2001). Many of these observations are likely to reflect the roles of integrin receptors in regulating activation of MAP kinase, Rac, and other signaling pathways (Campos *et al.*, 2004).

In part because they were characterized long before the identification of the major families of axon guidance molecules, such as the netrins, semaphorins, and ephrins, early studies on axon outgrowth and guidance focused on integrins and the cadherin and immunoglobulin families of cell adhesion molecules. In these studies, it was demonstrated that virtually all process extension on ECM substrates by neurons requires integrin function and that neuronal growth cones could distinguish between different ECM proteins and respond to orientation or gradients of ECM proteins by directed growth (Dubey *et al.*, 1999; McKenna and Raper, 1988). In addition, integrins have been shown to interact with several of the axon guidance systems. For example, semaphoring 7A-dependent promotion of axon growth requires integrin activity, while semaphoring-mediated activation of plexin signaling reduces integrin-based adhesion (Barberis *et al.*, 2004; Pasterkamp *et al.*, 2003). The A and B ephrins also control integrin activation (Davy and Robbins, 2000; Nakada *et al.*, 2005). Integrins interact genetically with the Slit-Robo pathway in *Drosophila* (Stevens and Jacobs, 2002). Evidence also suggests that two integrins may serve as receptors for netrins in epithelia (Yebra *et al.*, 2003). Despite these intriguing observations, ablation of either $\beta 1$ or αV integrins appear to have only minor effects on axon guidance in the brain although perturbations are most significant in the peripheral nervous system (Blaess *et al.*, 2004; Graus-Porta *et al.*, 2001; McCarty *et al.*, 2005; Proctor *et al.*, 2005). This is likely the result of neurons interacting with many different types of substrates, many of which do not require integrin function.

Despite the absence of major effects on axon guidance, integrin deletion affects many aspects of forebrain and cerebellar development. First, loss of $\beta 1$ integrins results in disruptions of the basal lamina that separates the brain from the overlying mesenchyme (Graus-Porta *et al.*, 2001). As a result, the migration of neurons is perturbed, resulting in abnormal lamination of the cortex and cerebellum. Similar phenotypes are observed in mice lacking another ECM receptor dystroglycan as well as in humans and mice with mutations in basal lamina–encoding genes, such as the laminin $\alpha 5$ subunit (Gleeson and Walsh, 2000). Although some evidence indicates that integrins modulate neuronal interactions with radial glia—which provide the substrate for the tangential migrations that establish the cortical lamination pattern (Sanada *et al.*, 2004; Schmid *et al.*, 2005)—the major phenotype observed in these mutants appears to stem from disruption of signaling pathways controlling neuronal migration that require integrity of the basal lamina (Beggs *et al.*, 2003).

Integrins have a number of additional actions that modulate brain development. Of particular interest, they have been shown to control survival and proliferation of some populations of neural stem cells (Leone *et al.*, 2005). Within the external granule cell layer of the cerebellum, integrin binding to laminin enhances the proliferative responsiveness of granule cells to sonic hedgehog (SHH), probably because association of SHH with laminin facilitates SHH activation of its receptor Smoothened (Blaess *et al.*, 2004).

Integrins have a number of potent, but poorly understood, effects on synaptic function and plasticity. In cell culture, interactions of astroglia with neurons mediated by the glial integrin $\alpha V \beta 3$ results in PKC activation in individual neurons that facilitates excitatory synaptogenesis (Hama *et al.*, 2004). The receptor on neurons mediating PKC activation is not known, but neurons express many proteins, including the ADAMs, L1, and amyloid precursor protein, which are known to interact with integrins and are therefore candidates to mediate this signaling pathway (Mechtersheimer *et al.*, 2001; Wright *et al.*, 2006; Yang *et al.*, 2006). In *Drosophila*, integrins control localization of postsynaptic proteins at the neuromuscular junction through a CamKII-mediated signaling cascade (Burgess *et al.*, 2002). At the vertebrate neuromuscular junction, expression of integrins in muscle, but not nerve, is required for synapse formation (Schwander *et al.*, 2004), possibly through interactions with agrin or promotion of basal lamina assembly/organization (Burgess *et al.*, 2002; Burkin *et al.*, 2000). Integrins have also been localized to the synaptic active zones of motor neuron axon terminals and mediate the enhancement of transmitter release caused by mechanical stretching of muscle fibers (Kashani *et al.*, 2001).

Although localization studies indicate that integrins are present at many synapses in the brain, genetic and pharmacological studies indicate that integrins are not required for synapse formation, but are required for normal synaptic plasticity. In a particularly elegant series of studies, the presence of integrins in the mushroom body of the *Drosophila* brain was shown to be required for short-term memory (Grotewiel *et al.*, 1998). Conditional expression of an integrin subunit in the adult mushroom body rescued the memory deficits, providing definitive evidence that this was an effect on function, not early development. Studies in the murine hippocampus have demonstrated that $\beta 1$ integrins are required for normal LTP (Chan *et al.*, 2006; Huang *et al.*, 2006). Studies of mice with reduced expression of individual $\beta 1$ integrin heterodimers have suggested that specific integrins have different functions at the synapse (Chan *et al.*, 2003). Acute pharmacological perturbations using inhibitory integrin reagents indicate that integrins are involved in regulation of both NMDA and AMPA receptor function and act through regulation of protein kinases and the actin cytoskeleton (Kramar *et al.*, 2003, 2006; Lin *et al.*, 2003). Clearly, much remains to be understood about interactions between integrins and the signaling pathways known to be fundamental in initiation and maintenance of LTP.

Integrins are also necessary for normal development of non-neuronal cells in the nervous system, including astroglia, oligodendrocytes, and Schwann cells. For example, the presence of $\beta 1$ integrins is required for normal morphological development of radial glia, and abnormalities in the morphology of these glia may underlie the neuronal lamination deficits observed in mice with mutants in genes encoding basal lamina constituents and their receptors, including both integrins and dystroglycan (Forster *et al.*, 2002; Gleeson and Walsh, 2000).

An abundance of studies in culture indicate that integrins have profound effects on the differentiation of the precursors to oligodendrocytes

(Colognato *et al.*, 2002, 2004). Despite these effects of integrin absence on oligodendrocyte development and myelination in the brain and spinal cord appear to be modest. In mice lacking $\beta 1$ integrins myelination appears to be completely normal during development and, following injury, in the adult (Benninger *et al.*, 2006). During development there is elevated apoptosis of premyelinating oligodendrocytes, but this does not prevent successful myelination. No obvious reduction in myelination is obvious in mice lacking either the αV or $\beta 8$ integrins, although this has been examined in the same detail as the studies performed using mice lacking the $\beta 1$ integrin subunit (McCarty *et al.*, 2005; Proctor *et al.*, 2005).

$\beta 1$ integrins are clearly essential for many steps in peripheral nerve development, including several aspects of Schwann cell differentiation. $\beta 1$ gene ablation in the neural crest cell lineage delayed the migration of Schwann cell precursors along embryonic peripheral nerves without detectable effects on proliferation or apoptosis (Pietri *et al.*, 2004). At later times, the absence of these integrins interfered with normal sorting of sensory axons with Schwann cells and reduced their myelination (Feltri *et al.*, 2002; Pietri *et al.*, 2004). Integrin deficiency in the neural crest cell lineage also interfered with basal lamina assembly in peripheral nerves and prevented normal differentiation of the neuromuscular junction. No obvious phenotypes in the peripheral nervous system have been reported in mice lacking the αV or $\beta 8$ integrins.

Absence of either αV or $\beta 8$ integrins does, however, have a profound influence on brain development. Expression of αV$\beta 8$ is required for normal vascular development in the brain. In its absence, vascular development is severely perturbed, resulting in massive embryonic hemorrhage. Cell-specific targeting has shown that this integrin must be expressed in the neuroepithelial, not the endothelial lineage (McCarty *et al.*, 2005; Proctor *et al.*, 2005). Intriguingly, αV$\beta 8$ has been shown to promote the activation of TGFβ through binding to an RGD sequence in the TGFβ latency-associated peptide (Mu *et al.*, 2002). Similarities between the phenotypes of the αV and $\beta 8$ and mutations in the TGFβ signaling pathway suggest that the vascular abnormalities may reflect, at least in part, a reduction in TGFβ activation. Surprisingly, in animals that survive hemorrhage, the vasculature recovers so that hemorrhage is no longer visible in young adults. These animals, however, do develop motor deficits and die prematurely, most likely as a result of neurodegeneration in the central nervous system (CNS).

Integrins are also of interest to neuroscientists because they have been implicated in several neurodegenerative disorders. Of particular importance, because integrins regulate the transit of lymphocytes, macrophage, and other cells across the blood–brain barrier in response to inflammatory stimuli, anti-integrin reagents are of great interest as therapeutic agents to control demyelinating diseases, such as multiple sclerosis that involve the immune system and inflammatory responses (Bartt, 2006; Kanwar *et al.*, 2000).

The involvement of integrins in inflammatory responses has created interest in the possibility of using anti–integrin reagents to alleviate several neurodegenerative disorders, including Parkinson's and Alzheimer's diseases (Austin *et al.*, 2006). Of special interest, several integrins appear to interact with amyloid precursor protein and these are postulated to mediate deposition or toxic actions of Aβ and amyloid formation (Bozzo *et al.*, 2004; Koenigsknecht and Landreth, 2004; Sondag and Combs, 2006). Thus future studies on integrin functions in the normal and diseased brain are likely to provide interesting insights, some of which may have practical applications.

2. NEURONAL CELL ADHESION AND NEURITE OUTGROWTH ASSAYS

2.1. Preparation of substrates

2.1.1. Standard substrate preparation

When substrates such as laminin, fibronectin and the collagens, are available in abundant quantities, substrates are typically prepared in the same way as for other cell types by incubation of substrata with ECM proteins in solution, washing, and blocking, and then neuronal adhesion or neurite outgrowth assays. For example, the laboratory has frequently used the following protocol (Hall *et al.*, 1987) in which sterile Linbro 96–well, flat-bottom tissue culture plates are coated with 100 μl per well of laminin, collagen IV, other ECM protein or antibody diluted in calcium– and magnesium–free phosphate buffered saline (CMF-PBS; 200 mg/liter KCl, 200 mg/liter K_2SO_4, 8.0 g/liter NaC1, and 2.16 g/liter Na_2HPO_4-$7H_2O$, pH 7.36 to 7.45) at a concentration typically of 10 to 100 μg/ml. Fibronectin is applied in 100 μl per well of sterile 0.1-M cyclohexylaminopropane sulfonic acid buffer, pH 9.0. Plates are incubated with proteins overnight in the cold room. After rinsing three to five times with sterile CMF-PBS, the wells are blocked by incubation with 1% BSA for at least 2 h at room temperature. Plates are again rinsed three to five times with sterile CMF-PBS and 100 μl of culture medium is added to each well. At this time antibodies are added to the wells, and plates are stored in the incubator, at 37° and 5% CO_2, until the cells are ready, typically about 1 h, but plates can be stored for several days at 4°.

2.1.2. Modified substrate preparation using limiting reagents

Many of the integrin–binding proteins in the nervous system are difficult to purify or present for other reasons in only limiting quantities. In this case, we recommend first coating the Linbro Titertek 96–well plastic dishes with nitrocellulose (Lagenaur and Lemmon, 1987). In this procedure, 5 cm^2 of nitrocellulose type-BA 85 (Schleicher and Schuell, Keene, NH) is dissolved in 6 ml of methanol. Aliquots are spread over the surface of microwells and allowed to dry under a laminar flow hood. Test protein samples are applied at 50 μl per well at concentrations approximately 100-fold lower than the concentrations used for standard substrate preparation. The presence of nitrocellulose on the substrate results in capture of virtually all of the protein in solution. Our laboratory has used this procedure to examine integrin-mediated interactions of neurons with tenascin and purified fragments of tenascin (Varnum-Finney et al., 1995).

2.2. Cell adhesion assays

For acute assays, neurons can be isolated and tested in serum-free conditions. In contrast, media conditions are quite important for assays involving long-term survival and differentiation of neurons in culture and conditions for maintenance of several different neuronal populations are described below. For acute assays using freshly isolated neuronal populations, our laboratory performs cell adhesion assays using procedures similar to those used for other cell types with cell substrata precoated with various ECM substrates and single-cell suspensions of neurons (for preparation of individual neuron populations, see below) centrifuged onto these substrates in 96–well plates. For cell attachment assays, after a time of typically 1 h (which can be varied), unattached cells are removed by gentle swirling of the medium and medium removal. Because neurons are typically present in comparatively small numbers, our assays typically measure adhesion by counting of attached neurons although for cell types present in large quantities, such as chick retinal neurons, assays using dyes and optical density measurements may be used instead. Cell counts are performed manually using randomly chosen fields of view, typically at least 15 fields at 200× magnification. It is important that fields of view be spaced evenly across the culture substrate surface and that wells be checked to ensure that there is a reasonably uniform distribution of neurons across the entire surface. In these assays, neurons can usually be distinguished from other cells on morphological criteria. Our laboratory also used preplating protocols on which cell suspensions are incubated with tissue culture plastic wells to remove preferentially non-neuronal cells, which adhere preferentially in these conditions.

2.3. Neurite outgrowth assays

Cells on coverslips are fixed with 4% paraformaldehyde and mounted on slides with gelvatol (50% glycerol plus polyvinyl chloride). Cultures are viewed using a Zeiss inverted LSM 5 Pascal confocal microscope using a 10×, 0.3 n.a. or 20×, 0.5 n.a. Plan Neufluoar objective. All images are captured under the same conditions in each type of experiment. After conversion to TIFF format, the selected images are exported to Image J

(National Institutes of Health software) for drawing neuronal images and final quantification.

To determine the percentage of cells with neurites, at least 100 cells are analyzed per culture. The number of cells with a process longer than one cell diameter is determined and compared to the total number of cells counted. In some instances, we have used more stringent criteria to define a neurite, such as requiring an extension longer than two cell diameters.

In some instances, it is important to determine the length and branching properties of neurons grown in conditions where their axons and dendrites are intermingled with those of other neurons. In these instances, we have transfected neurons with plasmids containing a fluorescent protein (e.g., Venus, EGFP, dsRED).

2.4. Axons versus dendrite quantification

For many neuronal populations, it is frequently important to distinguish between axons and dendrites. In this case, we have collected and fixed neurons as described above, but have stained cultures with antibodies to tau to identify axons or antibodies to MAP2 to identify dendrites (Rico *et al.*, 2004).

3. Neuronal Culture Procedures

Procedures used by our laboratory for isolation and culture of several different types of neurons are described below. These assays typically give enriched, not absolutely pure populations of neurons. For studies where highly purified populations of neurons are required, antibodies specific for different cell types have been used in sequence to deplete cell mixtures of unwanted cell types and to select for specific subpopulations of neurons, such as retinal ganglion cells (Barres and Raff, 1999; Goldberg *et al.*, 2002; Mi and Barres, 1999; Ullian *et al.*, 2004). These procedures have been crucial for advancing our understanding of the roles of cell interactions in differentiation and the functions of ECM proteins in processes such as synaptogenesis (Christopherson *et al.*, 2005). In instances where integrin functions in a defined population of neurons will be examined, readers are referred to the papers cited immediately above for descriptions of useful techniques.

3.1. Rat PC12 pheochromocytoma cell culture

PC12 cells are grown in standard culture flasks in Dulbecco's modified Eagle's medium with 4.5 g/liter of glucose (DME H-21) supplemented with 10% heat-inactivated horse serum, 5.0% newborn calf serum, 2 mM of glutamine, and 100 U/ml penicillin and streptomycin. Cells are passaged by incubating for 5 to 10 min with 0.5 mM EDTA in PBS (8 mM Na$_2$HPO$_4$, 2 mM KH$_2$PO$_4$, 2 mM KCl, 0.136 M NaCl, pH 7.4). For priming with NGF, PC12 cells are passaged as described above onto fresh plates at a low density (10^4 cells/cm^2) and cultured for 5 to 7 days in DME H-21 supplemented with 1.0% heat-inactivated horse serum, 5.0% newborn calf serum, 2 mM of glutamine, and 100 U/ml of penicillin and streptomycin. In addition, 50 ng/ml NGF is added to prime the cells. PC12 cells attach well to and grow on substrata coated with EHS laminin or collagen I, but exhibit only poor attachment and neurite outgrowth on laminin (Tomaselli *et al.*, 1990; Varnum-Finney and Reichardt, 1994).

3.2. Rodent sympathetic neuron cultures

Superior cervical ganglia (SCGs) are dissected from Sprague-Dawley neonatal (P0) rat pups. SCGs are dissociated by manually removing the outer sheath and then digesting the ganglia for 1 h in 1 mg/ml of collagenase/dispase in CMF-PBS (DeFreitas *et al.*, 1995; Hawrot and Patterson, 1979). Digested SCGs are washed in CMF-PBS and triturated to dissociate the cells. After two washes in CMF-PSS, the cells are resuspended in serum-free L-15 complete medium as described (Hawrot and Patterson, 1979). Alternatively, the N1 serum-free additives (Bottenstein *et al.*, 1980) can be substituted for rat serum. NGF is added to a final concentration of 100 ng/ml, glutamine to a concentration of 2 mM, and BSA to a final concentration of 1 mg/ml. Cells are grown in a 5% CO$_2$ atmosphere at 37°. Alternatively, to minimize growth of non-neuronal cells, cultures can be grown in L-15 air medium (no bicarbonate) and incubated in a humidified air incubator at 37° (Hawrot and Patterson, 1979).

3.3. Rodent DRG sensory neuron cultures

Newborn rat and mouse dorsal root ganglia (DRG) are dissected into Ca^{2+}-Mg^{2+}-free Hank's Balanced Salt Solution and incubated with 0.1% trypsin for 30 min at 37°. Ganglia are dissociated with trituration through a flame-polished, siliconized Pasteur pipette in HBSS containing 10% fetal calf serum (FCS). Dissociated cells are depleted of non-neuronal cells by preplating for 1 h at 37° in 60-mm tissue culture dishes in DRG growth medium consisting of Dulbecco's Modified Eagle's Medium (DMEM) (4.5 g/liter of glucose) with 10% FCS, 50 ng/ml 2.5s nerve growth factor, and 100 U/ml of penicillin/streptomycin. Unattached neurons are decanted, washed once in growth medium, and plated in growth medium at about 500,000 cells/35-mm dish precoated with 10 μg/ml EHS laminin (or other chosen substrate). After 3 days, cultures are treated twice for 48 h each with 10 μM cytosine arabinoside in growth medium, followed by 24 h in growth medium alone. Following treatment the cultures should contain more than 95% neurons, as assessed using morphological criteria (Tomaselli *et al.*, 1993).

3.4. Rodent trigeminal sensory neuron cultures

Dissociated trigeminal or DRG sensory neurons are cultured using a published protocol (Buchman and Davies, 1993). Trigeminal or dorsal root ganglia are dissected from E12.5 embryos or P0 newborn mice, and trypsinized for 5 min at 37° with 0.05% trypsin in calcium- and magnesium-free HBSS. After removal of the trypsin solution, the ganglia are washed twice with 10 ml of Hams F12 medium containing 10% HIHS, and are gently triturated with a siliconized Pasteur pipette to give a single-cell suspension. Dissociated neurons are then plated on 16-well chamber slides coated with poly (DL)ornithine (0.5 mg/ml) and laminin (10 μg/ml) in triplicate, at a density of 2000 cells/well in defined F-14 medium, containing thyroxine (400 ng/ml), triiodothyronine (340 ng/ml), progesterone (60 ng/ml), sodium selenite (38 ng/ml), 0.35% bovine serum albumin, and N2 supplement in the presence of 10 ng/ml NGF. For cultures in which it is desired to maintain other populations of neurons, this medium can be supplemented by 10 ng/ml BDNF, 10 ng/ml NT-3, or 50 ng/ml GDNF, optimal concentrations for each of these factors. Cultures are maintained in 5% CO_2 at 37° in a humidified incubator (Huang *et al.*, 1999).

3.5. Rodent hippocampal neuron cultures

Hippocampi are obtained from E-17 to P-2-rat or mouse embryos by dissection using scissors and needles. Dissected hippocampi are incubated for 15 to 20 min in a solution of 2 mg of trypsin per milliliter of Hank's balanced salts solution (HBSS) containing 2.4 g/liter of N-[2-hydroxyethyl]piperazine-N9-[2-ethanesulfonic acid] (HEPES) and 10 mg/liter of gentamicin, pH 7.2, at a concentration of two rat hippocampi per milliliter (three to four mouse hippocampi per milliliter). The hippocampi are then rinsed three times in 10 ml of HBSS, followed by a 5-min incubation in a solution of 1 mg of trypsin inhibitor per milliliter of HBSS, and finally rinsed three times with 10 ml of HBSS. Cells are then dissociated by trituration through the narrowed bore of a fired-polished Pasteur pipette. Neurons are plated on poly-L-lysine–coated cover glasses or tissue culture plastic at the density of 130 cells/mm^2 in MEM containing 10% horse serum, 1 mM of pyruvate, penicillin-streptomycin and 0.6% glucose ("plating medium"). After 4 h, the plating medium was replaced by neurobasal media, penicillin-streptomycin, and B27 supplements ("maintenance medium") (Elia *et al.*, 2006; Rico *et al.*, 2004; Xie *et al.*, 2000).

3.6. Rodent cortical neuron cultures

These neurons are cultured using the same procedures as used for hippocampal neuron cultures. Typically, cortices are easier to dissociate with trypsin and neurons are more sensitive to trypsin (Xie *et al.*, 2000).

3.7. Chick DRG sensory neuron culture

Embryonic chick dorsal root ganglia at Day 7 or 8 are dissected from the vicinity of the spinal cord manually using fine forceps or glass needles. Ganglia are dissociated into single cells by incubation in 0.05% trypsin in 0.2% versene, 0.10% glucose, 0.02% EDTA, 0.058% $NaHCO_2$ for 10 min at 37° followed by trituration. Dissociated cells are collected by centrifugation and resuspended in F12 containing 10% fetal bovine serum. To enrich for neuronal cells, cell suspensions are plated onto 60-mm tissue culture dishes (Falcon) for 1 to 3 hr. Neurons are pipetted from the culture dishes, centrifuged, and resuspended in DRG growth medium (F-12 containing 2% BSA (Serva), and 100 ng/ml nerve growth factor) at an appropriate density (1×10^3 cells per well). 50 μl of this cell suspension is added to the well of a 96-well culture dish coated with 10 μg/ml EHS laminin (or other chosen substrate). Neurons are gently centrifuged onto the dish and incubated at 37° in a 5% CO_2 atmosphere (Varnum-Finney and Reichardt, 1994; Wehrle and Chiquet, 1990).

3.8. Chick ciliary neuron culture

Our laboratory has used procedures developed by others (Nishi and Berg, 1977, 1981; Weaver *et al.*, 1995). E7.5 (stages 31 to 33, Hamburger and Hamilton, 1992) chick ciliary ganglia are dissected manually, after which they are incubated in 0.1% trypsin in Ca^{2+}-Mg^{2+}-free PBS (CMF-PBS) and dispersed by trituration through a fire-polished pipette. Neurons are cultured on plates precoated with laminin (collagen or other ECM substrates can be substituted) in 50% Eagle's MEM with Earl's balanced salts and 50% Ham's F-12 plus 2% chick embryo extract.

3.9. Chick motor neuron culture

Motor neuron cultures are prepared from stage 19 embryos. The dorsal part of the embryo at the brachial level is dissected by removing the dorsal aorta and flanking mesenchyme. This fragment is washed in PBS, and incubated in pancreatin (Gibco, Grand Island NY) for 2 min. Embryos are then placed in cold L-15 medium containing 5 mg/ml bovine serum albumin (BSA) (Sigma, St. Louis MO), and neural tubes are isolated by dissection from remaining sclerotomal tissue. Dorsal and ventral neural tube fragments are separated with a glass needle. Ventral fragments are then placed in 0.25% trypsin for 5 min at 37°, are washed in L-15 containing 10 mg/ml BSA, centrifuged, and resuspended in complete motor-neuron growth medium (F12 medium containing 0.5 mg/ml BSA (Serva, Germany). Brain-derived neurotrophic factor (BDNF) (10 ng/ml) and 10 ng/ml bFGF (Boehringer Mannheim, Indianapolis IN). These ventral neural tube tissue fragments are then gently triturated four times. Dissociated cells are centrifuged, resuspended in growth medium to an appropriate density (1×10^3 cells/well), plated in 100 μl growth medium, and incubated at 37° with 5% CO_2 on appropriate ECM substrates, such as laminin-coated or laminin and poly-D-lysine–coated substrata (Varnum-Finney and Reichardt, 1994; Wehrle and Chiquet, 1990).

3.10. Chick retinal neuron culture

Retinas are dissected from E6 or E12 chick embryos and are incubated for 6 min at 37° in 0.1% trypsin (in Ca^{2+}-Mg^{2+}-free PBS [CMF-PBS]). Digestion is stopped by adding 0.2 volumes of heat-inactivated fetal calf serum. Pellets are washed once in F12 nutrient mixture, and triturated in F12 containing 0.002% DNase I. For cell attachment assays, Linbro/Titer plates (Flow Laboratories, Inc., Maclean, VA) are coated overnight with 20 μg/ml of LN or collagen IV in CMF-PBS. Coated and uncoated wells are incubated for 2 h at room temperature with 1% BSA in CMF-PBS. Wells are washed twice with CMF-PBS, and about 100,000 retinal cells were added to each well, after preincubation in a sterile tube for 20 min at room temperature in F12 medium with additives (5 mg/ml insulin, 30 nM selenium, 25 mg/ml human transferrin, 100 U/ml penicillin and streptomycin, according to [Bottenstein et al., 1980]). Cells are sedimented to the bottom of wells by centrifugation, and incubated for 1 h at 37° in 5% CO_2 atmosphere in the same medium used during the preincubation. Unattached cells are removed by the brisk addition of warm medium followed by gentle vacuum suction. For neurite outgrowth assays, cells are cultured at 37° in a 5% CO_2 atmosphere, and cultures are examined in the microscope for neurite outgrowth after overnight or longer incubation (de Curtis and Reichardt, 1993; Neugebauer et al., 1991).

4. BIOCHEMICAL STUDIES USING CULTURED NEURONS

Purified neuronal populations are typically present in limited quantities, so sensitive assays are required to detect integrins and other proteins. It is absolutely essential to assess the purity of a culture before making conclusions about cell type–specific expression of integrins or other proteins because non-neuronal cells (both astroglia and endothelial cells that contaminate CNS cultures and Schwann cells and fibroblasts, contaminants of peripheral neuronal cultures) express many integrins. Ideally, neurons should be purified by dye labeling, antibody panning, or other procedures for biochemical assays (Barres and Raff, 1999; Christopherson et al., 2005; Meyer-Franke et al., 1998).

4.1. Surface labeling of integrins and other neuronal glycoproteins (used for DRG and ciliary ganglion neurons)

4.1.1. ^{125}I labeling

Dissociated neurons are cultured for 24 h on 100 mm^2 of Falcon tissue culture dishes previously coated with a purified ECM substrate dissolved in PBS. After removal of growth medium and washing with PBS, cultures are surface labeled by adding lactoperoxidase, H_2O_2 and ^{125}I in PBS for 15 min (Tomaselli et al., 1993). Lysates are prepared by adding 1.2 ml of lysis buffer containing 1% Triton-X-100 in PBS and protease inhibitors to the cultures. Lysed material is then scraped from the dish with a cell scraper and centrifuged at 20,000×g for 20 min at 4°. After preclearing twice with 100 μl of protein A-Sepharose, supernatants are incubated for 4 h with specific antibodies coupled to protein A-Sepharose (50 μl beads per culture, Pierce, Rockford IL). Antibodies are coupled to protein A-Sepharose with dimethylpimelimidate using the standard procedure (Harlow and Lane, 1988). Beads are then washed six times by adding 1.0 ml of 1% Triton-X-100 in PBS and pelleting beads at 134×g for 10 sec. Immunoprecipitated proteins are then separated with 6% SDS-PAGE in nonreducing conditions. Gels are dried and exposed to Kodak X Omat R film or are alternatively scanned in a molecular imager (Fujii, Molecular Dynamics).

4.1.2. Biotin labeling

At this juncture, 0.8 mg of sulfo-NHS-biotin (#21217, Pierce, Rockford, IL) is added and rocked with the cells at 4° for 1 h. Cells are then washed three times in TBS and extracted in RIPA (0.01 M Tris-Cl, pH 7.2, 0.15 M NaCl, 1% Na-deoxycholate, 1% Triton X-100, 0.1% SDS, 1% aprotinin, 4 mM PMSF and 10 μg/ml each antipain, leupeptin, pepstatin, and chymostatin) for 1 h at 4°. The extracts are centrifuged at 9170×g for 15 min and DNase is added to the soluble fraction at a final concentration of 100 ng/ml (Weaver et al., 1995).

For immunoprecipitations, all steps are carried out at 4°. Extracts are precleared once with Sepharose CL-4B followed by protein A-Sepharose (each 50 μl/lane for 45 min), and then incubated overnight at 4° with the appropriate antibody coupled directly to protein A-Sepharose (Harlow and Lane, 1988). Beads from all precipitation steps are collected, washed three times each in RIPA, two times in TBS with 0.5% Tween-20, and once in

TBS with 0.5% Tween-20 and 0.1% ovalbumin. Precipitates are boiled in nonreducing sample buffer, subjected to SDS-PAGE, and transferred to nitrocellulose (Schleicher and Schuell, Keene, NH). Transfer membranes are blocked for 1 h at room temperature in PBS with 10% BSA, 0.05% Tween-20, then incubated in PBS with 1% BSA, 0.05% Tween-20 (reaction buffer) with streptavidin-HRP (1:4000; Zymed, South San Francisco, CA) for 1 h at room temperature. After a brief rinse in TBS, the transfers are washed twice in 1% Triton X-100, 0.1% SDS, 0.5% Na-deoxycholate in TBS, for 5 min each, and then twice more in TBS. The transfers are then processed for chemiluminescent detection of HRP reaction product according to the manufacturer's specifications.

4.2. Immunocytochemistry

Cultured neurons are extremely fragile on coverslips. Standard procedures and antibodies can be used to visualize integrins and other proteins, but it is critical that the coverslips be handled delicately and not subjected to shear forces during application or removal of reagents. In our laboratory, coverslips are not dipped into reagents, but reagents are added gently to coverslips in horizontal position. Reagents are removed as slowly as possible during washes and coverslips are always left under liquid. Coverslips are carefully transferred to a humidified staining chamber. Cells are fixed with 4% paraformaldehyde for 10 min, washed with PBS, and are then incubated with PBS containing 5% horse or donkey serum and 0.2% Triton-X-100 for 10 min. Coverslips are then incubated overnight at 4° with either a monoclonal antibody or polyclonal antibody in PBS containing 5% horse or donkey serum. Coverslips are then washed with PBS and coverslips previously treated with a monoclonal antibody are incubated with biotinylated horse anti-mouse IgG diluted at 1:200 for 30 min. The coverslips are then washed in PBS and incubated for 30 min with fluorescein-labeled streptavidin-fluorescein diluted at 1:200. Coverslips previously treated with a polyclonal antibody are washed and incubated for 30 min with Texas-red labeled donkey anti-rabbit. Coverslips are mounted on slides with gelvatol and viewed and photographed with a Nikon photomicroscope, using the appropriate filters.

REFERENCES

Austin, S. A., Floden, A. M., Murphy, E. J., and Combs, C. K. (2006). Alpha-synuclein expression modulates microglial activation phenotype. *J. Neurosci.* **26,** 10558–10563.
Barberis, D., Artigiani, S., Casazza, A., Corso, S., Giordano, S., Love, C. A., Jones, E. Y., Comoglio, P. M., and Tamagnone, L. (2004). Plexin signaling hampers integrin-based adhesion, leading to Rho-kinase independent cell rounding, and inhibiting lamellipodia extension and cell motility. *FASEB J.* **18,** 592–594.

Barres, B. A., and Raff, M. C. (1999). Axonal control of oligodendrocyte development. *J. Cell Biol.* **147,** 1123–1128.

Bartt, R. E. (2006). Multiple sclerosis, natalizumab therapy, and progressive multifocal leukoencephalopathy. *Curr. Opin. Neurol.* **19,** 341–349.

Beggs, H. E., Schahin-Reed, D., Zang, K., Goebbels, S., Nave, K. A., Gorski, J., Jones, K. R., Sretavan, D., and Reichardt, L. F. (2003). FAK deficiency in cells contributing to the basal lamina results in cortical abnormalities resembling congenital muscular dystrophies. *Neuron* **40,** 501–514.

Benninger, Y., Colognato, H., Thurnherr, T., Franklin, R. J., Leone, D. P., Atanasoski, S., Nave, K. A., Ffrench-Constant, C., Suter, U., and Relvas, J. B. (2006). Beta1-integrin signaling mediates premyelinating oligodendrocyte survival but is not required for CNS myelination and remyelination. *J. Neurosci.* **26,** 7665–7673.

Blackmore, M., and Letourneau, P. C. (2006). L1, beta1 integrin, and cadherins mediate axonal regeneration in the embryonic spinal cord. *J. Neurobiol.* **66,** 1564–1583.

Blaess, S., Graus-Porta, D., Belvindrah, R., Radakovits, R., Pons, S., Littlewood-Evans, A., Senften, M., Guo, H., Li, Y., Miner, J. H., Reichardt, L. F., and Muller, U. (2004). Beta1-integrins are critical for cerebellar granule cell precursor proliferation. *J. Neurosci.* **24,** 3402–3412.

Bottenstein, J. E., Skaper, S. D., Varon, S. S., and Sato, G. H. (1980). Selective survival of neurons from chick embryo sensory ganglionic dissociates utilizing serum-free supplemented medium. *Exp. Cell Res.* **125,** 183–190.

Bozzo, C., Lombardi, G., Santoro, C., and Canonico, P. L. (2004). Involvement of beta(1) integrin in betaAP-induced apoptosis in human neuroblastoma cells. *Mol. Cell Neurosci.* **25,** 1–8.

Bronner-Fraser, M. (1994). Neural crest cell formation and migration in the developing embryo. *FASEB J.* **8,** 699–706.

Buchman, V. L., and Davies, A. M. (1993). Different neurotrophins are expressed and act in a developmental sequence to promote the survival of embryonic sensory neurons. *Development* **118,** 989–1001.

Burgess, R. W., Dickman, D. K., Nunez, L., Glass, D. J., and Sanes, J. R. (2002). Mapping sites responsible for interactions of agrin with neurons. *J. Neurochem.* **83,** 271–284.

Burkin, D. J., Kim, J. E., Gu, M., and Kaufman, S. J. (2000). Laminin and alpha7beta1 integrin regulate agrin-induced clustering of acetylcholine receptors. *J. Cell Sci.* **113,** 2877–2886.

Campos, L. S., Leone, D. P., Relvas, J. B., Brakebusch, C., Fassler, R., Suter, U., and Ffrench-Constant, C. (2004). Beta1 integrins activate a MAPK signalling pathway in neural stem cells that contributes to their maintenance. *Development* **131,** 3433–3444.

Chan, C. S., Weeber, E. J., Kurup, S., Sweatt, J. D., and Davis, R. L. (2003). Integrin requirement for hippocampal synaptic plasticity and spatial memory. *J. Neurosci.* **23,** 7107–7116.

Chan, C. S., Weeber, E. J., Zong, L., Fuchs, E., Sweatt, J. D., and Davis, R. L. (2006). Beta 1-integrins are required for hippocampal AMPA receptor-dependent synaptic transmission, synaptic plasticity, and working memory. *J. Neurosci.* **26,** 223–232.

Christopherson, K. S., Ullian, E. M., Stokes, C. C., Mullowney, C. E., Hell, J. W., Agah, A., Lawler, J., Mosher, D. F., Bornstein, P., and Barres, B. A. (2005). Thrombospondins are astrocyte-secreted proteins that promote CNS synaptogenesis. *Cell* **120,** 421–433.

Clegg, D. O., Wingerd, K. L., Hikita, S. T., and Tolhurst, E. C. (2003). Integrins in the development, function and dysfunction of the nervous system. *Front. Biosci.* **8,** d723–d750.

Colognato, H., Baron, W., Avellana-Adalid, V., Relvas, J. B., Baron-Van Evercooren, A., Georges-Labouesse, E., and Ffrench-Constant, C. (2002). CNS integrins switch growth factor signalling to promote target-dependent survival. *Nat. Cell Biol.* **4,** 833–841.

Colognato, H., Ramachandrappa, S., Olsen, I. M., and Ffrench-Constant, C. (2004). Integrins direct Src family kinases to regulate distinct phases of oligodendrocyte development. *J. Cell Biol.* **167,** 365–375.

Condic, M. L. (2001). Adult neuronal regeneration induced by transgenic integrin expression. *J. Neurosci.* **21,** 4782–4478.

Davy, A., and Robbins, S. M. (2000). Ephrin-A5 modulates cell adhesion and morphology in an integrin-dependent manner. *EMBO J.* **19,** 5396–5405.

de Curtis, I., and Reichardt, L. F. (1993). Function and spatial distribution in developing chick retina of the laminin receptor alpha 6 beta 1 and its isoforms. *Development* **118,** 377–388.

DeFreitas, M. F., Yoshida, C. K., Frazier, W. A., Mendrick, D. L., Kypta, R. M., and Reichardt, L. F. (1995). Identification of integrin alpha 3 beta 1 as a neuronal thrombospondin receptor mediating neurite outgrowth. *Neuron* **15,** 333–343.

Dubey, N., Letourneau, P. C., and Tranquillo, R. T. (1999). Guided neurite elongation and schwann cell invasion into magnetically aligned collagen in simulated peripheral nerve regeneration. *Exp. Neurol.* **158,** 338–350.

Ekstrom, P. A., Mayer, U., Panjwani, A., Pountney, D., Pizzey, J., and Tonge, D. A. (2003). Involvement of alpha7beta1 integrin in the conditioning-lesion effect on sensory axon regeneration. *Mol. Cell Neurosci.* **22,** 383–395.

Elia, L. P., Yamamoto, M., Zang, K., and Reichardt, L. F. (2006). p120 catenin regulates dendritic spine and synapse development through Rho-family GTPases and cadherins. *Neuron* **51,** 43–56.

Feltri, M. L., Graus Porta, D., Previtali, S. C., Nodari, A., Migliavacca, B., Cassetti, A., Littlewood-Evans, A., Reichardt, L. F., Messing, A., Quattrini, A., Mueller, U., and Wrabetz, L. (2002). Conditional disruption of beta 1 integrin in Schwann cells impedes interactions with axons. *J. Cell Biol.* **156,** 199–209.

Forster, E., Tielsch, A., Saum, B., Weiss, K. H., Johanssen, C., Graus-Porta, D., Muller, U., and Frotscher, M. (2002). Reelin, disabled 1, and beta 1 integrins are required for the formation of the radial glial scaffold in the hippocampus. *Proc. Natl. Acad. Sci. USA* **99,** 13178–13183.

Gleeson, J. G., and Walsh, C. A. (2000). Neuronal migration disorders: From genetic diseases to developmental mechanisms. *Trends Neurosci.* **23,** 352–359.

Goldberg, J. L., Espinosa, J. S., Xu, Y., Davidson, N., Kovacs, G. T., and Barres, B. A. (2002). Retinal ganglion cells do not extend axons by default: Promotion by neurotrophic signaling and electrical activity. *Neuron* **33,** 689–702.

Graus-Porta, D., Blaess, S., Senften, M., Littlewood-Evans, A., Damsky, C., Huang, Z., Orban, P., Klein, R., Schittny, J. C., and Muller, U. (2001). Beta1-class integrins regulate the development of laminae and folia in the cerebral and cerebellar cortex. *Neuron* **31,** 367–379.

Grotewiel, M. S., Beck, C. D., Wu, K. H., Zhu, X. R., and Davis, R. L. (1998). Integrin-mediated short-term memory in *Drosophila*. *Nature* **391,** 455–460.

Haack, H., and Hynes, R. O. (2001). Integrin receptors are required for cell survival and proliferation during development of the peripheral glial lineage. *Dev. Biol.* **233,** 38–55.

Hall, D. E., Neugebauer, K. M., and Reichardt, L. F. (1987). Embryonic neural retinal cell response to extracellular matrix proteins: Developmental changes and effects of the cell substratum attachment antibody (CSAT). *J. Cell Biol.* **104,** 623–634.

Hama, H., Hara, C., Yamaguchi, K., and Miyawaki, A. (2004). PKC signaling mediates global enhancement of excitatory synaptogenesis in neurons triggered by local contact with astrocytes. *Neuron* **41,** 405–415.

Hamburger, V., and Hamilton, H. L. (1992). A series of normal stages in the development of the chick embryo. *Dev. Dyn.* **195,** 231–272.

Harlow, E., and Lane, D. (1988). "Antibodies: A Laboratory Manual." Laboratory Press, Cold Spring Harbor.

Hawrot, E., and Patterson, P. H. (1979). Long-term culture of dissociated sympathetic neurons. *Methods Enzymol.* **58,** 574–584.

Huang, E. J., Zang, K., Schmidt, A., Saulys, A., Xiang, M., and Reichardt, L. F. (1999). POU domain factor Brn-3a controls the differentiation and survival of trigeminal neurons by regulating Trk receptor expression. *Development* **126,** 2869–2882.

Huang, Z., Shimazu, K., Woo, N. H., Zang, K., Muller, U., Lu, B., and Reichardt, L. F. (2006). Distinct roles of the beta1-class integrins at the developing and the mature hippocampal excitatory synapse. *J. Neurosci.* **26,** 11208–11219.

Kanwar, J. R., Harrison, J. E., Wang, D., Leung, E., Mueller, W., Wagner, N., and Krissansen, G. W. (2000). Beta7 integrins contribute to demyelinating disease of the central nervous system. *J. Neuroimmunol.* **103,** 146–152.

Kashani, A. H., Chen, B. M., and Grinnell, A. D. (2001). Hypertonic enhancement of transmitter release from frog motor nerve terminals: Ca2+ independence and role of integrins. *J. Physiol.* **530,** 243–252.

Kil, S. H., Krull, C. E., Cann, G., Clegg, D., and Bronner-Fraser, M. (1998). The alpha4 subunit of integrin is important for neural crest cell migration. *Dev. Biol.* **202,** 29–42.

Kloss, C. U., Werner, A., Klein, M. A., Shen, J., Menuz, K., Probst, J. C., Kreutzberg, G. W., and Raivich, G. (1999). Integrin family of cell adhesion molecules in the injured brain: Regulation and cellular localization in the normal and regenerating mouse facial motor nucleus. *J. Comp. Neurol.* **411,** 162–178.

Koenigsknecht, J., and Landreth, G. (2004). Microglial phagocytosis of fibrillar beta-amyloid through a beta1 integrin-dependent mechanism. *J. Neurosci.* **24,** 9838–9846.

Kramar, E. A., Bernard, J. A., Gall, C. M., and Lynch, G. (2003). Integrins modulate fast excitatory transmission at hippocampal synapses. *J. Biol. Chem.* **278,** 10722–10730.

Kramar, E. A., Lin, B., Rex, C. S., Gall, C. M., and Lynch, G. (2006). Integrin-driven actin polymerization consolidates long-term potentiation. *Proc. Natl. Acad. Sci. USA* **103,** 5579–5584.

Lagenaur, C., and Lemmon, V. (1987). An L1-like molecule, the 8D9 antigen, is a potent substrate for neurite extension. *Proc. Natl. Acad. Sci. USA* **84,** 7753–7757.

Le Douarin, N. M., and Dupin, E. (2003). Multipotentiality of the neural crest. *Curr. Opin. Genet. Dev.* **13,** 529–536.

Leone, D. P., Relvas, J. B., Campos, L. S., Hemmi, S., Brakebusch, C., Fassler, R., Ffrench-Constant, C., and Suter, U. (2005). Regulation of neural progenitor proliferation and survival by beta1 integrins. *J. Cell Sci.* **118,** 2589–2599.

Li, G., and Pleasure, S. J. (2005). Morphogenesis of the dentate gyrus: What we are learning from mouse mutants. *Dev. Neurosci.* **27,** 93–99.

Lin, B., Arai, A. C., Lynch, G., and Gall, C. M. (2003). Integrins regulate NMDA receptor-mediated synaptic currents. *J. Neurophysiol.* **89,** 2874–2878.

Mattern, R. H., Read, S. B., Pierschbacher, M. D., Sze, C. I., Eliceiri, B. P., and Kruse, C. A. (2005). Glioma cell integrin expression and their interactions with integrin antagonists. *Cancer Ther.* **3A,** 325–340.

McCarty, J. H., Lacy-Hulbert, A., Charest, A., Bronson, R. T., Crowley, D., Housman, D., Savill, J., Roes, J., and Hynes, R. O. (2005). Selective ablation of alphav integrins in the central nervous system leads to cerebral hemorrhage, seizures, axonal degeneration and premature death. *Development* **132,** 165–176.

McKenna, M. P., and Raper, J. A. (1988). Growth cone behavior on gradients of substratum bound laminin. *Dev. Biol.* **130,** 232–236.

Mechtersheimer, S., Gutwein, P., Agmon-Levin, N., Stoeck, A., Oleszewski, M., Riedle, S., Postina, R., Fahrenholz, F., Fogel, M., Lemmon, V., and Altevogt, P. (2001). Ectodomain shedding of L1 adhesion molecule promotes cell migration by autocrine binding to integrins. *J. Cell Biol.* **155,** 661–673.

Meyer-Franke, A., Wilkinson, G. A., Kruttgen, A., Hu, M., Munro, E., Hanson, M. G., Jr., Reichardt, L. F., and Barres, B. A. (1998). Depolarization and cAMP elevation rapidly recruit TrkB to the plasma membrane of CNS neurons. *Neuron* **21**, 681–693.

Mi, H., and Barres, B. A. (1999). Purification and characterization of astrocyte precursor cells in the developing rat optic nerve. *J. Neurosci.* **19**, 1049–1661.

Mu, D., Cambier, S., Fjellbirkeland, L., Baron, J. L., Munger, J. S., Kawakatsu, H., Sheppard, D., Broaddus, V. C., and Nishimura, S. L. (2002). The integrin alpha(v) beta8 mediates epithelial homeostasis through MT1-MMP–dependent activation of TGF-beta1. *J. Cell Biol.* **157**, 493–507.

Nakada, M., Niska, J. A., Tran, N. L., McDonough, W. S., and Berens, M. E. (2005). EphB2/R-Ras signaling regulates glioma cell adhesion, growth, and invasion. *Am. J. Pathol.* **167**, 565–576.

Neugebauer, K. M., Emmett, C. J., Venstrom, K. A., and Reichardt, L. F. (1991). Vitronectin and thrombospondin promote retinal neurite outgrowth: Developmental regulation and role of integrins. *Neuron* **6**, 345–358.

Nishi, R., and Berg, D. K. (1977). Dissociated ciliary ganglion neurons *in vitro*: Survival and synapse formation. *Proc. Natl. Acad. Sci. USA* **74**, 5171–5175.

Nishi, R., and Berg, D. K. (1981). Two components from eye tissue that differentially stimulate the growth and development of ciliary ganglion neurons in cell culture. *J. Neurosci.* **1**, 505–513.

Pasterkamp, R. J., Peschon, J. J., Spriggs, M. K., and Kolodkin, A. L. (2003). Semaphorin 7A promotes axon outgrowth through integrins and MAPKs. *Nature* **424**, 398–405.

Pietri, T., Eder, O., Breau, M. A., Topilko, P., Blanche, M., Brakebusch, C., Fassler, R., Thiery, J. P., and Dufour, S. (2004). Conditional beta1-integrin gene deletion in neural crest cells causes severe developmental alterations of the peripheral nervous system. *Development* **131**, 3871–3883.

Pinkstaff, J. K., Detterich, J., Lynch, G., and Gall, C. (1999). Integrin subunit gene expression is regionally differentiated in adult brain. *J. Neurosci.* **19**, 1541–1556.

Pinkstaff, J. K., Lynch, G., and Gall, C. M. (1998). Localization and seizure-regulation of integrin beta 1 mRNA in adult rat brain. *Brain Res. Mol. Brain Res.* **55**, 265–276.

Polman, C. H., O'Connor, P. W., Havrdova, E., Hutchinson, M., Kappos, L., Miller, D. H., Phillips, J. T., Lublin, F. D., Giovannoni, G., Wajgt, A., Toal, M., Lynn, F., *et al.* (2006). A randomized, placebo-controlled trial of natalizumab for relapsing multiple sclerosis. *N. Engl. J. Med.* **354**, 899–910.

Proctor, J. M., Zang, K., Wang, D., Wang, R., and Reichardt, L. F. (2005). Vascular development of the brain requires beta8 integrin expression in the neuroepithelium. *J. Neurosci.* **25**, 9940–9948.

Reichardt, L. F., and Tomaselli, K. J. (1991). Extracellular matrix molecules and their receptors: Functions in neural development. *Annu. Rev. Neurosci.* **14**, 531–570.

Rico, B., Beggs, H. E., Schahin-Reed, D., Kimes, N., Schmidt, A., and Reichardt, L. F. (2004). Control of axonal branching and synapse formation by focal adhesion kinase. *Nat. Neurosci.* **7**, 1059–1069.

Sanada, K., Gupta, A., and Tsai, L. H. (2004). Disabled-1–regulated adhesion of migrating neurons to radial glial fiber contributes to neuronal positioning during early corticogenesis. *Neuron* **42**, 197–211.

Schmid, R. S., Jo, R., Shelton, S., Kreidberg, J. A., and Anton, E. S. (2005). Reelin, integrin and DAB1 interactions during embryonic cerebral cortical development. *Cereb. Cortex* **15**, 1632–1636.

Schwander, M., Shirasaki, R., Pfaff, S. L., and Muller, U. (2004). Beta1 integrins in muscle, but not in motor neurons, are required for skeletal muscle innervation. *J. Neurosci.* **24**, 8181–8191.

Sondag, C. M., and Combs, C. K. (2006). Amyloid precursor protein cross-linking stimulates beta amyloid production and pro-inflammatory cytokine release in monocytic lineage cells. *J. Neurochem.* **97,** 449–461.

Stevens, A., and Jacobs, J. R. (2002). Integrins regulate responsiveness to slit repellent signals. *J. Neurosci.* **22,** 4448–4455.

Tomaselli, K. J., Doherty, P., Emmett, C. J., Damsky, C. H., Walsh, F. S., and Reichardt, L. F. (1993). Expression of beta 1 integrins in sensory neurons of the dorsal root ganglion and their functions in neurite outgrowth on two laminin isoforms. *J. Neurosci.* **13,** 4880–4888.

Tomaselli, K. J., Hall, D. E., Flier, L. A., Gehlsen, K. R., Turner, D. C., Carbonetto, S., and Reichardt, L. F. (1990). A neuronal cell line (PC12) expresses two beta 1-class integrins—alpha 1 beta 1 and alpha 3 beta 1—that recognize different neurite outgrowth-promoting domains in laminin. *Neuron* **5,** 651–662.

Tucker, R. P. (2004). Antisense knockdown of the beta1 integrin subunit in the chicken embryo results in abnormal neural crest cell development. *Int. J. Biochem. Cell Biol.* **36,** 1135–1139.

Ullian, E. M., Christopherson, K. S., and Barres, B. A. (2004). Role for glia in synaptogenesis. *Glia* **47,** 209–216.

Varnum-Finney, B., and Reichardt, L. F. (1994). Vinculin-deficient PC12 cell lines extend unstable lamellipodia and filopodia and have a reduced rate of neurite outgrowth. *J. Cell Biol.* **127,** 1071–1084.

Varnum-Finney, B., Venstrom, K., Muller, U., Kypta, R., Backus, C., Chiquet, M., and Reichardt, L. F. (1995). The integrin receptor alpha 8 beta 1 mediates interactions of embryonic chick motor and sensory neurons with tenascin-C. *Neuron* **14,** 1213–1222.

Verdier, Y., and Penke, B. (2004). Binding sites of amyloid beta-peptide in cell plasma membrane and implications for Alzheimer's disease. *Curr. Protein Pept. Sci.* **5,** 19–31.

Wallquist, W., Zelano, J., Plantman, S., Kaufman, S. J., Cullheim, S., and Hammarberg, H. (2004). Dorsal root ganglion neurons up-regulate the expression of laminin-associated integrins after peripheral but not central axotomy. *J. Comp. Neurol.* **480,** 162–169.

Wang, A. G., Yen, M. Y., Hsu, W. M., and Fann, M. J. (2006). Induction of vitronectin and integrin alphav in the retina after optic nerve injury. *Mol. Vis.* **12,** 76–84.

Weaver, C. D., Yoshida, C. K., de Curtis, I., and Reichardt, L. F. (1995). Expression and *in vitro* function of beta 1–integrin laminin receptors in the developing avian ciliary ganglion. *J. Neurosci.* **15,** 5275–5285.

Wehrle, B., and Chiquet, M. (1990). Tenascin is accumulated along developing peripheral nerves and allows neurite outgrowth *in vitro. Development* **110,** 401–415.

Wright, S., Malinin, N. L., Powell, K. A., Yednock, T., Rydel, R. E., and Griswold-Prenner, I. (2006). alpha2beta1 and alphaVbeta1 integrin signaling pathways mediate amyloid-beta–induced neurotoxicity. *Neurobiol. Aging, January 18.*

Xie, C., Markesbery, W. R., and Lovell, M. A. (2000). Survival of hippocampal and cortical neurons in a mixture of MEM+ and B27-supplemented neurobasal medium. *Free Radic. Biol. Med.* **28,** 665–672.

Yang, P., Baker, K. A., and Hagg, T. (2006). The ADAMs family: Coordinators of nervous system development, plasticity and repair. *Prog. Neurobiol.* **79,** 73–94.

Yebra, M., Montgomery, A. M., Diaferia, G. R., Kaido, T., Silletti, S., Perez, B., Just, M. L., Hildbrand, S., Hurford, R., Florkiewicz, E., Tessier-Lavigne, M., and Cirulli, V. (2003). Recognition of the neural chemoattractant netrin-1 by integrins alpha6beta4 and alpha3beta1 regulates epithelial cell adhesion and migration. *Dev. Cell* **5,** 695–707.

CHAPTER ELEVEN

METHODS FOR IDENTIFYING NOVEL INTEGRIN LIGANDS

Denise K. Marciano,[*,‡] Sumiko Denda,[†] *and* Louis F. Reichardt[‡]

Contents

Abstract

Integrins are cell adhesion receptors that have many important roles in organ development and tissue integrity, functioning to mediate interactions between cells and the ECM. The entire repertoire of integrins is vast, and the specific roles of each are determined by unique integrin–ligand interactions. These interactions allow for dynamic regulation of multiple processes. Despite intense efforts to elucidate individual integrin ligands, existing methods have been limiting. In this chapter, we describe methods developed in our laboratory to identify new integrin ligands that should be useful for characterizing novel integrin functions. These methods are applicable for studies on a variety of integrins, and may be extended to other cell surface receptors as well.

[*] Department of Medicine, Division of Nephrology, University of California, San Francisco, San Francisco, California
[†] Shiseido Research Center 2, Kanazawa-ku, Yokohama, Japan
[‡] Department of Physiology, Neuroscience Program, Howard Hughes Medical Institute, University of California, San Francisco, San Francisco, California

Methods in Enzymology, Volume 426

ISSN 0076-6879, DOI: 10.1016/S0076-6879(07)26011-2

© 2007 Elsevier Inc.

All rights reserved.

1. INTRODUCTION

Integrins comprise a family of heterodimeric cell surface receptors that play important roles in embryonic development, wound healing, hematopoiesis, hemostasis, and immunity (Geiger *et al.*, 2001; Hynes, 2002). Each consists of an α-subunit and a β-subunit, and in many cases, an individual α-subunit is able to pair with more than one β-subunit. To date, more than 20 different integrin heterodimers have been identified. In addition to their promiscuous partnering, most integrins have more than one extracellular ligand. Conversely, many matrix proteins, such as laminin, collagen, fibronectin, and vitronectin, have been shown to bind more than one integrin. This vast and intricate network of receptor–ligand complexes allows integrins to participate in a variety of cellular processes, such as cellular adhesion, migration, proliferation, differentiation, and survival (Geiger *et al.*, 2001; Howe *et al.*, 1998).

Integrins and their ligands have the capability of mediating bidirectional signaling (Takagi *et al.*, 2002). In outside-in signaling, an extracellular matrix (ECM) ligand or other ligand binds to its integrin receptor, transducing a signal to cytoskeletal and other cytoplasmic proteins to activate intracellular signaling pathways. In inside-out signaling, signals from cell surface receptors transduce intracellular signaling that changes the affinity of integrin for its extracellular ligand (Han *et al.*, 2006). Many of the signaling molecules have been identified (for review see Giancotti and Tarone, 2003), although the complex signaling pathways are still being elucidated.

While ECM proteins are one class of integrin ligands, the other major category of ligands are members of the immunoglobulin superfamily, such as vascular cell adhesion molecule (VCAM-1), intracellular adhesion molecules (ICAMs), and mucosal addressin cell adhesion molecule (MadCAM-1). In addition to these two ligand classes, E-cadherin, ADAM family members, and TGF-β have all been identified as integrin ligands. E-cadherin has been shown to be a ligand of αEβ7 in that an E-cadherin-Fc fusion protein binds to recombinant αEβ7 and solubilized αEβ7 from T lymphocytes (Higgins *et al.*, 1998). Several members of the ADAM (a disintegrin and metalloprotease) family of proteins have been shown to bind integrins via their disintegrin domains. Integrins known to interact with ADAM disintegrin-like domains include α4β1, α4β7, α5β1, α6β1, α9β1, and αvβ3 (White, 2003). Much of these data have been elucidated through cell-binding assays. The *in vivo* significance of these interactions remains unclear, although recently, cell culture experiments have demonstrated that specific ADAM–integrin interactions can alter integrin-mediated cell migration (Huang *et al.*, 2005). It has been postulated that ADAM–integrin interactions may promote the ectodomain shedding function of ADAMs, or alternatively, perturb integrin–ECM ligand interactions (White, 2003).

Several families of cell surface proteins also interact with integrins in *cis*, regulating their adhesive functions. Most notably, several tetraspanin proteins have been shown to associate with integrins and regulate their affinities (Lammerding *et al.*, 2003). In addition to regulating integrin affinity for ligands, the tetraspanins function as scaffolds for recruitment of cytoplasmic signaling proteins, including PKC (Chattopadhyay *et al.*, 2003). Urokinase-type plasminogen activator receptors (uPARs) have also been shown to bind multiple integrins in *cis*, including $\alpha M \beta 2$ (Mac-1), $\alpha 3 \beta 1$, and $\alpha 5 \beta 1$. As with ADAMs proteins, uPARs have been shown to alter integrin-mediated cell migration in a nonproteolytic fashion (Kugler *et al.*, 2003).

The process of identifying physiologically relevant integrin ligands has been arduous. Subsequent determination of the individual biological functions of each integrin–ligand pair has also been difficult, and in many instances cell culture models have substituted for *in vivo* models. Several factors have served as obstacles for ligand identification. First, in many conditions, integrins exist in multiple conformational states, with low-, intermediate-, and high-affinity conformations coexisting (Takagi *et al.*, 2002). Structural studies have shown that an integrin existing as a bent conformer in the extracellular domain has a low affinity for ligands, an extended conformer with a closed headpiece has an intermediate affinity, and an extended conformer with an open headpiece has a high affinity for ligand. Even in the presence of Mn^{2+}, a known activator of integrins, all three conformers of soluble $\alpha v \beta 3$ are present and exist in equilibrium. However, the presence of Mn^{2+} does shift the equilibrium to the high-affinity state so that a higher percentage of conformers bind ligand. In addition to Mn^{2+}, several other molecules and proteins, such as talin, are known to modulate the affinity state of integrins (Han *et al.*, 2006; Tadokoro *et al.*, 2003). Overexpression of talin in cells producing LFA-1 ($\alpha L \beta 2$) increases the ability of LFA-1 to bind its ICAM ligand (Kim *et al.*, 2003), a prime example of inside-out signaling. Talin also has been shown to increase the proportion of cells in the high-affinity state, inducing separation of the cytoplasmic α_L and β_2 domains (Kim *et al.*, 2003). Another example is the uPAR-$\alpha 5 \beta 1$ interaction. This interaction with uPAR changes $\alpha 5 \beta 1$ integrin conformation and the site at which $\alpha 5 \beta 1$ binds fibronectin from an RGD site to a non-RGD site (Wei *et al.*, 2005). Thus, integrin affinity and specificity for its ligand are modulated by multiple factors.

Since an integrin's conformation may equilibrate from high affinity to low affinity during purification or other procedures, it is possible that a bound ligand will be released. If only a fraction of ligand remains bound to its integrin, this will hinder *in vivo* ligand identification. Furthermore, even in the high-affinity state, many ligand-binding assays may not be sensitive enough to detect bound ligand. For example, the known dissociation constant of $\alpha L \beta 2$ for ICAM is 133 nM for the high-affinity state and 10,000-fold lower for the low-affinity state (Labadia *et al.*, 1998; Shimaoka *et al.*, 2001).

Given these limitations, most ECM integrin ligands have been identified using *in vitro* assays for *presumed* ligands rather than taking an unbiased approach. For example, integrin ligands have been identified using cell attachment assays in which cells expressing a particular integrin are tested for adherence to ECM-coated wells at increasing concentrations of purified ECM components (Tomaselli *et al.*, 1987). Specificity in this assay is often determined by blocking cell attachment with an integrin-blocking antibody or RGD peptide, or by using cells that either lack or contain a mutated integrin. Second, in cell-spreading assays, cells expressing an integrin heterodimer are assessed for their ability to extend processes on various ECM ligands (Denda *et al.*, 1998b). Third, gel filtration has been used to identify fibronectin, fibrinogen, and other proteins as ligands. For example, purified integrin and fibronectin are incubated together and then shown to co-elute from a gel filtration column (Horwitz *et al.*, 1986). Fourth, affinity purification over columns coupled with either ligand or integrin has been successful in providing evidence for direct interactions between several integrins and specific ligands. For example, this approach has been used to demonstrate direct binding of collagen to $\alpha1\beta1$ and $\alpha2\beta1$ (Pfaff *et al.*, 1994) and of fibronectin to $\alpha5\beta1$ (Cheng and Kramer, 1989). Fifth, surface plasmon resonance—in which purified soluble, truncated integrin is injected into flow cells that contain a particular purified ECM protein—has been used to characterize several previously identified integrin–ligand pairs (Pfaff *et al.*, 1994). Finally, in some instances, the affinity of integrin–ligand interactions is high enough to survive immunoprecipitation and washing procedures. For example, we have co-immunoprecipitated integrin $\alpha8\beta1$ from kidney extracts using an antibody to one of its ligands, nephronectin (Marciano, unpublished results).

All of these approaches have the limitation of testing only proposed ligands, and thus may overlook ligands that an unbiased approach would reveal. There are several instances in which identification of an integrin ligand has been delayed by the lack of an unbiased approach. For example, the identification of latent TGF-β as a ligand for integrins was very surprising as the TGF-β receptor complexes had been well characterized as ligand-activated protein kinases (Derynck and Zhang, 2003; Massague, 1987). Since latent TGF-β contains an RGD sequence and is located in the ECM, cell adhesion assays with a variety of integrins were performed. Surprisingly, the N-terminus of TGF-β (also known as the latency-associated peptide, or LAP) binds to $\alpha v\beta6$ in the adhesion assay, a finding that was confirmed using LAP-affinity column chromatography (Munger *et al.*, 1998). This interaction was shown later to be crucial in keratinocytes and airway epithelial cells for latent TGF-β activation, serving to induce matrix deposition and downregulate the inflammatory process after injury (Munger *et al.*, 1999).

As with TGF-β, the presence of an RGD sequence in a particular protein has stimulated studies to investigate the possibility that it functions as an integrin ligand. However, not all integrins recognize the RGD sequence, and some integrins recognize sites consisting of amino acids from more than one protein subunit, such as sequences in the triple helical domain of collagen (Sacca and Moroder, 2002). In addition to overlooking ligands, several of the abovementioned assays may identify ligands that are not relevant at physiological concentrations of integrin and ligand.

Several unbiased approaches could be used to discover novel protein–protein interactions that have not yet been used for integrin–ligand identification. One approach is to immunoprecipitate a specific integrin, and then evaluate co-eluted ligands by SDS-PAGE and subsequent mass spectrometry of gel fragments (Coon et al., 2005). In this manner, an integrin–ligand interaction could be identified with only trace amounts of ligand (approximately at femtomole level). A variant of this approach, using mass spectrometry to identify ligands from gel filtration peaks, might also yield novel interactions. Another approach, peptide display by filamentous phage, has been used successfully to identify peptide ligands, such as integrin–ligand mimetics (Koivunen et al., 1994, 1995), but has not identified physiologically relevant ligands. However, this may not be an ideal approach since the binding site of ligands is often derived from multiple segments of the protein that may not be contiguous (Sacca and Moroder, 2002).

We have developed a unique, unbiased approach to identify integrin ligands that has resulted in the discovery of two novel $\alpha8\beta1$ integrin ligands, nephronectin and osteopontin. Nephronectin was not known previously to be an ECM constituent (Brandenberger et al., 2001; Denda et al., 1998b). Because many more ECM proteins exist than have been characterized biochemically (Lander et al., 2001), there is a pressing need to develop methods to identify their receptors and other interactions. Using a soluble, truncated integrin heterodimer, we employed an expression cloning strategy and overlay assays to identify ligands. This approach had been used previously in tyrosine kinase receptors, namely c-kit, Mek-4, and Sek, to successfully identify ligands (Cheng and Flanagan, 1994; Flanagan and Leder, 1990), but had not been used with heterodimeric receptors such as integrins. In this approach, we engineered a soluble, truncated integrin heterodimer, consisting of the $\beta1$-subunit extracellular domain genetically fused to placental alkaline phosphatase, and the $\alpha8$-subunit extracellular domain genetically fused to a His6-myc tag. The His6 tag was inserted for ease of purification, and the alkaline phosphatase tag allowed us to quantify and trace the protein. When co-expressed in cultured cells, this heterodimeric probe ($\alpha8^t\beta1$-AP) yielded a soluble protein that retained its ability to bind known ligands, and could be used to detect and characterize novel ligands.

Once expressed, we purified $\alpha8^t\beta1$-AP heterodimer via its His6 tag with Ni^{++}-affinity chromatography and determined the functionality and specificity of $\alpha8^t\beta1$-AP using binding assays with known $\alpha8\beta1$ ECM ligands such as fibronectin and vitronectin. Binding was inhibited in the presence of RGD peptide and $\beta1$ monoclonal antibody, demonstrating its specificity.

After determining that the chimeric protein could bind known ligands, we used this soluble probe in an overlay assay (Far Western) to look for novel ligands, and found that that $\alpha8^t\beta1$-AP bound an additional band migrating at 70 to 90 kD that did not correspond to known ligands. To elucidate its identity we screened a cDNA expression library. We selected a lambda phage expression library (UNIZAP®) from murine embryonic heart since it lacks vitronectin, another $\alpha8\beta1$ ligand. Two million plaques were screened with $\alpha8^t\beta1$-AP, identifying both fibronectin and a novel protein, nephronectin.

To confirm that nephronectin was indeed a ligand for $\alpha8\beta1$, we performed cell adhesion assays and overlay assays. In the overlay assays, $\alpha8^t\beta1$-AP bound to blots containing kidney extract immunoprecipitated with anti-nephronectin antibody. There was no $\alpha8^t\beta1$-AP binding to blots that had been depleted of nephronectin. Our *in vivo* data from nephronectin null mice indicates that these mice have a similar phenotype as the $\alpha8$ null mice, further supporting the biological interaction of $\alpha8\beta1$ and nephronectin (Linton, J. *et al.*, 2007; Muller *et al.*, 1997).

The $\alpha8^t\beta1$-AP fusion protein also was used to localize sites of ligand binding directly in murine tissue using a procedure that has been named RAP (receptor affinity probe or receptor alkaline phosphatase) *in situ* by Cheng and Flanagan (1994). This is essentially an overlay assay on histological sections. $\alpha8^t\beta1$-AP binding in embryonic kidney was colocalized with nephronectin by anti-nephronectin antibody staining.

The ease with which $\alpha8^t\beta1$-AP was used to identify a novel ligand is due in part to its solubility, in comparison to full-length integrins, which are integral membrane proteins. Another factor may be the absence of cytoplasmic domains. Based on structural studies, it is well known that the α and β cytoplasmic tails are in close apposition in the low-affinity state, and they separate from each other in the high-affinity state (Kim *et al.*, 2003). It is possible, although we have not confirmed this by experimentation, that the absence of these membrane and cytoplasmic domains affects the conformational equilibrium, favoring the high-affinity state.

Discussed in subsequent sections are the materials and methods for $\alpha8^t\beta1$-AP construction, purification, cDNA expression cloning screen, and overlay assays for blots and histochemistry. We also describe methods for using intact, detergent-solubilized integrins for ligand identification and suggest possibilities for extension of these methods.

2. Production of Soluble $\alpha8^T\beta1$-AP

2.1. Construction of secreted integrin expression vectors

cDNA clones encoding the murine $\alpha8$ and $\beta1$ integrin subunits, including signal sequences, were modified by PCR reactions to produce soluble protein subunits lacking their cytoplasmic and transmembrane domains. Specific details, including primer sequences, have been published (Denda *et al.*, 1998a). The truncated $\alpha 8$ subunit was fused to six histidine residues followed by a Myc epitope followed by a stop codon (VIWATPNVSHHHHHHHGEQK-LISEEDL-stop). This was expressed in pCR3 (Invitrogen). A truncated integrin $\beta 1$ subunit ($\beta 1^t$) was generated by introducing a stop codon after the end of the extracellular domain (amino acid sequence number 728) by PCR. An integrin $\beta 1$ extracellular domain–alkaline phosphatase (AP) chimera ($\beta 1$-AP) was generated by isolating a modified pCR3-$\beta 1^t$ in which the stop codon at the end of the extracellular domain of $\beta 1^t$ was eliminated by PCR. This was fused with a fragment containing the alkaline phosphatase sequence from Aptag-1 and was expressed in pCR3 (Flanagan and Leder, 1990). Although very successful for expressing the $\alpha 8 \beta 1$ heterodimer, these truncations theoretically could prevent stable association of other α and β subunits because interactions between their transmembrane and cytoplasmic domains are lost. In such instances, fusion of the AP tag to the cytoplasmic domain of either subunit, followed by purification of the detergent-solubilized intact integrin heterodimer, may provide a useful reagent (see methods described in following sections).

2.2. Purification of soluble truncated integrin heterodimers

COS-7 cells were grown in 15-cm diameter dishes and transiently cotransfected using LipofectAmine Reagent (Gibco BRL) with the two plasmids, encoding extracellular domains of the α- and β-subunits plus indicated C-terminal tags ($\alpha 8^t$ and $\beta 1^t$ to express $\alpha 8^t$, β^t, $\alpha 8^t$, and $\beta 1$-AP to express $\alpha 8^t \beta 1$-AP). After treatment with the lipofection mixture for 5 h, the medium was changed to DMEM supplemented with Nutridoma HU (Boehringer, Indianapolis, IN) and antibiotics. The conditioned medium was collected every 2 to 4 days for 1 wk. After addition of $MgCl_2$ (at final concentration of 1 mM), phenylmethanesulfonyl fluoride (PMSF, 1 mM), and sodium azide (0.02%), the conditioned medium was filtered (paper number 1, Whatman, Maidstone, UK) and concentrated 10- to 20-fold using a YM100 membrane (Amicon, Beverly, MA). The concentrated medium was supplemented with Tris-Cl, pH 8.0 (at final concentration of 20 mM), imidazole (10 mM), and PMSF (0.5 mM), and was then incubated at 4° for 2 h in batch with Ni-NTA beads (Qiagen).

The beads were transferred into an empty column and washed with buffer A (20 mM Tris-Cl, pH 7.5, 300 mM NaCl, 1 mM $MgCl_2$, 0.02% sodium azide, 20 mM imidazole, 0.5 mM PMSF). After washing, bound proteins were eluted with elution buffer (20 mM Tris-Cl, pH 7.5, 50 mM NaCl, 1 mM $MgCl_2$, 0.02% sodium azide, 100 mM imidazole, 0.5 mM PMSF). To remove imidazole, $\alpha 8^t$ monomer, and nonspecifically bound contaminants of low molecular weight, the buffer of the eluate was exchanged by repeating concentration and dilution with the elution buffer without imidazole, using a Centricon 100 or Centriplus 100 filter apparatus

(Amicon, Beverly, MA). Protein concentration of the purified integrins was determined by both Coomassie Plus Protein Assay Reagent (Pierce, Rockford, IL) and silver staining of proteins after fractionation by SDS-PAGE. The purified heterodimers retained activity for at least 5 mo when stored at either 4° or −80°.

 ## 3. SOLID-PHASE BINDING ASSAYS WITH SOLUBLE INTEGRIN HETERODIMERS

Ninety-six–well plates (Maxisorp, Nunc, Rochester, NY) were coated with indicated concentrations of substrate proteins in TBS (25 mM Tris-Cl, pH 7.5, and 100 mM NaCl) at 4° overnight, blocked with 1% BSA (RIA grade, Sigma, St. Louis, MO) in TBS, and washed with TBS containing 1 mM MnCl$_2$ (TBS-Mn). Extracellular matrix ligands, including thrombospondin, laminin, fibronectin, osteopontin, and vitronectin were coated in the presence of 1 mM CaCl$_2$. As negative and positive controls, wells were coated with 1% BSA or 10 μg/ml FN120, respectively. One hundred microliters of $\alpha 8^t \beta 1$-AP (5 μg/ml in TBS-Mn) were then added to each well, and then incubated for 2 h at room temperature. After washing wells five times with TBS-Mn^{2+}, 100 μl of AP substrate (12 mM p-nitrophenyl phosphate, 0.5 mM MgCl$_2$, and 1 M diethanolamine, pH 9.8) were added and incubated at room temperature for an appropriate time. Integrin binding was quantified by measuring absorbance at 405 nm. $\alpha 8^t \beta 1^t$ biotinylated with NHC-LC-biotin (Pierce, Rockford, IL) was also used in the binding assay, and binding was detected as described above (DeFreitas *et al.*, 1995).

When working with scarce proteins or protein fragments, the plates were coated first with nitrocellulose and then with proteins as described (Lagenaur and Lemmon, 1987). The presence of nitrocellulose on the plate results in capture of virtually all of the protein in solution. In this case, we recommend first coating the Linbro Titertek 96-well plastic dishes with nitrocellulose (Lagenaur and Lemmon, 1987). In this procedure, 5 cm^2 of nitrocellulose type BA 85 (Schleicher and Schuell) is dissolved in 6 ml of methanol. Aliquots are spread over the surface of microwells and allowed to dry under a laminar flow hood. Test protein samples are applied at 50 μl per well at concentrations approximately 100-fold lower than the concentrations used for standard substrate preparation. Our laboratory has used this procedure to examine integrin-mediated interactions of neurons with tenascin and purified fragments of tenascin (Varnum–Finney *et al.*, 1995).

4. HISTOCHEMISTRY WITH SOLUBLE INTEGRIN HETERODIMERS

C57Bl/6 mouse embryos were fixed in 4% paraformaldehyde, embedded in paraffin, sectioned at 7 μm, stained with antibodies, and counterstained as described (Jones et al., 1994). Staining with $\alpha 8^t \beta 1$-AP was performed as described (Cheng and Flanagan, 1994) with some modifications. Sections were (1) blocked with 1% BSA, 25 mM Tris-Cl, pH 7.5, and 100 mM NaCl; (2) incubated for 4 to 12 h at room temperature with 7 μg/ml $\alpha 8^t \beta 1$-AP in 20 mM Tris-Cl, pH 7.5, 50 mM NaCl, 1 mM MnCl$_2$, 0.05% BSA, and 0.02% sodium azide; (3) washed with the same buffer; fixed with 60% acetone, 3% formaldehyde, and 20 mM HEPES, pH 7.0; washed with 20 mM HEPES, pH 7.0, and 150 mM NaCl; (4) heated at 65° for 1 h in this buffer; (5) rinsed in AP buffer (100 mM Tris-Cl, pH 9.5, 100 mM NaCl, 5 mM MgCl$_2$); and (6) incubated at room temperature with AP substrate solution (0.33 mg/ml nitroblue tetrazolium and 0.17 mg/ml 5-bromo-4-chloro-3-indoylphosphate in AP buffer). Sections were counterstained with methyl green (Zymed, South San Francisco, CA) and mounted with GVA mount (Zymed).

5. FAR WESTERN BLOTTING USING INTEGRIN HETERODIMERS

Protein extracts were obtained by homogenizing mouse embryos and kidneys in ice-cold extraction buffer (50 mM Tris-Cl, pH 7.5, 50 mM octylglucoside, 20 mM NaCl, 1 mM MgCl$_2$, 1 mM sodium vanadate, 1 mM sodium molybdate, 1 mM PMSF, 10 mg/ml leupeptin, 3 mg/ml pepstatin A) followed by centrifugation to remove debris. Protein concentrations were determined by Coomassie assays (Pierce). Far Western blotting was performed as described (Hildebrand et al., 1995) with some modifications. Proteins were separated by electrophoresis on SDS-polyacrylamide gels, transferred to nitrocellulose membranes, and renatured. Membranes were blocked at 4° with 10% BSA, 20 mM HEPES, pH 7.5, 75 mM KCl, 0.1 mM EDTA, 2.5 mM MgCl$_2$, and 0.05% NP40 for 1 h, and then 1% BSA, 25 mM Tris-Cl, pH 7.5, 50 mM NaCl, and 0.05% NP-40 for 1 h. Membranes were (1) washed in 0.1% BSA, 25 mM Tris-Cl, pH 7.5, 50 mM NaCl, 1 mM MnCl$_2$, and 0.05% NP-40; (2) incubated with 0.3 mg/ml $\alpha 8^t \beta 1$-AP in the same solution for 2 h at room temperature, and washed with this solution; (3) rinsed in AP buffer containing 0.5 mM MnCl$_2$; and (4) incubated with AP-substrate solution containing 0.5 mM MnCl$_2$.

In our experience of analyzing serum samples, the presence of fibronectin could obscure the presence of other proteins of similar molecular weight. When this was a problem, extracts were depleted of fibronectin by passing them three times through an anti-fibronectin antibody column. Alternatively, when we wished to assay interactions with a known protein, the extracts were immunoprecipitated with antiserum specific for this protein and protein A beads before Far Western analysis to detect binding to this protein in the absence of fibronectin (Denda *et al.*, 1998a).

6. LIGAND DETECTION USING INTACT INTEGRIN RECEPTORS

As described in the introduction, we used a lambda UNIZAP embryonic heart library. The lambda phage contains inserts whose expression is under the control of an isopropyl-β-D-thiogalactoyranoside (IPTG)–inducible promoter, and IPTG was used to screen for expressed proteins in infected bacteria. Numerous (2×10^6) plaques were screened with the $\alpha 8^t \beta 1$-AP.

Although the methods described above have worked exceedingly well in our laboratory, it is not clear that they provide methodology useful for all integrin heterodimers. The association of the two subunits of an integrin heterodimer is promoted not only by extracellular interactions mediated by the integrin α- and β-subunit head domains, but also by interactions mediated through their transmembrane and cytoplasmic domains. For many heterodimers, deletion of these domains, as was done in construction of the $\alpha 8 \beta 1$ chimeras described above, may result in subunit dissociation, making ligand detection impracticable (Hemler *et al.*, 1987). Although the introduction of interaction motifs, such as leucine zippers, may, in principle, stabilize integrin heterodimers, integrin activation is mediated through changes in the relationship between the α and β subunit stalks (Takagi *et al.*, 2002). In at least one instance, disulfide bonding of the integrin stalks has frozen it in its inactive form, which is clearly not appropriate for detecting functional interactions (Luo *et al.*, 2004). In section 7, we describe protocols used successfully by our laboratory to characterize a novel interaction between the integrin $\alpha 3 \beta 1$ and thrombospondin, which utilized an intact integrin solubilized in detergent (DeFreitas *et al.*, 1995). This approach seems likely to be more generally useful for characterizing integrin receptor ligands and can be further modified using recently developed procedures for incorporating membrane proteins into nanodiscs (Civjan *et al.*, 2003; Leitz *et al.*, 2006). It does require identification of a reasonable source for purification of the integrin, but this can be from a tumor or other cell line modified to ensure high expression of the integrin of interest. Alternatively, the assay can be modified by fusion of AP to the cytoplasmic C-terminus of the $\beta 1$ subunit.

7. $\alpha 3\beta 1$ Immunolabeling and Purification

As a first step, we label integrins on the surface of cells using biotinylation, and purify the integrin using an anti-cytoplasmic peptide antibody coupled to Sepharose. For integrin biotinylation on the surfaces of intact cells, cells were grown to confluency in 75-cm^2 tissue culture flasks. Rugli cells were used as a source for purification of the integrin $\alpha 3\beta 1$. Biotinylation was performed by a modification of the procedure described (Miyake *et al.*, 1991). Briefly, conflu-ent 75-cm^2 tissue culture flasks of Rugli cells were washed twice with warm PBS, after which 2 ml of PBS/0.1 M Na-HEPES, pH 8.0, was added to each flask or well with 80 μl of 10 mg/ml sulfo-NHS-biotin (Pierce Chemical Company, Rockford, IL) in PBS. After rocking for 1 h at room temperature, cells were washed twice with PBS, removed from the flask or well with CMF-PBS with 10 mM EDTA, washed again with CMF-PBS with 10 mM EDTA, and washed once with PBS to remove unreacted sulfo-NHS-biotin.

To immunopurify $\alpha 3\beta 1$, we used an affinity-purified antibody to the integrin $\alpha 3$A cytoplasmic domain (CYEAKGQKAEMRIQPSETER-LIDDY), which was coupled via the N-terminal cysteine residue to keyhole limpet hemocyanin (KLH), using the water-soluble hetero-bifunctional cross-linking reagent m-maleimidobenzoyl sulphosuccinimide ester according to the manufacturer's instructions (Pierce Chemical Co., Rockford, IL). Rabbit antisera to the peptide–KLH conjugates were raised in New Zealand white rabbits by a commercial vendor. The specific antibodies were affinity purified from the serum using as an affinity reagent the same cytoplasmic peptide coupled to thiopropyl-Sepharose (Pharmacia, Piscataway, NJ) via its amino-terminal cysteine residue. Antibodies were eluted with 0.1 M glycine-CI, pH 2.3, and dialyzed against CMF-PBS. Ten milligrams of affinity-purified, anti-α 3 cytoplasmic domain antibody was cross-linked to 3 ml of protein A-Sepharose (Pharmacia) with 20 mM dimethylpimelimidate (Pierce) in 0.2 M Na-borate, pH 9.0, as described (Harlow and Lane, 1988), and blocked with 0.2 M ethanolamine, pH 8.0. Biotinylated or nonbiotinylated Rugli cells from four confluent flasks were extracted in buffer B (50 mM N-octyl-β-D-glucopyranoside, 50 mM Tris-Cl, pH 7.5, 15 mM NaCI, 1 mM PMSF, CLAP) with 1 mM CaCl$_2$ and 1 mM MgCl$_2$, centrifuged at 10,000×g for 15 min, and incubated with CL-4B Sepharose (Pharmacia). The protease inhibitor cocktail, CLAP, was prepared as a 1000× stock with chymostatin, leupeptin, antipain, and pepstatin, each at 7 mg/ml in dimethylsulfoxide. The supernatant was then incubated with anti-$\alpha 3$-Sepharose for at least 4 h. Following six washes in buffer B, $\alpha 3\beta 1$ was eluted for 1 h with the $\alpha 3$A cytoplasmic domain peptide (1 mg/ml in buffer B).

While our laboratory has used this procedure only to purify the integrin $\alpha 3\beta 1$, we have prepared specific antibodies to almost all of the cytoplasmic domains of the integrins and believe this procedure is generally useful for purification of individual integrin heterodimers. Our laboratory has also solubilized these receptors through use of the detergent N-octyl-β-D-glucopyranoside. Recently, exciting procedures for incorporating membrane proteins into soluble nanoscale nanodiscs have been described that prevent aggregation and preserve function through use of a membrane scaffold protein (Chan *et al.*, 2003; Leitz *et al.*, 2006). This methodology seems likely to be particularly useful for examination of integrin functions.

8. Receptor-Binding Assays

For receptor-binding assays, substrate (75 μl per well) was incubated in 96-well dishes overnight at 4°, and then blocked with 1% BSA in PBS for 1 h at room temperature. Receptor purified as described above was diluted in buffer B with 1 mM CaCl$_2$ and 1 mM MgCl$_2$ at 1:10 to 1:50, respectively. The solutions were added to the substrata-blocked dishes and allowed to bind at room temperature for 2 h. The wells were then washed five times with buffer C (25 mM N-octyl-β-D-glucopyranoside, 50 mM Tris-CI, pH 7.5, 15 mM NaCl, 1 mM PMSF, CLAP) with 1 mM CaCl$_2$ and 1 mM MgCl$_2$ and incubated with streptavidin-conjugated horseradish peroxidase (Zymed) diluted in the same buffer, for 1 h at 40°. After washing five times more with buffer C with 1 mM CaCl$_2$ and 1 mM MgCl$_2$, binding was assayed by developing with 3,3′,5,5′-tetramethylbenzidine reagent (Kirkegaard and Perry, Gaithersburg, MD) and stopping the reaction with an equal volume of 1 M H$_3$PO$_4$. Absorbance at 450 nm was read in a microtiter plate reader (Flow Laboratories). Results were zeroed to wells with no receptor added. For experiments testing various divalent cations, 1 mM CaCl$_2$ and 1 mM MgCl$_2$ were replaced by the divalent cation tested. For experiments with antibodies added, receptor was used at a 1:50 dilution with antibody at 150 μg/ml. The receptor and antibody were incubated at room temperature for 1 h before the binding assay.

REFERENCES

Brandenberger, R., Schmidt, A., Linton, J., Wang, D., Backus, C., Denda, S., Müller, U., and Reichardt, L. F. (2001). Identification and characterization of a novel extracellular matrix protein nephronectin that is associated with integrin alpha8beta1 in the embryonic kidney. *J. Cell Biol.* **154,** 447–458.

Chan, C. S., Weeber, E. J., Kurup, S., Sweatt, J. D., and Davis, R. L. (2003). Integrin requirement for hippocampal synaptic plasticity and spatial memory. *J. Neurosci.* **23,** 7107–7116.

Chattopadhyay, N., Wang, Z., Ashman, L. K., Brady-Kalnay, S. M., and Kreidberg, J. A. (2003). alpha3beta1 integrin-CD151, a component of the cadherin-catenin complex, regulates PTPmu expression and cell–cell adhesion. *J. Cell Biol.* **163,** 1351–1362.

Cheng, H. J., and Flanagan, J. G. (1994). Identification and cloning of ELF-1, a developmentally expressed ligand for the Mek4 and Sek receptor tyrosine kinases. *Cell* **79,** 157–168.

Cheng, Y. F., and Kramer, R. H. (1989). Human microvascular endothelial cells express integrin-related complexes that mediate adhesion to the extracellular matrix. *J. Cell. Physiol.* **139,** 275–286.

Civjan, N. R., Civjan, N. R., Bayburt, T. H., Schuler, M. A., and Sligar, S. G. (2003). Direct solubilization of heterologously expressed membrane proteins by incorporation into nanoscale lipid bilayers. *Biotechniques* **35,** 556–560, 562–563.

Coon, J. J., Syka, J. E. P., Shabanowitz, J., and Hunt, D. F. (2005). Protein identification using sequential ion/ion reactions and tandem mass spectrometry. *Proc. Natl. Acad. Sci. USA* **102,** 9463–9468.

DeFreitas, M. F., Yoshida, C. K., Frazier, W. A., Mendrick, D. L., Kypta, R. M., and Reichardt, L. F. (1995). Identification of integrin alpha 3 beta 1 as a neuronal thrombospondin receptor mediating neurite outgrowth. *Neuron* **15,** 333–343.

Denda, S., Müller, U., Crossin, K. L., Erickson, H. P., and Reichardt, L. F. (1998a). Utilization of a soluble integrin-alkaline phosphatase chimera to characterize integrin alpha 8 beta 1 receptor interactions with tenascin: Murine alpha 8 beta 1 binds to the RGD site in tenascin-C fragments, but not to native tenascin-C. *Biochemistry* **37,** 5464–5474.

Denda, S., Reichardt, L. F., and Müller, U. (1998b). Identification of osteopontin as a novel ligand for the integrin alpha8 beta1 and potential roles for this integrin-ligand interaction in kidney morphogenesis. *Mol. Biol. Cell* **9,** 1425–1435.

Derynck, R., and Zhang, Y. E. (2003). Smad-dependent and Smad-independent pathways in TGF-beta family signalling. *Nature* **425,** 577–584.

Flanagan, J. G., and Leder, P. (1990). The kit ligand: A cell surface molecule altered in steel mutant fibroblasts. *Cell* **63,** 185–194.

Geiger, B., Bershadsky, A., Pankov, R., and Yamada, K. M. (2001). Transmembrane crosstalk between the extracellular matrix—cytoskeleton crosstalk. *Nat. Rev. Mol. Cell Biol.* **2,** 793–805.

Giancotti, F. G., and Tarone, G. (2003). Positional control of cell fate through joint integrin/receptor protein kinase signaling. *Annu. Rev. Cell Dev. Biol.* **19,** 173–206.

Han, J., Lim, C. J., Watanabe, N., Soriani, A., Ratnikov, B., Calderwood, D. A., Puzon-McLaughlin, W., Lafuent,e, E. M., Boussiotis, V. A., Shattil, S. J., and Ginsberg, M. H. (2006). Reconstructing and deconstructing agonist-induced activation of integrin alphaIIbbeta3. *Curr. Biol.* **16,** 1796–1806.

Harlow, E., and Lane, D. (1988). "Antibodies: A Laboratory Manual." Cold Spring Harbor Press, Cold Spring Harbor, NY.

Hemler, M. E., Huang, C., Takada, Y., Schwarz, L., Strominger, J. L., and Clabby, M. L. (1987). Characterization of the cell surface heterodimer VLA-4 and related peptides. *J. Biol. Chem.* **262,** 11478–11485.

Higgins, J. M., Mandlebrot, D. A., Shaw, S. K., Russell, G. J., Murphy, E. A., Chen, Y. T., Nelson, W. J., Parker, C. M., and Brenner, M. B. (1998). Direct and regulated interaction of integrin alphaEbeta7 with E-cadherin. *J. Cell Biol.* **140,** 197–210.

Hildebrand, J. D., Schaller, M. D., and Parsons, J. T. (1995). Paxillin, a tyrosine phosphorylated focal adhesion-associated protein binds to the carboxyl terminal domain of focal adhesion kinase. *Mol. Biol. Cell* **6,** 637–647.

Horwitz, A., Duggan, K., Buck, C., Beckerle, M. C., and Burridge, K. (1986). Interaction of plasma membrane fibronectin receptor with talin—a transmembrane linkage. *Nature* **320,** 531–533.

Howe, A., Aplin, A. E., Alahari, S., and Juliano, R. L. (1998). Integrin signaling and cell growth control. *Curr. Opin. Cell Biol.* **10,** 220–231.

Huang, J., Bridges, L. C., and White, J. M. (2005). Selective modulation of integrin-mediated cell migration by distinct ADAM family members. *Mol. Biol. Cell* **16,** 4982–4991.

Hynes, R. O. (2002). Integrins: Bidirectional, allosteric signaling machines. *Cell* **110,** 673–687.

Jones, K. R., Fariñas, I., Backus, C., and Reichardt, L. F. (1994). Targeted disruption of the BDNF gene perturbs brain and sensory neuron development but not motor neuron development. *Cell* **76,** 989–999.

Kim, M., Carman, C. V., and Springer, T. A. (2003). Bidirectional transmembrane signaling by cytoplasmic domain separation in integrins. *Science* **301,** 1720–1725.

Koivunen, E., Wang, B., and Ruoslahti, E. (1994). Isolation of a highly specific ligand for the alpha 5 beta 1 integrin from a phage display library. *J. Cell Biol.* **124,** 373–380.

Koivunen, E., Wang, B., and Ruoslahti, E. (1995). Phage libraries displaying cyclic peptides with different ring sizes: Ligand specificities of the RGD-directed integrins. *Bio./Technol.* **13,** 265–270.

Kugler, M. C., Wei, Y., and Chapman, H. A. (2003). Urokinase receptor and integrin interactions. *Curr. Pharm. Des.* **9,** 1565–1574.

Labadia, M. E., Jeanfavre, D. D., Caviness, G. O., and Morelock, M. M. (1998). Molecular regulation of the interaction between leukocyte function-associated antigen-1 and soluble ICAM–1 by divalent metal cations. *J. Immunol.* **161,** 836–842.

Lagenaur, C., and Lemmon, V. (1987). An L1-like molecule, the 8D9 antigen, is a potent substrate for neurite extension. *Proc. Natl. Acad. Sci. USA* **84,** 7753–7757.

Lammerding, J., Kazarov, A. R., Huang, H., Lee, R. T., and Hemler, M. E. (2003). Tetraspanin CD151 regulates alpha6beta1 integrin adhesion strengthening. *Proc. Natl. Acad. Sci. USA* **100,** 7616–7621.

Lander *et al.*, (2001). Initial sequencing and analysis of the human genome. *Nature* **409.** 860–921

Leitz, A. J., Bayburt, T. H., Barnakov, A. N., Springer, B. A., and Sligar, S. G. (2006). Functional reconstitution of Beta2-adrenergic receptors utilizing self-assembling Nanodisc technology. *Biotechniques* **40,** 601–602, 604, 606, passim.

Luo, B. H., Takagi, J., and Springer, T. A. (2004). Locking the beta3 integrin I-like domain into high and low affinity conformations with disulfides. *J. Biol. Chem.* **279,** 10215–10221.

Massague, J. (1987). Identification of receptors for type-beta transforming growth factor. *Methods Enzymol.* **146,** 174–195.

Miyake, K., Weissman, I. L., Greenberger, J. S., and Kincade, P. W. (1991). Evidence for a role of the integrin VLA-4 in lympho-hemopoiesis. *J. Exp. Med.* **173,** 599–607.

Muller, U., Wang, D., Denda, S., Meneses, J. J., Pedersen, R. A., and Reichardt, L. F. (1997). Integrin alpha8beta1 is critically important for epithelial-mesenchymal interactions during kidney morphogenesis. *Cell* **88,** 603–613.

Munger, J. S., Harpel, J. G., Giancotti, F. G., and Rifkin, D. B. (1998). Interactions between growth factors and integrins: Latent forms of transforming growth factor-beta are ligands for the integrin alphavbeta1. *Mol. Biol. Cell* **9,** 2627–2638.

Munger, J. S., Harpel, J. G., Giancotti, F. G., and Rifkin, D. B. (1999). The integrin alpha v beta 6 binds and activates latent TGF beta 1: A mechanism for regulating pulmonary inflammation and fibrosis. *Cell* **96,** 319–328.

Pfaff, M., Göhring, W., Brown, J. C., and Timpl, R. (1994). Binding of purified collagen receptors (alpha 1 beta 1, alpha 2 beta 1) and RGD-dependent integrins to laminins and laminin fragments. *Eur. J. Biochem.* **225,** 975–984.

Sacca, B., and Moroder, L. (2002). Synthesis of heterotrimeric collagen peptides containing the alpha1beta1 integrin recognition site of collagen type IV. *J. Peptide Sci.* **8,** 192–204.

Shimaoka, M., Ferzly, M., Oxvig, C., Takagi, J., and Springer, T. A. (2001). Reversibly locking a protein fold in an active conformation with a disulfide bond: integrin alphaL I domains with high affinity and antagonist activity *in vivo*. *Proc. Natl. Acad. Sci. USA* **98,** 6009–6014.

Tadokoro, S., Shattil, S. J., Eto, K., Tai, V., Liddington, R. C., de Pereda, J. M., Ginsberg, M. H., and Calderwood, D. A. (2003). Talin binding to integrin beta tails: A final common step in integrin activation. *Science* **302,** 103–106.

Takagi, J., Petre, B. M., Walz, T., and Springer, T. A. (2002). Global conformational rearrangements in integrin extracellular domains in outside-in and inside-out signaling. *Cell* **110,** 599–611.

Tomaselli, K. J., Damsky, C. H., and Reichardt, L. F. (1987). Interactions of a neuronal cell line (PC12) with laminin, collagen IV, and fibronectin: Identification of integrin-related glycoproteins involved in attachment and process outgrowth. *J. Cell Biol.* **105,** 2347–2358.

Varnum-Finney, B., Venstrom, K., Muller, U., Kypta, R., Backus, C., Chiquet, M., and Reichardt, L. F. (1995). The integrin receptor alpha 8 beta 1 mediates interactions of embryonic chick motor and sensory neurons with tenascin-C. *Neuron* **14,** 1213–1222.

Wei, Y., Czekay, R. P., Robillard, L., Kugler, M. C., Zhang, F., Kim, K. K., Xiong, J. P., Humphries, M. J., and Chapman, H. A. (2005). Regulation of alpha5beta1 integrin conformation and function by urokinase receptor binding. *J. Cell Biol.* **168,** 501–511.

White, J. M. (2003). ADAMs: Modulators of cell–cell and cell–matrix interactions. *Curr. Opin. Cell Biol.* **15,** 598–606.

CHAPTER TWELVE

Analysis of Integrin Functions in Peri-Implantation Embryos, Hematopoietic System, and Skin

Eloi Montanez,* Aleksandra Piwko-Czuchra,* Martina Bauer,* Shaohua Li,[†] Peter Yurchenco,[‡] *and* Reinhard Fässler*

Contents

* Max Planck Institute of Biochemistry, Department of Molecular Medicine, Martinsried, Germany
† Department of Surgery, Robert Wood Johnson Medical School, New Brunswick, New Jersey
‡ Department of Pathology, Robert Wood Johnson Medical School, Piscataway, New Jersey

Methods in Enzymology, Volume 426
ISSN 0076-6879, DOI: 10.1016/S0076-6879(07)26012-4

© 2007 Elsevier Inc.
All rights reserved.

Abstract

Integrins mediate cell adhesion, permit traction forces important for cell migration, and cross-talk with growth factor receptors to regulate cell proliferation, cell survival, and cell differentiation. The plethora of functions explains their central role for development and disease. The progress in mouse genetics and the ease with which the mouse genome can be manipulated enormously contributed to our understanding of how integrins exert their functions at the molecular level. In the present chapter, we describe tests that are routinely used in our laboratory to investigate embryos, organs, and cells (peri-implantation embryos, hematopoietic system, epidermis, and hair follicles) that lack the expression of integrins or integrin-associated proteins.

1. INTRODUCTION

Integrins are α/β heterodimeric, type I cell surface receptors that mediate cell–extracellular matrix (ECM) and cell–cell interactions. The binding of integrins to their ligands depends on an activation mechanism that is not fully understood and triggered by an inside–out signaling. When activated integrins have bound their ligands they in turn trigger an outside-in signaling including actin reorganizations and activation of various intracellular signaling pathways. Therefore, it is not astounding that genetic and pharmacological studies firmly established their fundamental importance for multicellular life.

The integrin family with more than 20 members is found on blastomeres and on all cells of the developing and adult mouse. Gene deletions and modifications in mice have enormously contributed to our understanding of how integrins, integrin-associated proteins, and signaling pathways work *in vivo*. Some deletions arrest development and lead to embryonic lethality.

Embryonic lethal phenotypes are overcome with tissue-specific or temporally restricted gene deletion, allowing the study of integrin functions in specific cell types at specific time points in development or in adulthood. The *Cre/loxP* system enables spatiotemporal gene ablations by removing DNA sequences that are flanked by 34-nucleotide-long *loxP* sites with the recombinase *Cre* from the bacteriophage P1. The expression of *Cre* under the control of a specific tissue promoter allows spatial gene ablations, while the use of inducible promoters that drive *Cre* expression provide a temporal control of the *Cre*-mediated gene recombination. The consequence of the *in vivo* gene deletion analysis can be complemented with the establishment of primary cells, which can be immortalized and used to molecularly dissect the signaling pathways regulated by integrins.

In the present chapter, we describe protocols that are routinely used in our laboratory to investigate the role of integrins and integrin-associated proteins during the peri-implantation stage using embryoid bodies (EBs), in the hematopoietic system and in the skin and skin appendages.

2. Analysis of Integrin Functions during Peri-Implantation Development

Deletion of $\beta1$ integrin as well as several of its cytoplasmic-associated proteins leads to embryonic lethality at the peri-implantation stage (Fassler and Meyer, 1995; Legate *et al.*, 2006). The difficulty of obtaining sufficient numbers of embryos at this early stage of mouse development complicates the elucidation of the cellular and molecular defects that underlie lethality. In order to circumvent this problem, we generate and analyze EBs from pluripotent mouse embryonic stem cells (ESCs).

ESCs derived from the inner cell mass of 3.5-day-old (E3.5) embryos called blastocysts, can be differentiated *in vitro* into a variety of cell types, and thus provide a powerful model system to investigate cellular and molecular events occurring during differentiation into various cell lineages. Suspension aggregates of ESCs spontaneously form multicellular three-dimensional structures called EBs. They closely recapitulate the *in vivo* cell differentiation programs occurring during the peri-implantation stage of development and follow a well-defined temporal and molecular program taking place during implantation of mammalian embryos. Initially the outer layer of the ESC aggregate differentiates into primitive endoderm cells (or hypoblast), which lay down the first embryonic basement membrane (BM). The BM, in turn, triggers differentiation of the adjacent ESCs into a columnar epithelium (epiblast or primitive ectoderm), while the remaining ESCs, which fail to contact the BM undergo apoptosis leading to the formation of a central cavity (Fig. 12.1A). The generation of EBs from ESCs has become

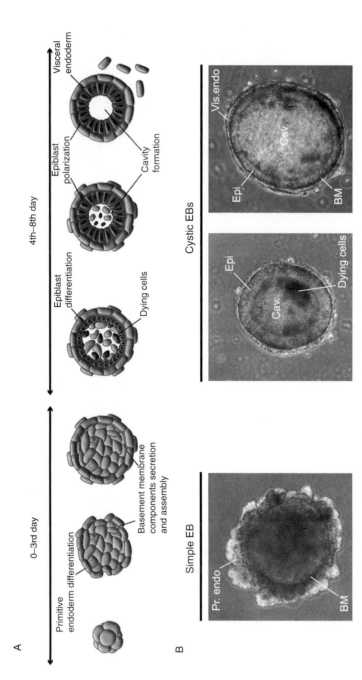

Figure 12.1 Stage of EB development. (A) Schematic representation of the major events that take place during EB differentiation. (B) Phase-contrast pictures of simple and cystic EBs. BM, basement membrane; cav, cavity; EB, embryoid bodies; Epi, epiblast; Pr. endo, primitive endo-derm; Vis. endo, visceral endoderm. (See color insert.)

an attractive model to unravel the cellular and molecular properties of genes involved in peri-implantation mouse development.

2.1. Generation of EBs from ESC aggregates

Although there are many different protocols to produce EBs, all of them share a final step of suspension culture of ESC aggregates. The most critical and variable step of all protocols is the generation of ESC aggregates. The protocol described here is based on one developed in the Yurchenco laboratory. Major steps include the (1) culture of undifferentiated ESCs on feeder cells, (2) generation of ESC aggregates called EB precursors, (3) removal of feeder cells, and (4) culture of EB precursors in suspension for 7 to 8 days. Steps are summarized in Fig. 12.2. Ideally, EB precursors are formed by three to four cells, since larger cell clusters rapidly degenerate in suspension and single cells poorly differentiate into EBs.

All cell culture steps of this protocol are performed in a cell culture incubator at 37° in a humidified 5% CO_2 atmosphere.

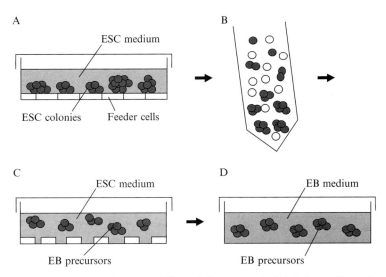

Figure 12.2 Main steps of the EBs differentiation protocol. (A) Culture of undifferentiated ESCs to a 70 to 80% confluence on a monolayer of growth-inactivated feeder cells. (B) Generation ESC clusters, called EB precursors, by repeated pipetting. (C) Removal of feeder cells by plating cell suspension in a tissue culture dish for 30 to 40 min. (D) Culture of EB precursors in suspension in bacterial dishes for 7 to 8 days (1×10^3 clusters/dish). EB, embryoid bodies; ESC, embryonic stem cells.

2.1.1. Materials

Glass fine-tip Pasteur pipettes: Pull glass Pasteur pipettes over the flame to obtain a fine tip, approximately 2 μm in diameter. Ideally, the liquid should flow through the tip of the pipette as a very fine stream. The pulled Pasteur pipettes are then autoclaved.

Feeder cell medium: High-glucose Dulbecco's Modified Eagle Medium (DMEM) with sodium pyruvate (Gibco) supplemented with 10% heat-inactivated fetal bovine serum (FBS) (Gibco), 2 mM L-glutamine (Gibco), 100 U/ml penicillin/100 μg/ml streptomycin (100×, stock PAA).

ESC medium: High-glucose DMEM with sodium pyruvate (Gibco) supplemented with 20% heat-inactivated FBS, 2 mM L-glutamine (Gibco), 0.1 mM 2-β-mercaptoethanol (Merck), 1× nonessential amino acids from a 100× stock solution (Gibco), 1000 U/ml leukemia inhibitory factor (LIF) (Gibco), 100 U/ml penicillin/100 μg/ml streptomycin (100×, stock PAA).

EB medium: ESC medium lacking LIF.

ESC lines: We typically use R1 and R1-derived mutant ESC lines. The D3 ESC line is also commonly used in our laboratory (Aumailley *et al.*, 2000).

Feeder cells: We use irradiated primary mouse embryonic fibroblasts (MEFs). STO cells, which are immortalized 3T3-like fibroblasts, have also been successfully used (Li *et al.*, 2002).

2.1.2. Culture of undifferentiated ESCs

Mouse ESCs can be maintained in an undifferentiated, pluripotent state indefinitely by culturing them on a monolayer of growth-inactivated MEFs or on gelatin coats in the presence of LIF, an inhibitor of ESC differentiation. MEFs are isolated from E13.5 embryos and growth inactivated by γ-irradiation according to an established protocol (Talts *et al.*, 1999). Growth-inactivated MEFs can be kept in culture for 10 to 14 days and have to be seeded either the day before plating of ESCs or together with them.

Since high-density cultures of ESC have the tendency to differentiate, they should never reach more than 80% confluence. Furthermore, the medium should be replaced daily. A suitable ESC culture for EB formation consists of spherical-shape ESC colonies growing on top of a monolayer of feeder cells. However, some mutants ESCs such as, for example, β1 integrin-deficient ESCs, have severe adhesion defects. These cells have to be handled very carefully to allow them to settle by gravity before changing the medium.

Procedure

1. The day before plating the ESCs, the feeder cells are seeded in feeder cell medium onto a 10-cm tissue culture dish in a subconfluent density (90 to 95%).
2. The ESCs are rapidly thawed at 37°, transferred into a 15-ml conical tube containing 10 ml of ESC medium and spun down for 5 min at 300×g.

3. The medium is sucked off and the ESC pellet is resuspended in 10 ml of fresh ESC medium.
4. The feeder cell medium is removed from the feeder cells and rinsed once with 10 ml of prewarmed (37°) PBS.
5. The ESCs are seeded onto the feeder cell layer and grown until they reach 70 to 80% confluence (typically 2 days after seeding).

2.1.3. Generation of EB precursors
Procedure

1. The ESC medium is removed and the culture dish with ESCs is rinsed twice with 10 ml of prewarmed (37°) PBS.
2. The ESC colonies are detached from feeders by adding 3 ml of 0.25% Trypsin/0.53 mM EDTA for 2 to 3 min at room temperature (RT). Cell detachment should be monitored under the microscope and can be facilitated by tapping the tissue culture dish.
3. When ESC colonies start to separate from feeder cells, the trypsin is immediately inactivated by adding 6 ml of feeder cell medium.
4. After repeated pipetting with a 5-ml plastic pipette, the cell suspension is collected in a 15-ml conical tube. ESC aggregates are then settled by gravity for 10 min at RT.
5. The medium (containing most of the feeder cells and single ESCs) is discarded and the pellet (containing ESC clusters), is subsequently resuspend in 2 ml of EB medium.
6. Cells within ESC colonies are tightly attached to each other. To obtain EB precursors, ESC clusters are broken by passing them through a fine-tip Pasteur pipette with a 2-μm diameter. The size is carefully checked under the microscope and the process is repeated until almost all ESC clusters display a suitable size (three to four cells).

2.1.4. Removal of feeder cells
Residual feeder cells will interfere with EB differentiation and therefore, have to be removed as complete as possible. Since feeder cells attach faster to the tissue culture dish than ESCs, the EB precursors suspension can be efficiently depleted of feeder cells by replating the culture suspension for 30 to 45 min. This panning can also be used to remove most of the remaining large ESC aggregates.

Procedure

1. The EB medium is added to the suspension with the EB precursors (Step 6 in Section 2.1.3.1) to obtain a final volume of 8 ml and incubated in a new fresh tissue culture dish.
2. After 10 to 15 min of incubation, the main part of the feeder cells is reattached to the tissue culture dish and large ESC aggregates have settled

down by gravity. The medium containing EB precursors is carefully aspirated and transferred into a fresh tissue culture dish.

3. The procedure is repeated around three times to ensure complete feeder cell depletion.

2.1.5. Culture of EB precursors in suspension

1. EB differentiation is initiated by suspending the EB precursors in EB medium and subsequently transferring them (from Step 3, Section 2.1.4.1) into bacterial-grade Petri dishes containing EB medium to obtain a final volume of around 10 ml. The density of the culture should not be higher than 1×10^3 EB precursors/dish. This day is considered as day 0 of EB differentiation.

2. The EB precursors are cultured for 3 days without changing the medium. To avoid fusion between EB precursors, the bacterial dishes are not moved during this critical culture period. After 3 days, an outer layer of primitive endoderm cells and a BM can be distinguished in many of the aggregates by phase-contrast microscopy. At this stage of differentiation cell aggregates are termed simple EBs (Fig. 12.1B).

3. On the 3rd day of EB differentiation, the tissue culture dish is tilted for 10 min to allow cell aggregates to settle by gravity. Three to four volumes of the medium are carefully removed and replaced with 10 ml of fresh EB medium. During this step it is important to carefully suck off the culture medium to avoid removing too many of the cell aggregates.

4. The cell aggregates are cultured for another 4 to 5 days. The medium is changed every 2nd day. At the 7th day of EB differentiation, a high percentage of simple EBs will develop an inner layer of columnar epithelial cells (epiblast or primitive ectoderm) and a fluid-filled cavity. These structures are called cystic EBs (Fig. 12.1B)

2.2. Immunocytochemical characterization of EBs

Embryoid bodies can be collected, fixed, embedded, and cut for characterization at all time points throughout the culture period. Sections can be stained with specific antibodies that allow the analysis of cell biological processes such as cell differentiation, secretion and assembly of the BM, cell survival, cell proliferation, cell polarization, and so on.

2.2.1. Materials

Fixation solution: 4% paraformaldehyd (PFA) in PBS, always prepared fresh. PFA is available as a powder (Sigma) or as a liquid (Merck). The dissolution of the PFA powder can be facilitated by stirring the solution at 60°.
Permeabilization solution: 0.1% TritonX-100 in PBS.
Blocking solution: 3% BSA in PBS.

DAPI solution: 4′,6-diamidino-2-phenylindole (Sigma), 1 ml of stock solution of 1 mg/ml in Millipore water is prepared and stored at −20° in the dark. For staining, a 0.1 μg/ml solution is prepared in PBS.

Primary and secondary (Table 12.1) antibody solutions: The antibodies are diluted in 3% BSA in PBS to a dilution recommended by the manufacturer.

2.2.2. Embedding EBs into cryomatrix and obtaining cryosections
Procedure

1. The EBs are transferred from the bacterial dishes into 15-ml conical tubes and allowed to settle by gravity for 10 min.
2. The supernatant is removed and the EBs-pellet is then resuspended in 10 ml of PBS.
3. The EBs are settled again by gravity for 10 min and again resuspended in 10 ml of PBS.

Table 12.1 List of antibodies used to stain cryosections of EBs

Protein	Application	Source
Anti-cleaved-caspase-3	Apoptosis marker	Cell signalling
Tec3	Proliferation marker	Dako
Anti-gata-4	Endoderm marker	Santa Cruz
Anti-α-feto protein	Endoderm marker	Santa Cruz
Anti-talin	Integrin associated protein	Sigma
Anti-integrin α6	Adhesion receptor	PharMingen
Anti-integrin β1, clone MB1.2	Adhesion receptor	Chemicon
Anti-integrin β1 9EG7 epitope	Active integrin	Sigma
Anti-β-dystroglycan	ECM-cell receptor	Tebu-bio
Anti-laminin α1	BM component	Chemicon
Anti-collagen IV	BM component	Chemicon
Anti-nidogen	BM component	SantaCruz
Anti-perlecan	BM component	Chemicon
Anti-fibronectin	BM component	Chemicon
GM130	Polarization marker	BD biosciences
Anti-E-cadherin	Cell-cell adhesion	Zymed
Anti-ZO-1	Cell-cell adhesion	Zymed

ECM, extracellular matrix; BM, basement membrane.
Note: TRICT- and FITC-conjugated phalloidin is used to detect F-actin. Cy3- and FITC-conjugated antibodies specific for mouse IgG, mouse IgM, rabbit IgG, and rat IgG (Jackson Immunochemicals) are used as secondary antibodies.

4. The EBs are spun down at 150×*g* and incubated in fixation solution for 30 min at RT.
5. The fixing solution is removed, and EBs are subsequently incubated with 7.5% sucrose in PBS for 3 h at RT and then with 15% sucrose in PBS overnight at 4°.
6. The 15% sucrose-PBS solution is removed and replaced by 1 ml of cryomatrix (Shandon Cryomatrix^{TM}, Thermo).
7. The cryomatrix (containing EBs) is transferred into a standard cryo mold (Tissue-Tek®, Sakura), and placed on a cooled copper plate on dry ice.
8. When the cryomatrix is completely frozen 8 μm-thick sections are cut using a cryostate (Microm). Frozen sections are collected onto positively charged SuperFrost®Plus glass microscope slides (Microm). The slides can either be immediately stained or stored at −80° for several weeks.

2.2.3. Immunofluorescence staining of EB cryosections
Procedure

1. The slides are rinsed three times with PBS.
2. The sections are subsequently incubated with permeabilization solution for 2 min on ice (4°).
3. The sections are then incubated with blocking solution for 45 min at RT.
4. The blocked sections are incubated with the primary antibody solution (Table 12.1) overnight at 4° in a humidified chamber.
5. The slides are washed three times for 5 min with PBS.
6. The sections are incubated with secondary antibody solution (Table 12.1) diluted in PBS for 1 h at RT in a humidified chamber. From this step on the sections should be protected from light.
7. The antibody solutions are removed from the slides and then they are incubated with DAPI solution for 5 min at RT.
8. The slides are rinsed three times with PBS and carefully dried around the sample. The sections are mounted with an antifading medium (Elvanol) and air-dried for 30 min. The slides can be immediately analyzed or stored in the dark for several days at 4°.

During mammalian development, the first wave of apoptosis occurs when the embryo cavitates. This process is reproduced in EBs where the inner cells of the simple EBs undergo apoptosis. Integrins and integrin-associated proteins modulate cell survival and thus cavity formation. Deletion of integrin-linked kinase (ILK) leads to a cavitation defect (Sakai *et al.*, 2003), whereas deletion of PINCH-1 (particularly interesting Cys-His-rich protein-1) triggers apoptosis of endoderm cells (Li *et al.*, 2005). Therefore, the analysis of apoptosis is an important topic in the study of EBs formation that can be easily done in cryosections by using the *In Situ* Cell Death Detection Kit, POD (TUNEL Technology, Roche).

3. ANALYSIS OF INTEGRIN FUNCTIONS IN BLOOD

The leukocyte-specific $\beta2$ and $\beta7$ integrins as well as almost all members of the $\beta1$ integrin subfamily play important roles in the hematopoietic system (Sixt et al., 2006). The integrins are required for the homing of hematopoietic stem cells to the fetal liver and bone marrow (Hirsch et al., 1996; Potocnik et al., 2000), hematopoiesis (Bouvard et al., 2001), extravasation of leukocytes at sites of inflammation (Mizgerd et al., 1997; Yednock et al., 1992), and the formation of lymphatic organs such as Peyer's patches, and so on. Consequently, the analysis of integrin function in the hematopoietic system will undoubtedly contribute to our understanding of how the various blood cells move from one site to another, how the cells communicate with each other, and how they exert their functional properties. A serious problem of genetic studies of blood cells can be early embryonic lethality, or the loss of gene expression in several non-hematopoietic cell types relevant for blood cells, such as endothelial cells or stromal cells. This problem can be elegantly circumvented with conditional and inducible gene deletions. Numerous, well-characterized *Cre* mouse lines are available that restrict the gene ablation in a constitutive or temporal manner to specific cell types or compartments of the hematopoietic system. This rich resource makes genetic studies of hematopoiesis very attractive and doable (Table 12.2).

The *Mx1-Cre* mouse strain represents an excellent option to temporally ablate genes in the hematopoietic system (Kuhn et al., 1995). The *Mx1* promoter is activated by interferon-α or -β, whose expression in turn can readily be induced by the intraperitoneal injection of the synthetic double-stranded RNA polyinosinic-polycytidylic acid (pI–pC). Since the *Mx1* promoter is highly active in hematopoietic stem cells, pI–pC injections lead to a rapid and efficient deletion of *loxP*-flanked (also–called floxed) genes in all hematopoietic cells. Unfortunately, a large number of nonhematopoietic cells, such as hepatocytes and endothelial cells, among others, also express the interferon α/β receptors and hence also lose the floxed gene. To restrict the deletion to the hematopoietic compartment, the *loxP/Mx1-Cre* system must be combined with the generation of bone marrow chimeras (Fig. 12.3), which is a rather easy task. This task is usually done with C57BL/6 mice that express the Ly-5.2 and B6SJL that express the Ly-5.1 surface antigen on all leukocytes. The two mouse strains provide an elegant system to generate bone marrow chimeras between C57BL/6 donors (floxed gene/Mx1-Cre bone marrow) and lethally irradiated, wild-type B6SJL hosts, thereby providing a marker to easily exclude remaining cells of the irradiated host from the analysis.

Bone marrow chimeras cannot be generated when a gene mutation leads to embryonic lethality. In such a case, it may still be possible to isolate and transfer hematopoietic stem cells from the fetal liver cells or the adreno-gonadal-mesonephros region into lethally irradiated hosts (Gribi et al., 2006).

Table 12.2   Cre lines for gene deletions in the hematopoietic system

Promoter	Expression pattern	Special notes/reference
Lck	T cells	Onset of deletion during the DN1 stage of thymocyte development (Lee *et al.*, 2001)
CD4	T cells	Onset of deletion during the double positive stage of thymocyte development (Lee *et al.*, 2001)
mb1	B cells	Onset of deletion in early pro-B cells, low frequency of deletion in T cells (Hobeika *et al.*, 2006)
CD19	B cells	Onset of deletion during the pre-B cell stage (Rickert *et al.*, 1997)
CD21	B cells	Onset of deletion in mature B cells, high deletion in the ovary (Kraus *et al.*, 2004)
hCD2	T cells and B cells	Onset of deletion before the DN4 stage of thymocyte development and in early pro-B cells, mosaic expression in testis (de Boer *et al.*, 2003)
Vav	All cells of hematopoietic system	Expression in ovaries and testis (de Boer *et al.*, 2003), endothelial cells, and testis (Georgiades *et al.*, 2002)
CD11c	Dendritic cells (DC)	Cre expression is induced by tamoxifen, only a few CD11c$^+$ DCs are expressing Cre (Probst *et al.*, 2003)
GATA1	erythroid cells and mast cells	Onset of deletion at the time of Ter119 expression, deletion also in megakaryocytes and eosinophils (Jasinski *et al.*, 2001)
LysM	granulocytes	Partial deletion in macrophages and splenic CD11c$^+$ DCs (Clausen *et al.*, 1999)
CD11b	peritoneal macrophages, mature osteoclasts	Partial deletion in macrophages and granulocytes, deletion in B and T cell subsets (Ferron and Vacher, 2005)
Pf4	megakaryocytes	Complete and specific deletion in the megakaryocytic lineage (Tiedt *et al.*, 2006)

3.1. Generation of bone marrow chimeras

3.1.1. Materials

Borgal: 7.5% antibiotic solution for small animals (commercially available from Hoechst).

pI-pC: Poly(I)-poly(C) (Amersham Biosciences); the stock solution with a final concentration of 2 mg/ml is prepared with PBS according to manufacturer's instructions and stored at −20°.

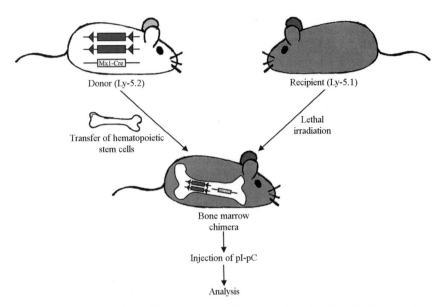

Figure 12.3 Generation of bone marrow chimeras using the *Mx1*-Cre/loxP system. Bone marrow from a Ly-5.2–positive donor mouse carrying the target gene flanked by *loxP sites*, and the *Mx1-Cre* transgene is transferred into an irradiated, Ly-5.1–positive wild-type recipient mouse. Four weeks after bone marrow transfer the hematopoietic system of the recipient mouse is fully reconstituted and the knockout can be induced by repeated pI-pC injections.

3.1.2. Procedure

1. The ideal recipients are 8- to 12-week-old B6SJL mice that are sex-matched with the donor mice (to prevent an immune response against donor cells). The recipient mice are γ-irradiated at 1000 rad with an X-ray machine.
2. Single-cell suspensions of bone marrow cells are prepared (see Section 3.2), resuspended in PBS, and adjusted to 5×10^6 cells/ml PBS.
3. The recipient mice are immobilized in custom-made conical metal chambers (Fig. 12.4) and the tails of the recipient mice are warmed for 30 sec in a 50° water bath (to increase blood flow and vessel size). Around 200 μl of the bone marrow cell suspension (1×10^6 cells/mouse) are injected with a 30G needle into the lateral tail vein.
4. The first 2 weeks after the transplantation, 1 ml of Borgal solution is added to 500 ml of drinking water; this will prevent infections in the irradiated, immunocompromised mice.
5. Four weeks after the bone marrow transplantation, the hematopoietic system of the recipient mouse is fully reconstituted with donor cells (this can be easily checked by determining the number of Ly-5.1–positive

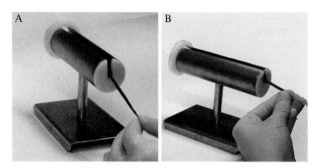

Figure 12.4 Injection of bone marrow cells into a recipient mouse. (A) The mouse is immobilized in a conical metal chamber. (B) Bone marrow is injected into the lateral tail vein of the prewarmed tail with a 30G needle.

cells in the peripheral blood using the flow cytometer). To induce the *Mx1-Cre*–mediated gene ablation, the pI-pC stock solution is diluted to 0.5 mg/ml with PBS, and 500 μl is then injected intraperitoneally. The pI-pC injection is repeated 2 and 4 days later.

6. It is necessary to wait at least 4 weeks before attempting *in vivo* analysis of the hematopoietic system to exclude effects caused by the pI-pC–triggered interferon production.

3.2. Standard flow cytometric analysis of cell surface receptors

Analysis of the cellular composition of hematopoietic tissues/compartments is done by determining the expression of cell surface receptors unique for distinct subsets of blood cells with the fluorescence-activated cell sorter (FACS).

3.2.1. Materials

Isoflurane: Isoba, a ready-to-use solution is commercially available (Essex Tierarznei)

Heparin: 20 U/ml heparin (Sigma) in TBS (50 mM Tris, 10 mM NaCl, pH 7.5).

ACK buffer: Ammonium chloride potassium phosphate buffer: 150 mM NH$_4$Cl, 1 mM KHCO$_3$, 0.1 mM EDTA, pH 7.3.

FACS-PBS: PBS supplemented with 1% BSA.

70-μm cell strainer: sterile 70-μm nylon cell strainer (BD Falcon).

Propidium–iodide: Propidium–iodide (Sigma) is prepared as a stock solution of 50 μg/ml in PBS, which can be stored at 4° in the dark.

3.2.2. Isolation of peripheral blood leukocytes
Procedure

1. The mouse is anesthetized by putting it into a beaker that contains a tissue drenched with 1 to 2 ml of isoflurane. When movements cease, the mouse should be immediately removed from the beaker. Extended exposure to isoflurane stops the heart beat and makes blood collection impossible. Around 50 μl of blood are collected with a heparin-soaked glass capillary from the retro-orbital venous plexus. The blood is resuspended in 500 μl PBS and centrifuged for 5 min at 400×g. The supernatant is carefully removed and discarded.
2. The cell pellet is resuspended in 500-μl ACK buffer, incubated for 5 min at RT, and centrifuged again. The supernatant, which should appear reddish due to the lysis of the erythrocytes, should be carefully removed and discarded. The cell pellet is resuspended in 100 μl of PBS. The amount of cells should be sufficient for two to three flow cytometry stainings.

3.2.3. Isolation of bone marrow cells
Procedure

1. The femurs of a mouse are carefully dissected and the muscles are removed by rubbing each femur with a Kleenex tissue. If cells should be sterile (e.g., for generation of dendritic cells), the femurs must stay intact during the removal of the muscle tissue and are incubated after muscle removal for 2 min in 70% ethanol and then rinsed in PBS.
2. Both ends of the femur are abscised with a sharp scalpel and the bone is subsequently flushed from both ends with 10-ml ice-cold PBS using a 10-ml syringe and a 23G needle. The isolated cell clumps are broken up into a single cell suspension by vigorous pipetting and then transferred through a 70-μm cell strainer into a 50-ml tube.

3.2.4. Isolation of cells from lymphatic tissues
Procedure

1. The organs of interests (such as thymus, spleen, lymph nodes, or Peyer's patches) are isolated, freed from surrounding connective tissue as much as possible, rinsed with PBS to remove blood and put into a cell strainer, which is placed in a Petri dish together with 10 ml of ice-cold PBS. To keep the cell strainer at 4°, the Petri dish is kept on ice.
2. Each organ is homogenized by gently squeezing it with the piston of a plastic syringe through the cell strainer. The cell suspension is transferred from the Petri dish into a 50-ml tube by filtering it again through the cell strainer. To obtain all remaining cells floating in the Petri dish, the Petri dish is rinsed again with 10 ml of ice-cold PBS and the suspension is combined with the other cells in the 50-ml tube.

3.2.5. Immunofluorescence staining of suspended hematopoietic cells

Procedure

1. The cells are pelleted by centrifugation at $300 \times g$, resuspended in an appropriate volume (2 to 10 ml) of ice-cold PBS, and counted.
2. If a centrifuge capable of centrifuging 96-well plates is available, the staining is most conveniently performed in 96-well round-bottom plates. For each staining, 1×10^6 cells or 100 μl of peripheral blood suspension are pipetted into each well. The plate is centrifuged for 5 min at $300 \times g$. The supernatant is removed and the pellet resuspended in 50 μl of FACS-PBS with the first antibody.
3. After a 30-min incubation at $4°$ in the dark, 200 μl FACS-PBS are added to each well. Subsequently, the cells are pelleted by centrifugation ($300 \times g$). Upon removal of the supernatant, the cell pellet is resuspended in 50-μl of FACS-PBS containing the secondary antibody.
4. The cell suspension is incubated for 15 min at $4°$ in the dark, and then diluted by adding 200 μl of FACS-PBS and finally centrifuged again at $300 \times g$ and RT. The supernatant is removed and the cell pellet resuspended in 200 μl FACS-PBS. Immediately before measuring, the fluorescent signal with the flow cytometer 10 μl of propidium-iodide (50 μg/ml) is added to each sample, which allows identification of dead cells. After brief vortexing, the fluorescence is measured with a cytometer.

Table 12.3 lists a number of surface markers that are commonly used singly or in combination to analyze the hematopoietic system. Please note that many of the markers are also useful for immunostaining of tissue sections of lymphatic organs, thereby also allowing assessment of the spatial distribution of blood cell types.

3.3. Flow cytometric lacZ staining of hematopoietic cells

The bacterial *lacZ* gene is an elegant means to examine the expression of a gene of interest. If knockout constructs are engineered to replace the disrupted gene with a *lacZ* gene, *Cre*-driven excisions can easily be monitored histochemically. Expression of the *lacZ* gene results in the production of β-galactosidase (β-gal), which can be readily detected by a number of chromogenic or fluorogenic β-gal substrates (Rotman *et al.*, 1963). This allows visualization of the gene deletion by histochemically staining tissue sections (see Section 4.4) or whole-mount embryos. Importantly, lacZ-positive cells can also be detected by flow cytometry. In the protocol described here, the lacZ measurement can be combined with immunofluorescent staining allowing to precisely determining the identity of the lacZ-positive cells.

Table 12.3 Surface antigens supporting analysis of the hematopoietic system

Surface markers	Cell type
Spleen	
$CD4^+$	T helper cells
$CD8^+$	Cytotoxic T cells
IgD^+	B cells
$NK1.1^+$	NK and NKT cells
Ter-119	Erythroblasts
$B220^+$ CD21high CD23low	Marginal zone B cells
CD23high CD21low	Follicular B cells
Bone marrow	
$B220^+$ IgM^-	Pre–pro B cells and later developmental stages
$CD19^+$	Pro B cells and later developmental stages
B220low IgM^+	Immature B cells
B220high IgM^+ IgD^+	All mature B cells
$Mac-1^+$ $Gr-1^+$	Granulocytes
$Mac-1^+$ $Gr-1^-$	Monocytes
Ter-119	Erythroblasts
$NK1.1^+$	NK and NKT cells
$CD4^+$	T helper cells
$CD8^+$	Cytotoxic T cells
Lin^- $c-kit^+$ Sca-1high	Hematopoietic stem cells
Lymph nodes	
$B220^+$ $CD19^+$	B cells
$CD4^+$	T helper cells
$CD8^+$	Cytotoxic T cells
Thymus	
Lin^-	$B220^-$, $CD4^-$, $CD8^-$, $Mac1^-$, $Gr-1^-$, $Ter119^-$
Lin^- $CD25^-$ $CD44^+$	DN1
Lin^- $CD25^+$ $CD44^+$	DN2
Lin^- $CD25^+$ $CD44^-$	DN3
Lin^- $CD25^-$ $CD44^-$	DN4
$CD4^-$ $CD8^-$	Double-negative T cells
$CD4^+$ $CD8^+$	Double-positive T cells
$CD4^+$	CD4 single-positive T cells
$CD8^+$	CD8 single-positive T cells
Peyer's patches	
$CD4^+$	T-helper cells
$CD8^+$	Cytotoxic T cells

3.3.1. Materials

FDG: Fluorescein Di-(B–D–galactopyranoside) (Sigma). Dissolve 5 mg of
FDG in a mixture of 304.8 μl of H_2O, 38.1 μl of dimethyl sulfoxide
(DMSO) (Sigma), and 38.1 μl of ethanol (8:1:1) to obtain a 20-mM
solution. Add 3429 μl of H_2O (10:1) as soon as the FDG is completely
dissolved to obtain a 2-mM working solution. Store aliquots of around
300 μl at $-20°$ in the dark.

Procedure

1. If immunofluorescence staining is combined with lacZ staining, the cells
   are stained as described (see Section 3.2.5). Use around 3×0^6 cells per
   staining, since the described method can lead to cell mortality and hence
   loss of cells. After staining, the cells are pelleted by centrifugation at
   $300 \times g$ at RT and resuspended in 20 μl of FACS-PBS.
2. The cell suspension is admixed with 20 μl FDG to each sample, and
   incubated for 75 sec in a 37° water bath to allow FDG uptake. The
   uptake is then quickly terminated by the addition of 200 μl of ice-cold
   FACS-PBS. Since exceeding the 75-sec incubation time with FDG
   significantly increases cell death, avoid handling more than five samples
   at the same time.
3. The cell suspensions are incubated for 2 h on ice in the dark and then
   measured in the flow cytometer. Excluding dead cells by the addition of
   propidium–iodide is recommended (see Section 3.2.5).

Flow cytometry and histology allow determination of the number, size,
distribution, and differentiation of distinct cell populations. Additional
assays further help to pinpoint potential functional defects of blood cell
lineages. Standard procedures include *in vitro* colony formation assays with
stem/progenitor cells, which permit evaluation of the formation of ery-
throid and myeloid cells; co-culture assays of bone marrow precursors with
stromal cells that permit determination of differentiation and proliferation of
B cells; and bone marrow precursor cell differentiation into dendritic cells,
which permits determination of T-cell proliferation and dendritic cell
analysis *in vitro*.

3.4. Generation of dendritic cells from bone marrow

Bone marrow–derived dendritic cells represent an elegant tool to study
antigen presentation and T-cell proliferation. Furthermore, dendritic cells
are also used to study cell migration, polarity, phagocytosis, adhesion, and
podosome formation (Calle *et al.*, 2006), among other processes. Our

protocol for generating dendritic cells is a slightly modified procedure originally described by Lutz *et al.* (1999).

3.4.1. Materials

R10 medium: RPMI-1640 (Gibco) supplemented with 100 U/ml of penicillin, 100 μg/ml of streptomycin, 2 mM of L-glutamin (all from PAA), and 10% heat-inactivated FCS (Gibco).

GM-CSF: rmGM-CSF (Peprotech)—20 ng/ml correspond to 200 U/ml. Alternatively, cell culture supernatant collected from Ag8653 myeloma cells transfected with the murine GM-CSF cDNA can be used (Zal *et al.*, 1994).

Freezing medium: Heat-inactivated FCS is supplemented with 10% DMSO. Freezing medium is always freshly prepared.

LPS: Lipopolysaccharides from *Escherichia coli* (Sigma). A stock solution of 1 mg/ml is prepared with PBS.

Procedure

1. At day 0, bone marrow cells are isolated as described in Section 3.2.3, resuspended in 10 ml of R10 medium, and counted.
2. Around 2.5 $\times$ 10^6 cells in a total volume of 10 ml of R10 medium containing 20 ng/ml GM-CSF or 10% GM-CSF supernatant are transferred to a 10-cm Petri dish. Dendritic cells should be cultured in bacterial-grade Petri dishes, since they strongly adhere to the plastic surface of cell culture dishes, which prevents their differentiation.
3. Ten milliliters of R10 containing 20 ng/ml GM-CSF or 20% GM-CSF supernatant are added at day 3 to the bone marrow culture. At day 6, 10 ml of the medium are carefully removed by tilting the plate slightly and slowly sucking off the medium. Stirring and shaking should be avoided when taking the plates out of the incubator. The removed medium is replaced with 10 ml of fresh R10 containing 20 ng/ml of GM-CSF or 20% GM-CSF supernatant.
4. At day 8, one of the following two possibilities is selected. First, the immature dendritic cells can be frozen for later usage: The cells from a 10-cm cell culture dish are collected by gentle pipetting, centrifugation at 300$\times$*g*, and the cell pellet is resuspended in 1 ml of freezing medium. The 1-ml suspension is transferred into a freezing tube and quickly transferred to a $-80°$ freezer. Second, the immature dendritic cells are brought to maturation: The cells from a 10-cm culture dish are collected as described above, resuspended in 10 ml of R10 containing 20 ng/ml GM-CSF or 10% GM-CSF and 200 ng/ml LPS, and cultured overnight in a 6-cm cell culture dish.

3.5. *In vitro* T-cell proliferation assay

Several possibilities are available to trigger and determine T-cell proliferation. First, T cells can be treated with ionophores and phorbol esters (e.g., ionomycin and phorbol-12-myristate-13-acetate [PMA]). Ionomycin increases the intracellular calcium concentration by facilitating calcium transport through the plasma membrane. Using this procedure, protein kinase C (PKC) can be activated in a phospholipase C–independent manner. Additionally, phorbol esters such as PMA activate PKC by mimicking the action of diacylglycerol (DAG). Second, T-cell proliferation can also be induced by cross-linking T-cell receptors (TCR) with antibodies to the CD3 T-cell co-receptor. This treatment mimics antigen-dependent TCR cross-linking, and thereby evokes signaling from the TCR. Third, antigen-dependent T-cell stimulation can be rather easily examined by crossing the integrin knockouts with a mouse strain expressing a transgenic TCR. A transgenic MHC class II–restricted TCR is expressed in the OT-II.2 mouse strain, which generates $CD4^+$ T cells specific for the $OVA_{323-339}$ peptide from chicken ovalbumin (Barnden *et al.*, 1998). Upon intercrossing gene-targeted mice with OT-II.2 transgenic mice, mutant $CD4^+$ T cells can be stimulated by co-culture with $OVA_{323-339}$–loaded dendritic cells. A similar system exists with the OT-I mice for $CD8^+$ T cells (Clarke *et al.*, 2000).

We usually monitor division of the stimulated T cells by determining the dilution of the carboxy-fluorescein diacetate succinimidyl ester (CFSE). CFSE diffuses freely into cells where it is converted by esterases into a membrane-impermeant dye, which becomes covalently bound to cellular proteins and is then capable to emit a fluorescence signal that can be assessed with a flow cytometer equipped with 488-nm excitation and emission filters. The CFSE fluorescence signal is halved during each cell division.

An alternative procedure to monitor cell division is measuring the incorporation of H^3-thymidine (Krishnamoorthy *et al.*, 2006), which is more sensitive but does not permit the analysis of the proliferation on a single-cell basis.

3.5.1. Materials

Magnetic sorting of T cells: $CD4^+$ T Cell Isolation Kit, LS MACS columns, and magnetic separation unit (all from Miltenyi).
ACK buffer: 150 mM NH_4Cl, 1 mM $KHCO_3$, 0.1 mM EDTA, pH 7.3.
MACS buffer: PBS containing 0.5% bovine serum albumin and 2 mM EDTA. The MACS buffer is degassed by applying vacuum or sonication.
anti-CD3e: Purified hamster anti-mouse CD3e monoclonal antibody (BD Biosciences, clone 145–2C11).
CFSE: Carboxyfluorescein diacetate succinimidyl ester (CFDA, SE) (Molecular Probes). A 6-mM stock solution is prepared in DMSO and stored at $-20°$ in the dark.

$OVA_{323-339}$ peptide: The sequence of the $OVA_{323-339}$ peptide is H-ISQAVH AHAEINEAGR-OH. A stock of 2 mg/ml is prepared in PBS and 100 μl aliquots are stored at $-20°$.

Dendritic cells: 1.5 × 10^6 bone marrow–derived dendritic cells matured overnight with 200 ng/ml LPS (see Section 3.4) are resuspended in R10 medium to a final concentration of 0.4 × 10^6/ml.

PMA: Phorbol-12-myristate-13-acetate (Calbiochem). A 1-mg/ml stock is prepared in DMSO, and 20-μl aliquots are stored at $-20°$. A 1:300 dilution in R10 medium is the working dilution and is always freshly prepared.

Ionomycin: Ionomycin calcium salt from *Streptomyces conglobatus* (Calbiochem). A 1-mg/ml stock is prepared in DMSO and 20-μl aliquots are stored at $4°$ in the dark. A 1:50 dilution in R10 medium is the working dilution and is always freshly prepared.

3.5.2. Magnetic sorting of T cells
For high sorting purities, it is critical to cool the cells and solutions on ice.

Procedure

1. The spleen is dissected and a single cell suspension prepared (see Section 3.2.4). The cell suspension is pelleted by centrifugation at $300 \times g$, and then thoroughly resuspended in 5 ml of ACK buffer and incubated at RT for 5 min. Afterward, 10 ml of PBS is added to the cell suspension and centrifuged.

2. After centrifugation, the cell pellet is resuspended in 10 ml of MACS buffer, counted, and 4 × 10^7 cells are centrifuged. The cells are labeled and sorted using the $CD4^+$ T Cell Isolation Kit according to the manufacturer's instructions. With this kit, non-$CD4^+$ T cells are labeled indirectly magnetically with a cocktail of biotin-conjugated antibodies and magnetic anti-biotin beads, and are subsequently depleted. Therefore, the isolated $CD4^+$ T cells remain untouched.

3.5.3. CFSE staining of T cells
Procedure

1. The sorted $CD4^+$ T cells are resuspended in PBS in a concentration of 1 × 10^7/ml. The CFSE stock solution is diluted 1:240 with PBS and added 1:50 to the cell suspension to reach a final concentration of 0.5 μM. The mixture is briefly vortexed to ensure even distribution. The T-cell suspension is incubated for 10 min at RT in the dark.

2. The reaction is stopped by the addition of 5 ml of ice-cold R10 medium. The cells are centrifuged at $300 \times g$, and then taken up in R10 medium to obtain a final concentration of 1 × 10^6/ml.

3.5.4. Proliferation assay
Procedure

1. Fifty microliters of 10 μg/ml of anti-CD3e in PBS is added to three of the positive control wells (Fig. 12.5) of a 96-well, round-bottom plate, and the plate is then incubated for 1 h at 37°.
2. The dilutions of the OVA peptide are prepared according to Fig. 12.5 in a 12-well plate. Each dilution step must be thoroughly mixed before proceeding to the next dilution.
3. One hundred microliters of the peptide dilution are added to the respective wells (Fig. 12.5). The wells in row 5, which lack OVA peptide, serve as negative controls. Next, 50 μl of dendritic cells (DCs) (0.4×10^6/ml) and 50 μl of CFSE-labeled T cells (1×10^6/ml) are then added to each well.
4. The anti-CD3e–coated wells are washed carefully two times with 200 μl of PBS. Then 100 μl of R10 medium are supplemented with 2 μl of PMA (1:300 dilution) and 100 μl of CFSE-labeled T cells (1×10^6/ml) are added to each well. This will result in a final PMA concentration of 33 ng/ml. To the remaining "positive control wells," 100 μl of R10 medium is supplemented with 2 μl of PMA (1:300 dilution) and 2 μl of ionomycin (1:50) dilution, and 100 μl CFSE-labeled T cells (1×10^6/ml) is added. This will result in a final ionomycin concentration of 200 ng/ml.
5. The plate is incubated for 72 h at 37° in a regular cell-culture incubator. For each measurement, the cells from three wells are collected and pooled by thorough pipetting. The CFSE signal is determined in the fluorescein channel of a flow cytometer. It is often convenient to analyze additional parameters such as activation markers, for example. In this case, an immunofluorescence staining of the cells with antibodies against markers such as CD25, CD44, and CD69 antibodies must be performed as described in Section 3.2.5 before the flow cytometric analysis.

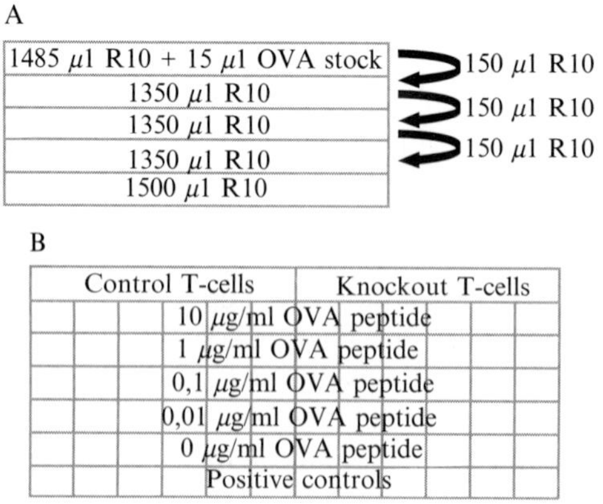

A

1485 μl R10 + 15 μl OVA stock	150 μl R10
1350 μl R10	150 μl R10
1350 μl R10	150 μl R10
1350 μl R10	
1500 μl R10	

B

		Control T-cells		Knockout T-cells		
		10 μg/ml OVA peptide				
		1 μg/ml OVA peptide				
		0,1 μg/ml OVA peptide				
		0,01 μg/ml OVA peptide				
		0 μg/ml OVA peptide				
		Positive controls				

Figure 12.5 Pipetting scheme for peptide dilutions and T-cell proliferation assays.

3.6. *In vivo* T-cell proliferation assay

The intravenous injection of T cells together with antigen–loaded dendritic cells into mice is the most physiological way to examine antigen–dependent proliferation of TCR transgenic T cells. The two injected cell types will move to the spleen where they interact in a physiological environment and trigger T-cell proliferation.

3.6.1. Materials

Magnetic sorting of T cells, ACK buffer, MACS buffer, CFSE, and OVA$_{323-339}$ peptide. See Section 3.5.
Dendritic cells: 2.5×10^6 bone marrow–derived dendritic cells/mouse, matured overnight with 200 ng/ml of LPS. They are prepared as described in Section 3.4.

3.6.2. Magnetic sorting of T cells

Magnetic sorting of the cells is performed as described in Section 3.5. Around 3×10^6 T cells are needed per mouse. Therefore, 5×10^7 splenocytes should be sorted to obtain enough cells for three experiments.

3.6.3. CFSE staining of T cells
Procedure

1. The sorted CD4$^+$ T cells are resuspended in PBS to obtain a concentration of 2×10^7/ml. The CFSE stock solution is diluted 1:600 with PBS and added 1:1 to the cell suspension to obtain a final concentration of 5 μM. The mixture is vortexed briefly to ensure even distribution. The T-cell suspension is incubated for 10 min at 37°.
2. The reaction is stopped by the addition of 10 ml of ice-cold R10 medium. The cells are then centrifuged at $300 \times g$ and taken up in PBS to obtain a final concentration of 15×10^6/ml.

3.6.4. Proliferation assay
Procedure

1. Around 5 to 15×10^6 matured dendritic cells are resuspended in 3 ml of R10 medium containing 20 ng/ml of GM-CSF or 10% GM-CSF supernatant and 20 μg/ml of Ova peptide. They are transferred in a 50-ml tube and incubated for 4 h in a cell culture incubator. A control batch of dendritic cells should be incubated without the peptide. After the incubation, the dendritic cells are washed twice with 10 ml of PBS, and are taken up in PBS after a subsequent centrifugation to obtain a dendritic cell concentration of 12.5×10^6/ml.

2. About 3×10^6 of CFSE-labeled OT-II T cells resuspended in 200 μl PBS are injected into the tail vein of a mouse (see Section 3.1). Sex-matched mice of 6 to 10 weeks of age are used as recipients. Around 4 h later, 2.5×10^6 of Ova-loaded dendritic cells resuspended in 200 μl of PBS are injected into the same recipients.

3. Three days later the spleens are dissected and a single cell suspension is prepared as described in Section 3.2.4. To analyze T-cell proliferation by flow cytometry, 5×10^6 splenocytes are stained with the desired markers and measured in a flow cytometer. Care should be taken to adjust the gains and compensations of the cytometer to avoid leakage of the CFSE signal into other channels. Furthermore, it is important to include a control with unloaded dendritic cells to determine the fluorescence intensity of nondividing T cells.

3.7. Induction of experimental autoimmune encephalomyelitis in C57/BL6 mice

Experimental autoimmune encephalomyelitis (EAE) is a very well-established animal model representing the human disease multiple sclerosis (MS). Over the years it has become increasingly clear that integrins play an important role for the development of both EAE and MS (Yednock *et al.*, 1992). Nonetheless, the exact function of integrins during the disease-relevant processes of T-cell activation and proliferation and extravasation into the brain are still not clear. Aspects like T-cell adhesion to endothelial cells can be analyzed *in vitro* in adhesion assay (Reiss *et al.*, 1998), as well as *in vivo* by intravital microscopy (Vajkoczy *et al.*, 2001). But to analyze the overall contribution of genetically altered hematopoietic cells to the development of EAE, it is necessary to induce EAE in knockout mice. This has been done in many mutant mouse strains that are deficient in adhesion molecules with a potential role in inflammatory conditions (Kanwar *et al.*, 2000; Kerfoot *et al.*, 2006).

3.7.1. Materials

Complete Freund's Adjuvant (CFA): 50 mg of *Mycobacterium tuberculosis* H37 Ra (Difco Laboratories) is first thoroughly ground with 1 ml of Incomplete Freund's Adjuvant (Sigma) using mortar and pestle, and then suspended with an additional 9 ml of Incomplete Freund's Adjuvant to obtain a total volume of around 10 ml, and finally stored at 4°. The suspension must always be homogenized with a syringe prior to usage.

MOG$_{35-55}$ peptide: The sequence is H-MEVGWYRSPFSRVVHLYRNGK-OH. A stock of 2 mg/ml is prepared in PBS and 500-μl aliquots are stored at $-20°$.

Syringes, adaptor, and needles: $2\times$ 1-ml glass tuberculin syringes with Luer lock tip and a female Luer lock to female Luer lock syringe adaptor (Sigma), and disposable 23G and 26G needles.

Pertussis toxin: Pertussis toxin (List Biological Laboratories). The vial containing pertussis toxin powder is reconstituted according to the manufacturer's instructions to a 100 ng/μl-stock solution and stored at 4°.

3.7.2. Procedure

1. For every mouse, 100 μl MOG in PBS and 100 μl CFA are mixed using two syringes that are connected with an adaptor. The two solutions are mixed until a white emulsion forms, which is hard to push from one syringe to the other. The mixed emulsion is kept on ice. The mouse is anesthetized with Isoflurane and 100 μl of the emulsion are injected with a 23G needle subcutaneously into the tail base as well as the neck of the mouse.
2. To prepare 400 ng of pertussis toxin, add 4 μl of the stock solution to 200 μl PBS. The toxin is injected intraperitoneally using a 26G needle.
3. The mice are assessed daily for clinical EAE symptoms and weight loss. Clinical scoring should be performed blinded if possible, and according to the established classification as, for example, described by Krishnamoorthy et al. (2006).

3.8. Mononuclear cell isolation from the central nervous system

The first step to analyze the diseased mice is often histological analysis of the central nervous system (CNS). With H&E and luxol fast blue staining of CNS sections, infiltration of immune cells and demyelination of nerves can be assessed (Krishnamoorthy et al., 2006). Additionally, mononuclear cells can also be isolated from the CNS by a Percoll density-gradient centrifugation, and subsequently be analyzed by flow cytometry.

3.8.1. Materials

Percoll: Medium for density centrifugation with a density of 1.130 g/ml (Fluka).
10× PBS: 80 g NaCl, 2 g KCl, 2 g KH2PO4 and 14.4 g Na2HPO4 × 2 H$_2$O are dissolved in 1 liter of distilled water.
Medium: DMEM cell culture medium (Gibco), supplemented with 2% FCS and 2.5% HEPES (1-M stock solution, pH 7.0).

Procedure

1. For each gradient, 9 ml of Percoll are mixed with 1 ml of 10× PBS to obtain an isotonic Percoll stock solution (IPSS) with a density of 1.123 g/ml. The IPSS can be stored at 4° for about 2 weeks. For each gradient, 4 ml of IPSS are mixed with 2.23 ml of PBS to obtain Percoll with a density of 1.08 g/ml.

2. All glass- and plasticware should be coated with serum-containing medium by wetting them once. This prevents cells sticking to surfaces.

3. The brain is carefully dissected out from the skull. The spinal column is transected in the cervical and lumbar part. The spinal cord is flushed out with 10 ml of PBS by inserting the end of a truncated 18G needle connected to a 10-ml syringe into the spinal column and applying gentle pressure with the syringe. Both the brain and spinal cord are rinsed with PBS to remove blood and put into a cell strainer, which is placed in a Petri dish together with 5 ml of ice-cold medium. To maintain the tissue at 4°, the Petri dish is kept on ice.

4. The brain is homogenized by gently squeezing it with the piston of a plastic syringe through the cell strainer. The piston should be only pressed, not ground, since grinding will result in increased mortality of T lymphocytes. The cell suspension is transferred from the Petri dish into a 15-ml tube by filtering it again through the cell strainer. To obtain all remaining cells floating in the Petri dish, the Petri dish is rinsed again with 5 ml of ice-cold medium, and then the suspension is combined with the other cells in the 15-ml tube. The volume is adjusted to 12.5 ml with the medium, and the suspension is mixed with 5.4 ml of IPSS.

5. Five milliliters of Percoll with a density of 1.08 g/ml are carefully under-layed by putting the tip of a 5-ml plastic pipette filled with 5.5 ml of the latter to the very bottom of the tube and slowly releasing the liquid. The last 0.5 ml should not be released from the pipette since this would often result in disturbance of the layering.

6. The gradient is centrifuged at $1200 \times g$ for 30 min at 20°. Afterward, the myelin debris, which is on top, is sucked away. The mononuclear cells that accumulated at the interface of the two solutions are sucked away with a medium-coated Pasteur pipette and transferred into a medium-coated 50-ml Falcon tube. The recovered suspension is filled up to 50 ml with medium and centrifuged at $300 \times g$ for 10 min at 4°. The pelleted cells can now be resuspended—for instance in FACS-PBS—and analyzed by flow cytometry as described in Section 3.2.5.

4. ANALYSIS OF INTEGRIN FUNCTIONS IN SKIN

In normal undamaged skin, the expression of integrins (with an exception of $\alpha v \beta 8$) is restricted to the basal keratinocyte layer of the epidermis and the outer root sheath (ORS) of the hair follicle (HF) (Fig. 12.6A to C). Some of the integrins such as $\alpha 3 \beta 1$ or $\alpha 6 \beta 4$ are expressed constitutively, while others like $\alpha 5 \beta 1$ or $\alpha v \beta 6$ are upregulated upon wounding, pathological conditions such as inflammation, or in culture (Watt, 2002).

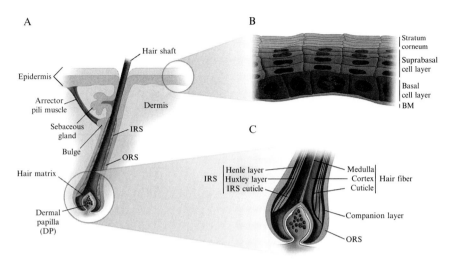

Figure 12.6 (A) Scheme of the mammalian skin and anagen hair follicle. (B) Close-up of the epidermis and the underlying BM separating the epidermal from the dermal compartment. The basal cell layer contains stem cells and transiently amplifying cells, which ensure the self-renewal of the epidermis. After the transiently amplifying cells cease proliferation, they differentiate and mature into suprabasal keratinocytes (the suprabasal cell layer). During this transition, the keratinocytes lose their contact with the BM and undergo an apoptosis-related process termed terminal differentiation. Finally, they die and become cornified squames (the stratum corneum) and are shed from the surface of the skin. (C) The hair follicle is made up of eight concentric epithelial sheaths: the ORS, which is continuous with the basal keratinocyte layer of the epidermis, the companion layer, three layers of the IRS, and three layers of hair-producing cells. BM, basement membrane; IRS, inner root sheath; ORS, outer root sheath. (See color insert.)

The HF is an important appendage of the skin that develops during embryogenesis and constantly renews throughout the lifetime of a mouse. The HF renewal or cycle is divided into the growth or anagen phase, regression or catagen phase, resting or telogen phase, and shedding or exogen phase that may not occur in all HF cycles (Alonso and Fuchs, 2006) (Fig. 12.7). The morphogenesis of HFs critically depends on the expression of $\beta1$ integrins. During HF cycling, the follicular stem cells have to migrate from the bulge to the hair matrix, proliferate, and differentiate into various layers of the inner root sheath (IRS) and ORS (Fig. 12.6C). This makes the HF an excellent *in vivo* model for studying the cellular mechanisms that regulate the balance between self-renewal and differentiation of adult stem cells (Watt *et al.*, 2006).

The variety of cellular processes occurring in the epidermis as well as in the HF, the accessibility of stem cells, the possibility to isolate and culture primary keratinocytes *in vitro*, and the availability of well-established skin tumor induction models (such as DMBA-TPA induced carcinogenesis

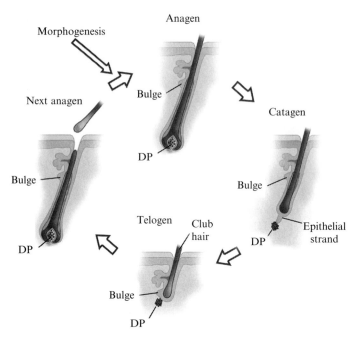

Figure 12.7 The hair follicle cycle. The hair-follicle morphogenesis begins during embryogenesis and is completed with the formation of the anagen hair follicle, which is made of an upper permanent part starting above the arrector pili muscle insertion, and a lower part, which undergoes cycling changes with a phase of growth (anagen), regression (catagen), and rest (telogen). The first postnatal anagen (and every subsequent one) starts when the mesenchymal cells of the DP induce the stem cells located in the bulge region of the hair follicle to proliferate and migrate into the hair matrix region of the follicle, where they further differentiate into the eight epithelial lineages. DP, dermal papilla. (See color insert.)

[DiGiovanni, 1992; Massoumi *et al.*, 2006]) and *in vivo* wound-healing protocols (DiPietro and Burns, 2003; Grose *et al.*, 2002) make the skin a perfect model to study the role of integrins under both physiological and pathological conditions.

With the improvement of the gene-targeting technology, most of the skin-specific integrins and several integrin cytoplasmic domain-associated proteins (such as ILK) have been deleted either in a constitutive fashion or using a conditional approach in combination with the expression of the Cre recombinase under skin-specific promoters, such as Keratin 5-Cre (Ramirez *et al.*, 2004) or Keratin 14-Cre (Vasioukhin *et al.*, 1999). Loss of $\alpha2$, $\alpha9$, $\beta5$, and $\beta6$ integrin subunits has no or only minor effects on skin development or wound healing, while deletion of $\alpha3$, $\alpha6$, $\beta1$, and $\beta4$ reveal striking defects in skin development and maintenance. The absence of the latter integrin chains in the basal keratinocytes of the skin leads to skin blistering of various severities due to the defective adhesion of the basal keratinocytes to the BM. Furthermore, skin-restricted deletions of the $\beta1$ integrin subunit or ILK reveals additional defects in keratinocytes such as defective proliferation, interfollicular terminal differentiation, HF maintenance, and keratinocytes migration both *in vivo* and *in vitro* (Lorenz *et al.*, 2007; Watt, 2002). The severe skin and HF defects occurring upon the deletion of the $\beta1$ integrin gene triggered further investigations of more subtle gene modifications in which point mutations were introduced into the cytoplasmic domain of the $\beta1$ integrin aiming at the functional dissection of distinct signaling motifs in this domain (Czuchra *et al.*, 2006). The introduction of such subtle mutations and the analysis of the mutant epidermis and HFs led to the important conclusion that skin and HF development and maintenance require integrin activation, but not the tyrosine phosphorylation of the $\beta1$-integrin cytoplasmic domain.

4.1. Isolation of mouse skin

To investigate the role of integrins for development and maintenance of the epidermis and the cycling and morphology of the HF, skin samples from various pre- and post-natal stages have to be collected and analyzed. Because skin morphology greatly differs by body region, skin samples must always be isolated from the same location (e.g., back skin). The HF morphogenesis and the first postnatal catagen, telogen, and anagen development follow a rather precise time scale (Table 12.4), while later HF cycles are not synchronized. Therefore, in comparative studies care must be taken to ensure that skin tissues are derived from the same hair cycle phase. Furthermore, possible defects occurring later in life (e.g., fibrosis) demand skin sampling and analysis from adult unsynchronized animals. However, the correct choice of the appropriate developmental stage for skin analysis ultimately depends solely on the phenotype observed.

Table 12.4 Time-scale for hair follicle morphogenesis and cycling in C57BL/6 mice

Developmental stage	Main stages of hair follicle morphogenesis and cycling
~E15	Beginning of follicular morphogenesis
P9	Completion of follicular morphogenesis
P16	First postnatal catagen
P19	First postnatal telogen
P28	First postnatal anagen
P42	Second postnatal catagen
P49	Second postnatal telogen
~P84	Second postnatal anagen

Sources: Muller-Rover *et al.* (2001) and Paus *et al.* (1999).

4.1.1. Procedure

1. Mice are killed by decapitating (newborn mice) or cervical dislocation. Prior to skin sampling from adult mice, the fur is carefully removed with a shaver (Fig. 12.8A and B).

2. The back skin (approximately 1 to 1.5 × 3 cm) is dissected down to the subcutis level and placed with the subcutis down on a nylon membrane (Hybond-XL^{TM}, Amersham Biosciences). Both ends of the skin sample are stretched and carefully smoothed (Fig. 12.8C to E). PVDF membranes should be avoided since they might detach from the skin during the preparation of paraffin blocks.

3. The skin is then cut into two similarly sized stripes parallel to the vertebral line using a sharp scalpel. The side where the skin was cut is marked with a pen on the back side of the nylon membrane (Fig. 12.8F to H).

4. The overextending nylon membrane is removed with a scalpel, and the skin stripe is cut into two pieces perpendicular to the first skin separation (Fig. 12.8I). The tissue pieces are directly used to prepare cryoblocks (Fig. 12.9) or paraffin blocks (Fig. 12.10).

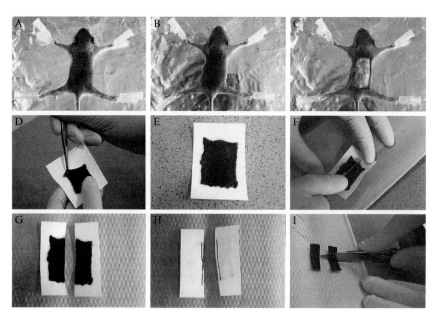

Figure 12.8 Isolation of mouse skin. For a detailed description, see Section 4.1 (See color insert.)

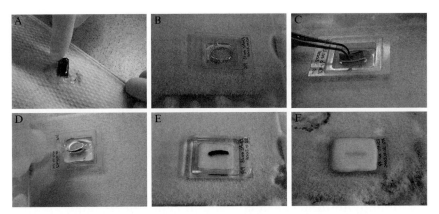

Figure 12.9   Preparation of cryoblocks. For a detailed description, see Section 4.2.

Figure 12.10   Preparation of paraffin blocks. For a detailed description, see Section 4.5. (See color insert.)

4.2. Embedding of skin in cryomatrix and preparation of cryosections

4.2.1. Procedure

1. The embedding medium (Shandon Cryomatrix^{TM}, Thermo) is rubbed into the remaining hairs (Fig. 12.9A).
2. A tissue mold (Tissue-Tek$^{®}$, Sakura) containing a small amount of an embedding medium is placed on a copper plate cooled on dry ice (Fig. 12.9B). When the embedding medium starts to freeze, immediately position the skin piece with the marked side down into the mold and cover the remaining tissue with cryomatrix (wait until the skin piece is immobilized in the embedding medium (Fig. 12.9C through E). When

the embedding matrix is completely frozen, transfer the block to −80°
(Fig. 12.9F).

3. The skin samples embedded in the cryomatrix are cut into 10-μm slices
at −20° using a regular cryostate (Microm). Ensure that the blade cuts
parallel to the longitudinal axis of the HF and avoid cross-sections (if
necessary change the orientation of the blade). The frozen sections are
then transferred to positively charged SuperFrost®Plus glass microscope
slides (Microm), air-dried for 30 min at RT, and stored at −80°. Repeti-
tive thawing of the cryosections should be avoided since this leads to
freeze-thaw artifacts.

4.3. Immunofluorescence staining of skin cryosections

To characterize the mutant skin in more detail, skin cryosections prepared
according to the above protocol can be used to perform immunofluorescence
staining. Table 12.5 lists antibodies helpful for the analysis of HF and epidermal
differentiation, integrin expression, BM organization, and cell–cell
interactions.

4.3.1. Materials

Fixation solution: 4% PFA in PBS, can be aliquoted and stored at −20°.
Permeabilization solution: 0.1% TritonX-100 in PBS. Should be always
    freshly prepared.
Blocking solution: 3% BSA in PBS.
DAPI solution: 4′,6-diamidino-2-phenylindole (Sigma). One-milliliter
    stock solution of 1 mg/ml in Millipore water is prepared and stored at
    −20° in darkness. For staining, a 0.1-μg/ml solution is prepared in PBS.
Primary and secondary (Table 12.5) antibody solutions: The antibodies are
    diluted according to the manufacturer's instructions in 1% BSA in PBS.

Procedure

1. Cryosections are warmed up for 1 h at RT, and incubated for 20 min
   with the fixation solution in a glass container at RT. (For some antibodies, skin tissue fixed with reagents other than PFA, such as methanol,
   might be required.)

2. The slides with skin tissue sections are washed with PBS three times for
   5 min each, and if necessary incubated with the permeabilization solution for 5 min on ice. Afterward, the slides are rinsed once with PBS.

3. The skin sections are incubated with the blocking solution for 30 min at
   RT or overnight at 4°, and subsequently encircled with a Pap-Pen to
   create a water-repellent area, thus minimizing the quantity of staining
   solution needed. The Pap-Pen applied in this step must be chosen

Table 12.5 List of antibodies commonly used to stain cryosections of skin

Antibody	Stained compartment	Source
Anti-integrin $\beta 1$, clone MB1.2	Basal keratinocytes, ORS	Chemicon
Anti-CD49f (integrin $\alpha 6$ chain), clone GoH3	Basal keratinocytes (concentrated at BM, ORS)	BD biosciences
Anti-CD104 (integrin $\beta 4$ chain), clone 346–11A		
Anti-laminin 5 (γ2LE4–6)	BM component	Described in Brakebusch et al., 2000
Anti-nidogen		Santa Cruz
Anti-collagen IV		Chemicon
Anti-perlecan		
Anti-cytokeratin 10 (DE-K10)	Expressed in all suprabasal cells, but absent in basal cells	Covance
Anti-keratin 6	Companion layer of the hair follicle, aberrantly expressed in the hyper- or hypo-proliferative epidermis	
Anti-keratin 14 (AF 64)	Basal keratinocytes, ORS	
Anti-loricrin	Keratohyalin component typical for terminally differentiated keratinocytes	
Anti-K6irs1	Specific for IRS (all layers)	Described in Langbein et al., 1999, 2001, 2003

(continued)

Table 12.5 (*continued*)

Antibody	Stained compartment	Source
Anti-K6irs2	Specific for IRS (IRS cuticle)	
Anti-K6irs3	Specific for IRS (IRS cuticle)	
Anti-K6irs4	Specific for IRS (Huxley layer)	
Anti-hHb2	Specific for hair fiber (HS cuticle)	
Anti-hHa4	Specific for hair fiber (upper cortex of HS)	
Anti-hHa5	Specific for hair fiber (HS matrix)	
Anti-E-cadherin, clone ECCD-2	Component of the adherens junctions; found in several HF compartments, and in basal and suprabasal layers of epidermis	Zymed
Anti-β-catenin		Sigma
Anti-α-catenin		
Anti-desmoplakin	Component of the desmosomes; found in several HF compartments, basal and suprabasal layers of epidermis	Research diagnostics

BM, basement membrane; HF, hair follicle; HS, hair shaft; IRS, inner root sheath; ORS, outer root sheath.
Note: Cy3-and FITC-conjugated antibodies specific for mouse IgM, rabbit IgG and rat IgG (Jackson Immunochemicals) are used as secondary antibodies.

carefully because some are leaky and/or interfere with the DAPI staining, and thus may destroy the section. (A Pap-Pen from G. Kisker-Products for Biotechnology is recommended.) However, when performing the DAPI staining, we encircle the sections with solid wax warmed up to 37°, which is otherwise routinely used to embed tissue (see Section 4.5).

4. The nylon membrane is carefully removed from the skin sections to prevent folding back of the membrane and thereby masking parts of the tissue.

5. The samples are incubated with 50 to 100 μl of the primary antibody solution for 90 min in a humidified chamber.

6. The slides are washed with PBS three times at 5 min each, and subsequently incubated with 50 to 100 μl of the secondary antibody solution for 1 h in a humidified chamber. From this step on, the slides should be shielded from light.

7. The antibody solution is discarded and sections are incubated with 100 μl of the DAPI solution for 5 min (optional step).

8. The slides are washed with PBS three times at 5 min each and carefully dried around the skin section using a Kleenex tissue. The sections are mounted with anti-fading medium (Elvanol) and dried for 30 min. Slides can be analyzed immediately or stored in the dark at 4° for several days or at −20° for several months.

4.4. LacZ staining on cryosections

As already introduced in Section 3.3, the bacterial *lacZ* gene can be engineered into the targeting construct so that it is activated upon *Cre*-mediated gene disruption. Consequently, it can then be used to examine the expression of the disrupted gene. This approach was used in our laboratory to monitor the *K5-Cre*-driven excision of the β1 integrin gene (Brakebusch *et al.*, 2000). The protocol described below uses the β-gal substrate 5-bromo-4-chloroindoxyl-beta-D-galactopyranoside (X-gal) that produces blue precipitates that are insoluble in fixed cryosections.

4.4.1. Materials

10× Solution A: A 1 M K$_2$HPO$_4$ solution of pH 7.4 (adjusted with HCl) is prepared and stored at RT. A suitable volume is 1 liter.

4× Solution B: 26 mM EDTA and 8 mM MgCl$_2$ are prepared in 4× of solution A, the pH is adjusted to 7.4, and the solution stored at RT. A suitable volume is 1 liter.

Fixation solution: 0.2% glutaraldehyde is always freshly prepared with 1× of solution B.

2× Washing solution: A solution of 0.02% Na-deoxycholate, 0.04% NP-40, is prepared with 2× of solution B and stored at RT. A suitable volume is 1 liter.

Staining solution: A 1× solution of 10 mM $K_3Fe(CN)_6$, 10 mM K_4Fe (CN)$_6$, and 0.5 mg/ml X-gal is prepared with 1× of washing solution, stored at RT in the dark, and always filtered before use.

4.4.2. Procedure

1. Cryosections are warmed up for 1 h at RT, encircled with a Pap-Pen (see Section 4.3) to create a water-repellent area, and incubated with 100 μl of the fixation solution for 5 min at RT.
2. The sections are washed in a jar with 1× of washing solution three times at 5 min each, and incubated with 100 μl of staining solution overnight at RT in the dark.
3. Optionally, sections can be counterstained with H&E (see Section 4.6) and mounted with Elvanol. After staining, slides can be stored at 4° for several months.

4.5. Embedding and cutting of skin for paraffin sections

4.5.1. Procedure

1. The skin piece is placed together with the nylon membrane into an embedding cassette and incubated with fixation solution (see Section 4.3) at 4° overnight (Fig. 12.10A).
2. The skin is then dehydrated at RT in a graded alcohol series consisting of 50%, 70%, 80%, 90%, and three times 100% ethanol for 60 min for each dilution.
3. The dehydrated skin is incubated with xylol two times at 15 min each, and then placed together with the embedding cassettes in molten paraffin (Paraplast embedding media, Parablast X-tra, Sigma) at 60° with two changes for 90 min each.
4. A tissue mold containing a small amount of hot wax is placed on a cooling plate (−7°) of an embedding machine (Shandon). As soon as the paraffin starts to solidify, the skin piece is positioned with the side of the marked nylon membrane down into the mold (cut surface facing the bottom surface of the mold, Fig. 12.10B to E). After the skin piece is immobilized in the embedding medium, the mold is covered with the lid of the embedding cassette and filled with wax. When the paraffin is completely solidified, the paraffin block is removed from the mold and transferred to 4° (Fig. 12.10F).
5. Approximately 6- to 8-μm thick sections are cut using a microtome (Microm) and transferred into a 40° water bath where they are smoothed. In order to avoid cross-sections of HFs, it is important to ensure that the blade cuts parallel to the longitudinal axis of HFs (if necessary change the orientation of the blade). The sections are

subsequently collected onto the surface of positively charged glass microscope slides, and dried at 37° overnight. Paraffin sections are stored at 4°.

Paraffin sections can subsequently be stained with H&E or be used for specific epitope detection in immunohistochemistry. In addition, paraffin sections can be used to detect bromodeoxyuridine (BrdU) incorporated into the DNA of proliferating cells, which is a widely used technique to investigate the proliferation of interfollicular and follicular keratinocytes, as well as for *in vivo* epidermal stem-cell labeling.

4.6. H&E staining on paraffin sections

4.6.1. Materials

Hematoxylin: Meyer's hemalaum solution (Merck) is diluted 1:5 with distilled water and always filtered before use. The hematoxylin solution can be stored at RT.

Eosin: 1% Eosin (Merck) is diluted in 0.05% CH_3COOH and can be stored at RT.

Graded ethanol series: Prepared in jars, separately for hydratation and dehydrataion of the tissue, can be stored at RT and reused.

4.6.2. Procedure

1. The sections are deparaffinized and incubated with xylol two times at 5 min each, and then hydrated with a graded ethanol/water dilution series of 100% (two times), 95%, 90%, 80% and 70%, respectively, for 3 min each dilution. Finally, the tissue slides are incubated with water for 5 min. All incubation steps are performed at RT.
2. The tissue slides are stained for 5 to 10 min in hematoxylin. The staining intensity is controlled by regular microscopic inspections.
3. When the appropriate hematoxylin staining intensity is achieved, the reaction is stopped by washing the slides for 10 min in running tap water. Subsequently, the sections are stained with eosin solution for 1 to 3 min (the staining intensity is also controlled microscopically), and when the proper intensity is achieved, the sections are washed for 3 min in tap water.
4. The slides are dehydrated with a graded series of ethanol/water dilutions starting at 70% and increasing to 90%, 95%, and finally, two times at 100% for 3 min each dilution.
5. The sections are washed twice with xylol, and the nylon membrane is carefully removed from the skin sections to prevent folding back of the membrane (and thus avoiding masking tissue). Finally, the sections are mounted with Entellan (Entellan new, Merck). The slides can be stored at 4° for several weeks.

4.7. Evaluation of keratinocyte proliferation by BrdU incorporation method

4.7.1. Materials

BrdU (Roche) solution: 5 mg/ml stock is freshly prepared in sterile PBS. The solution is prepared just prior to use and is kept until use on ice.

Trypsin solution: 0.1% trypsin (Gibco) and 0.1% $CaCl_2$ are diluted in water and can be aliquoted and stored at $-20°$. For a glass jar, approximately 75 ml of the solution are needed.

Blocking solution A: 1% H_2O_2 in methanol is freshly prepared.

Blocking solution B: 0.5% BSA and 0.1% Tween20 are diluted in PBS and can be stored at $4°$.

BrdU-POD (Roche) antibody solution: The antibody is diluted 1:40 in blocking solution B. The solution is prepared just prior to use and is kept until use on ice.

Substrate solution: Prepared just before use by mixing 5 ml of 15 mM 3,3′-diaminobenzidine tetrahydrochloride (DAB, Sigma) dissolved in water with 95 ml of 0.5 M Tris-HCl, pH 7.6, and 120 μl of 5% H_2O_2.

Hematoxylin: Meyer's hemalaum solution (Merck) is diluted 1:5 with distilled water and always filtered before use. The hematoxylin solution can be stored at RT.

4.7.2. Procedure

1. Mice are injected with 10 μl of BrdU solution per gram of body weight and sacrificed 2 h later.
2. Sections from injected mice are deparaffinzed following the same procedure as described in the hematoxylin/eosin staining protocol (see Section 4.6).
3. The slides are washed with distilled water for 5 min and then incubated in a jar containing 4N HCl for 20 min. Afterward, the slides are washed with distilled water for 5 min.
4. The sections are incubated in a jar containing prewarmed trypsin solution. The jar is placed in a $37°$ water bath. Afterward, the slides are washed with water for 5 min.
5. The slides are then incubated in a jar with blocking solution A for 10 min at RT, and subsequently washed with PBS three times at 5 min each.
6. The slides are treated with blocking solution B three times, at 5 min each, and the tissue is encircled with a Pap-Pen (see Section 4.3) to minimize use of staining solution.
7. The slides are incubated with 50 μl of the BrdU-POD antibody solution overnight at $4°$. The following day the sections are washed three times with PBS.

8. The sections are incubated in a jar with substrate solution, and the ongoing reaction is regularly microscopically monitored. The reaction is stopped, at the latest, after 10 min by incubating the slides for 10 min in water.

9. The sections are counterstained with hematoxylin 5× for 10 sec (with careful inspections of staining intensity) and washed with tap water for 10 min.

10. The slides are finally dehydrated and mounted following the same procedure as described in H&E staining protocol (see Section 4.6).

5. IN VITRO SKIN ANALYSIS

The *in vivo* skin analysis should be complemented with state-of-the-art *in vitro* analysis of freshly isolated keratinocytes. The primary keratinocytes can then be used to analyze integrin expression and activation by FACS and/or Western blot assay, integrin-dependent cell adhesion with adhesion assays, spreading and migration using microscopic methods, and various biochemical analyses of integrin-dependent signaling pathways (Brakebusch *et al.*, 2000; Czuchra *et al.*, 2006; Grose *et al.*, 2002).

5.1. Isolation of primary keratinocytes

Our protocol for isolation of epidermal keratinocytes is a modified procedure from that described by Romero *et al.* (1999), and can be used to isolate keratinocytes from newborn or adult mice. However, attention should be paid to the hair cycle stage since at anagen, the separation of the epidermis from the dermis is highly inefficient. Therefore, we either isolate keratinocytes from P0-P5 and P21-P23 mice or animals older than 2 mo. The average keratinocyte yield from P0-P5 mice is 10×10^6 cells, from P21-P23 mice 20×10^6, and from adult animals (older than 2 mo) up to 50×10^6 cells.

5.1.1. Materials

Chelex-treated FCS: 200 g chelex resin (BioRad) is dissolved in 5 liters of water. The pH is adjusted with concentrated HCl to 7.4 (the pH adjusts very slowly), and the solution is then filtered through a folded paper filter (Schleicher & Schuell). Subsequently, 500 ml of heat-inactivated FCS (Gibco) is stirred at 4° overnight, together with the chelex resin, and then filtered once again through the folded paper filter. To sterilize the chelex-treated FCS, it is subsequently filtered through a 0.2-μm filter, aliquoted in 40-ml portions, and stored at −20°. This chelex treatment is essential, and removes the calcium ions from the serum, which prevents differentiation of keratinocytes.

Keratinocyte growth medium (KGM): Minimal Essential Medium (MEM, from Sigma) is complemented with 5 μg/ml insulin, 10 ng/ml EGF, 10 μg/ml transferrin, 10 μM phosphorylethanolamine, 10 μM ethanolamine (all from Sigma), 0.36 μg/ml hydrocortisone (Calbiochem), 2 mM L-Glutamine (Invitrogen), 1× Penicillin/Streptomycin (PAA), 45 μM CaCl$_2$, and 8% chelex-treated FCS.

Antibiotic solution: Penicillin/streptomycin (PAA), nystatin (Sigma), and fungizone (Invitrogen) are diluted to a final concentration of 2× in PBS and stored at 4°.

Trypsin solution: 0.8% trypsin (Invitrogen) diluted in PBS and stored at 4°.

DNAse medium: 0.25 mg/ml DNAse (Sigma) is always freshly prepared with MEM (Sigma) and supplemented with 8% chelex-treatedd FCS and kept on ice.

Coating medium: MEM (Sigma) buffered with 20 mM Hepes, pH 7.3 and supplemented with 0.1 mg/ml BSA, 30 μg/ml collagen I (Vitrogen 100 collagen, Cohesion), 10 μg/ml fibronectin (Sigma), and 1 mM CaCl$_2$. Plastic plates are coated for 2 h and glass surfaces for 4 h. Coated and parafilm-sealed plates can be stored at 4° for several weeks.

5.1.2. Procedure

1. Mice are killed by ether (newborn mice) or cervical dislocation, and then the entire skin is shaved to remove the fur. The mouse is then rinsed with water.

2. The mouse is dipped for 1 min in 70% ethanol (to sterilize the epidermis) and then rinsed again with water. Subsequent steps must be carried out under sterile conditions if the keratinocytes are to be cultured later.

3. The limbs, tail, and ears are removed (bacteriological dishes are used as a working area).

4. The skin is peeled off from the trunk and the subcutaneous fat layer is then removed with a sharp scalpel. This step is very crucial since incomplete removal of the subcutaneous fat layer will prevent efficient trypsin digestion. While skin derived from adult mice is quite resistant to mechanical forces, the skin of young mice is fragile, and therefore, the scrapping should be carried out with great care (bacteriological dishes are used as a working area).

5. The skin is cut into several pieces and put with the dermis into a 10-cm tissue culture dish filled with 20 ml of antibiotic solution and incubated for at least 10 min on ice.

6. The skin pieces are then transferred with the epidermis side up into a 10-cm tissue culture dish filled with 20 ml of trypsin solution and incubated for 50 min at 37° (in a tissue culture incubator).

7. The skin pieces are then transferred to a new bacteriological dish, and the epidermis can be separated from the dermis by scraping it off with a

small, curved forceps. The epidermis should come off very easily; if not the mouse was either in anagen phase or the trypsin treatment was insufficient.

8. The epidermis is transferred into a 50-ml Falcon tube containing 25 ml of ice-cold DNAse medium. Shake for 30 min in a 37° water bath and subsequently pipet up and down several times.

9. The cell suspension is filtered through a 70-μm cell strainer (Becton Dickinson) and then spun down for 5 min at 300×g.

10. The cells are washed once with 10 to 20 ml of keratinocyte growth medium, and immediately used for an experiment or further plated onto tissue culture dishes (4 to 6 × 10^6 cells per 10-cm tissue culture dish) precoated with coating medium.

11. The KGM medium is changed the next day, and then every second day. The cells are cultured in a humidified chamber at 34 to 37°, 5% CO_2. Since keratinocytes rapidly differentiate at high confluence and frequent passaging can lead to immortalization, all experiments should be carried out when the keratinocyte culture has reached 80 to 90% confluence.

5.2. Detachment of keratinocytes from cell culture dishes

Since keratinocytes adhere strongly to cell culture plates coated with ECM proteins, a special trypsinization procedure is required to detach them from the plate surface.

5.2.1. Materials

Chelating solution: 0.05 M EDTA prepared in PBS.
Trypsinization solution: 2× Trypsin-EDTA (10× stock, Gibco) diluted with PBS and stored at 4°. The trypsinization solution must be warmed to 37° before use.
KGM medium: As described in section 5.1.

5.2.2. Procedure

1. The keratinocyte layer is washed with PBS and then incubated with 4 ml of chelating solution for 5 to 10 min at 37°. The cells are frequently monitored under the microscope, and when cells start to round up, the chelating solution is quickly sucked off.

2. The cells are then incubated with 1 ml of trypsinization solution for 2 to 5 min at 37°. The trypsin reaction is stopped with KGM, the cells collected in a 15-ml tube, spun down for 5 min at 300×g, and resuspended in an appropriate volume of KGM.

5.3. Analysis of expression and activation of integrins on freshly isolated or cultured keratinocytes by flow cytometry

To determine the surface levels of integrins and their activation state, freshly isolated or cultured keratinocytes are immunostained with antibodies recognizing keratinocyte-specific integrins and activation-specific epitopes and subsequently measured by flow cytometry. Table 12.6 lists antibodies useful for such analysis.

The protocol for the staining procedure is the same as described in section 3.2.5, with the exception that the cell numbers required for one staining differ, and range between 0.5 to 1×10^6 of freshly isolated or 0.3 to 0.5×10^6 of cultured keratinocytes, respectively.

5.4. Analysis of fibronectin fragment binding by primary keratinocytes

An alternative method to the measurement of the binding of activation-specific antibodies for determining integrin activation on keratinocytes is the flow cytometric analysis of manganese-induced ligand binding to fluorescently labeled ECM proteins. In Czuchra *et al.* (2006), we determined binding efficiency of an His-tagged FN fragment containing the central cell binding domain comprised of the type III repeats 7 to 10 (FNIII7–10)

Table 12.6 List of antibodies commonly used in the analysis of integrin expression and activation on primary keratinocytes

Antibody[a]	Comments
Anti-CD29 (integrin β1 chain), clone Ha2/5	Detected on freshly isolated and cultured keratinocytes
Anti-CD104 (integrin β4 chain), clone 346–11A	
Anti-CD49b (integrin α2 chain), clone Ha1/29	
Anti-CD49f (integrin α6 chain), clone GoH3	
Anti-CD49e (integrin α5 chain), clone 5H10–27	Detected on cultured keratinocytes
Anti-CD51 (integrin α_V chain), clone RMV-7	
Anti-CD29 (integrin β1 chain), clone 9EG7	Activation-specific antibody

[a] All antibodies are from BD Biosciences.

(Ohashi and Erickson, 2005). This FN fragment can be expressed in large quantities in *Escherichia coli*, purified with Ni-based affinity chromatography, and then conjugated with a fluorescent dye (Alexa Fluor 647). Since the $\alpha5\beta1$ integrin is not expressed on freshly isolated keratinocytes, only cultured cells, which have induced expression of $\alpha5\beta1$, can be used for the FN binding assay.

5.4.1. Materials

FACS-TRIS (for measuring basal activation of integrins): 24 mM Tris-HCl, pH 7.4, supplemented with 137 mM NaCl and 2.7 M KCl. A suitable volume is 1 liter.

FACS-TRIS-Mn (for measuring activated integrins): FACS-TRIS supplemented with 5 mM Mn^{2+}.

FACS-TRIS-EDTA (for measuring inactivated integrins): FACS-TRIS supplemented with 2 mM EDTA.

FN-Alexa 647 solution: FNIII7–10 fragment labeled with Alexa 647 diluted in the appropriate FACS-TRIS (with or without manganese or EDTA). Recommended concentration range up to 0.7 μM.

Propidium iodide: Propidium iodide (Sigma) is diluted to a stock solution of 50 μg/ml in PBS and can be stored at 4° in the dark. A suitable volume is 50 ml.

5.4.2. Procedure

Except for the incubation step with the FN fragment, all steps must be performed on ice.

1. Approximately 3×10^5 keratinocytes are resuspended in ice-cold FACS-TRIS, in a volume not exceeding 200 μl per sample, transferred to a round-bottom 96-well plate and centrifuged for 5 min at 300×g.

2. The cell pellet is resuspended in 50 μl FNIII7–10-Alexa 647 solution and incubated 40 min at RT.

3. The cells are washed with 200 μl of the ice-cold FACS-TRIS, FACS-TRIS-Mn, or FACS-TRIS-EDTA (the same FACS-TRIS solution is applied that was used to dilute the FN fragment in the given sample), pelletted by centrifugation for 5 min at 300×g, and resuspended again in 200 μl of the ice-cold FACS-TRIS used for the dilution of the FN fragment.

4. The labeled keratinocytes are transferred into 1.2-ml microtubes. To exclude dead cells, 10 μl of a 50-μg/ml stock solution of propidium iodide is added to each sample. Subsequently the bound FNIII7–10-Alexa 647 signal is measured with the flow cytometer.

5.5. Assessment of integrin-dependent adhesion to ECM proteins

5.5.1. Materials

Recombinant laminin $\alpha3A\beta3\gamma2$ (laminin 332): Laminin 332 (generously provided to us by Dr. Monique Aumailley, University of Cologne, Germany) (Rousselle and Aumailley, 1994). Serial dilutions are prepared with PBS containing 1% BSA. Recommended concentration range: up to 5 μg/ml.

FN (Sigma): Serial dilutions are prepared with PBS containing 1% BSA. Recommended concentration range: up to 50 μg/ml.

Collagen I (Vitrogen 100 collagen, Cohesion): Serial dilutions are prepared with PBS containing 1% BSA. Recommended concentration range: up to 15 μg/ml. We routinely use collagen I, which is inexpensive and widely available, present in an active conformation, and recognized by the same integrins that also bind collagen IV. The natural ligand of keratinocytes is collagen IV, which, however, is expensive and sold by only a few companies.

Blocking solution: PBS supplemented with 1% BSA.

Assay MEM (AMEM): MEM supplemented with 2 mM L-glutamine, 1× penicillin/streptomycin and 45 μM CaCl$_2$. The solution can be stored at 4°.

Substrate buffer: 7.5 mM NPAG (Sigma) is dissolved at 37 to 40° in 0.1 M sodium citrate, pH 5.0, and mixed 1:1 with 0.5% Triton X-100 diluted in water. The substrate buffer can be stored at −20°.

Stop buffer: 200 ml 0.05 M glycine, pH 10.4, is mixed with 2 ml of 5 M EDTA and stored at RT.

5.5.2. Procedure

1. Flat-bottom 96-well plates are coated with 50 μl of increasing concentrations of integrin ligands at 4° overnight. Since the proteins are highly adhesive to plastic, the serial dilutions are prepared in glass tubes. Avoid vortexing.

2. The plates are washed with PBS and incubated with 100 μl of the blocking solution at 37° for 90 min or at 4° overnight.

3. Keratinocytes are detached from the cell culture dish (see Section 5.2) and washed with 10 ml of AMEM.

4. The 96-well plate is washed with 200 μl of PBS per well, and then 4 × 10^4 cells in 100 μl of AMEM are seeded per well. The cells are allowed to adhere for 30 to 45 min. Keep 12 × 10^4 cells in a 1.5-ml tube as "loading control." The loading control is needed to determine the relative number of adherent cells, compared to the number of cells seeded in each well.

5. The adherent cells are washed three times with PBS (to remove non-adherent cells), and then 50-μl substrate buffer is added to each well. The cells are incubated with substrate buffer at 37° overnight.
6. Spun-down cells kept as loading controls and resuspended in 150-μl substrate buffer. Aliquot 50 μl onto three wells of the 96-well plate and incubate overnight at 37°.
7. Add 75 μl stop buffer to each well and measure the absorbance at 405 nm.
8. Determine the relative number of attached cells compared to the number of cells in the loading control.

5.6. Analysis of integrin-mediated spreading of primary keratinocytes

5.6.1. Materials
Coating medium and KGM: Described in Section 5.1.

5.6.2. Procedure

1. Cells are detached from tissue culture dish and resuspended in KGM.
2. Approximately 2.5 to 3 $\times$ 10^5 keratinocytes are seeded in a volume not to exceed 2 ml on precoated six-well plates, and are started to be monitored when the first cells begin to attach to the bottom surface of the plate so that the contrast can be properly adjusted.
3. The cells are monitored using live-cell imaging microscopy or by picturing the cells with a digital camera connected to a phase contrast microscope every 10 to 30 min for 4 h. This prolonged time, when compared to other cell types, such as fibroblasts, is due to the fact that keratinocytes generally require more time to fully spread. In addition, primary keratinocytes are not homogenous, and single cells differ greatly in their spreading kinetics.
4. The cell area of keratinocytes is measured at selected time points using suitable software, such as MetaMorph 6.0.

5.7. *In vitro* wound-healing assay with primary keratinocytes

5.7.1. Materials

Coating medium and KGM can be found in Section 5.1.
Mitomycin C (Sigma): A 0.5-mg/ml stock is prepared in water and can be stored at −20°. Mitomycin C, a proliferation inhibitor, is added to the cells, to exclude differences in proliferative properties of the cells, which could possibly influence the results of the *in vitro* wound healing assay.

5.7.2. Procedure

1. Around 1.5 to 2×10^6 freshly isolated keratinocytes are seeded onto precoated six-well plates and cultured until they reach confluence.
2. The KGM is removed and keratinocytes are incubated with fresh KGM containing 4 μg/ml of mitomycin C at 37° for 4 h. Subsequently, the cells are washed three times with PBS and further analyzed in fresh KGM medium.
3. The cell monolayer is wounded by scraping with a yellow tip across the monolayer. The cells migrating into the scratch area are monitored either using live-cell imaging microscopy or by picturing the scratched cells with a digital camera connected to a phase contrast microscope every 0.5 to 2 h. The dynamics of the wound closure is calculated by measuring the width of the wound at selected time points. The wound of approximately 500 μm is closed by wild-type primary keratinocytes within 12 h.

5.8. Trans-well migration assay with primary keratinocytes

5.8.1. Materials

Coating medium, KGM, and EGF can be found in Section 5.1.
AMEM and laminin 332 can be found in Section 5.5.
Staining and fixation solution: 0.1% crystal violet (to stain cells) is resuspended in 20% methanol (to fix the cells).

5.8.2. Procedure

1. The bottom surface of trans-well filters (8-μm pore size, Becton Dickinson) is coated with 100 μl of coating medium or 5 μg/ml of recombinant laminin 322 at 37° for 2 h and then washed once with PBS.
2. The keratinocytes are detached from the cell culture dish and washed with AMEM. In the meantime, the filters are placed in a 24-well plate containing 600 μl of KGM, with or without additional 50 ng/ml of EGF.
3. Around 4×10^4 cells in 200 μl of AMEM are placed on the upper surface of the filter and allowed to migrate at 37° for 14 h to the lower filter surface.
4. The cells are removed from the upper surface of the filter using a cotton swab. The filter is rinsed once with PBS, and subsequently incubated for 5 min with the staining and fixing solution. Keep a "positive filter control" where cells are not wiped off, and place the control filter in the staining and fixation solution. The positive filter control is needed to determine the relative number of migrated cells, compared to the number of cells seeded in each filter.
5. The filter is rinsed by repeated dipping in a beaker with water and then allowed to dry.

6. The relative number of migrated cells is determined by counting the attached cells from five randomly chosen microscopic fields on the filters and comparing it to the number of cells found in five randomly chosen microscopic fields on positive control filters.

5.9. Immunfluorescence on primary keratinocytes

To investigate whether primary keratinocytes can form normal integrin adhesion sites and develop a well-organized actin cytoskeleton, adherent cells are fixed and immunostained with a variety of antibodies recognizing proteins found in focal complexes and focal adhesions (Zaidel-Bar *et al.*, 2004), and stained with fluorescently labeled phalloidin to visualize the F-actin cytoskeleton.

5.9.1. Materials

Fixation, permeabilization, and DAPI solutions can be found in Section 4.3.
Blocking solution: PBS supplemented with 5% BSA.
Primary (Table 12.7) and secondary antibodies: Diluted according to manufacturer's instructions in PBS supplemented with 2% BSA.

5.9.2. Procedure

1. Around 1 to 2×10^5 freshly isolated keratinocytes are plated per well in an eight-well glass chamber slide (LAB-TEK) precoated with coating medium.
2. At 36 to 48 h after seeding (at this time point, most of the cells are fully spread), the slides are washed once with PBS and incubated with 200 μl

Table 12.7 Selection of antibodies staining the focal adhesions of cultured mouse keratinocytes

Antibody	Source
Anti-β1 integrin, clone MB1.2	Chemicon
Anti-FAK	Upstate
Anti-paxillin	BD transduction laboratories
Anti-FAK[pY397] phosphospecific antibody	Biosource international
Anti-paxillin[pY31] phosphospecific antibody	
Anti-paxillin[pY118] phosphospecific antibody	
Anti-vinculin	Sigma
Anti-talin	

Notes: TRICT- and FITC-conjugated phalloidin is used to detect F-actin. Cy3- and FITC-conjugated antibodies specific for mouse IgG, mouse IgM, rabbit IgG, and rat IgG (Jackson Immunochemicals) are used as secondary antibodies.

of the fixation solution (warm up to RT before use) for 20 min at RT. If necessary, keratinocytes should be treated with 200 μl of the permeabilizing solution for 5 min on ice and then rinsed with PBS.

3. The cells are then incubated with 200 μl of the blocking solution at 4° overnight.
4. The cells are incubated with 100 μl of the primary antibody solution at RT for 90 min.
5. Subsequently, the cells are washed with PBS three times for 10 min each, and then incubated with 100 μl of the secondary, fluorophore-conjugated antibody at RT for 60 min. From this step on, the cells should be protected from light.
6. The antibody solution is removed from the chamber slides. If required, the cells are incubated with DAPI solution at RT for 5 min.
7. The slides are washed with PBS three times for 10 min each, and finally, the chamber grid is carefully removed. The cells are mounted with a drop of anti-fading medium (Elvanol) and dried at RT for 30 min. The slides can be immediately analyzed or stored in the dark at 4° for several days or at −20° for several weeks.

REFERENCES

Alonso, L., and Fuchs, E. (2006). The hair cycle. *J. Cell Sci.* **119**, 391–393.

Aumailley, M., Pesch, M., Tunggal, L., Gaill, F., and Fassler, R. (2000). Altered synthesis of laminin 1 and absence of basement membrane component deposition in β1 integrin-deficient embryoid bodies. *J. Cell Sci.* **113**, 259–268.

Barnden, M. J., Allison, J., Heath, W. R., and Carbone, F. R. (1998). Defective TCR expression in transgenic mice constructed using cDNA-based α- and β-chain genes under the control of heterologous regulatory elements. *Immunol. Cell Biol.* **76**, 34–40.

de Boer, J., Williams, A., Skavdis, G., Harker, N., Coles, M., Tolaini, M., Norton, T., Williams, K., Potocnik, A. J., and Kioussis, D. (2003). Transgenic mice with hematopoietic and lymphoid specific expression of Cre. *Eur. J. Immunol.* **33**, 314–325.

Bouvard, D., Brakebusch, C., Gustafsson, E., Aszódi, A., Bengtsson, T., Berna, A., and Fässler, R. (2001). Functional consequences of integrin gene mutations in mice. *Circ. Res.* **89**, 211–223.

Brakebusch, C., Grose, R., Quondamatteo, F., Ramirez, A., Jorcano, J. L., Pirro, A., Svensson, M., Herken, R., Sasaki, T., Timpl, R., Werner, S., and Fässler, R. (2000). Skin and hair follicle integrity is crucially dependent on β1 integrin expression on keratinocytes. *EMBO J.* **19**, 3990–4003.

Calle, Y., Burns, S., Thrasher, A., and Jones, G. (2006). The leukocyte podosome. *Eur. J. Cell Biol.* **85**, 151–157.

Clarke, S. R., Clarke, S. R., Barnden, M., Kurts, C., Carbone, F. R., Miller, J. F., and Heath, W. R. (2000). Characterization of the ovalbumin-specific TCR transgenic line OT-I: MHC elements for positive and negative selection. *Immunol. Cell Biol.* **78**, 110–117.

Clausen, B. E., Burkhardt, C., Rieth, W., Renkawitz, R., and Forster, I. (1999). Conditional gene targeting in macrophages and granulocytes using LysMcre mice. *Transgenic Res.* **V8**, 265–277.

Czuchra, A., Meyer, H., Legate, K. R., Brakebusch, C., and Fässler, R. (2006). Genetic analysis of β1 integrin "activation motifs" in mice. *J. Cell Biol.* **174**, 889–899.

DiGiovanni, J. (1992). Multistage carcinogenesis in mouse skin. *Pharmacol. Ther.* **54**, 63–128.

DiPietro, L. A., and Burns, A. L., eds. (2003). "Wound Healing: Methods and Protocols." Humana Press, Totowa, NJ.

Fassler, R., and Meyer, M. (1995). Consequences of lack of $\beta 1$ integrin gene expression in mice. *Genes Dev.* **9**, 1896–1908.

Ferron, M., and Vacher, J. (2005). Targeted expression of Cre recombinase in macrophages and osteoclasts in transgenic mice. *Genesis* **41**, 138–145.

Georgiades, P., Ogilvy, S., Duval, H., Licence, D. R., Charnock-Jones, D. S., Smith, S. K., and Print, C. G. (2002). VavCre transgenic mice: A tool for mutagenesis in hematopoietic and endothelial lineages. *Genesis* **34**, 251–256.

Gribi, R., Hook, L., Ure, J., and Medvinsky, A. (2006). The differentiation program of embryonic definitive hematopoietic stem cells is largely $\alpha 4$ integrin independent. *Blood* **108**, 501–509.

Grose, R., Hutter, C., Bloch, W., Thorey, I., Watt, F. M., Fässler, R., Brakebusch, C., and Werner, S. (2002). A crucial role of $\beta 1$ integrins for keratinocyte migration *in vitro* and during cutaneous wound repair. *Development* **129**, 2303–2315.

Hirsch, E., Iglesias, A., Potocnik, A. J., Hartmann, U., and Fässler, R. (1996). Impaired migration but not differentiation of haematopoietic stem cells in the absence of $\beta 1$ integrins. *Nature* **380**, 171–175.

Hobeika, E., Thiemann, S., Storch, B., Jumaa, H., Nielsen, P. J., Pelanda, R., and Rethet, M. (2006). Testing gene function early in the B cell lineage in mb1-cre mice. *Proc. Natl. Acad. Sci. USA* **103**, 13789–13794.

Jasinski, M., Keller, P., Fujiwara, Y., Orkin, S. H., and Bessler, M. (2001). GATA1-Cre mediates Piga gene inactivation in the erythroid/megakaryocytic lineage and leads to circulating red cells with a partial deficiency in glycosyl phosphatidylinositol-linked proteins (paroxysmal nocturnal hemoglobinuria type II cells). *Blood* **98**, 2248–2255.

Kanwar, J. R., Harrison, J. E., Wang, D., Leung, E., Mueller, W., Wagner, N., and Krissansen, G. W. (2000). $\beta 7$ integrins contribute to demyelinating disease of the central nervous system. *J. Neuroimmunol.* **103**, 146–152.

Kerfoot, S. M., Norman, M. U., Lapointe, B. M., Bonder, C. S., Zbytnuik, L., and Kubes, P. (2006). Reevaluation of P-selectin and $\alpha 4$ integrin as targets for the treatment of experimental autoimmune encephalomyelitis. *J. Immunol.* **176**, 6225–6234.

Kraus, M., Alimzhanov, M.B, Rajewsky, N., and Rajewsky, K. (2004). Survival of resting mature B lymphocytes depends on BCR signaling via the Igα/β heterodimer. *Cell* **117**, 787–800.

Krishnamoorthy, G., Lassmann, H., Wekerle, H., and Holz, A. (2006). Spontaneous opticospinal encephalomyelitis in a double-transgenic mouse model of autoimmune T cell/B cell cooperation. *J. Clin. Invest.* **116**, 2385–2392.

Kuhn, R., Schwenk, F., Aguet, M., and Rajewsky, K. (1995). Inducible gene targeting in mice. *Science* **269**, 1427–1429.

Langbein, L., Rogers, M. A., Praetzel, S., Winter, H., and Schweizer, J. (2003). K6irs1, K6irs2, K6irs3, and K6irs4 represent the inner-root-sheath-specific type II epithelial keratins of the human hair follicle1. *J. Invest. Dermatol.* **120**, 512–522.

Langbein, L., Rogers, M. A., Winter, H., Praetzel, S., Beckhaus, U., Rackwitz, H. R., and Schweizer, J. (1999). The catalog of human hair keratins. I. Expression of the nine type I members in the hair follicle. *J. Biol. Chem.* **274**, 19874–19884.

Langbein, L., Winter, H., Schweizer, J., and Grosshans, E. (2001). The catalog of human hair keratins. II. Expression of the six type II members in the hair follicle and the combined catalog of human type I and II keratins. *J. Biol. Chem.* **276**, 35123–35132.

Lee, P. P., Fitzpatrick, D. R., Beard, C., Jessup, H. K., Lehar, S., Makar, K. W., Perez-Melgosa, M., Sweetser, M. T., Schlissel, M. S., Nguyen, S., Cherry, S. R., Tsai, J. H., *et al.* (2001). A critical role for Dnmt1 and DNA methylation in T cell development, function, and survival. *Immunity* **15**, 763–774.

Legate, K. R., Montañez, E., Kudlacek, O., and Fässler, R. (2006). ILK, PINCH and parvin: The tIPP of integrin signalling. *Nat. Rev. Mol. Cell Biol.* **7,** 20–31.

Li, S., Bordoy, R., Stanchi, F., Moser, M., Braun, A., Kudlacek, O., Wewer, U. M., Yurchenco, P. D., and Fässler, R. (2005). PINCH1 regulates cell–matrix and cell–cell adhesions, cell polarity and cell survival during the peri-implantation stage. *J. Cell Sci.* **118,** 2913–2921.

Li, S., Harrison, D., Carbonetto, S., Fassler, R., Smyth, N., Edgar, D., and Yurchenco, P. D. (2002). Matrix assembly, regulation, and survival functions of laminin and its receptors in embryonic stem cell differentiation. *J. Cell Biol.* **157,** 1279–1290.

Lorenz, K., Grashoff, C., Torka, R., Sakai, T., Langbein, L., Bloch, W., Aumailley, M., and Fässler, R. (2007). ILK is required for epidermal and hair follicle morphogenesis. *J.Cell Biol.* **177,** 501–513.

Lutz, M. B., Kukutsch, N., Ogilvie, A. L., Röbner, S., Koch, F., Romani, N., and Schuler, G. (1999). An advanced culture method for generating large quantities of highly pure dendritic cells from mouse bone marrow. *J. Immunol. Methods* **223,** 77–92.

Massoumi, R., Chmielarska, K., Hennecke, K., Pfeifer, A., and Fässler, R. (2006). Cyld inhibits tumor cell proliferation by blocking Bcl–3-dependent NF-κB signaling. *Cell* **125,** 665–677.

Mizgerd, J. P., Kubo, H., Kutkoski, G. J., Bhagwan, S. D., Scharffetter-Kochanek, K., Beaudet, A. L., and Doerschuk, C. M. (1997). Neutrophil emigration in the skin, lungs, and peritoneum: Different requirements for CD11/CD18 revealed by CD18-deficient mice. *J. Exp. Med.* **186,** 1357–1364.

Muller-Rover, S., Handjiski, B., van der Veen, C., Eichmuller, S., Foitzik, K., McKay, I. A., Stenn, K. S., and Paus, R. (2001). A comprehensive guide for the accurate classification of murine hair follicles in distinct hair cycle stages. *J. Invest. Dermatol.* **117,** 3–15.

Ohashi, T., and Erickson, H. P. (2005). Domain unfolding plays a role in superfibronectin formation. *J. Biol. Chem.* **280,** 39143–39151.

Paus, R., Muller-Rover, S., Van Der Veen, C., Maurer, M., Eichmuller, S., Ling, G., Hofmann, U., Foitzik, K., Mecklenburg, L., and Handjiski, B. (1999). A comprehensive guide for the recognition and classification of distinct stages of hair follicle morphogenesis. *J. Invest. Dermatol.* **113,** 523–532.

Potocnik, A. J., Brakebusch, C., and Fässler, R. (2000). Fetal and adult hematopoietic stem cells require β1 integrin function for colonizing fetal liver, spleen, and bone marrow. *Immunity* **12,** 653–663.

Probst, H. C., Lagnel, J., Kollias, G., and van den Broek, M. (2003). Inducible transgenic mice reveal resting dendritic cells as potent inducers of CD8+ T cell tolerance. *Immunity* **18,** 713–720.

Ramirez, A., Page, A., Gandarillas, A., Zanet, J., Pibre, S., Vidal, M., Tusell, L., Genesca, A., Whitaker, D. A., Melton, D. W., and Jorcano, J. L. (2004). A keratin K5Cre transgenic line appropriate for tissue-specific or generalized Cre-mediated recombination. *Genesis* **39,** 52–57.

Reiss, Y., Hoch, G., Deutsch, U., and Engelhardt, B. (1998). T cell interaction with ICAM-1-deficient endothelium *in vitro*: Essential role for ICAM-1 and ICAM-2 in transendothelial migration of T cells. *Int. Immunol.* **28,** 3086–3099.

Rickert, R. C., Roes, J., Roes, J., and Rajewsky, K. (1997). B lymphocyte-specific, Cre-mediated mutagenesis in mice. *Nucleic Acids Res.* **25,** 1317–1318.

Romero, M. R., Carroll, J. M., and Watt, F. M. (1999). Analysis of cultured keratinocytes from a transgenic mouse model of psoriasis: Effects of suprabasal integrin expression on keratinocyte adhesion, proliferation and terminal differentiation. *Exp. Dermatol.* **8,** 53–67.

Rotman, B., Zderic, J. A., and Edelstein, M. (1963). Fluorogenic substrates for beta-D-galactosidases and phosphatases derived from fluorescein (3,6-dihydroxyfluoran) and its monomethyl ether. *Proc. Natl. Acad. Sci. USA* **50,** 1–6.

Rousselle, P., and Aumailley, M. (1994). Kalinin is more efficient than laminin in promoting adhesion of primary keratinocytes and some other epithelial cells and has a different requirement for integrin receptors. *J. Cell Biol.* **125,** 205–214.

Sakai, T., Li, S., Docheva, D., Grashoff, C., Sakai, K., Kostka, G., Braun, A., Pfeifer, A., Yurchenco, P. D., and Fässler, R. (2003). Integrin-linked kinase (ILK) is required for polarizing the epiblast, cell adhesion, and controlling actin accumulation. *Genes Dev.* **17,** 926–940.

Sixt, M., Bauer, M., Lammermann, T., and Fassler, R. (2006). $\beta1$ integrins: Zip codes and signaling relay for blood cells. *Curr. Opin. Cell Biol.* **18,** 482–490.

Talts, J., Brakebusch, C., and Fässler, R. (1999). Integrin gene targeting. *Integrin Protocols* **129,** 153–188.

Tiedt, R., Schomber, T., Hao-Shen, H., and Skoda, R. C. (2006). Pf4-Cre transgenic mice allow generating lineage-restricted gene knockouts for studying megakaryocyte and platelet function *in vivo. Blood* **109,** 1503–1506.

Vajkoczy, P., Laschinger, M., and Engelhardt, B. (2001). α4-integrin-VCAM-1 binding mediates G protein-independent capture of encephalitogenic T cell blasts to CNS white matter microvessels. *J. Clin. Invest.* **108,** 557–565.

Vasioukhin, V., Degenstein, L., Wise, B., and Fuchs, E. (1999). The magical touch: Genome targeting in epidermal stem cells induced by tamoxifen application to mouse skin. *Proc. Natl. Acad. Sci. USA* **96,** 8551–8556.

Watt, F. M. (2002). Role of integrins in regulating epidermal adhesion, growth and differentiation. *EMBO J.* **21,** 3919–3926.

Watt, F. M., Lo Celso, C., and Silva-Vargas, V. (2006). Epidermal stem cells: An update. *Curr. Opin. Genet. Dev.* **16,** 518–524.

Yednock, T. A., Cannon, C., Fritz, L. C., Sanchez-Madrid, F., Steinman, L., and Karin, N. (1992). Prevention of experimental autoimmune encephalomyelitis by antibodies against α4β1 integrin. *Nature* **356,** 63–66.

Zaidel-Bar, R., Cohen, M., Addadi, L., and Geiger, B. (2004). Hierarchical assembly of cell–matrix adhesion complexes. *Biochem. Soc. Trans.* **32,** 416–420.

Zal, T., Volkman, A., and Stockinger, B. (1994). Mechanisms of tolerance induction in major histocompatibility complex class II–restricted T cells specific for a blood-borne self-antigen. *J. Exp. Med.* **180,** 2089–2099.

CHAPTER THIRTEEN

IDENTIFICATION AND MOLECULAR CHARACTERIZATION OF MULTIPLE PHENOTYPES IN INTEGRIN KNOCKOUT MICE

Chun Chen *and* Dean Sheppard

Contents

Abstract

Each of the 24 known integrin subunits has now been inactivated in mice, and a growing number of conditional null lines are becoming available. Lines of mice expressing null mutations in integrin subunit genes have taught us a great deal about the remarkably diverse functions that integrins perform *in vivo* in mammals. Thorough evaluation of the phenotypes manifested by these lines has also revealed a number of previously unexpected integrin ligands and signaling partners. In this article, we review approaches that can contribute to optimal use of this valuable resource.

Lung Biology Center, Department of Medicine, University of California, San Francisco, San Francisco, California

Methods in Enzymology, Volume 426
ISSN 0076-6879, DOI: 10.1016/S0076-6879(07)26013-6
© 2007 Elsevier Inc.
All rights reserved.

1. INTRODUCTION

The basic structure of integrin α and β subunits is highly conserved through millions of years of evolution. However, this family is quite small in invertebrates, which generally express between 2 and 5 integrin heterodimers. The complexity of the family dramatically increased with the evolution of vertebrates, so that mammals express at least 24 distinct heterodimers. This increase in structural complexity has been accompanied by a dramatic increase in the number of distinct cellular functions performed by integrins. Whereas in invertebrates, integrins are principally involved in the maintenance of cell adhesion and structural tissue integrity, in vertebrates they have been recruited to mediate arrest of rapidly moving cells, regulation of cell growth, survival, and differentiation and even activation of latent growth factors stored in the extracellular space. With the evolution of numerous organs in vertebrates, members of the integrin family have also taken on a variety of distinct organ-specific functions. Much of what is known about these new roles of integrins has come from evaluation of phenotypic abnormalities in integrin knockout mice.

2. STRATEGIES FOR GENERATING INTEGRIN KNOCKOUT MICE

Early integrin knockouts were generally made by replacing a critical exon in an integrin subunit gene, with the gene encoding a neomycin resistance cassette to allow selection of targeted clones by treatment with a neomycin analogue. Because of the complex tertiary structure of integrin extracellular domains and the relatively few examples of alternative splicing, it is likely that deletion of almost any exon within the extracellular domain would result in an effective null mutation. However, demonstration of the absence of integrin subunit protein and/or mRNA $3'$ of the targeted region is essential for meaningful interpretation of experiments with these early knockout lines. Another problem with this strategy is the possibility that the strong promoter in the inserted neomycin-resistance cassette will exert long-range effects on the expression of nearby genes on the same chromosome, leading to erroneous assignment of a mutant phenotype to the effects of the deleted integrin subunit. To circumvent this problem, more recent knockout lines have been generated with the selection maker cassette flanked by LoxP or FRT sites to allow removal by expression of the DNA recombinases Cre or Flp in either the targeted ES cells or *in vivo* using lines expressing these recombinases in the germline.

Because of the important roles played by many integrins in development, several unconditional integrin knockouts lead to embryonic or early postnatal mortality (Bader *et al.*, 1998; Fassler and Meyer, 1995; Georges-Labouesse *et al.*, 1996; Kreidberg *et al.*, 1996; Muller *et al.*, 1997; Murgia *et al.*, 1998; Stephens *et al.*, 1995; Yang *et al.*, 1993, 1995; Zhu *et al.*, 2002). Much has been learned about the contributions of these subunits to developing adult cells and tissues by the analysis of chimeric mice generated from a mixture of wild-type and knockout ES cells (Arroyo *et al.*, 1999; Hirsch *et al.*, 1996). More recently, lines have been generated expressing conditional alleles in which one or more critical exon is flanked by loxP sites to allow cell-specific knockout (McCarty *et al.*, 2005; Potocnik *et al.*, 2000; Proctor *et al.*, 2005; Scott *et al.*, 2003) and additional conditional integrin knockout lines have been generated but not yet described in published manuscripts.

3. VALIDATION OF OBSERVED PHENOTYPES BY TRANSGENIC RESCUE *IN VIVO*

To demonstrate that an observed phenotype is actually explained by the loss of the specific targeted integrin subunit, *in vivo* rescue by transgenic overexpression of the deleted subunit in a relevant cell type should be attempted whenever possible. One strategy for this purpose is the generation of transgenic mice expressing the deleted subunit under the control of a tissue-specific promoter. This approach has been used to demonstrate the role of the integrin $\beta 6$ subunit in negatively regulating lung inflammation and the induction of pulmonary emphysema, and to show that expression of this subunit in a subset of alveolar epithelial cells is sufficient to reverse each of these phenotypes (Huang *et al.*, 1998b; Morris *et al.*, 2003). In addition to proving that an observed phenotype is due to the loss of the deleting subunit, *in vivo* rescue allows the performance of structure–function analysis *in vivo*. This approach demonstrated that interaction of the $\alpha v\beta 6$ integrin with ligand was required for suppression of both pulmonary inflammation and the development of emphysema, since transgenic expression of a point mutant in the β subunit MIDAS motif that was shown to be normally expressed but to abrogate ligand binding (Huang *et al.*, 1995) was unable to rescue either phenotype (Morris *et al.*, 2003). In contrast, a mutant that lacked the last 11 amino acids of the carboxy terminus, which was shown to be incapable of supporting integrin-mediated proliferation *in vivo* (Agrez *et al.*, 1994), fully rescued both phenotypes, demonstrating that neither was dependent on this function of the integrin.

4. *In Vitro* Transgenic Rescue

Rescue experiments can also be productively employed using cultured cells from integrin-subunit knockout mice if convincing and informative *in vitro* assays can be generated that are directly relevant to an *in vivo* phenotype. This approach has recently been employed to confirm that the severe neutropenia seen in newborn mice lacking the integrin α9 subunit is due to the loss of α9 in bone marrow cells (Chen *et al.*, 2006). In that case, the human α9 subunit was heterologously expressed by retroviral transfer in bone marrow derived from α9 knockout mice. α9 expression substantially rescued the production of neutrophils in response to the neutrophil-inducing cytokine G-CSF in *in vitro* colony-forming assays. A similar approach was used to demonstrate that the protection of integrin β5–subunit knockout mice from increased vascular permeability in a model of acute lung injury was due to loss of this integrin from pulmonary endothelial cells (Su *et al.*, 2007). In that case, lines of immortalized pulmonary vascular endothelial cells were generated from β5 knockout mice and shown to be completely resistant to the induction of increased monolayer permeability in response to three different edemagenic agonists. Reconstitution of these cells with the human β5 subunit completely restored the response to all three agonists, and this effect could be completely inhibited by a blocking monoclonal antibody to αvβ5.

5. Generation of Blocking Mouse Monoclonal Antibodies that Recognize Murine Integrins

Validation of *in vivo* phenotypes in knockout mice is often hampered by the lack of specific blocking antibodies that recognize murine proteins and are suitable for *in vivo* experiments. Knockout lines that survive into early adulthood provide a great opportunity to generate such reagents, since these mice will not recognize the inactivated protein as self and can therefore be immunized to generate high-affinity blocking monoclonal antibodies targeting the knocked-out protein. In our experience, the optimal immunogen to generate such antibodies is a murine cell line expressing high levels of the knocked-out integrin. Ideally, this line should express the murine integrin, but we have also been successful using lines expressing the human paralog of the knocked-out subunit as a heterodimer with its murine partner. Because all other proteins on these murine cells are recognized as self, most hybridomas generated with this strategy recognize the integrin that has been inactivated. Furthermore, because most integrin subunits are 80 to 90% identical among mammals, most of the antibodies generated with this approach will recognize the inactivated integrin from all mammalian species. We have successfully used this strategy to generate mouse mono-clonal antibodies against $\alpha v\beta 5$ (Su *et al.*, 2006) and $\alpha v\beta 6$ (Huang *et al.*, 1998a), and have used these to validate the role of $\alpha v\beta 5$ in multiple models of acute lung injury (Su *et al.*, 2007) and the roles of $\alpha v\beta 6$ in both acute lung injury and pulmonary fibrosis (Pittet *et al.*, 2001). These antibodies can also serve as the platform for humanized versions that could be used as thera-peutics for diseases in which a role for integrins is suggested by studies in knockout mice.

Identical phenotypes in integrin knockout mice and wild-type mice treated with integrin-blocking antibodies provide strong support for the specificity of the knockout phenotypes. However, it is important to keep in mind that loss of an integrin throughout growth and development, and blockade of integrin ligation in adult animals, need not result in identical phenotypes. Informative examples of this problem come from studies of the roles of the $\alpha v\beta 3$ and $\alpha v\beta 5$ integrins in angiogenesis. In that case, convinc-ing evidence from multiple model systems suggests that blockade of either integrin can potently inhibit angiogenesis induced by either specific growth factors or by tumors (Brooks *et al.*, 1994a,b, 1995). However, mice homo-zygous for null mutations of the $\beta 3$ subunit, the $\beta 5$ subunit, or both, do not have impaired angiogenic responses to any stimuli that have been studied (Reynolds *et al.*, 2002). In fact, $\beta 3$ and $\beta 5$ double-knockout mice demon-strate the opposite effect, an enhancement of angiogenesis. There are two explanations for the apparent discrepancies in these results. First, mice lacking the $\beta 3$ subunit throughout life undergo significant genetic compen-sation for this loss, with a marked upregulation of the proangiogenic receptor, VEGF-R2. In addition, the mechanism of impaired angiogenesis in response to inhibitors that interfere with ligation of these integrins does not appear to be the equivalent of loss of the integrin, but is rather due to proapoptotic signals that are initiated when the integrins are present but not ligated (Stupack and Cheresh, 2003).

6. SELECTION OF INTEGRIN KNOCKOUT LINES TO STUDY

Integrin subunits do not randomly associate with one another. Rather, they can assemble to form a restricted number of heterodimers as shown in Fig. 13.1. As is apparent from this figure, most integrin subunits are present in only a single heterodimer, whereas a few (most notably $\beta 1$, $\beta 2$, and αv) can form heterodimers with multiple partners. If an investigator has a hypothesis to test about a specific integrin heterodimer, it is clearly advan-tageous to use mice in which the subunit that has been inactivated is only present in a single heterodimer. For this reason, we have chosen only to

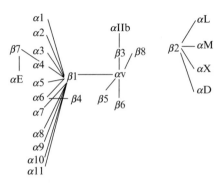

Figure 13.1 Integrin family map. Lines connect α and β subunits that have been identified as integrin heterodimers. Note that a few subunits (i.e., β1, αv, and β2) can form multiple heterodimers, but most subunits are present only in a single unique heterodimer.

make knockouts of such subunits. Much has been learned about the general roles of broad subfamilies of integrins based on inactivation of promiscuous subunits such as αv (Bader *et al.*, 1998) or β1 (Fassler and Meyer, 1995; Stephens *et al.*, 1995), but in principle results from these lines should be interpreted with caution because of the potential for competing effects of various heterodimers that share the same subunit. Clearly, distinguishing which integrins are responsible for any phenotypes observed with these lines depends on follow-up studies with mice expressing null mutations of subunits that are restricted to a single heterodimer.

 7. SELECTION OF PHENOTYPIC ASSAYS

One major challenge in optimally mining the information that is potentially available from integrin knockout mice is deciding what organs to evaluate and which assays to perform. Because identification of many integrin functions requires carefully timed evaluation of specific organs and cell types and/or use of these knockouts in specific disease models, it is likely that investigators will continue to identify additional phenotypes in these mice over the next several years. As a guide for how to decide where and how to look for phenotypes, we will review some informative examples from our own experience.

Of course, the first step in evaluation of any new knockout line is determining whether the knockout results in an increase in mortality during development or the immediate postnatal period. This is usually best accomplished by initially generating heterozygote intercrosses and genotyping on the order of 100 offspring at 2 to 3 weeks of age. With 100 offspring, it is easy to determine whether homozygous knockouts are present at the expected

25% frequency. As soon as it becomes apparent that there is a loss of some or all homozygous null offspring, it is advisable to initiate timed matings and to check mating females at least twice daily for the presence of vaginal plugs. Toward the end of each expected pregnancy, it is then important to view the pregnant mothers at least twice daily and to frequently count and carefully observe all of the delivered pups to look for signs of distress or newborn loss. Distressed or dead pups should be photographed, genotyped, carefully examined, and dissected, and all major organs should be embedded, sectioned, and stained for morphologic examination (initially by hematoxylin and eosin staining). As an example, we have used this approach to identify respiratory failure from congenital chylothorax as the cause of death in all mice homozygous for a knockout of the integrin $\alpha9$ subunit (Huang *et al.*, 2000b). No $\alpha9$ null mice initially survived until genotyping at p14, but all were observed to develop respiratory distress and bilateral milky pleural fluid collections between 5 and 10 days after birth. This phenotype was due to leaking from the thoracic duct, the large lymphatic vessels that returns lymph to the systemic circulation, and this phenotype identified the $\alpha9\beta1$ integrin as an important determinant of normal lymphatic development. A similar strategy has identified early postnatal mortality from a variety of causes in mice lacking the $\alpha3$ (Kreidberg *et al.*, 1996), $\alpha6$ (Georges-Labouesse *et al.*, 1996), $\alpha8$ (Muller *et al.*, 1997), αv (Bader *et al.*, 1998), $\beta4$ (Murgia *et al.*, 1998), and $\beta8$ (Zhu *et al.*, 2002) subunits.

In the absence of postnatal loss or distress (which implies prenatal mortality), the best approach is then to sacrifice pregnant mothers at various time points after the onset of pregnancy, and to genotype and carefully observe the morphology of a sufficient number of embryos to identify the precise time of the onset of embryonic loss. Examination of whole-mount embryos and stained sections from these embryos usually provides clues to the mechanisms underlying embryonic lethality (Yang *et al.*, 1993, 1995). For the most severe integrin knockout, loss of the $\beta1$ subunit that is contained in 12 different integrin heterodimers, the knockout produced a defect in gastrulation that could only be detected by the presence of empty implantation sites at E5.5 (Fassler and Meyer, 1995; Stephens *et al.*, 1995).

8. IDENTIFICATION OF ADDITIONAL PHENOTYPES IN INTEGRIN KNOCKOUTS THAT SURVIVE EMBRYONIC DEVELOPMENT

Because of the divergent roles that integrins play in multiple organs and in multiple cell types, several integrin knockout lines have been found to have multiple phenotypes. Many of these phenotypes will become apparent over time with careful observation and the use of targeted functional assays, but many others will remain silent in the absence of specific interventions. Some of these phenotypes can be intelligently hypothesized based on previously published *in vitro* observations or *in vivo* studies with integrin antagonists, but many are surprising.

8.1. Pattern of distribution

One obvious starting point for generating hypotheses about potential phenotypes in integrin knockout mice is knowledge of the cells and tissues in which the integrin is normally expressed. For example, our earlier observation that the $\alpha9\beta1$ integrin was highly expressed in human neutrophils (Taooka *et al.*, 1999) led to evaluation of neutrophils in newborn mice prior to their death from chylothorax. Our finding that neutrophil counts were dramatically reduced in 7-day-old mice led us to identify a previously unsuspected role for this integrin in enhancing signaling from the G-CSF receptor that is responsible for the normal differentiation and proliferation of neutrophils (Chen *et al.*, 2006). In that case, serendipity also played an important role. We were fortunate that $\alpha9$ knockout mice die soon after birth, because older mice appear to compensate for this loss of integrin signaling. When we generate with a conditional null allele of the $\alpha9$ subunit that we crossed to mice expressing cre recombinase in leukocytes (vav1-cre), these mice have a similar reduction in neutrophil counts at 7 days of age, but the mice recover gradually and have normal neutrophil counts as adults (unpublished observations).

Knowledge about the pattern of distribution was also very helpful in identifying the phenotype of mice lacking the integrin $\beta6$ subunit. This subunit was initially cloned from airway epithelial cells (Sheppard *et al.*, 1990) and was found by *in situ* hybridization and tissue staining to be largely restricted to epithelial cells in a subset of epithelial organs (Breuss *et al.*, 1993, 1995). This information helped us to target our initial morphologic evaluation of $\beta6$ knockout mice to epithelial organs such as the lungs and skin and accelerated our observations of exaggerated environmentally induced inflammation in these organs.

However, some integrins are so widely expressed that the pattern of distribution is not particularly informative. This has been the case for the integrin $\beta5$ subunit, which is ubiquitously expressed. Despite this ubiquitous expression, mice lacking $\beta5$ looked remarkably normal to us for the first 5 years after we generated this line (Huang *et al.*, 2000a). In this case, identification of specific phenotypes was highly dependent on guesses made based on prior *in vitro* studies, which suggested roles for $\alpha v\beta5$ in specific biological processes (see below). In other cases, there are no good reagents available to adequately assess the pattern of distribution of a particular integrin subunit. One way to get around this problem would be to generate specific integrin subunit–reporter mice, but no such reagents have yet been made for most integrin subunits.

9. IDENTIFICATION OF INTEGRIN KNOCKOUT PHENOTYPES BASED ON EDUCATED GUESSES FROM *IN VITRO* EXPERIMENTS

Many of the roles that have been suggested for integrins based on studies with primary cell lines or cultured cells have not been validated in integrin knockout mice. One important use of these mice can be identifying which of the many suggestions made from *in vitro* studies actually have *in vivo* relevance. In a few notable cases, clues from *in vitro* studies have been critical in identifying knockout phenotypes that would not otherwise have been apparent. Two examples from $\beta 5$ subunit knockout mice are informative in this regard. We had these mice in our facility for several years and had no idea that they develop age-related blindness. In retrospect, this is not surprising, since mice obtain enough clues from other sensory inputs to function normally in the protective environment of a caged facility without requiring normal vision. However, Finnemann *et al.* (1997) had found that retinal pigment epithelial cells use $\alpha v \beta 5$ for the required daily phagocytosis of shed pigment, and suspected that in the absence of this integrin mice might gradually accumulate shed pigment. When these authors carefully evaluated the retinas of the $\beta 5$ knockout mice, their hypothesis was clearly correct, which led them to perform careful evaluation of vision in aging mice (Nandrot *et al.*, 2004).

Another example is the role played by $\alpha v \beta 5$ in regulating rapid increases in vascular permeability. Cheresh and colleagues (in Friedlander *et al.*, 1995) had previously shown that inhibition of this integrin played a particularly important role in blocking angiogenesis induced by the growth factor VEGFA. These authors therefore suspected that $\beta 5$ knockout mice would be protected from *in vivo* effects of this growth factor. As noted above, loss or blockade of $\alpha v \beta 5$ does not result in a decrease in VEGFA-induced angiogenesis. However, $\beta 5$ knockout mice were protected from another important effect of VEGF, increased vascular permeability (Eliceiri *et al.*, 2002). The resistance of these mice to VEGF-induced increases in vascular permeability also resulted in protection in a model of ischemic stroke. Based on these results, we further evaluated the role of $\alpha v \beta 5$ in two models of acute lung injury, a disorder characterized by increased pulmonary vascular permeability, and found that loss or blockade of this integrin led to dramatic protection (Su *et al.*, 2006). Interestingly, in contrast to the systemic circulation, in the pulmonary circulation *in vivo* and in cultured pulmonary endothelial cells studied *in vitro*, $\alpha v \beta 5$ appears to be required for induction of increased permeability in response to multiple stimuli in addition to VEGFA.

10. Use of Expression Microarrays to Suggest Possible Phenotypic Assays

One of the major goals in studies of knockout mice is the identification of roles for the inactivated genes in pathways that would not have been suspected based on previous *in vitro* studies. We have found whole-genome expression microarrays to be an especially valuable tool for suggesting such roles. One way to use this strategy is to perform microarrays from organs that have a clear-cut but unexplained phenotype. For example, when we first generated $\beta6$ knockout mice, we observed exaggerated pulmonary inflammation, but protection from induced pulmonary fibrosis. These phenotypes suggested, but did not prove, that the cytokine-transforming growth factor β might play an important role downstream of the integrin, since loss of TGFβ1 leads to widespread tissue inflammation in mice, and TGFβ (Shull *et al.*, 1992) is a well-known central regulator of tissue fibrosis. Microarrays from the lungs of wild-type and $\beta6$ knockout mice at baseline and after treatment with the fibrogenic drug, bleomycin, identified a group of more than 50 known TGFβ-inducible genes that were expressed at higher levels after bleomycin in the lungs of wild-type mice than in the lungs of $\beta6$ knockout animals, providing strong support for a role for TGFβ in this pathway (Kaminski *et al.*, 2000). *In vitro* data showed that $\alpha v\beta6$ is actually capable of binding to and activating TGFβ (Munger *et al.*, 1999), which we now strongly suspect is the explanation for each of the phenotypic features we observed.

Closer analysis of the pattern of gene expression at various time points after treatment with bleomycin using self-organizing maps identified a small cluster of TGFβ-inducible genes that were clearly induced within 5 days after bleomycin treatment, a time point well in advance of the onset of pulmonary fibrosis, but the peak time point for development of another response to bleomycin pulmonary edema from acute lung injury. This finding suggested that $\alpha v\beta6$-mediated TGFβ activation might modulate the development of acute lung injury, an effect that we have demonstrated both for bleomycin and for lung injury in response to intratracheal endotoxin (Pittet *et al.*, 2001) and large tidal volume ventilation (Jenkins *et al.*, 2006).

Further analysis of the difference in gene expression at baseline in the lungs of wild-type and $\beta6$ knockout mice identified a metalloprotease, MMP12, that was expressed at 200-fold higher levels in the lungs of $\beta6$ knockout mice than in wild-type mice (Morris *et al.*, 2003). Previous studies had suggested that MMP12 might be responsible for the pulmonary emphysema that can be induced in mice by long-term exposure to cigarette smoke (Hautamaki *et al.*, 1997). However, no previous studies had identified any roles for integrins or for TGFβ in regulating the development of emphysema. We reasoned that if MMP12 was actually the cause of emphysema in response to cigarette smoke, then b6 knockout mice with such high baseline levels of MMP12 expression should develop emphysema. However, in mice

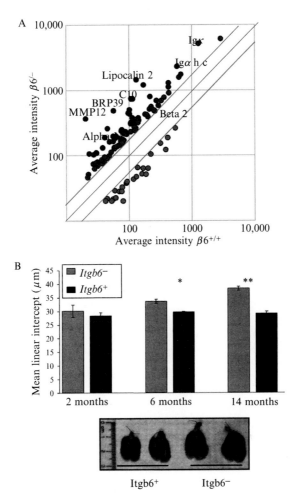

Figure 13.2 Microarray identification of increased MMP12 expression and age-related emphysema in Itgb6 null mice. (A) Microarray analysis of whole-lung gene-expression profiling using Afymetrix mu6500. The names of selected highly induced genes are labeled. MMP12 was the single most highly induced gene in the lungs of Itgb6 null mice. (B) Upper panel shows mean linear intercepts (micrometers plus/minus standard error of the mean) of alveolar septae measured in the lungs of five integrin $\beta6$ wildtype and five integrin $\beta6$ null mice at 2, 6, and 14 months of age. $\star p = 0.0006$, $\star\star p = 0.0002$. Lower panel shows lungs from control or Itgb6 null mice inflated at a constant filling pressure. Itgb6 null mice develop progressive alveolar enlargement with age.

we had studied before that time, which we routinely euthanized by 2 months after birth, we had not seen any evidence of emphysema. Based on the microarray results, we simply let the mice get older and then carefully performed morphometric evaluations of alveolar size at 6, 8, and 14 months of age. By 6 months of age, $\beta6$ knockout mice had mild, but significant emphysema, which became progressively more apparent with age (Morris et al., 2003) (Fig. 13.2). This important phenotype, which we showed was due to the loss of TGFβ activation, would never have been noticed without strong suggestive evidence from expression microarrays.

 11. SUMMARY

Integrin knockout mice are a powerful tool for identifying *in vivo* roles for members of the integrin family and novel ligands (e.g., TGFβ) and downstream signaling pathways (e.g., GCSFR signaling). In this brief review, we have described some of the approaches that can be used to optimize the information obtained from studies using these mice. Using mice with inactivation of integrin subunits that are only present in a single heterodimer, and combining carefully timed morphologic evaluation with expression microarrays of selected organs, it has thus far been possible to identify a number of unexpected roles for specific integrin heterodimers. Additional insights have been provided by targeted evaluations or interventions based on functions suggested from *in vitro* studies. Given the thousands of papers that have suggested a myriad of roles for integrins *in vitro*, many hypotheses clearly remain to be examined in these mice. With the increasing availability of mice expressing conditional null alleles, it should also be possible to identify multiple additional roles for integrins whose unconditional loss leads to embryonic or perinatal mortality.

REFERENCES

Agrez, M., Chen, A., Cone, R. I., Pytela, R., and Sheppard, D. (1994). The alpha v beta 6 integrin promotes proliferation of colon carcinoma cells through a unique region of the beta 6 cytoplasmic domain. *J. Cell Biol.* **127,** 547–556.

Arroyo, A. G., Yang, J. T., Rayburn, H., and Hynes, R. O. (1999). Alpha4 integrins regulate the proliferation/differentiation balance of multilineage hematopoietic progenitors *in vivo*. *Immunity* **11,** 555–566.

Bader, B. L., Rayburn, H., Crowley, D., and Hynes, R. O. (1998). Extensive vasculogenesis, angiogenesis, and organogenesis precede lethality in mice lacking all alpha v integrins. *Cell* **95,** 507–519.

Breuss, J. M., Gallo, J., DeLisser, H. M., Klimanskaya, I. V., Folkesson, H. G., Pittet, J. F., Nishimura, S. L., Aldape, K., Landers, D. V., Carpenter, W., Gillett, N., Sheppard, D., *et al.* (1995). Expression of the beta 6 integrin subunit in development, neoplasia and tissue repair suggests a role in epithelial remodeling. *J. Cell Sci.* **108,** 2241–2251.

Breuss, J. M., Gillett, N., Lu, L., Sheppard, D., and Pytela, R. (1993). Restricted distribution of integrin beta 6 mRNA in primate epithelial tissues. *J. Histochem. Cytochem.* **41,** 1521–1527.

Brooks, P. C., Clark, R. A., and Cheresh, D. A. (1994a). Requirement of vascular integrin alpha v beta 3 for angiogenesis. *Science* **264,** 569–571.

Brooks, P. C., Montgomery, A. M., Rosenfeld, M., Reisfeld, R. A., Hu, T., Klier, G., and Cheresh, D. A. (1994b). Integrin alpha v beta 3 antagonists promote tumor regression by inducing apoptosis of angiogenic blood vessels. *Cell* **79,** 1157–1164.

Brooks, P. C., Stromblad, S., Klemke, R., Visscher, D., Sarkar, F. H., and Cheresh, D. A. (1995). Antiintegrin alpha v beta 3 blocks human breast cancer growth and angiogenesis in human skin. *J. Clin. Invest.* **96,** 1815–1822.

Chen, C., Huang, X., Atakilit, A., Zhu, Q. S., Corey, S. J., and Sheppard, D. (2006). The integrin alpha9beta1 contributes to granulopoiesis by enhancing granulocyte colony-stimulating factor receptor signaling. *Immunity* **25,** 895–906.

Eliceiri, B. P., Puente, X. S., Hood, J. D., Stupack, D. G., Schlaepfer, D. D., Huang, X. Z., Sheppard, D., and Cheresh, D. A. (2002). Src-mediated coupling of focal adhesion kinase to integrin alpha(v)beta5 in vascular endothelial growth factor signaling. *J. Cell Biol.* **157,** 149–160.

Fassler, R., and Meyer, M. (1995). Consequences of lack of beta 1 integrin gene expression in mice. *Genes Dev.* **9,** 1896–1908.

Finnemann, S. C., Bonilha, V. L., Marmorstein, A. D., and Rodriguez-Boulan, E. (1997). Phagocytosis of rod outer segments by retinal pigment epithelial cells requires alpha(v) beta5 integrin for binding but not for internalization. *Proc. Natl. Acad. Sci. USA* **94,** 12932–12937.

Friedlander, M., Brooks, P. C., Shaffer, R. W., Kincaid, C. M., Varner, J. A., and Cheresh, D. A. (1995). Definition of two angiogenic pathways by distinct alpha v integrins. *Science* **270,** 1500–1502.

Georges-Labouesse, E., Messaddeq, N., Yehia, G., Cadalbert, L., Dierich, A., and Le Meur, M. (1996). Absence of integrin alpha 6 leads to epidermolysis bullosa and neonatal death in mice. *Nat. Genet.* **13,** 370–373.

Hautamaki, R. D., Kobayashi, D. K., Senior, R. M., and Shapiro, S. D. (1997). Requirement for macrophage elastase for cigarette smoke-induced emphysema in mice. *Science* **277,** 2002–2004.

Hirsch, E., Iglesias, A., Potocnik, A. J., Hartmann, U., and Fassler, R. (1996). Impaired migration but not differentiation of haematopoietic stem cells in the absence of beta1 integrins. *Nature* **380,** 171–175.

Huang, X., Griffiths, M., Wu, J., Farese, R. V., Jr., and Sheppard, D. (2000a). Normal development, wound healing, and adenovirus susceptibility in beta5-deficient mice. *Mol. Cell Biol.* **20,** 755–759.

Huang, X., Wu, J., Spong, S., and Sheppard, D. (1998a). The integrin alphavbeta6 is critical for keratinocyte migration on both its known ligand, fibronectin, and on vitronectin. *J. Cell Sci.* **111,** 2189–2195.

Huang, X., Wu, J., Zhu, W., Pytela, R., and Sheppard, D. (1998b). Expression of the human integrin beta6 subunit in alveolar type II cells and bronchiolar epithelial cells reverses lung inflammation in beta6 knockout mice. *Am. J. Respir. Cell Mol. Biol.* **19,** 636–642.

Huang, X. Z., Chen, A., Agrez, M., and Sheppard, D. (1995). A point mutation in the integrin beta 6 subunit abolishes both alpha v beta 6 binding to fibronectin and receptor localization to focal contacts. *Am. J. Respir. Cell Mol. Biol.* **13,** 245–251.

Huang, X. Z., Wu, J. F., Ferrando, R., Lee, J. H., Wang, Y. L., Farese, R. V., Jr., and Sheppard, D. (2000b). Fatal bilateral chylothorax in mice lacking the integrin alpha9-beta1. *Mol. Cell Biol.* **20,** 5208–5215.

Jenkins, R. G., Su, X., Su, G., Scotton, C. J., Camerer, E., Laurent, G. J., Davis, G. E., Chambers, R. C., Matthay, M. A., and Sheppard, D. (2006). Ligation of protease-activated receptor 1 enhances alpha(v)beta6 integrin-dependent TGF-beta activation and promotes acute lung injury. *J. Clin. Invest.* **116,** 1606–1614.

Kaminski, N., Allard, J. D., Pittet, J. F., Zuo, F., Griffiths, M. J., Morris, D., Huang, X., Sheppard, D., and Heller, R. A. (2000). Global analysis of gene expression in pulmonary fibrosis reveals distinct programs regulating lung inflammation and fibrosis. *Proc. Natl. Acad. Sci. USA* **97,** 1778–1783.

Kreidberg, J. A., Donovan, M. J., Goldstein, S. L., Rennke, H., Shepherd, K., Jones, R. C., and Jaenisch, R. (1996). Alpha 3 beta 1 integrin has a crucial role in kidney and lung organogenesis. *Development* **122,** 3537–3547.

McCarty, J. H., Lacy-Hulbert, A., Charest, A., Bronson, R. T., Crowley, D., Housman, D., Savill, J., Roes, J., and Hynes, R. O. (2005). Selective ablation of αv integrins in the central nervous system leads to cerebral hemorrhage, seizures, axonal degeneration and premature death. *Development* **132,** 165–176.

Morris, D. G., Huang, X., Kaminski, N., Wang, Y., Shapiro, S. D., Dolganov, G., Glick, A., and Sheppard, D. (2003). Loss of integrin alpha(v)beta6–mediated TGF-beta activation causes Mmp12-dependent emphysema. *Nature* **422,** 169–173.

Muller, U., Wang, D., Denda, S., Meneses, J. J., Pedersen, R. A., and Reichardt, L. F. (1997). Integrin alpha8beta1 is critically important for epithelial-mesenchymal interactions during kidney morphogenesis. *Cell* **88,** 603–613.

Munger, J. S., Huang, X., Kawakatsu, H., Griffiths, M. J., Dalton, S. L., Wu, J., Pittet, J. F., Kaminski, N., Garat, C., Matthay, M. A., Rifkin, D. B., and Sheppard, D. (1999). The integrin alpha v beta 6 binds and activates latent TGF beta 1: A mechanism for regulating pulmonary inflammation and fibrosis. *Cell* **96,** 319–328.

Murgia, C., Blaikie, P., Kim, N., Dans, M., Petrie, H. T., and Giancotti, F. G. (1998). Cell cycle and adhesion defects in mice carrying a targeted deletion of the integrin beta4 cytoplasmic domain. *EMBO J.* **17,** 3940–3951.

Nandrot, E. F., Kim, Y., Brodie, S. E., Huang, X., Sheppard, D., and Finnemann, S. C. (2004). Loss of synchronized retinal phagocytosis and age-related blindness in mice lacking alphavbeta5 integrin. *J. Exp. Med.* **200,** 1539–1545.

Pittet, J. F., Griffiths, M. J., Geiser, T., Kaminski, N., Dalton, S. L., Huang, X., Brown, L. A., Gotwals, P. J., Koteliansky, V. E., Matthay, M. A., and Sheppard, D. (2001). TGF-beta is a critical mediator of acute lung injury. *J. Clin. Invest.* **107,** 1537–1544.

Potocnik, A. J., Brakebusch, C., and Fassler, R. (2000). Fetal and adult hematopoietic stem cells require beta1 integrin function for colonizing fetal liver, spleen, and bone marrow. *Immunity* **12,** 653–663.

Proctor, J. M., Zang, K., Wang, D., Wang, R., and Reichardt, L. F. (2005). Vascular development of the brain requires beta8 integrin expression in the neuroepithelium. *J. Neurosci.* **25,** 9940–9948.

Reynolds, L. E., Wyder, L., Lively, J. C., Taverna, D., Robinson, S. D., Huang, X., Sheppard, D., Hynes, R. O., and Hodivala-Dilke, K. M. (2002). Enhanced pathological angiogenesis in mice lacking beta3 integrin or beta3 and beta5 integrins. *Nat. Med.* **8,** 27–34.

Scott, L. M., Priestley, G. V., and Papayannopoulou, T. (2003). Deletion of {alpha}4 integrins from adult hematopoietic cells reveals roles in homeostasis, regeneration, and homing. *Mol. Cell. Biol.* **23,** 9349–9360.

Sheppard, D., Rozzo, C., Starr, L., Quaranta, V., Erle, D. J., and Pytela, R. (1990). Complete amino acid sequence of a novel integrin beta subunit (beta 6) identified in epithelial cells using the polymerase chain reaction. *J. Biol. Chem.* **265,** 11502–11507.

Shull, M. M., Ormsby, I., Kier, A. B., Pawlowski, S., Diebold, R. J., Yin, M., Allen, R., Sidman, C., Proetzel, G., Calvin, D., Annunziata, N., and Doetschrman, T. (1992). Targeted disruption of the mouse transforming growth factor-beta 1 gene results in multifocal inflammatory disease. *Nature* **359,** 693–699.

Stephens, L. E., Sutherland, A. E., Klimanskaya, I. V., Andrieux, A., Meneses, J., Pedersen, R. A., and Damsky, C. H. (1995). Deletion of beta 1 integrins in mice results in inner cell mass failure and peri-implantation lethality. *Genes Dev.* **9,** 1883–1895.

Stupack, D. G., and Cheresh, D. A. (2003). Apoptotic cues from the extracellular matrix: Regulators of angiogenesis. *Oncogene* **22,** 9022–9029.

Su, G., Hodnett, M., Wu, N., Atakilit, A., Kosinski, C., Godzich, M., Huang, X. Z., Kim, J. K., Frank, J. A., Matthay, M. A., Sheppard, D., and Pittet, J. F. (2007). Integrin $\alpha v \beta 5$ regulates lung vascular permeability and pulmonary endothelial barrier function. *Am. J. Respir. Cell Mol. Biol.* In press.

Taooka, Y., Chen, J., Yednock, T., and Sheppard, D. (1999). The integrin a9b1 mediates adhesion to activated endothelial cells and transendothelial neutrophil migration through interaction with vascular cell adhesion molecule-1. *J. Cell Biol.* **145,** 413–420.

Yang, J. T., Rayburn, H., and Hynes, R. O. (1993). Embryonic mesodermal defects in alpha 5 integrin-deficient mice. *Development* **119,** 1093–1105.

Yang, J. T., Rayburn, H., and Hynes, R. O. (1995). Cell adhesion events mediated by alpha 4 integrins are essential in placental and cardiac development. *Development* **121,** 549–560.

Zhu, J., Motejlek, K., Wang, D., Zang, K., Schmidt, A., and Reichardt, L. F. (2002). beta8 integrins are required for vascular morphogenesis in mouse embryos. *Development* **129,** 2891–2903.

CHAPTER FOURTEEN

PURIFICATION, ANALYSIS, AND CRYSTAL STRUCTURE OF INTEGRINS

Jian-Ping Xiong,* Simon L. Goodman,[†] *and* M. Amin Arnaout*

Contents

Abstract

Integrins are large modular cell-surface receptors that regulate almost every aspect of cellular function through bidirectional signals transmitted across the lipid bilayer. Regulation of integrin activity is accomplished by complex and still incompletely understood biochemical pathways that modify integrin ligand binding, clustering, trafficking, and signaling functions. The dynamic tertiary and quaternary changes required to channel some of these activities have hampered, until recently, the crystal structure determination of these heterodimeric receptors. In this chapter, we review the methods used to purify and characterize these proteins biophysically and functionally, and to derive their three-dimensional structures.

* Structural Biology Program, Leukocyte Biology and Inflammation Program, Nephrology Division, Massachusetts General Hospital and Harvard Medical School, Charlestown, Massachusetts
[†] Preclinical Oncology Research, Merck KGaA, Darmstadt, Germany

Methods in Enzymology, Volume 426
ISSN 0076-6879, DOI: 10.1016/S0076-6879(07)26014-8

© 2007 Elsevier Inc.
All rights reserved.

1. INTRODUCTION

Integrins are composed of α- and β-subunits, each a type I membrane protein, which form a noncovalent heterodimer complex (Hynes, 1992). In mammals, eight distinct but homologous β-subunits have been identified, each of which can associate with one or more of the 18 α subunits, forming 24 distinct receptors. Nine of the α-subunits contain an additional von Willebrand factor (VWF) type A domain (known as αA or αI-domain in integrins) in the extracellular segment, which has been shown to mediate the divalent-cation–dependent ligand binding in this αA-containing integrin subgroup (Arnaout, 2002). The αA-lacking integrin subgroup binds ligand in part through a highly homologous αA-like domain (known as βA- or βI-domain) present in every β-subunit (Xiong *et al.*, 2002). An invariant feature of integrin ligands, most of which are extracellular matrix (ECM) proteins (Humphries *et al.*, 2006), is the presence of a solvent-exposed acidic residue, which forms the core of the ligand-binding interface with the integrin (Lee *et al.*, 1995d).

Integrin-mediated cell–cell and cell–matrix interactions are complex, highly regulated processes that play critical roles in fundamental cellular functions including cell motility, proliferation, differentiation, and survival. Integrins physically link the cell's cytoskeleton to the ECM or to other cells, allowing the transfer of mechanical forces and biochemical cues across the plasma membrane, thus affecting many cell signaling pathways including G-protein- and c-AMP signaling, ion fluxes, and protein phosphorylation (Ingber, 2006). These integrin-mediated activities involve different states of integrin affinity (inactive, intermediate, active) that are regulated allosterically and in a bidirectional manner (Arnaout *et al.*, 2005). The structural basis of integrin affinity modulation and its interaction with ligand have been the subject of recent reviews (Arnaout *et al.*, 2005; Ginsberg *et al.*, 2005; Humphries *et al.*, 2003; Luo and Springer, 2006). In this chapter, we describe the purification and biophysical characterization of integrins and the methods used to derive their three-dimensional (3D) crystal structures.

2. PURIFICATION AND ANALYSIS OF INTEGRIN HETERODIMERS

Integrins have a number of challenging biochemical properties that make their isolation in an active native state technically demanding. First, integrins are large proteins that form obligate noncovalent α/β heterodimers, with both subunits being type I membrane proteins and thus are associated with membrane lipids (Hynes, 1987). Second, each subunit binds divalent cations, which is vital for ligand binding and for stability and association of the subunits (Howard *et al.*, 1982). Third, each subunit is heavily glycosylated with complex multicatenary sugar trees, which may affect subunit association (Isaji *et al.*, 2006). Fourth, integrins also undergo profound, rapid, low-energy dependent conformational changes during their biological function (Calvete, 2004; Hynes, 1992). Fifth, integrins form small or large clusters and interact with the cytoskeleton, signaling proteins, and with other transmembrane proteins, which are vital for their functionality (Ingber, 2006).

2.1. Integrin isolation

Many integrins have been isolated in an active state from nonionic detergent extracts of cultured cells or fresh tissues using affinity chromatography, or they have been produced in recombinant forms. The choice of strategy depends on your goals, resources, and the amounts of integrin you need. Low amounts of several native human integrins are commercially available ($\alpha 1\beta 1$, $\alpha 3\beta 1$, $\alpha 5\beta 1$, $\alpha V\beta 3$, and $\alpha V\beta 5$, Chemicon, Temecula, CA). For isolating and analyzing larger quantities of purified functionally active integrins, we describe some current technology, particularly focusing on $\alpha v\beta 3$. The strategies involved, however, are very similar between integrins, so the methods can be readily translated to other molecules. As integrins tend to be low- or moderate-copy number surface receptors, large quantities of cells are needed to get biochemical amounts. Common sources are human placenta or outdated platelet concentrates, with their accompanying biohazards. This and the limited range of integrins accessible in these sources have driven a move toward using recombinant integrins. Due to similar mass and biochemistry, only affinity chromatography techniques can be relied on to yield reasonably pure integrins from native sources. Several recombinant integrin forms, usually lacking transmembrane and cytoplasmic domains, have been successfully engineered and purified.

2.1.1. Extraction and purification of native integrins

Large masses of viable cells can be used as sources for isolating native integrins; the majority of biochemical preparations have used outdated platelet concentrates or placenta. Antibody- and ligand-affinity chromatography can be effective for isolation, but may select or selectively activate a subpopulation of the target integrin, and confuse the subsequent interpretation of data. As moderate–copy number integrins (i.e., between 10^4 and 10^5 copies per cell), with each cell expressing several or many types of integrins, and with a rather similar biochemistry, even affinity techniques give low yields: 0.01 to 0.1 μg of a particular integrin is present in 10^6 cells (Kurzinger and Springer, 1982). Accordingly, traditional sources of native integrins have been placenta, a source of integrins $\alpha 5\beta 1$, $\alpha V\beta 3$, $\alpha V\beta 5$, $\alpha 1\beta 1$, and $\alpha 2\beta 1$, outdated human platelet concentrates, a source of $\alpha 2\beta 1$ and $\alpha IIb\beta 3$ (Bennett *et al.*, 1983), and leukocyte concentrates, a source of $\alpha L\beta 2$ (CD11a/CD18) (Bennett *et al.*, 1983) and $\alpha M\beta 2$ (CD11b/CD18) (Pierce *et al.*, 1986). The ability to isolate these integrins usually depends on the availability of high-quality antibodies as affinity reagents. Platelet membranes are a special case, and classical chromatography yields integrin $\alpha IIb\beta 3$ at reasonable purity from them (Eirin *et al.*, 1986; Muller *et al.*, 1993). Although the integrins resist many proteases that accompany extraction, the affinity columns are less tolerant, and there is rapid column degradation. This can make the procedures very costly, lengthy, and variable. Whatever the reasons for their attractiveness in the first instance, we strongly recommend that an effort be made to produce a recombinant form of the integrin of interest.

2.1.2. Purification strategies

Several strategies have yielded good-quality native integrins. Classically, peptide affinity chromatography, especially on Arg-Gly-Asp, (RGD)-based peptide affinity columns were used to isolate analytical amounts of $\alpha 5\beta 1$, $\alpha IIb\beta 3$, and $\alpha V\beta 3$ (Pytela *et al.*, 1985a,b, 1986). This approach also works well for generating larger amounts. Ligand affinity chromatography based on other ECM components is effective, but more difficult than peptide affinity, due to the complicated biochemistry necessary to isolate these molecules and fragments. Collagen or its fragments has been used to isolate $\alpha 1\beta 1$ and $\alpha 2\beta 1$ (Pfaff *et al.*, 1994; Vandenberg *et al.*, 1991), and $\alpha 3\beta 1$ and $\alpha 7\beta 1$ have been isolated using laminin or its fragments (Gehlsen *et al.*, 1988; von der Mark *et al.*, 1991). Fibronectin fragments have also been used to capture $\alpha 5\beta 1$ and $\alpha 4\beta 1$ (Pfaff *et al.*, 1994). Many high-quality integrin-binding antibodies are available, and depending on the tissue source, subunit, or complex, specific affinity columns can be produced. Unfortunately, the availability of antibodies that cross-react with other species is more limited. Antibody affinity chromatography has been used to isolate several integrins in bulk, including $\alpha V\beta 3$ (Smith and Johnson, 1988), $\alpha V\beta 5$ (Smith *et al.*, 1990), and $\alpha 5\beta 1$ (Mould *et al.*, 1996).

In general, the integrins are rather stable, and quite resistant to protease activity (Tarone *et al.*, 1982), so conventional protease cocktails are quite adequate to protect them during purification. Pre-extractions with high-salt buffers or weak detergents are also well tolerated—the original methodologies with only minor modifications work well. Depending on available resources, the pre-extraction washing steps can be repeated *ad libitum* to reduce the debris passing over the affinity column, and as the amounts of material used are large (a placenta weighs between 500 and 750 g), typical 10:1 (extraction buffer: tissue) ratios are difficult for a research laboratory to manipulate. Lowering the extraction volumes and repeating the steps work well.

Divalent cations can stabilize α/β heterodimers, and usually millimolar Ca^{2+} or Mg^{2+} are added to maintain native structure during the isolation. Integrin stability is also enhanced by moderate (nonbasic) pH conditions (Hemler *et al.*, 1987; Hynes, 1992). This may conflict with purification, especially during elution from antibody affinity columns. Finally, integrins are conformationally active, and in at least some cases (e.g., $\alpha IIb\beta 3$) the ligand mimetics used for affinity chromatography can alter the integrin activation state (Du *et al.*, 1991).

Depending on the degree of enrichment in the tissue source, the affinity steps can be more or less sophisticated. For $\beta1$ integrins, invasin columns are invaluable (Eble *et al.*, 1998), while many integrins bind with acceptable affinity to Arg-Gly-Asp (RGD)-based peptide affinity columns (Pytela *et al.*, 1985a,b, 1986). But for most integrins, only antibody affinity columns provide the necessary purity of native product. For peptide affinity columns, competing peptide, as classically used, or EDTA, with fraction collection into high divalent cation buffer can be used (Pfaff *et al.*, 1994; Pytela *et al.*, 1986). For antibody columns, high or, preferably, low pH elution can be used. The divalent cations needed for integrin function and to stabilize integrin structure form more or less insoluble hydroxides at pH over 7.2 to 7.4, so correction of the pH of acidic column elutions should be performed both rapidly and cautiously.

Purification from human term placenta   The method basically follows the strategy described by Smith and Cheresh (1988) and Mould *et al.* (1996), as modified by Mitjans *et al.* (1995). All operations are performed at between 0 and 4°. Term placenta,[1] chopped into convenient pieces and then cleaned of blood and debris by mincing into 0.05% (w/v) digitonin, 2 mM CaCl$_2$, 2 mM PMSF, and 25 mM HEPES, pH 7.4,[2] and after low-speed centrifugation (2000g 30 min), the pellet is re-extracted and centrifuged in the same buffer. Exact extraction volumes for this wash are unimportant, and depend on centrifuge capacity. The integrins are solubilized from the resulting pellet by homogenization and stirring for 1 h in buffer A (100 mM Octyl-β-D-glucopyranoside (OG),[3] 1 mM CaCl$_2$, 1 mM MgCl$_2$, 2 mM PMSF, 25 mM HEPES pH 7.4). After mid-speed centrifugation (10,000g × 1 h), the supernatant is recirculated for 16 h over a 15-ml affinity[4] column. Many

---

[1] Take the appropriate biohazard precautions. The whole procedure until completion of the extraction is likely to produce aerosols, so an appropriate safety cabinet with controlled airflow is recommended. It is prudent to assume pathological viral contamination until the affinity column has been extensively washed with detergent-containing buffers, and even then to be cautious until the purified material is 0.2 μM filtered.

[2] Tris-HCl buffer and 150 mM NaCl without digitonin can be used here. Pfaff *et al.* (1994) obtained good results using 1 M NaCl in neutral buffer as buffer A.

[3] Depending on resources, NP-40 or even Triton-X100 can be used as primary extraction detergent (Mould *et al.*, 1996; Pfaff *et al.*, 1994). Once the integrin has been captured on the affinity column, it is possible to perform detergent exchange by washing the column in the detergent of choice. This can reduce the amounts of more costly detergents (e.g., octylglucoside) needed for preparation and does not seem to affect quality of the preparation.

[4] Affigel 10 matrix conjugated to 2.5 mg of antibody per milliliter of gel, according to manufacturers instructions. In essence, 30 ml of Affigel 10 is washed on a sintered glass filter with 200 ml of ice-cold PBS, pH 7.4, before being gently transferred to a 50-ml conical plastic tube containing 15 ml of 2.5 mg/ml monoclonal antibody LM609 (directed against the $\alpha V\beta3$ heterodimer). The tube is rotated at 4° overnight and allowed to sediment; supernatant is removed (for protein assay to determine coupling efficiency) and the conjugate is blocked by incubation with 1M ethanolamine-HCL, pH 8.0. After packing into a 2.5-cm–diameter column and washing with buffer B (buffer A containing 25 mM OG), the column is stored in buffer D plus 0.02% sodium azide.

integrins can be captured from this supernatant. For example, immobilized LM609 captures $\alpha V\beta 3$ (Smith and Cheresh, 1988); P1F6 has been used to capture $\alpha V\beta 5$; LM142 (all available from Chemicon, Temecula, CA) (Smith et al., 1990) or 14D9 (Calbiochem) capture all αV integrins (Mitjans et al., 1995); and mAb 13 or the 140 kDa fragment of fibronectin (Pfaff et al., 1994) is used to capture $\alpha 5\beta 1$ (Mould et al., 1996). Collagen IV peptide CB3 can be used to capture $\alpha 1\beta 1$ (Pfaff et al., 1994). If ligand affinity rather than antibody affinity is used, adding 1 mM MnCl$_2$ to the solubilization buffers may improve yields. After washing the column with buffer B (buffer A containing 25 mM OG) till OD280 reaches background, integrins are eluted from antibody columns at low pH, using buffer C (Buffer B, but buffered with 10 mM Na-acetate, pH 3.1). The eluant is neutralized with 3-M Tris (pH 8.8), dialyzed against buffer B, and concentrated approximately 10-fold to 1 mg/ml in Ultra-4 centrifugal concentrator units (Millipore). The purified receptor is aliquoted and may be stored at $-80°$. Ligand affinity columns (e.g., RGD-columns) are eluted using buffer B plus 10 mM EDTA, without divalent cations, and the eluant immediately neutralized by adding MgCl$_2$ (1M), dialyzed against buffer B, and concentrated and stored as above.

Purification from thrombocyte concentrates Outdated human thrombocytes[5] are a rich source of αIIb$\beta 3$ and $\alpha 2\beta 1$. Outdated thrombocyte concentrates from the local blood bank are washed by dilution in one volume of Tyrodes buffer, centrifuged at 1200g and the pellet lysed in buffer A for 1 h on ice with stirring. After high-speed centrifugation (35,000g $\times$ 1 h), the supernatant is recirculated over a GRGDSPK peptide column (Pfaff et al., 1994; Pytela et al., 1986) pre-equilibrated with buffer B. The column is washed, and αIIb$\beta 3$ eluted with 10 mM GRGDSPK in buffer D (buffer B, without divalent cations). The eluant is dialyzed, concentrated, and stored as described for placental integrins above (Mitjans et al., 1995).

2.1.3. Extraction and purification of recombinant integrin heterodimers

Detergent extracts of placenta and blood cells generate overwhelming amounts of contaminant molecules that rapidly degrade affinity columns. Thus, such procedures are for the rich and resourceful—and, even more importantly, still leave many integrins inaccessible to biochemistry. However, in recent years, several integrins have been produced by recombinant DNA technologies. This must now be considered the method of choice, especially as the integrins can be engineered to ease purification, and as

---

[5] Take the appropriate biohazard precautions when working with thrombocytes.

elegantly demonstrated in recent years, to resolve previously inaccessible structure–activity relationships.

We routinely use the Baculovirus system (Abrams *et al.*, 1994; Kraft *et al.*, 1999; Mehta *et al.*, 1998), but Schneider cells (Eble *et al.*, 1998), 3T3 (Briesewitz *et al.*, 1993; Weinacker *et al.*, 1994), COS (Dada *et al.*, 1998; Gulino *et al.*, 1995), HEK293 (Higgins *et al.*, 1998), and CHO (Mould *et al.*, 2003a) expression systems have also been used to produce full-length and partially truncated integrins. The choice of system depends on what the integrin is needed for. For structural studies using NMR, full-length integrins are still challengingly large, but fragments that do not need heterodimerization for fidelity have been expressed (Denda *et al.*, 1998), including the A-domains (Lee *et al.*, 1995b,c; Michishita *et al.*, 1993; Qu and Leahy, 1995), and the short cytoplasmic domains (Li *et al.*, 2001; Vinogradova *et al.*, 2000). Insect cells can give high yields, but the final glycosylation is nonmammalian; there, terminal processing is dominated by high-mannose trees and leads to a homogenous glycosylation that may be advantageous for structural studies (Hsieh and Robbins, 1984; Sissom and Ellis, 1989). Insect culture media contain high concentrations of Mg^{2+}, a divalent cation that stabilizes many integrins, and they are heterologous, so that if antibody-affinity chromatography is used in the purification, the levels of cross-reaction are low. Finally, insect cells are defective for dibasic cleavage (Bruinzeel *et al.*, 2002). As a subset of integrins has dibasic, site-dependent, cleaved α-subunits, this can be a disadvantage. To avoid the complexities of restoring the appropriate furin system in insect cells (Bruinzeel *et al.*, 2002), we found that mild trypsin treatment rapidly completed the cleavage of the αV subunit in the heterodimer into its heavy and light chain segments (MAA, unpublished observations).

It was a pleasant surprise to discover that several integrins, when engineered as deletion mutants lacking transmembrane and cytoplasmic domains, were secreted as a stable, functional heterodimer, including αMβ2 (CD11b/CD18) (Dana *et al.*, 1991), α1β1 (Briesewitz *et al.*, 1993), αIIbβ3 (Wippler *et al.*, 1994), αVβ3 (Mehta *et al.*, 1998), αVβ6, (Kraft *et al.*, 1999; Weinacker *et al.*, 1994), α4β1 (Clark *et al.*, 2000), α5β1 (Coe *et al.*, 2001), and α8β1 (Denda *et al.*, 1998). Such truncated soluble molecules are ideal for many studies, including ligand binding and structural investigations. Secreted soluble proteins are easier to purify than integral membrane components—and cellular contaminants and detergents are avoided. In addition, several tagging and chain-stabilization techniques have been used to ease purification, and to enhance the stability or detectability of the heterodimers, including Fc-fusions (Coe *et al.*, 2001), Jun-Fos dimerization domains (Eble *et al.*, 1998), and alkaline phosphatase chimeras (Denda *et al.*, 1998). Such tagging can save months of subsequent effort, and open the way to getting integrins from nonhuman organisms. As divalent cations are intimately involved in integrin function, it may not be wise to attempt His-tag protocols. Although soluble engineered integrins can be active and specific, and retain defining epitopes

(Clark *et al.*, 2000; Mehta *et al.*, 1998), as with all tagged recombinants, possible effects of the tags should of course be kept firmly in mind when interpreting the resulting data: the C-termini of the integrin subunits evidently realign during physiological activation and signal transduction (Kim *et al.*, 2003), and so manipulating this region of the molecule may lead to artifacts.

The yields from both eukaryotic and insect cell cultures are not dramatic (100 to 1000 μg/liter), but, being recombinant, they are scalable in a way that placentas tend not to be. In addition, the biohazard level is minimal in comparison to the human material. Purified recombinant integrins are stable, and have a reasonable storage life when stored aseptic at $4°$. Snap-freezing and storage at $-80°$ in normal physiological salt solutions in the absence of additional cryoprotectants works well, both for native and recombinant integrins, and leaves the integrins active in ligand-binding assays for years.

Purification using baculovirus systems For integrins still containing transmembrane regions, cells expressing the recombinant proteins can be harvested from the culture medium by low-speed centrifugation, detergent extraction, and purification protocols as described above for isolating integrins from platelets as detailed here and elsewhere (Kraft *et al.*, 1999; Mehta *et al.*, 1998). Naturally, the affinity chromatography techniques are greatly simplified, and group-specific (pan-β1, invasin, pan-αV) reagents can be used rather than subunit- or complex-specific reagents, as only one integrin is present in the recombinant, and the affinity columns also have a dramatically extended life span.

cDNA encoding individual integrin subunits engineered as desired, for example by PCR-based mutation to generate cleavage and truncation sites, are sub-cloned into appropriate Baculovirus vectors, such as pBacPAK8 and pBacPAK9 (Clontech), and high-titer viral clones are produced and propagated in Sf9 cells. The resulting recombinant Baculovirus is used to infect High Five cells (Invitrogen) at a multiplicity of infection between 5 and 10 (Clark *et al.*, 2000; Kraft *et al.*, 1999; Mehta *et al.*, 1998). After 2 h of infection, the medium is cautiously replaced with fresh Express Five medium (Invitrogen) and the infected cells are cultured for 48 to 64 h, before harvesting cells (full length) or supernatant (secreted proteins). Cells are centrifuged ($1000g$, 10 min), and full-length integrins are then prepared by detergent extraction of the cell pellet, following the protocol described above for platelets, and using the affinity column of choice. For soluble integrins, the medium is centrifuged and filtered to remove any cellular material, and concentrated approximately 10-fold before refiltration and passage (but in the absence of detergents) over antibody affinity columns conjugated with an appropriate antibody (see above), essentially as described for placental integrins (Kraft *et al.*, 1999; Mehta *et al.*, 1998). Following elution and dialysis with buffer E (150 mM NaCl, 1 mM MgCl$_2$, 1 mM CaCl$_2$, 25 mM Tris-HCl, pH 7.4), the proteins are sterile-filtered and are stable for months

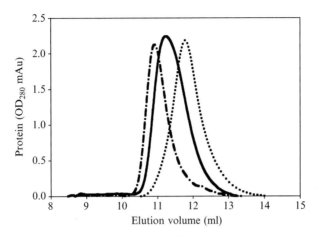

Figure 14.1 Recombinant ectodomain of $\alpha V \beta 3$ Mehta *et al.* (1998) resolved on an S200 10/300 GL molecular sieve column (GE Healthcare), in neutral physiological salt containing 1 μM calcium (dotted), 200 mM manganese (solid), or manganese plus 200 nM cilengitide, an RGD peptide (bar-dot-bar). Note the relatively symmetrical peaks and the cation-dependent change in elution point, which reflects alterations in molecular form in solution.

when stored at 4°, or for longer periods when stored at −80°. Apart from the recommended presence of divalent cations, the storage buffers are rather flexible, so Tris, MES pH 6.0, HEPES, pH 7.4, or Na-acetate, pH 6.0, all maintain stable integrins. Bearing in mind that integrin activity involves their specific and extensive clustering at the cell surface, it is perhaps remarkable that soluble integrins prepared in Baculovirus can be concentrated using, for example, centrifugal concentrators (Millipore Ultra-4 system) at ∼5 to 8 mg/ml in neutral salt solutions, yet remain monodispersed as judged by gel filtration on S200 molecular sieve columns (Fig. 14.1).

2.2. Integrin characterization

The characterization of membrane proteins in detergent solutions is notoriously difficult. But many years of study have demonstrated that native, recombinant, or recombinant-engineered integrins prepared as described are suitable for biochemical studies, and retain many biological activities and specificities that match their activities in a cellular environment (Clark *et al.*, 2000; Kraft *et al.*, 1999; Mehta *et al.*, 1998). Functional assays have been based on ligand binding measured in equilibrium, ELISA-like configurations, or by surface plasmon resonance.

Luckily, integrins have evolved a structure in which ligand binding, and transmembrane and cytoplasmic signaling functions are spatially independent (Chan *et al.*, 1992; Xiong *et al.*, 2002). It thus seems valid to examine the function and structure of the extracellular and intracellular domains in isolation. Data obtained with purified integrins have confirmed and extended data from cells. As integrins are sensitive to conformational activation and modulation, it is always wise to compare data obtained from the isolated receptor assays with the same integrins expressed *in situ*, and some more advanced model systems using liposomes have also been used to avoid possible artifacts of adsorbance to plastic (Muller *et al.*, 1993; Parise and Phillips, 1985). However, for many applications, simple ELISA-like assays are highly satisfactory (Clark *et al.*, 2000; Eble *et al.*, 1998; Kraft *et al.*, 1999; Mehta *et al.*, 1998; Nachman and Leung, 1982; Pfaff *et al.*, 1994; Smith *et al.*, 1990).

2.2.1. Ligand-binding assays

Integrin ligand-binding assays function in a variety of configurations; the choice depends on the amounts of integrin and ligand available, buffer requirements, and goal of the assay. An early binding assay used an ELISA format (Nachman and Leung, 1982), native platelet $\alpha IIb\beta 3$, and immobilized fibrinogen. Interaction was detected with polyclonal anti-$\alpha IIb\beta 3$ antibodies; we term this configuration an "inverted assay" (Fig. 14.2A). The assays follow conventional ELISA protocols (Clark *et al.*, 2000; Coe *et al.*, 2001; Kraft *et al.*, 1999; Mehta *et al.*, 1998; Pfaff *et al.*, 1994; Smith *et al.*, 1990), with

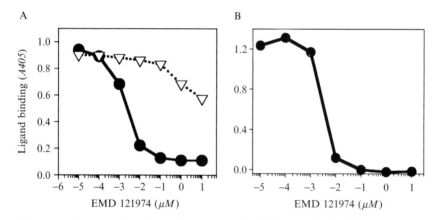

Figure 14.2 Recombinant integrin ectodomain in ligand-binding inhibition assays function in several configurations, including inverted assay (A) and antibody capture (B). In the inverted assay, fibronectin fragments of type III domains 9 to 10 are adsorbed in physiological saline plus 1 mM Ca^{2+} and 1 mM Mn^{2+}, and recombinant soluble ectodomains of $\alpha V\beta 3$ (Mehta *et al.*, 1998) (circles) or $\alpha V\beta 6$ (Kraft *et al.*, 1999) (triangles) plus cilengitide (EMD 121974) added. Bound integrins were detected with biotinylated antibody LM142 (Cheresh and Spiro, 1987). In the antibody capture, plates were coated with 20H9 (Mitjans *et al.*, 1995), a noninhibitory, beta-3 chain–binding antibody. The $\alpha V\beta 3$ ectodomain was allowed to bind, and then biotinylated vitronectin was added in the presence of cilengitide. In both assays, binding was detected using antibiotin peroxidase conjugates. Note the similar IC_{50} obtained in these assays for $\alpha V\beta 3$–ligand interactions. $\alpha V\beta 6$ is much less sensitive to cilengitide.

carrier protein and weak nonionic detergents (i.e., 0.05% Tween-20) being incorporated into the incubation and blocking buffers if necessary to reduce background binding. Signal-to-noise ratios higher than 10 are readily attained when using peroxidase or alkaline phosphatase conjugates. The assays are rather tolerant of salt and buffer conditions, incubation times, and temperatures and blocking conditions, and ligand detection methods; the presence of divalent cations and neutral pH are two critical requirements. A basic protocol is described below. Such assays have been successfully used to analyze the role of cations and pharmacological inhibitors of integrin activities.

As detergents can modulate the adsorption of proteins to plastic, such inverted assays are useful for integrins in detergent solution—even then one must be aware that detergent can strip ligands off plastic substrates, and the system should be controlled for such events. But if there is no concern that the substrate will modify the integrin, it is often more convenient to adsorb the integrin, and detect the subsequent ligand binding (Kraft et al., 1999; Mehta et al., 1998; Smith et al., 1990). We have also successfully used capture-ELISA configurations, where noninhibitory antibodies are adsorbed to the ELISA plate, and with ligand or integrin captured on the immobilized antibody before interaction studies with the complementary binding partner (Fig. 14.2B). Such techniques can avoid adsorption artifacts. Similar capture techniques have been used for integrin $\alpha L\beta2$ in BIAcore studies (Bechard et al., 2001). The choice of configuration depends very much on the goal of the assay. For example, phage display screens are tedious unless the target is immobilized (Kraft et al., 1999; Pasqualini et al., 1995). As previously mentioned, liposomes have been used to position integrins into a more correct biological environment for binding; however, the need for this is debatable, as liposomes also only approximate a native lipid environment.

Several good-quality protein ligands of the integrins—for example, fibronectin, fibrinogen, collagens, vitronectin, and laminin I—are commercially available. In principle, the inverted assay, where the ligand is on the solid phase, might mimic the conformation assumed in the immobile environment of the ECM; however, there has been little comparison of the various assay formats. Integrin-binding recombinant ligands or protease-digested fragments of several of the ligands are easy to prepare (e.g., fibronectin, fibrinogen) (Leahy et al., 1996), and being small and monomeric, are convenient and stable for routine large-scale assays. The reader should be aware that the biological function of many integrin ligands involves their aggregation, like fibrillar collagens, or they are heterodispersed mixtures (like standard commercial vitronectin [Yatohgo et al., 1988]). Consequently, their storage and usage in these assays must be carefully assessed.

In inverted assays, integrins can be detected with one of the many well-defined and commercially available noninhibitory anti-integrin antibodies. It is prudent to confirm that the chosen detection antibodies do not in fact inhibit the integrin in question, such as by using a cell–attachment inhibition assay (Abrams *et al.*, 1994).

Both radioactive and nonradioactive detection have been used to detect integrin–ligand interactions. Biotin–antibiotin systems (based either on labeled avidin, or on antibiotin antibody) work well (Hantgan *et al.*, 1993). Ligands and recombinant integrins themselves can be directly biotinylated, and the biotinylated derivatives are stable for more than 6 mo at 4°.

Basic binding assays   These assays are in ELISA format and detection of binding occurs via enzyme-linked secondary antibodies. Purified recombinant or native $\alpha V\beta 3$ is diluted to 1 mg/ml in coating buffer F (10 μM MnCl$_2$, 1 mM MgCl$_2$, 1 mM CaCl$_2$, 150 mM NaCl, 20 mM Tris-HCl, pH 7.4), and allowed to adsorb to 96-well microtitre plates (2 to 16 h at 4°). After washing in TBS, the plates are blocked with 3% BSA in buffer F (2 h at 37°). Biotinylated ligand diluted (0.1 to 10 mg/ml) in buffer F containing 1 mg/ml BSA is added, and the plate incubated for 3 h (37°) before washing with TBS and detection of bound ligand with antibiotin phosphatase conjugated antibody. This assay can be modified to investigate divalent cation dependency and sensitivity to pharmacological inhibitors by simply adjusting buffer conditions (Haubner *et al.*, 1996). In the inverted assay, simply coat the ligand, and following blocking, incubate the integrin with the plate (Kraft *et al.*, 1999). Integrin binding is then detected by an appropriate noninhibiting subunit-specific reagent—for instance, 5C10 or 14G11 antibodies (Calbiochem) for $\alpha V\beta 3$—can be used as the primary antibody, followed by an enzyme-linked second-layer antibody.

BIAcore analysis   Surface plasmon resonance can be very useful for analyzing ligand receptor interactions. Recombinant soluble integrins function in this system, and particularly when in solution (apparently, the chip chemistry denatures the integrins during coupling). The integrin ligands fibrinogen and vitronectin conjugated to CM5 chips bind weakly to soluble $\alpha V\beta 3$ (Takagi *et al.*, 2002). Adenovirus proteins on the chip bind to $\alpha V\beta 5$ (Chiu *et al.*, 1999). Solving the technical obstacles in this system will be a valuable contribution.

2.2.2. Hydrodynamic characteristics

The size and shape of integrins varies depending on their activation and ligation status, and this is a topic of intense interest. However, popular indirect methods involving the exposure of antibody epitopes in response to a particular treatment protocol are not informative about what has actually been altered in the molecule. Once a soluble form of an integrin has been obtained, classical technologies can be used to study its solution properties using molecular sieve chromatography, equilibrium ultracentrifugation, and dynamic light scattering. When combined with imaging technologies, these can help provide a robust concept of the molecular dynamics.

Molecular sieve chromatography Molecular sieve chromato-graphy (MSC) estimates the hydrodynamic properties of an integrin in solution. The crystal structure can be used to compute the hydrodynamic radius (Garcia de la Torre *et al.*, 2003) compared to the MSC values. While this does not give provide such high resolution of information content as static imaging by electron microscopy or crystallography, it can provide valuable correlations to imaging data. For proteins in detergent solution, micelle formation makes MSC unreliable. But soluble integrins have allowed high-resolution MSC studies (Adair *et al.*, 2005; Takagi *et al.*, 2002) and the resolution of integrin-ligand complexes (Adair *et al.*, 2005). The hydrodynamic radius of the molecule shifts depend on ionic conditions. Mn^{2+} ions, which enhance ligand binding, cause an increase in the hydrodynamic radius of a soluble recombinant $\alpha V\beta 3$ (Takagi *et al.*, 2002), over Ca^{2+}, which does not enhance binding. For proteins that rapidly alter their size and shape in relationship to the run-time of the MSC, this technique gives an averaged value for hydrodynamic radius. The Superdex S200 HR-10/30 (now Superdex S200 10/300 GL) column (GE-Amersham Healthcare) resolves well at the appropriate molecular weight range of the integrin monomers, and using AKTA-FPLC systems, low-microgram amounts of protein can be analyzed in each run. A typical experiment would load 10 μg of protein salt exchanged by spin column (Centrisep, Princeton separations) into the appropriate column buffer, and centrifugally filtered (0.45 mM Ultrapore HV, Millipore), in a 50-μl probe volume, and run isocratically at 0.4 ml/min, with detection at OD280. This system can readily and robustly detect changes of 0.15-nm Stokes radius (Fig. 14.1).

Dynamic light scattering (DLS) Dynamic light scattering (DLS) measures scattering intensity coefficients with time, which are related to particle motion. The Stokes-Einstein equation is then applied to the measured distribution of diffusion coefficients to obtain absolute values for hydrodynamic parameters including translational diffusion coefficient— hence Stokes radius and molecular weight—from dilute protein solutions. The technique is exquisitely sensitive to aggregates in the solution, and to the degree of interaction of protein and solvent (via the second virial coefficient), which explains its routine use in crystallography laboratories. Only one team has reported the use of DLS on integrins, focusing on $\alpha IIb\beta 3$ to study conformation of the detergent-extracted native molecule (Hantgan *et al.*, 1993), changes in shape following inhibitor treatment (Hantgan *et al.*, 2002), and induction of oligomerization (Hantgan *et al.*, 2003). It appears that this technique has unexploited potential for integrin studies, being fast, compact, and sensitive. The need for extreme cleanliness in solution handling may be a limitation in working with detergent systems.

Equilibrium ultracentrifugation   The classical technique of analy-tical ultracentrifugation measures the sedimentation velocity of a molecule, and so can be used to derive its absolute weight-averaged molecular mass and translational frictional coefficients, which is determined by its average shape in solution. It has been used primarily on integrin $\alpha IIb\beta3$, both as a full-length detergent-extracted molecule and on ectodomain constructs (Hantgan *et al.*, 2004a; Rivas *et al.*, 1991a,b, 1996). Like DLS, but unlike MSC, ultra-centrifugation gives absolute values for molecular parameters, which can be combined with molecular modeling and imaging to determine integrin size, shape, and heterogeneity under various conditions (Garcia de la Torre *et al.*, 2003; Hantgan *et al.*, 2004b). Ultracentrifugation is lengthy but is a useful for measurements not only of molecular weight, but also the presence of integrin oligomers (Hantgan *et al.*, 2001, 2004a). It can be valuable to probe parameters that are difficult to investigate by other techniques; for example, it has been used to study the extent of oligomerization of the integrin transmembrane domains (Li *et al.*, 2001, 2004).

3. ATOMIC STRUCTURE OF INTEGRINS USING MACROMOLECULAR CRYSTALLOGRAPHY

Three-dimensional structure determinations of integrins have used X-ray crystallography, electron microscopy, NMR, and low-angle X-ray scattering. In this chapter, we will focus on the use of macromolecular crystallography to define the 3D structure of integrins, the interaction of integrins with ligands, and the nature of its conformational states. Methods describing the purification of the ligand-binding integrin αA domain have been described previously (Li and Arnaout, 1999) and will not be addressed here. In addition, the use of electron microscopy (Adair and Yeager, 2002; Adair *et al.*, 2005; Iwasaki *et al.*, 2005; Nermut *et al.*, 1988; Takagi *et al.*, 2002), NMR (Li *et al.*, 2001; Ulmer *et al.*, 2001; Vinogradova *et al.*, 2002; Wegener *et al.*, 2007; Weljie *et al.*, 2002), and low-angle X-ray solution scattering (Mould *et al.*, 2003b) in the structure determination of integrins is discussed elsewhere in this volume.

3.1. Crystal structure determination of the integrin ectodomain

3.1.1. Protein preparation
The availability of purified monodispersed integrins in sufficient quantities is essential for successful crystallization. As noted earlier, one of the major challenges in crystallization of integrins is that they assume different

conformational states in solution. Thus, keeping the purified integrin predominantly in one state is critical. Prior to crystallization trials, we buffer-exchange the soluble integrin ectodomain—purified as described above—in Ca^{2+}-containing buffer, using either a dialysis- or membrane-filtration device (Amicon). For the recombinant ectodomain of integrin $\alpha V\beta 3$, the protein buffer is 20 mM HEPES, pH 7.5, 100 mM NaCl, and 5 mM CaCl$_2$. Under these conditions, the protein typically assumes the "bent" conformation, as assessed by electron microscopy of negatively stained protein followed by image analysis (Adair *et al.*, 2005). The $\alpha V\beta 3$ ectodomain is also stable in the above buffer for several months at 4°.

3.1.2. Protein crystallization

We have used the hanging drop vapor diffusion method (Weber, 1997) for crystallization of the $\alpha V\beta 3$ ectodomain (Xiong *et al.*, 2001, 2002, 2004). Initial crystallization screening trials were carried out at room temperature (RT). The resulting crystals assumed clustered needle-like shapes. Switching crystal growth to 4° produced nice hexagonal protein crystals (Fig. 14.3). The hexagonal crystals were routinely obtained at room temperature when micro-seeding or macro-seeding techniques were used. The hexagonal crystals grow in a pH range from 4.5 to 6.0, in 10 to 14% polyethylene glycol 3350 or 4000, 0.6 M NaCl and 5 mM CaCl$_2$ at 4°. The protein concentration used can range from 5 to 10 mg/ml. The crystals appear within 2 days and grow to a dimension of $0.4 \times 0.3 \times 0.2$ mm within 1 week (Fig. 14.3). Although the same crystal morphology can be obtained from both 4° and RT, diffraction of the 4° crystals appears to be slightly better.

3.1.3. Cryoprotectant and heavy atom–derivative screening

The hexagonal crystals can diffract up to 4 Å using our in-house X-ray machine and to about 3 Å using the Advanced Photon Source (APS) synchrotron (Fig. 14.4). The crystal belongs to P3$_2$21 space group with unit cells

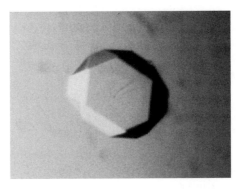

Figure 14.3   A hexagonal crystal of ectodomain of human integrin $\alpha V\beta 3$ grown at 4°. (See color insert.)

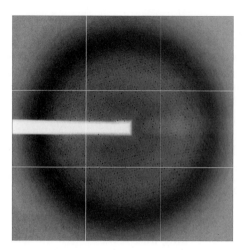

Figure 14.4 Diffraction pattern from the integrin αVβ3 crystal maintained at 100 K and collected at APS beamline ID19, 1.03321a degrees with Å crystal oscillation angle of 0.2 degrees over a 180-degree range, 2-sec exposure and at a distance of 400 mm from the CCD detector. The dark ring corresponds to ~3Å resolution.

of 130 Å, 130 Å, and 310 Å. The crystal decays very fast upon exposure to X-rays. To increase the crystal lifetime in the X-ray beam, diffraction data collection are done at low temperatures. Improper freezing however significantly increases mosaicity, making data collection very challenging. Trying a large number of the most commonly used cryoprotectants (Rodgers, 1997), we find that the best cryoprotectant is 20 to 24% glycerol. We also determined that the best protocol to transfer the integrin crystals from mother liquid to cryosolution is in a stepwise 2% glycerol increments, each time leaving the crystal for 3 to 5 minutes in the cryosolution to equilibrate. Once the crystal is transferred to the mother liquid containing 20 to 24% glycerol, it can be either flash frozen in liquid nitrogen or in cold stream. By using this approach, mosaicity can be limited to a 0.3- to 0.6-degree range.

Heavy atom screening was also very challenging. In most cases, soaking crystals in heavy-atom solutions either destroyed the crystal or dramatically decreased crystal diffraction. The best heavy-atom derivative we found was the nonlanthanide $K_2Pt(NO_2)_2$. Overnight soaking in 0.4-mM $K_2Pt(NO_2)$ resulted in a resolvable-difference Patterson map (Perutz, 1956) with only a modest decrease in crystal diffraction. The substitution of Ca^{2+}, which binds all integrins, with lanthanides is an excellent strategy to prepare heavy-atom derivatives. First, lanthanide-soaked crystals diffracted very similarly with wild-type crystals, allowing use of multiple wavelength anomalous diffraction (MAD) (Hendrickson, 1991) in structure determination. Second, it generates little disturbance in the crystal, resulting in minimal change in mosaicity compared to wild-type. Although $GdCl_3$ and $SmCl_3$ work equally well, we soaked crystals in $LuCl_3$ in order to maximize the phasing power and avoid the usage of long wavelengths in data collection. Generally, 2 days of soaking in 0.2 mM $LuCl_3$ is sufficient to generate full substitutions.

3.1.4. Data collection

To avoid considerable overlap in reflection spots and to minimize X-ray exposure time, it is ideal to orient the crystal such that its reciprocal axis c* is parallel to the spindle axis during data collection. Since the large face of integrin $\alpha V\beta 3$ crystal is perpendicular to the reciprocal axis c*, it is difficult to orient the crystals accordingly. The native data set was collected at a 1.03321-Å wavelength with frames acquired at 0.2-degree oscillation angles over an 180-degree range. For the derivative data sets, fluorescence XAFS spectra were first measured to obtain effective absorption edges (Fig. 14.5). The "inverse beam" method was used to ensure that Bijvoet difference could be measured as accurately as possible. Since the platinum (Pt) derivative decayed rapidly, only the data set at peak wavelength (1.07194 Å) was collected to obtain the large Bijvoet difference (Bijovet *et al.*, 1951). For the Lu derivative, the MAD data sets were collected at three wavelengths: inflection point (1.3412 Å), peak point (1.3407 Å), and remote point (1.2398 Å). All data sets were processed using the software HKL2000 (Otwinowski and Minor, 1997). In protein crystals produced in the presence of the pro-adhesive metal ion Mn^{2+}, the positions of Mn^{2+} were confirmed from anomalous-difference Fourier maps using data collected

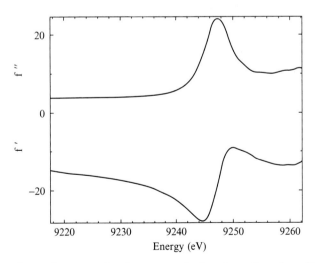

Figure 14.5 Anomalous scattering factors near the L_{III} absorption edge of Lu from the integrin $\alpha V\beta 3$ crystal. This spectrum was measured at the APS beam line ID19. The inflection point is chosen at the f′ minimum (for maximal dispersive difference), and the peak point is chosen at f″ maximum (for maximal Bijvoet difference). These tracings show strong binding of Lu^{3+} to the protein crystal.

at the wavelength 1.2398 Å, where Mn^{2+} has a reasonable anomalous contribution ($f'' = 1.96$ electrons).

3.1.5. Structural determination

The Pt position can be readily determined through difference Patterson maps (Blow, 1958; Perutz, 1956) (Fig. 14.6). The rest of the Pt sites are determined through difference Fourier maps (Stryer *et al.*, 1964). The Lu sites were found through cross-Fourier maps, such that both derivatives shared the same origin. Both positions, B-factors and occupancies, were refined using SHARP (De La Fortelle and Bricogne, 1997). Phases were calculated and improved using SOLOMON (Abrahams and Leslie, 1996). The resulting Fourier maps were used for peptide tracing. As most bulky side chains were not readily apparent, use of the multiple glycosylation sites and the unusually high number of disulfide bonds were key to the successful peptide tracing of such a large and multichain protein (Fig. 14.7A and B). The initial model was refined using CNS cells (Brunger *et al.*, 1998). After several cycles of iterative refinement and adjustment, individual isotropic B-factors were refined. The crystal structure of $\alpha V\beta 3$ ectodomain was determined in the presence of Ca^{2+} ($\alpha V\beta 3$-Ca) and Mn^{2+} ($\alpha V\beta 3$-Mn). The final model contained all four αV-subunit domains (named seven-bladed

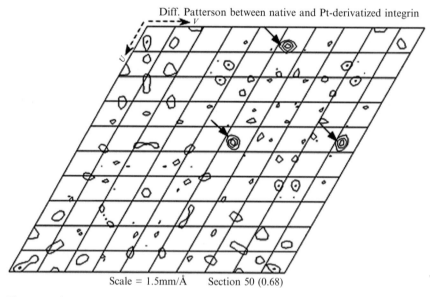

Diff. Patterson between native and Pt-derivatized integrin

Scale = 1.5mm/Å   Section 50 (0.68)

Figure 14.6 Difference Patterson map of the Pt derivative at the Harker section z = 0.667 with data from 20.0 to 6.0 Å. From the peaks (highest density, highlighted by arrows) on Harker sections, the Pt position can be calculated. The rest of the Pt sites are then determined using difference Fourier maps. The Pt data were first scaled onto the wild-type data using the FHSCAL program in CCP4, and then the difference Patterson map calculated using program FFT in CCP4.

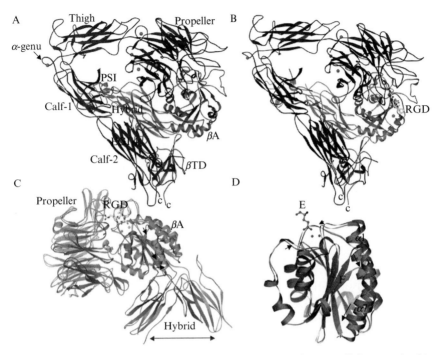

Figure 14.7 Integrin crystal structures and conformational states. All diagrams in this figure were made using ribbon models (Carson, 1987). Ribbon diagrams showing the crystal structure of the ectodomain of integrin $\alpha V \beta 3$ in its unliganded (A) and RGD-bound (B) (Xiong *et al.*, 2001, 2002, 2004) states. In (A), the protein is bent at a flexible region (α-genu, arrow), with an adjacent metal ion (orange sphere) (the β-genu is not shown). The four metal ions at the bottom of the propeller (orange spheres) and the ADMIDAS ion (purple sphere) are shown. In (B), two additional metal ions at MIDAS (cyan sphere) and LIMBS (gray sphere) are found, the former directly coordinating the Asp residue from the cyclic pentapeptide cilengitide, Arg-Gly-Asp-[D-Phe]-[N-methyl-Val-] (carbon, nitrogen, and oxygen atoms are in yellow, blue, and red, respectively). (C) Ribbon diagram of the propeller, βA, and hybrid domains from the liganded bent $\alpha V \beta 3$ ectodomain (red) and the αIIb$\beta 3$ integrin fragment (green) (Xiao *et al.*, 2004). Nonmoving parts of the Cα tracings are in gray. (D) Ribbon diagram of the crystal structures of open (red) and closed (green) αA domain from integrin CD11b (Lee *et al.*, 1995a,b). Major conformational differences are indicated by arrows that also show the direction of movement. The ligand Glu is in gold. (See color insert.)

β-propeller, an Ig-like thigh domain, and two large β-sandwich domains named calf-1 and calf-2) and five (an Ig-like hybrid domain, an αA-like βA domain, EGF domains 3 and 4, and a novel β-tail domain) of the eight β3-subunit domains, with three domains (the N-terminal PSI domain and EGF1 and EGF2) not included. We have recently added the structure of PSI to the ectodomain (Xiong *et al.*, 2004) (Fig. 14.7A), but the electron density of the remaining EGF1 and EGF2 was weak and so was excluded. (The crystal structure of EGF1 together with the hybrid and PSI domains from a fragment containing only these three domains has recently been determined [Shi *et al.*, 2005], which should help in resolving the structure of EGF1 within the whole ectodomain.)

The crystal structure of the $\alpha V\beta 3$ ectodomain is assembled into an ovoid head perched atop two "legs." The ligand-binding head is formed of an αV propeller domain and the βA domain from the $\beta 3$-subunit (Fig. 14.7A,B). These two domains resemble the $G\beta$- and $G\alpha$-subunits of G-proteins, respectively, and contact each other in a strikingly similar manner, suggesting a dynamic interface (Xiong *et al.*, 2001). Both $\alpha V\beta 3$-Ca and $\alpha V\beta 3$-Mn contain six metal ions each, four at the bottom of the propeller domain, one in the calf-2 domain, and one metal ion bound at the ADMIDAS (adjacent to metal ion-dependent adhesion *s*ite) of βA (Fig. 14.7A). The ADMIDAS metal ion links two conformationally sensitive segments of βA ($\alpha 1$ helix and F-$\alpha 7$ loop) helping to stabilize the integrin in the unliganded state (Arnaout *et al.*, 2005). An unexpected feature of the crystal structure is that it is severely bent at its α- and β-genu "knee" (the junctions between the thigh and calf-2 domains of αV and approximately EGF1 and EGF2 of $\beta 3$, respectively), such that the head contacts the lower leg segments of the same molecule (Fig. 14.7A). This feature is not a crystal artifact, as it has also been demonstrated in solution (Adair *et al.*, 2005; Takagi *et al.*, 2002), and appears to stabilize the unliganded state of the integrin (Xiong *et al.*, 2001).

The crystal structure of the binding site for the prototypical ligand peptide RGD in the integrin $\alpha V\beta 3$ ectodomain was determined by soaking wild-type protein crystals with a pentapeptide containing the RGD motif EMD 121974, cilengitide in the presence of Mn^{2+} (Xiong *et al.*, 2002) (Fig. 14.7B). The RGD-bound structure displayed a bent conformation (Fig. 14.7B), with small tertiary and quaternary changes that are linked to ligand binding (Arnaout *et al.*, 2005). It also contained two additional Mn^{2+} ions bound to the βA domain, one at MIDAS, which coordinates binding of the ligand aspartate in RGD to the integrin, and the other bound at LIMBS (ligand-associated metal binding site), which helps stabilize the MIDAS ion–ligand coordination (Fig. 14.7B) (Arnaout *et al.*, 2005). The original six metal ions found in the unliganded structure are retained at the same sites, but a functionally important switch in coordination of the ADMIDAS metal ion takes place such that it no longer links the $\alpha 1$ helix and F-$\alpha 7$ loop, while at the same time stabilizing the adjacent MIDAS ion in the liganded state (Xiong *et al.*, 2002).

A third conformation of a ligand-bound $\alpha IIb\beta 3$ fragment containing only the propeller and βA domains and a portion of the β-leg (hybrid + PSI domains) has also been determined using molecular replacement (Xiao *et al.*, 2004) (Fig. 14.7C). It differs from the crystal structure of the RGD-bound ectodomain of $\alpha V\beta 3$ mainly in a large ~ 70-degree outward opening at the hinge between the hybrid and βA domains that is associated with a downward movement in position of the activation-sensitive F-$\alpha 7$ loop and a one-turn downward and outward movement in the position of the c-terminal $\alpha 7$ helix of βA (Fig. 14.7C). It has been proposed, based largely on mutagenesis and EM structures of negatively stained integrins, that this structure represents the high-affinity state (Xiao et al., 2004). In this "switchblade" model (Takagi et al., 2002), the switch of native integrins to high affinity is initiated by physiologic inside-out activation signals that result in a greater than 100-Å separation of the cytoplasmic tails (Kim et al., 2003), causing head–leg separation, genu-straightening, and the opening of the βA/hybrid hinge. However, a recent EM study has shown that the bent conformation can bind the physiologic ligand fibronectin (domains 7 to 10) (Adair et al., 2005) in association with a very modest ∼11 ± 4-degree βA/hybrid hinge opening. As detailed in the study by Adair et al. (2005), this finding cannot be explained away by model bias, and is supported by recent studies in native α5β1 (Wei et al., 2005), αLβ2 (CD11a/CD18) (Chen et al., 2006), and αMβ2 (CD11b/CD18) (Gupta et al., 2007), Thus, genu straightening is not an obligate feature of the high-affinity conformation; much smaller quaternary changes appear sufficient. A recent molecular dynamics study (Puklin-Faucher et al., 2006) found that activation of the βA domain could be achieved with a very modest (∼20-degree) opening of the βA/hybrid hinge, well within the bent EM structure of the αVβ3-fibronectin complex (Adair et al., 2005). It thus appears that inside-out cytoplasmic tail separation triggers allosteric changes involving βA contact with the leg domains βTD (the deadbolt model) (Gupta et al., 2007) and perhaps calf-2 (Kamata et al., 2005) that lead to a small βTD/βA hinge opening within a bent conformation that is sufficient to switch βA to the high-affinity binding state (Gupta et al., 2007). The strain build-up induced by the binding of physiologic ligand to the bent high-affinity integrin is released by the additional opening of the βA/hybrid hinge (Fig. 14.7C) that likely leads to various degrees of genu extension, the latter perhaps representing the outside-in signaling state of the integrin (Gupta et al., 2007).

3.2. Crystal structure determination of αA domain

3.2.1. Protein preparation

The ligand-binding αA domain exists in low "closed" and high (open) affinity states (Li et al., 1998; Xiong et al., 2000). It can be overexpressed in Escherichia coli as either His-tagged or GST-tagged fusion proteins (Michishita et al., 1993; Qu and Leahy, 1995). Generally, the tags are removed after the affinity purification step. Prior to crystallization, the protein is normally passed through a gel filtration column. The CD11b αA domain protein is released from GST using on-gel overnight thrombin digestion at 4° on a slow, rotating platform. For the high-affinity form, the slow rotation is more critical to avoid protein aggregation. Released CD11b αA domain protein is buffer exchanged and purified on an SP cation ion exchange column. Both forms of CD11b αA domain protein are eluted from this column around 0.2 to 0.3 M NaCl.

The purified proteins are passed on to Superdex 75 column, and the respective fractions pooled and concentrated to about 10 mg/ml. Intermediate affinity forms have been obtained following stabilizing site-specific mutagenesis (Shimaoka *et al.*, 2003).

3.2.2. Crystallization and structure determination

The low- and high-affinity forms of the CD11b αA domains have been crystallized under PEG at room temperature using the hanging- or sitting-drop vapor diffusion method in Ca^{2+}, Mn^{2+}, or Mg^{2+}-containing buffer (Lee *et al.*, 1995c,d; Li *et al.*, 1998; Qu and Leahy, 1995; Xiong *et al.*, 2000). In our hands, the protein crystals form within 3 days to 3 weeks and are usually rod-shaped. The crystallization drop of the "open" form may form a sticky film cover, making crystal harvesting difficult. The "closed" crystal can diffract to about 2.3 Å using an in-house R-Axis IV detector and CuKa radiation from a Rigaku RU200 rotating anode. The "open"-form crystal requires a synchrotron beamline to collect high-resolution data. Crystal structure determination of the high- (Lee *et al.*, 1995c) and low-affinity (Lee *et al.*, 1995d; Qu and Leahy, 1995) forms utilized selenomethionine-labeled proteins (Harrison *et al.*, 1994), and MAD phases were determined at 1.7 Å to 1.8 Å resolution (Fig. 14.7D). Subsequently, molecular or multiple isomorphous replacement methods and multiple anomalous diffraction have been used to solve the crystal structures of other αA domains (Emsley *et al.*, 1997; Shimaoka *et al.*, 2003; Vorup-Jensen *et al.*, 2003) and A-domains found in nonintegrin proteins such as VWFA1 (Emsley *et al.*, 1998), anthrax receptors (Lacy *et al.*, 2004; Santelli *et al.*, 2004), Ku70 and Ku80 subunits of DNA repair protein (Walker *et al.*, 2001), the pre-budding complex components sec23 and sec34 (Bi *et al.*, 2002), the Ro autoantigen (Stein *et al.*, 2005), and complement component factors Bb and C2a (Bhattacharya *et al.*, 2004; Milder *et al.*, 2006).

4. Summary

Integrins play critical roles in the cell's sensation and response to the ever-changing chemical and mechanical signals in its complex microenvironment. Recent advances in the purification of water-soluble forms of integrins have resulted in the successful determination of the crystal structure of these large glycosylated heterodimers. The crystal structure has resolved at the atomic level the basis for the interaction of these receptors with ligands and explained the uniform requirement for an acidic residue and a metal ion in the binding of physiologic ligands to these receptors. Crystal structure determination has resulted in an expanding number of studies aimed at assessing structure–activity relationships in these receptors.

Although millions of years of evolution resulted in intricate regulatory pathways to control the activation state of these receptors, these pathways occasionally run amok, causing potentially life-threatening diseases ranging from heart attacks and stroke to cancer, arthritis, and multiple sclerosis. Several successes have been made in targeting integrins for drug development using antibody-based and peptidomimetic antagonists (Staunton *et al.*, 2006), but toxicities resulting from these reagents present ongoing challenges. The integrin crystal structures elucidate several allosteric sites, such as those at the $\alpha A/\beta A$, βA/hybrid, $\beta A/\beta TD$, and βTD/calf-2 interfaces (Arnaout *et al.*, 2005), which offer new targets that could circumvent the limitations of antibody-based or peptidomimetic antagonists.

ACKNOWLEDGMENTS

The authors acknowledge support from the National Institutes of Health (grants DK48549 and HL-70219).

REFERENCES

Abrahams, J. P., and Leslie, A. G. W. (1996). Methods used in the structure determination of bovine mitochondrial F1 ATPase. *Acta Cryst.* **D52**, 30–42.

Abrams, C., Deng, Y. J., Steiner, B., O'Toole, T., and Shattil, S. J. (1994). Determinants of specificity of a baculovirus-expressed antibody Fab fragment that binds selectively to the activated form of integrin alpha IIb beta 3. *J. Biol. Chem.* **269**, 18781–18788.

Adair, B. D., Xiong, J. P., Maddock, C., Goodman, S. L., Arnaout, M. A., and Yeager, M. (2005). Three-dimensional EM structure of the ectodomain of integrin $\alpha V\beta 3$ in a complex with fibronectin. *J. Cell Biol.* **168**, 1109–1118.

Adair, B. D., and Yeager, M. (2002). Three-dimensional model of the human platelet integrin alpha IIbbeta 3 based on electron cryomicroscopy and x-ray crystallography. *Proc. Natl. Acad. Sci. USA* **99**, 14059–14064.

Arnaout, M. A. (2002). Integrin structure: New twists and turns in dynamic cell adhesion. *Immunol. Rev.* **186**, 125–140.

Arnaout, M. A., Mahalingam, B., and Xiong, J. P. (2005). Integrin structure, allostery, and bidirectional signaling. *Annu. Rev. Cell Dev. Biol.* **11**, 2606–2621.

Bechard, D., Scherpereel, A., Hammad, H., Gentina, T., Tsicopoulos, A., Aumercier, M., Pestel, J., Dessaint, J. P., Tonnel, A. B., and Lassalle, P. (2001). Human endothelial-cell specific molecule-1 binds directly to the integrin CD11a/CD18 (LFA-1) and blocks binding to intercellular adhesion molecule-1. *J. Immunol.* **167**, 3099–3106.

Bennett, J. S., Hoxie, J. A., Leitman, S. F., Vilaire, G., and Cines, D. B. (1983). Inhibition of fibrinogen binding to stimulated human platelets by a monoclonal antibody. *Proc. Natl. Acad. Sci. USA* **80**, 2417–2421.

Bhattacharya, A. A., Lupher, M. L., Jr., Staunton, D. E., and Liddington, R. C. (2004). Crystal structure of the A domain from complement factor B reveals an integrin-like open conformation. *Structure* **12**, 371–378.

Bi, X., Corpina, R. A., and Goldberg, J. (2002). Structure of the Sec23/24-Sar1 pre-budding complex of the COPII vesicle coat. *Nature* **419**, 271–277.

Bijovet, J. M., Peerdeman, A. F., and van Bommel, A. J. (1951). Determination of the absolute configuration of optically active compounds by means of x-rays. *Nature* **168**, 271–272.

Blow, D. M. (1958). The structure of haemoglobin.7.determination of phase angles in the non-centrosymmetric[100] zone. *Proc. R. Soc. Lond.* **247**, 302–336.

Briesewitz, R., Epstein, M. R., and Marcantonio, E. E. (1993). Expression of native and truncated forms of the human integrin alpha 1 subunit. *J. Biol. Chem.* **268**, 2989–2996.

Bruinzeel, W., Yon, J., Giovannelli, S., and Masure, S. (2002). Recombinant insect cell expression and purification of human beta-secretase (BACE–1) for X-ray crystallography. *Protein Expr. Purif.* **26**, 139–148.

Brunger, A. T., Adams, P. D., Clore, G. M., DeLano, W. L., Gros, P., Grosse-Kunstleve, R. W., Jiang, J. S., Kuszewski, J., Nilges, M., Pannu, N. S., *et al.* (1998). Crystallography and NMR system: A new software suite for macromolecular structure determination. *Acta Cryst. D Biol. Cryst.* **54**, 905–921.

Calvete, J. J. (2004). Structures of integrin domains and concerted conformational changes in the bidirectional signaling mechanism of alphaIIbbeta3. *Exp. Biol. Med. (Maywood)* **229**, 732–744.

Carson, M. (1987). Ribbon models of macromolecules. *J. Mol. Graphics* **5**, 103–106.

Chan, B. M. C., Kassner, P. D., Schiro, J. A., Byers, R., Kupper, T. S., and Hemler, M. E. (1992). Distinct cellular functions mediated by different VLA integrin a subunit cytoplasmic domains. *Cell* **68**, 1051–1060.

Chen, J., Yang, W., Kim, M., Carman, C. V., and Springer, T. A. (2006). Regulation of outside-in signaling and affinity by the beta2 I domain of integrin alphaLbeta2. *Proc. Natl. Acad. Sci. USA* **103**, 13062–13067.

Cheresh, D. A., and Spiro, R. C. (1987). Biosynthetic and functional properties of an Arg-Gly-Asp-directed receptor involved in human melanoma cell attachment to vitronectin, fibrinogen, and von Willebrand factor. *J. Biol. Chem.* **262**, 17703–17711.

Chiu, C. Y., Mathias, P., Nemerow, G. R., and Stewart, P. L. (1999). Structure of adenovirus complexed with its internalization receptor, alphavbeta5 integrin. *J. Virol.* **73**, 6759–6768.

Clark, K., Newham, P., Burrows, L., Askari, J. A., and Humphries, M. J. (2000). Production of recombinant soluble human integrin alpha4beta1. *FEBS Lett.* **471**, 182–186.

Coe, A. P., Askari, J. A., Kline, A. D., Robinson, M. K., Kirby, H., Stephens, P. E., and Humphries, M. J. (2001). Generation of a minimal alpha5beta1 integrin-Fc fragment. *J. Biol. Chem.* **276**, 35854–35866.

Dana, N., Fathallah, D. F., and Arnaout, M. A. (1991). Expression of a soluble and functional form of the human b2 integrin CD11b/CD18. *Proc. Natl. Acad. Sci. USA* **88**, 3106–3110.

De La Fortelle, E., and Bricogne, G. (1997). Maximum-likelihood heavy-atom parameter refinement for multiple isomorphous replacement and multiwavelength anomalous diffraction methods. *Methods Enzymol.* **276**, 472–494.

Denda, S., Muller, U., Crossin, K. L., Erickson, H. P., and Reichardt, L. F. (1998). Utilization of a soluble integrin-alkaline phosphatase chimera to characterize integrin alpha 8 beta 1 receptor interactions with tenascin: Murine alpha 8 beta 1 binds to the RGD site in tenascin-C fragments, but not to native tenascin-C. *Biochemistry* **37**, 5464–5474.

Du, X., Plow, E. F., Frelinger, A. L., O'Toole, T. E., Loftus, J. C., and Ginsberg, M. H. (1991). Ligands "activate" integrin aIIbb3 (platelet GPIIb-IIIa). *Cell* **65**, 409–416.

Eble, J. A., Wucherpfennig, K. W., Gauthier, L., Dersch, P., Krukonis, E., Isberg, R. R., and Hemler, M. E. (1998). Recombinant soluble human alpha 3 beta 1 integrin: Purification, processing, regulation, and specific binding to laminin-5 and invasin in a mutually exclusive manner. *Biochemistry* **37**, 10945–10955.

Eirin, M. T., Calvete, J. J., and Gonzalez-Rodriguez, J. (1986). New isolation procedure and further biochemical characterization of glycoproteins IIb and IIIa from human platelet plasma membrane. *Biochem. J.* **240**, 147–153.

Emsley, J., Cruz, M., Handin, R., and Liddington, R. (1998). Crystal structure of the von Willebrand Factor A1 domain and implications for the binding of platelet glycoprotein Ib. *J. Biol. Chem.* **273**, 10396–10401.

Emsley, J., King, S. L., Bergelson, J. M., and Liddington, R. C. (1997). Crystal structure of the I domain from integrin alpha2beta1. *J. Biol. Chem.* **272**, 28512–28517.

Garcia de la Torre, J., Perez Sanchez, H. E., Ortega, A., Hernandez, J. G., Fernandes, M. X., Diaz, F. G., and Lopez Martinez, M. C. (2003). Calculation of the solution properties of flexible macromolecules: Methods and applications. *Eur. Biophys. J.* **32**, 477–486.

Gehlsen, K. R., Dillner, L., Engvall, E., and Ruoslahti, E. (1988). The human laminin receptor is a member of the integrin family of cell adhesion receptors. *Science* **241**, 1228–1229.

Ginsberg, M. H., Partridge, A., and Shattil, S. J. (2005). Integrin regulation. *Curr. Opin. Cell Biol.* **17**, 509–516.

Gulino, D., Martinez, P., Delachanal, E., Concord, E., Duperray, A., Alemany, M., and Marguerie, G. (1995). Expression and purification of a soluble functional form of the platelet alpha IIb beta 3 integrin. *Eur. J. Biochem.* **227**, 108–115.

Gupta, V., Gylling, A., Alonso, J. L., Sugimori, T., Ianakiev, P., Xiong, J. P., and Arnaout, M. A. (2007). The β-tail domain ({beta}TD) regulates physiologic ligand binding to integrin CD11b/CD18. *Blood* **109**(8), 3513–3520; Epub 2006 Dec. 14.

Hantgan, R. R., Braaten, J. V., and Rocco, M. (1993). Dynamic light scattering studies of alpha IIb beta 3 solution conformation. *Biochemistry* **32**, 3935–3941.

Hantgan, R. R., Gibbs, W., Stahle, M. C., Aster, R. H., and Peterson, J. A. (2004a). Integrin clustering mechanisms explored with a soluble alphaIIbbeta3 ectodomain construct. *Biochim. Biophys. Acta* **1700**, 19–25.

Hantgan, R. R., Lyles, D. S., Mallett, T. C., Rocco, M., Nagaswami, C., and Weisel, J. W. (2003). Ligand binding promotes the entropy-driven oligomerization of integrin alpha IIb beta 3. *J. Biol. Chem.* **278**, 3417–3426.

Hantgan, R. R., Stahle, M., Del Gaizo, V., Adams, M., Lasher, T., Jerome, W. G., McKenzie, M., and Lyles, D. S. (2001). AlphaIIb's cytoplasmic domain is not required for ligand-induced clustering of integrin alphaIIbbeta3. *Biochim. Biophys. Acta* **1540**, 82–95.

Hantgan, R. R., Stahle, M. C., Connor, J. H., Lyles, D. S., Horita, D. A., Rocco, M., Nagaswami, C., Weisel, J. W., and McLane, M. A. (2004b). The disintegrin echistatin stabilizes integrin alphaIIbbeta3's open conformation and promotes its oligomerization. *J. Mol. Biol.* **342**, 1625–1636.

Hantgan, R. R., Stahle, M. C., Jerome, W. G., Nagaswami, C., and Weisel, J. W. (2002). Tirofiban blocks platelet adhesion to fibrin with minimal perturbation of GpIIb/IIIa structure. *Thromb. Haemost.* **87**, 910–917.

Harrison, C. J., Bohm, A. A., and Nelson, H. C. M. (1994). Crystal structure of the DNA binding domain of the heat shock transcription factor. *Science* **263**, 224–227.

Haubner, R., Gratias, R., Diefenbach, B., Goodman, S. L., Jonczyk, A., and Kessler, H. (1996). Structural and functional aspects of RGD-containing cyclic pentapeptides as highly potent and selective integrin V3 antagonists. *J. Am. Chem. Soc.* **118**, 7461–7472.

Hemler, M. E., Huang, C., Takada, Y., Schwarz, L., Strominger, J. L., and Clabby, M. L. (1987). Characterization of the cell surface heterodimer VLA-4 and related peptides. *J. Biol. Chem.* **262**, 11478–11485.

Hendrickson, W. A. (1991). Determination of macromolecular strcutrues from anomalous diffraction of synchrotron radiation. *Science (Washington DC)* **254**, 51–58.

Higgins, J. M., Mandlebrot, D. A., Shaw, S. K., Russell, G. J., Murphy, E. A., Chen, Y. T., Nelson, W. J., Parker, C. M., and Brenner, M. B. (1998). Direct and regulated interaction of integrin alphaEbeta7 with E-cadherin. *J. Cell Biol.* **140**, 197–210.

Howard, L., Shulman, S., Sadanandan, S., and Karpatkin, S. (1982). Crossed immunoelectrophoresis of human platelet membranes. The major antigen consists of a complex of glycoproteins, GPIIb and GPIIIa, held together by Ca2+ and missing in Glanzmann's thrombasthenia. *J. Biol. Chem.* **257**, 8331–8336.

Hsieh, P., and Robbins, P. W. (1984). Regulation of asparagine-linked oligosaccharide processing. Oligosaccharide processing in Aedes albopictus mosquito cells. *J. Biol. Chem.* **259,** 2375–2382.

Humphries, J. D., Byron, A., and Humphries, M. J. (2006). Integrin ligands at a glance. *J. Cell Sci.* **119,** 3901–3903.

Humphries, M. J., McEwan, P. A., Barton, S. J., Buckley, P. A., Bella, J., and Paul Mould, A. (2003). Integrin structure: Heady advances in ligand binding, but activation still makes the knees wobble. *Trends Biochem. Sci.* **28,** 313–320.

Hynes, R. O. (1987). Integrins: A family of cell surface receptors. *Cell* **48,** 549–554.

Hynes, R. O. (1992). Integrins: Versatility, modulation and signaling in cell adhesion. *Cell* **69,** 11–26.

Ingber, D. E. (2006). Cellular mechanotransduction: Putting all the pieces together again. *FASEB J.* **20,** 811–827.

Isaji, T., Sato, Y., Zhao, Y., Miyoshi, E., Wada, Y., Taniguchi, N., and Gu, J. (2006). N-glycosylation of the beta-propeller domain of the integrin alpha5 subunit is essential for alpha5beta1 heterodimerization, expression on the cell surface, and its biological function. *J. Biol. Chem.* **281,** 33258–33267.

Iwasaki, K., Mitsuoka, K., Fujiyoshi, Y., Fujisawa, Y., Kikuchi, M., Sekiguchi, K., and Yamada, T. (2005). Electron tomography reveals diverse conformations of integrin alphaIIbbeta3 in the active state. *J. Struct. Biol.* **150,** 259–267.

Kamata, T., Handa, M., Sato, Y., Ikeda, Y., and Aiso, S. (2005). Membrane-proximal {alpha}/{beta} stalk interactions differentially regulate integrin activation. *J. Biol. Chem.* **280,** 24775–24783.

Kim, M., Carman, C. V., and Springer, T. A. (2003). Bidirectional transmembrane signaling by cytoplasmic domain separation in integrins. *Science* **301,** 1720–1725.

Kraft, S., Diefenbach, B., Mehta, R., Jonczyk, A., Luckenbach, G. A., and Goodman, S. L. (1999). Definition of an unexpected ligand recognition motif for alphav beta6 integrin. *J. Biol. Chem.* **274,** 1979–1985.

Kurzinger, K., and Springer, T. A. (1982). Purification and structural characterization of LFA-1, a lymphocyte function-associated antigen, and Mac-1, a related macrophage differentiation antigen associated with the type three complement receptor. *J. Biol. Chem.* **257,** 12412–12418.

Lacy, D. B., Wigelsworth, D. J., Scobie, H. M., Young, J. A., and Collier, R. J. (2004). Crystal structure of the von Willebrand factor A domain of human capillary morphogenesis protein 2: An anthrax toxin receptor. *Proc. Natl. Acad. Sci. USA* **101,** 6367–6372.

Leahy, D. J., Aukhil, I., and Erickson, H. P. (1996). A crystal structure of a four-domain segment of human fibronectin encompassing the RGD loop and synergy region. *Cell* **84,** 155–164.

Lee, J.-O., Anne-Bankston, L., Arnaout, M. A., and Liddington, R. C. (1995a). Two conformations of the integrin A-domain (I-domain): A pathway for activation? *Structure* **3,** 1333–1340.

Lee, J.-O., Rieu, P., Arnaout, M. A., and Liddington, R. (1995b). Crystal structure of the A-domain from the a-subunit of b2 integrin complement receptor type 3 (CR3, CD11b/CD18). *Cell* **80,** 631–638.

Lee, J. O., Bankston, L. A., Arnaout, M. A., and Liddington, R. C. (1995c). Two conformations of the integrin A-domain (I-domain): A pathway for activation? *Structure* **3,** 1333–1340.

Lee, J. O., Rieu, P., Arnaout, M. A., and Liddington, R. (1995d). Crystal structure of the A domain from the alpha subunit of integrin CR3 (CD11b/CD18). *Cell* **80,** 631–638.

Li, R., and Arnaout, M. A. (1999). Functional analysis of the beta 2 integrins. *Methods Mol. Biol.* **129,** 105–124.

Li, R., Babu, C. R., Lear, J. D., Wand, A. J., Bennett, J. S., and DeGrado, W. F. (2001). Oligomerization of the integrin alphaIIbbeta3: Roles of the transmembrane and cytoplasmic domains. *Proc. Natl. Acad. Sci. USA* **98**, 12462–12467.

Li, R., Gorelik, R., Nanda, V., Law, P. B., Lear, J. D., DeGrado, W. F., and Bennett, J. S. (2004). Dimerization of the transmembrane domain of Integrin alphaIIb subunit in cell membranes. *J. Biol. Chem.* **279**, 26666–26673.

Li, R., Rieu, P., Griffith, D. L., Scott, D., and Arnaout, M. A. (1998). Two functional states of the CD11b A-domain: Correlations with key features of two Mn2+-complexed crystal structures. *J. Cell Biol.* **143**, 1523–1534.

Luo, B. H., and Springer, T. A. (2006). Integrin structures and conformational signaling. *Curr. Opin. Cell Biol.* **18**, 579–586.

Mehta, R. J., Diefenbach, B., Brown, A., Cullen, E., Jonczyk, A., Gussow, D., Luckenbach, G. A., and Goodman, S. L. (1998). Transmembrane-truncated alphavbeta3 integrin retains high affinity for ligand binding: Evidence for an 'inside-out' suppressor? *Biochem. J.* **330**(Pt 2), 861–869.

Michishita, M., Videm, V., and Arnaout, M. A. (1993). A novel divalent cation-binding site in the A domain of the beta 2 integrin CR3 (CD11b/CD18) is essential for ligand binding. *Cell* **72**, 857–867.

Milder, F. J., Raaijmakers, H. C., Vandeputte, M. D., Schouten, A., Huizinga, E. G., Romijn, R. A., Hemrika, W., Roos, A., Daha, M. R., and Gros, P. (2006). Structure of complement component C2A: Implications for convertase formation and substrate binding. *Structure* **14**, 1587–1597.

Mitjans, F., Sander, D., Adan, J., Sutter, A., Martinez, J. M., Jaggle, C. S., Moyano, J. M., Kreysch, H. G., Piulats, J., and Goodman, S. L. (1995). An anti-alpha v-integrin antibody that blocks integrin function inhibits the development of a human melanoma in nude mice. *J. Cell Sci.* **108** (Pt 8), 2825–2838.

Mould, A. P., Akiyama, S. K., and Humphries, M. J. (1996). The inhibitory anti-beta1 integrin monoclonal antibody 13 recognizes an epitope that is attenuated by ligand occupancy. Evidence for allosteric inhibition of integrin function. *J. Biol. Chem.* **271**, 20365–20374.

Mould, A. P., Barton, S. J., Askari, J. A., McEwan, P. A., Buckley, P. A., Craig, S. E., and Humphries, M. J. (2003a). Conformational changes in the integrin beta A domain provide a mechanism for signal transduction via hybrid domain movement. *J. Biol. Chem.* **278**, 17028–17035.

Mould, A. P., Symonds, E. J., Buckley, P. A., Grossmann, J. G., McEwan, P. A., Barton, S. J., Askari, J. A., Craig, S. E., Bella, J., and Humphries, M. J. (2003b). Structure of an integrin-ligand complex deduced from solution x-ray scattering and site-directed mutagenesis. *J. Biol. Chem.* **278**, 39993–39999.

Muller, B., Zerwes, H. G., Tangemann, K., Peter, J., and Engel, J. (1993). Two-step binding mechanism of fibrinogen to alpha IIb beta 3 integrin reconstituted into planar lipid bilayers. *J. Biol. Chem.* **268**, 6800–6808.

Nachman, R. L., and Leung, L. L. (1982). Complex formation of platelet membrane glycoproteins IIb and IIIa with fibrinogen. *J. Clin. Invest.* **69**, 263–269.

Nermut, M. V., Green, N. M., Eason, P., Yamada, S. S., and Yamada, K. M. (1988). Electron microscopy and structural model of human fibronectin receptor. *EMBO J.* **7**, 4093–4099.

Otwinowski, Z., and Minor, W. (1997). "Processing of X-ray diffraction data collected in oscillation mode," Vol. 276. Academic Press, New York.

Parise, L. V., and Phillips, D. R. (1985). Reconstitution of the purified platelet fibrinogen receptor. Fibrinogen binding properties of the glycoprotein IIb-IIIa complex. *J. Biol. Chem.* **260**, 10698–10707.

Pasqualini, R., Koivunen, E., and Ruoslahti, E. (1995). A peptide isolated from phage display libraries is a structural and functional mimic of an RGD-binding site on integrins. *J. Cell Biol.* **130,** 1189–1196.

Perutz, M. F. (1956). Isomorphous replacement and phase determination in non-centrosymmetric space groups. *Acta Cryst.* **9,** 867–873.

Pfaff, M., Gohring, W., Brown, J. C., and Timpl, R. (1994). Binding of purified collagen receptors (alpha 1 beta 1, alpha 2 beta 1) and RGD-dependent integrins to laminins and laminin fragments. *Eur. J. Biochem.* **225,** 975–984.

Pierce, M. W., Remold-O'Donnell, E., Todd, R. F. I., and Arnaout, M. A. (1986). N-terminal sequence of human leukocyte glycoprotein M01: Conservation across species and homology to platelet IIb/IIIa. *Biochim. Biophys. Acta* **874,** 368–371.

Puklin-Faucher, E., Gao, M., Schulten, K., and Vogel, V. (2006). How the headpiece hinge angle is opened: New insights into the dynamics of integrin activation. *J. Cell Biol.* **175,** 349–360.

Pytela, R., Pierschbacher, M. D., Ginsberg, M. H., Plow, E. F., and Ruoslahti, E. (1986). Platelet membrane glycoprotein IIb/IIIa: Member of a family of Arg-Gly-Asp——specific adhesion receptors. *Science* **231,** 1559–1562.

Pytela, R., Pierschbacher, M. D., and Ruoslahti, E. (1985a). A 125/115-kDa cell surface receptor specific for vitronectin interacts with the arginine–glycine–aspartic acid adhesion sequence derived from fibronectin. *Proc. Natl. Acad. Sci. USA* **82,** 5766–5770.

Pytela, R., Pierschbacher, M. D., and Ruoslahti, E. (1985b). Identification and isolation of a 140 kd cell surface glycoprotein with properties expected of a fibronectin receptor. *Cell* **40,** 191–198.

Qu, A., and Leahy, D. J. (1995). Crystal structure of the I-domain from the CD11a/CD18 (LFA-1, aLb2) integrin. *Proc. Natl. Acad. Sci. USA* **92,** 10277–10281.

Rivas, G., Tangemann, K., Minton, A. P., and Engel, J. (1996). Binding of fibrinogen to platelet integrin alpha IIb beta 3 in solution as monitored by tracer sedimentation equilibrium. *J. Mol. Recognit.* **9,** 31–38.

Rivas, G. A., Aznarez, J. A., Usobiaga, P., Saiz, J. L., and Gonzalez-Rodriguez, J. (1991a). Molecular characterization of the human platelet integrin GPIIb/IIIa and its constituent glycoproteins. *Eur. Biophys. J.* **19,** 335–345.

Rivas, G. A., Usobiaga, P., and Gonzalez-Rodriguez, J. (1991b). Calcium and temperature regulation of the stability of the human platelet integrin GPIIb/IIIa in solution: An analytical ultracentrifugation study. *Eur. Biophys. J.* **20,** 287–292.

Rodgers, D. W. (1997). Practical cryocrystallography. *Methods Enzymol.* **276,** 183–203.

Santelli, E., Bankston, L. A., Leppla, S. H., and Liddington, R. C. (2004). Crystal structure of a complex between anthrax toxin and its host cell receptor. *Nature* **430,** 905–908.

Shi, M., Sundramurthy, K., Liu, B., Tan, S. M., Law, S. K., and Lescar, J. (2005). The crystal structure of the plexin-semaphorin-integrin domain/hybrid domain/I-EGF1 segment from the human integrin beta2 subunit at 1.8-A resolution. *J. Biol. Chem.* **280,** 30586–30593.

Shimaoka, M., Xiao, T., Liu, J. H., Yang, Y., Dong, Y., Jun, C. D., McCormack, A., Zhang, R., Joachimiak, A., Takagi, J., *et al.* (2003). Structures of the alphaL I Domain and Its Complex with ICAM-1 Reveal a Shape-Shifting Pathway for Integrin Regulation. *Cell* **112,** 99–111.

Sissom, J., and Ellis, L. (1989). Secretion of the extracellular domain of the human insulin receptor from insect cells by use of a baculovirus vector. *Biochem. J.* **261,** 119–126.

Smith, D. B., and Johnson, K. S. (1988). Single step purification of ploypeptides expressed in *Escherichia coli* as fusions with glutathione-S-transferase. *Gene* **67,** 31–40.

Smith, J. W., and Cheresh, D. A. (1988). The Arg-Gly-Asp binding domain of the vitronectin receptor. Photoaffinity cross-linking implicates amino acid residues 61–203 of the beta subunit. *J. Biol. Chem.* **263,** 18726–18731.

Smith, J. W., Vestal, D. J., Irwin, S. V., Burke, T. A., and Cheresh, D. A. (1990). Purification and functional characterization of integrin alpha v beta 5. An adhesion receptor for vitronectin. *J. Biol. Chem.* **265,** 11008–11013.

Staunton, D. E., Lupher, M. L., Liddington, R., and Gallatin, W. M. (2006). Targeting integrin structure and function in disease. *Adv. Immunol.* **91,** 111–157.

Stein, A. J., Fuchs, G., Fu, C., Wolin, S. L., and Reinisch, K. M. (2005). Structural insights into RNA quality control: The Ro autoantigen binds misfolded RNAs via its central cavity. *Cell* **121,** 529–539.

Stryer, L., Kendrew, J. C., and Watson, H. C. (1964). The mode of attachment of the azide ion to sperm whale metmyoglobin. *J. Mol. Biol.* **37,** 96–104.

Takagi, J., Petre, B. M., Walz, T., and Springer, T. A. (2002). Global conformational rearrangements in integrin extracellular domains in outside-in and inside-out signaling. *Cell* **110,** 599–611.

Tarone, G., Galetto, G., Prat, M., and Comoglio, P. M. (1982). Cell surface molecules and fibronectin-mediated cell adhesion: Effect of proteolytic digestion of membrane proteins. *J. Cell Biol.* **94,** 179–186.

Ulmer, T. S., Yaspan, B., Ginsberg, M. H., and Campbell, I. D. (2001). NMR analysis of structure and dynamics of the cytosolic tails of integrin alpha IIb beta 3 in aqueous solution. *Biochemistry* **40,** 7498–7508.

Vandenberg, P., Kern, A., Ries, A., Luckenbill-Edds, L., Mann, K., and Kuhn, K. (1991). Characterization of a type IV collagen major cell binding site with affinity to the alpha 1 beta 1 and the alpha 2 beta 1 integrins. *J. Cell Biol.* **113,** 1475–1483.

Vinogradova, O., Haas, T., Plow, E. F., and Qin, J. (2000). A structural basis for integrin activation by the cytoplasmic tail of the alpha IIb-subunit. *Proc. Natl. Acad. Sci. USA* **97,** 1450–1455.

Vinogradova, O., Velyvis, A., Velyviene, A., Hu, B., Haas, T., Plow, E., and Qin, J. (2002). A structural mechanism of integrin alpha(IIb)beta(3) "inside-out" activation as regulated by its cytoplasmic face. *Cell* **110,** 587–597.

von der Mark, H., Durr, J., Sonnenberg, A., von der Mark, K., Deutzmann, R., and Goodman, S. L. (1991). Skeletal myoblasts utilize a novel beta 1-series integrin and not alpha 6 beta 1 for binding to the E8 and T8 fragments of laminin. *J. Biol. Chem.* **266,** 23593–23601.

Vorup-Jensen, T., Ostermeier, C., Shimaoka, M., Hommel, U., and Springer, T. A. (2003). Structure and allosteric regulation of the alpha X beta 2 integrin I domain. *Proc. Natl. Acad. Sci. USA* **100,** 1873–1878.

Walker, J. R., Corpina, R. A., and Goldberg, J. (2001). Structure of the Ku heterodimer bound to DNA and its implications for double-strand break repair. *Nature* **412,** 607–614.

Weber, P. C. (1997). Overview of protein crystallization methods. *Methods Enzymol.* **276,** 13–22.

Wegener, K. L., Partridge, A. W., Han, J., Pickford, A. R., Liddington, R. C., Ginsberg, M. H., and Campbell, I. D. (2007). Structural basis of integrin activation by talin. *Cell* **128,** 171–182.

Wei, Y., Czekay, R. P., Robillard, L., Kugler, M. C., Zhang, F., Kim, K. K., Xiong, J. P., Humphries, M. J., and Chapman, H. A. (2005). Regulation of alpha5beta1 integrin conformation and function by urokinase receptor binding. *J. Cell Biol.* **168,** 501–511.

Weinacker, A., Chen, A., Agrez, M., Cone, R. I., Nishimura, S., Wayner, E., Pytela, R., and Sheppard, D. (1994). Role of the integrin alpha v beta 6 in cell attachment to fibronectin. Heterologous expression of intact and secreted forms of the receptor. *J. Biol. Chem.* **269,** 6940–6948.

Weljie, A. M., Hwang, P. M., and Vogel, H. J. (2002). Solution structures of the cytoplasmic tail complex from platelet integrin alpha IIb– and beta 3–subunits. *Proc. Natl. Acad. Sci. USA* **99,** 5878–5883.

Wippler, J., Kouns, W. C., Schlaeger, E. J., Kuhn, H., Hadvary, P., and Steiner, B. (1994). The integrin alpha IIb-beta 3, platelet glycoprotein IIb-IIIa, can form a functionally active heterodimer complex without the cysteine-rich repeats of the beta 3 subunit. *J. Biol. Chem.* **269,** 8754–8761.

Xiao, T., Takagi, J., Coller, B. S., Wang, J. H., and Springer, T. A. (2004). Structural basis for allostery in integrins and binding to fibrinogen-mimetic therapeutics. *Nature* **432,** 59–67.

Xiong, J. P., Li, R., Essafi, M., Stehle, T., and Arnaout, M. A. (2000). An isoleucine-based allosteric switch controls affinity and shape shifting in integrin CD11b A-domain. *J. Biol. Chem.* **275,** 38762–38767.

Xiong, J. P., Stehle, T., Diefenbach, B., Zhang, R., Dunker, R., Scott, D. L., Joachimiak, A., Goodman, S. L., and Arnaout, M. A. (2001). Crystal structure of the extracellular segment of integrin alpha Vbeta3. *Science* **294,** 339–345.

Xiong, J. P., Stehle, T., Goodman, S. L., and Arnaout, M. A. (2004). A Novel Adaptation of the Integrin PSI Domain Revealed from Its Crystal Structure. *J. Biol. Chem.* **279,** 40252–40254.

Xiong, J. P., Stehle, T., Zhang, R., Joachimiak, A., Frech, M., Goodman, S. L., and Arnaout, M. A. (2002). Crystal structure of the extracellular segment of integrin alphaV-beta3 in complex with an Arg-Gly-Asp ligand. *Science* **296,** 151–155.

Yatohgo, T., Izumi, M., Kashiwagi, H., and Hayashi, M. (1988). Novel purification of vitronectin from human plasma by heparin affinity chromatography. *Cell Struct. Funct.* **13,** 281–292.

CHAPTER FIFTEEN

ELECTRON MICROSCOPY OF INTEGRINS

Brian D. Adair* *and* Mark Yeager*,†

Contents

Abstract

Integrins are a family of heterodimeric, cell-surface receptors that mediate interactions between the cytoskeleton and the extracellular matrix. We have used electron microscopy and single-particle image analysis combined with molecular modeling to investigate the structures of the full-length integrin $\alpha_{IIb}\beta_3$ and the ectodomain of $\alpha_V\beta_3$ in a complex with fibronectin. The full-length integrin $\alpha_{IIb}\beta_3$ is purified from human platelets by ion exchange and gel filtration chromatography in buffers containing the detergent octyl-β-D-glucopyranoside, whereas the recombinant ectodomain of $\alpha_V\beta_3$ is soluble in aqueous buffer. Transmission electron microscopy is performed either in negative stain, where the protein is embedded in a heavy metal such as uranyl acetate, or in the frozen-hydrated state, where the sample is flash-frozen such that the buffer is vitrified and native conditions are preserved. Individual integrin particles are selected from low-dose micrographs, either by manual identification or an automated method using a cross-correlation search of the micrograph against

* Department of Cell Biology, The Scripps Research Institute, La Jolla, California
† Division of Cardiovascular Diseases, Scripps Clinic, La Jolla, California

Methods in Enzymology, Volume 426   © 2007 Elsevier Inc.
ISSN 0076-6879, DOI: 10.1016/S0076-6879(07)26015-X   All rights reserved.

a set of reference images. Due to the small size of integrin heterodimers (∼250 kDa) and the low electron dose required to minimize beam damage, the signal-to-noise level of individual particles is quite low, both by negative-stain electron microscopy and electron cryomicroscopy. Consequently, it is necessary to average many particle images with equivalent views. The particle images are subjected to reference-free alignment and classification, in which the particles are aligned to a common view and further grouped by statistical methods into classes with common orientations. Assessment of the structure from a set of two-dimensional averaged projections is often difficult, and a further three-dimensional (3D) reconstruction analysis is performed to classify each particle as belonging to a specific projection from a single 3D model. The 3D reconstruction algorithm is an iterative projection-matching routine in which the classified particles are used to construct a new, 3D map for the next iteration. Docking of known high-resolution structures of individual subdomains within the molecular envelope of the 3D EM map is used to derive a pseudoatomic model of the integrin complex. This approach of 3D EM image analysis and pseudoatomic modeling is a powerful strategy for exploring the structural biology of transmembrane signaling by integrins because it is likely that multiple conformational states will be difficult to crystallize, whereas the different states should be amenable to electron cryomicroscopy.

1. Introduction

Integrins are a family of heterodimeric transmembrane receptors that modulate cell adhesion, such as platelet aggregation (Shattil, 1993), as well as other important cellular processes such as migration, differentiation, proliferation, and programmed cell death (Hynes, 1992). They participate in fundamental biological and pathological processes, including development, angiogenesis, wound healing, and neoplastic transformation (Brooks et al., 1994; DeSimone, 1994; Howe et al., 1998; Price and Loscalzo, 1999). Integrin function is regulated by signaling events, which can originate via the binding of an extracellular ligand, often containing a canonical RGD sequence, thereby eliciting conformational changes at the integrin's cytoplasmic carboxy termini (so-called "outside-in signaling"). Alternatively, cytoskeletal proteins may interact with the integrin cytoplasmic domains to elicit conformational changes in the extracellular domains that change the affinity for external ligands (so-called "inside-out signaling") (Hynes, 1999).

Integrins are found in comparatively low levels in native cells with one exception: the integrin $\alpha_{IIb}\beta_3$ is found in high levels in platelets (Calvete, 1994; Phillips et al., 1988). Due to the presence of the transmembrane domains, detergents must be used to solubilize full-length integrins. The presence of detergent micelles complicates a number of structural techniques, including electron cryomicroscopy (cryo-EM). To date, no recombinant system is available for expression of full-length integrins to high levels. In contrast, comparatively high yields can be achieved by expression of the ectodomains without the transmembrane and cytoplasmic domains (Mehta *et al.*, 1998). Another advantage of studying the isolated ectodomains or cytoplasmic domains is that detergent-free buffers can be used for purification (Ostermeier and Michel, 1997). However, since signaling by integrins is mediated via the transmembrane domains, their absence necessarily removes an essential component of the system. In recent years, there has been increased interest in the use of electron microscopy (EM) and image analysis to examine the structure and conformational states of integrins (Takagi *et al.*, 2002, 2003). In this review, we will restrict ourselves to a discussion of our three-dimensional (3D) analyses of full-length $\alpha_{IIb}\beta_3$ purified from human platelets (Adair and Yeager, 2002) and the recombinant ectodomain of $\alpha_V\beta_3$ expressed in SF9 and High Five cells (Adair *et al.*, 2005).

2. PURIFICATION OF FULL-LENGTH $\alpha_{IIB}\beta_3$ FROM PLATELETS

After washing platelet membranes, detergent-solubilized $\alpha_{IIb}\beta_3$ is purified by successive steps of ion-exchange and gel filtration chromatography (Fig. 15.1). These conditions yield the low-affinity state of $\alpha_{IIb}\beta_3$ (Faull *et al.*, 1994; Phillips *et al.*, 1991), and 1 to 2 mg can be isolated in a typical experiment. (Affinity chromatography with an RGD peptide covalently attached to gel beads can be used to isolate the high-affinity state [Faull *et al.*, 1994; Pytela *et al.*, 1986]).

1. Obtain freshly outdated blood products from a local blood bank. The platelets should be no older than 7 days from the time of phlebotomy. A 200- to 250-ml volume of platelets (either a five-unit pack or a single pack isolated by plasmaphoresis) will arrive at room temperature, and the experiment should begin immediately. The phoresis pack will not be contaminated with red blood cells. However, in using platelet units, red blood cells should be removed by centrifugation (Beckman JA30-15 rotor, 1000 rpm (120×*g*), 20 min, 20°). The supernatants are saved and the pellets discarded.

2. Prepare 250 ml of TBS (25 m*M* Tris, 150 m*M* NaCl, pH 7.2) plus 5 m*M* EDTA. Isolate the platelets by centrifugation (Beckman JA30-15 rotor, 2800 rpm [950×*g*], 20 min, 20°). Gently resuspend the platelet pellets in TBS/EDTA, and repeat the centrifugation.

3. Prepare 500 ml of TBS/Ca (TBS plus 1 m*M* CaCl₂), resuspend the pellets in ~250 ml of TBS/Ca, and repeat the centrifugation.

4. Repeat Step 3. Resuspend the final washed pellets in ~60 ml of TBS/Ca with a Pasteur pipette and add protease inhibitors (25 μ*M* leupeptin

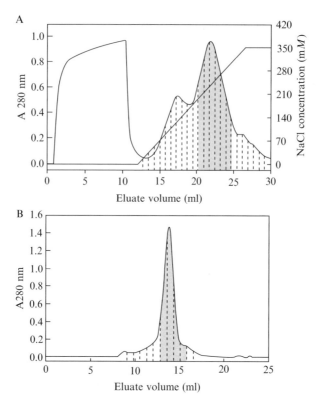

Figure 15.1   Purification of $\alpha_{IIb}\beta_3$ from resting platelets by ion exchange and gel filtration chromatography. (A) Elution profile for resource Q column. The absorbance profile at 280 nm and the NaCl gradient are indicated with black lines. $\alpha_{IIb}\beta_3$ is eluted at NaCl concentrations ranging from 200 to 300 mM. The shaded gray area identifies the collected fractions containing $\alpha_{IIb}\beta_3$ as assessed by gel electrophoresis. (B) Elution profile from superose 6 gel filtration column. $\alpha_{IIb}\beta_3$ purified with this procedure is in the low-affinity state.

and E64). Distribute the suspension into two plastic Beckman JA30 centrifuge tubes. With each tube on ice, sonicate the suspension at full power using a probe sonifier (Branson) equipped with a microtip. Deliver six 10-sec bursts to each tube, with a rest of 10 to 20 sec between each burst.

5. Transfer the suspensions to two Beckman SW28 tubes. Carefully layer 10 ml of 40% sucrose into the bottom of each tube using a syringe. Centrifuge the membrane suspension on the sucrose cushion (Beckman SW28 rotor, 25,000 rpm, 3 h, 30 min, 20°).

6. Use a Pasteur pipette to isolate the yellowish, cloudy band at the 0/40% interface, and dilute with 20 mM of Tris, pH 7.2, plus protease inhibitors in SW28 centrifuge tubes. The number of tubes (typically three) depends on the amount of sucrose that is recovered in the band. Isolate the pellets by centrifugation (Beckman SW28 rotor, 25,000 rpm, 1 h, 20°).

7. Prepare 10 ml of 5% octyl-β-D-glucopyranoside (βOG) (Calbiochem) in 20 mM Tris, pH 7.2, with protease inhibitors. Use this buffer and a Pasteur pipette to resuspend the pellets. Deliver a very brief (1 sec) burst of sonication if necessary to break up any aggregates. Stir the suspension overnight at room temperature.

8. Prepare 100 ml of 20 mM Tris (pH 7.2) with 1% βOG and 40 ml of 20 mM Tris, pH 7.2, with 350 mM NaCl.

9. Connect a 1-ml resource Q column to a Pharmacia Äkta FPLC system, and equilibrate the column with the low-salt buffer from Step 8.

10. Remove any undissolved material from the platelet suspension from Step 7 by centrifugation (Beckman SW40 [1 tube], 25,000 rpm, 45 min, 4°).

11. Ion exchange chromatography using the Resource Q column is performed at 2 ml/min with a 10- to 15-ml linear gradient from 0 to 350 mM NaCl. The absorbance at 280 nm is monitored, and 0.75-ml fractions are collected (Fig. 15.1A).

12. The peak fractions (Fig. 15.1A) are analyzed by sodium dodecyl sulfate polyacrylamide gel electrophoresis (Pharmacia Phast gel, 10 to 15% to ~70 Avh, 1-μl sample). The samples are placed in nonreducing sample buffer and not boiled prior to loading. The gel is stained with Coomassie Brilliant Blue, and fractions on the edge of the peak that show noticeable contamination are discarded.

13. Concentrate the pooled peak fractions (typically 2 to 3 ml) to ~0.5 ml using a centrifugal concentrator (Centricon 50, 3000 to 5000×g, room temperature).

14. Equilibrate a superose 6 10/300 GL column (10 mm × 300 mm, 24-ml bed volume, also connected to an Äkta system) with 20 mM Tris (pH 7.2) plus 1% βOG at a flow rate of 0.3 ml/min. The loading loop volume is 0.5 or 1.0 ml, and fractions of 0.75 ml are collected during chromatography (Fig. 15.1B).

15. The A280 peak fractions (Fig. 15.1B) are analyzed by Phast gel electrophoresis as in Step 12, and the fractions that display pure bands corresponding to the α_{IIb} and β_3 polypeptides are pooled and concentrated as in Step 13. Add additional protease inhibitors as a concentrated stock to pooled fractions. The concentrated $\alpha_{IIb}\beta_3$ is stable for several weeks at 4°.

3. PREPARATION OF SPECIMENS FOR NEGATIVE-STAIN EM

Our laboratory uses two types of transmission EM: negative-stain EM (Hayat and Miller, 1990), in which macromolecular complexes are surrounded by a heavy metal stain (Fig. 15.2), and cryo-EM (Fig. 15.3), in which protein complexes are suspended in a vitrified buffer (Chiu, 1993;

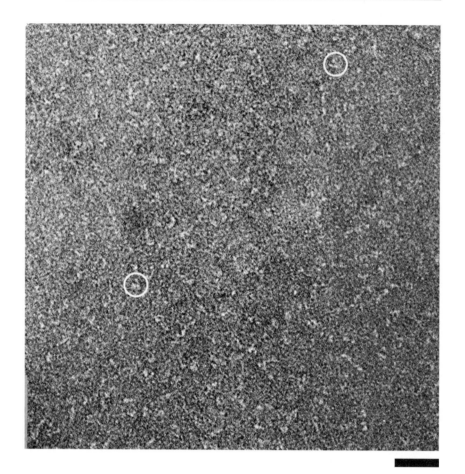

Figure 15.2   Negative-stain electron micrograph of the soluble ectodomain of $\alpha_V\beta_3$. The molecules are randomly distributed making recognition of views difficult. Two putative side views are circled, showing the characteristic "V" shape. Scale bar represents 500 Å.

Dubochet *et al.*, 1988). Negative staining is generally a first step in any EM structural study because it is a convenient method for assessing concentration, intactness, purity, and aggregation of the specimen. Negative staining has a number of advantages, the greatest being the high contrast provided by puddling of the heavy metal stain around the particles, which is of particular importance when trying to visualize small complexes such as ~250-kDa integrin heterodimers. Negative staining also requires less material compared with cryo-EM. A less important advantage is that negatively stained EM grids may be manipulated at room temperature and do not require specialized cryostages for maintaining the grid at the temperature of liquid nitrogen (about −180°). However, negative staining has several possible sources of artifacts. The sample is dried during preparation of the EM grid,

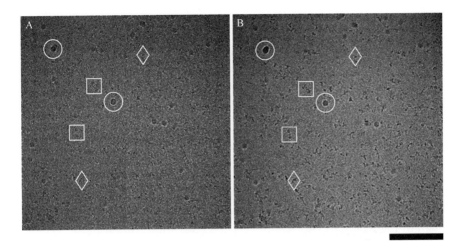

Figure 15.3   Cryo-electron micrographs of $\alpha_{IIb}\beta_3$ in detergent showing a pair of cryo-micrographs recorded at different defocus values. (A) The first image was recorded at 2.5-μm underfocus and an electron dose of 13 e$^-$/Å^2. (B) A second micrograph of the same area was recorded at 4.0-μm underfocus and an electron dose of 35 e$^-$/Å^2. The second, highly defocused micrograph accentuates the low-frequency terms that in turn accentuate the coarse features in the image, such as the particle boundary, thereby facilitating particle selection. Having selected the particles in the second micrograph, the same particle images can be boxed in the first micrograph, which contains higher-frequency terms for obtaining a higher resolution 3D structure. Areas of contamination have been circled in both micrographs. The squares indicate possible aggregates. The diamonds identify more compact particles, possibly a side view in the lower left (compare with Fig. 15.12A) and a top-down view in the upper right (compare with Fig. 15.12I). The scale bar is 1000 Å.

and the staining environment is nonphysiological (e.g., the pH of uranyl acetate is ~5.5). Drying may collapse and distort the complex, and the nonphysiological solution may cause denaturation and/or conformational changes. The formation of a stain shell can create mechanical stress, although this is more often a problem with larger, porous complexes such as ribosomes and virus particles. Negative staining procedures have also been known to disorder proteins, as has been observed in the head domains in myosin filaments (Craig and Woodhead, 2006; Zhao and Craig, 2003a). In this system, ordering in the heads was observed by X-ray diffraction, but not initially seen in diffraction from electron micrographs of negatively stained myosin filaments. Negative staining procedures have an additional limitation in that contrast is produced between the shell of precipitated stain and the protein. Thus, in general, only the surface of the complex is visualized. (In contrast, cryo-EM allows visualization of both external and internal structural details.) Furthermore, uneven puddling of the stain may not faithfully reflect the surface features of the complex. In spite of these reservations, negative staining is a surprisingly good mordant for preserving biological structure (Zhao and Craig, 2003b).

During the staining procedure, a sample that is fairly dilute (roughly 0.05 to 1 mg/ml) compared with samples for cryo-EM (roughly 1 to 5 mg/ml) is incubated on an EM grid covered with a thin, flat carbon layer ($\sim$200 Å thick). The stain solution is wicked off with filter paper, and the grid is allowed to dry thoroughly (Fig. 15.4B). As the residual liquid dries, the stain precipitates and coats the surface of the complex. One characteristic in the choice of a stain (in addition to molecular weight) is the tendency to precipitate with a defined grain size rather than as crystals of the metal salt. Uranyl acetate precipitates to form grains $\sim$15 Å in diameter, which limits the achievable resolution during image processing. Uranyl crystals may form in buffers, so that staining solutions should be prepared by dissolving the stain in high-purity H_2O. Uranyl crystals may still form during storage, so the solutions are filtered (0.2- to 0.45-μM pore size) immediately prior to use. Artifactual positive staining can also occur with uranyl acetate due to actual binding to the protein. Nevertheless, we have had more consistent results with uranyl acetate for staining integrins compared with tungstate stains such as phosphotungstic acid or methalaminetungstate. (However, an advantage of tungstate stains is that the pH of the solution can be adjusted. In our experience, more even staining is achieved when the pH is adjusted with KOH rather than NaOH.)

There are several equally successful procedures for staining specimens. The approach we describe has been optimized for the recombinant $\alpha_V\beta_3$ ectodomain in aqueous buffer. The sample concentration, amount of washing, and the duration of staining of the EM grids may vary for other specimens, in particular with full-length $\alpha_{IIb}\beta_3$ in detergent buffer. In general, buffers with detergent require more washing. The protein solution is typically 0.05 to 0.1 mg/ml. In the study of $\alpha_V\beta_3$ bound to a fragment of fibronectin (FN7-EDB-10), we employed concentrations of 1.0 mg/ml, due to concerns that dilution of the sample might lead to dissociation of the integrin/fibronectin complex (Adair et al., 2005).

1. Place a piece of Parafilm (4 to 6 in^2) on a flat work surface (Fig. 15.4A). Arrange drops as linear tracks, one track for each grid, moving from front to back. The first drop will be an aliquot (3 to 5 μl) of the sample. The next three drops ($\sim$25 μl) in the track are mQ H_2O (which may contain 200 μM $MnCl_2$), and the final three drops are the staining solution ($\sim$25 μl).
2. Using EM forceps (e.g., Dumont #5), carefully pick up a grid with the tips of the forceps just at the edge of the grid so that the carbon layer is not broken. Place the side of the grid with continuous carbon (i.e., the side opposite that with exposed grid bars) onto the sample drop and incubate for 3 to 5 min. (Alternatively, one can hold the grid between the tips of the tweezers and pipette 3 to 5 μl of sample directly onto the grid

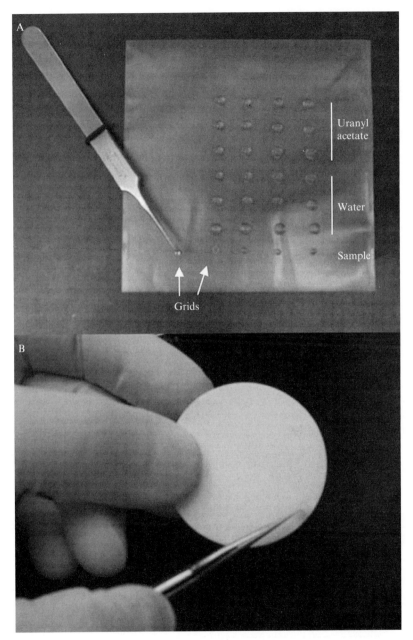

Figure 15.4 Negative staining. The top panel (A) shows a piece of Parafilm with four vertical tracks of drops for staining four grids. Sample drops are in the bottom row, and a grid has been applied to the drop in the lower left to allow adherence of the sample to the grid. The next three drops moving up are water for rinsing the grid. The top three drops are negative-stain solution. The first two drops of stain are used for rinsing the

[Fig. 15.4A, left]. Keep the droplet clear of the tweezer tips to prevent spreading of the droplet between the tips by capillary action.)

3. Carefully pick up the grid, again by the edge. (The sample droplet will often adhere to the grid, and if only a small amount of sample is available, a pipettor can be used to retrieve the sample, while leaving a residual liquid layer of sample on the grid.) Touch an edge of the grid to a piece of Whatman filter paper such that the sample liquid is wicked off. The edge of the grid is touched perpendicularly (i.e., with the grid making a 90-degree angle with the filter paper), so that the buffer is wicked off rather than blotted off (Fig. 15.4B). This tends to generate a more even spread of the sample on the grid. After wicking, clean the forceps by pinching the tips on the filter paper to remove any sample that may have migrated between the tips by capillary action.

4. Place the specimen side of the grid on the first water drop in the track. Position the grid as close to the drop (horizontally) as possible without touching, and drop the grid onto the droplet such that it floats on the top of the droplet. Clean any excess liquid from the forceps by briefly closing the forceps on the edge of the filter paper.

5. After a few seconds, pick up the grid from the droplet and repeat the wicking procedure in Step 3 and the procedure in Step 4 for the next two water droplets and the first two droplets of stain in the track. (In general, the water rinses remove salts, sugars, detergents, and glycerol in the sample buffer that may interfere with even puddling of the stain around the macromolecules.)

6. As in Steps 3 and 4, transfer the grid to the final droplet of stain and time the incubation (60 to 90 sec as optimized for the sample). Lift the grid with forceps and remove the stain solution by wicking as in Step 3. The grid is allowed to air dry and then transferred to a slot in a grid storage box (Ted Pella). If the grid is not going to be immediately examined, it is best to store the grid box in a vacuum dessicator.

4. PREPARATION OF SPECIMENS FOR CRYO-EM

The powerful advantage of cryo–EM over conventional EM methods is that the "native" structure of biological macromolecules can be examined in a defined physiological buffer without introducing potential artifacts commonly associated with preparation methods involving chemical fixation,

grid, and the top drop is used for final incubation in negative stain. An alternative way to apply the sample is shown at the left in which a drop of sample (3 to 5 μl) has been applied directly to an EM grid attached to the forceps. The bottom panel (B) shows the technique of wicking the droplet off the EM grid by touching the edge of the grid to a piece of filter paper that is held perpendicular to the grid.

dehydration, embedding, shadowing, or staining (Chiu, 1986; Dubochet et al., 1988; Unwin, 1986). Contrast in unstained, vitrified specimens arises from differences in scattering density among the protein, nucleic acid, and solvent components, and is also enhanced by judicious defocusing of the electron image (Jaffe and Glaeser, 1984; Milligan et al., 1984; Taylor and Glaeser, 1976). In single-particle image reconstruction, a large number of views of individual, unoriented molecules (>1000, and frequently >10,000) are combined to generate a 3D density map. The advantage of this technique is precisely that ordered samples (such as a crystal or helical filament) or highly symmetric single particles (such as icosahedral viruses) are not required. (The conical tilt method of single-particle reconstruction can be used when the particles have a preferred orientation on the continuous carbon layer of the EM grid [Frank et al., 1996].) However, assemblies with ordered symmetry such as two-dimensional (2D) crystals and helical tubes allow the examination of complexes much smaller than ~250 kDa, often to a resolution better than 10 Å, so that elements of secondary structure can be visualized. The ability to achieve higher resolution by cryo-EM and single-particle image analysis was demonstrated in three landmark studies that resolved the α-helices within the capsid shell of the hepatitis virus (Böttcher et al., 1997; Conway et al., 1997) and papillomavirus (Trus et al., 1997). Admittedly, the 60-fold symmetry afforded by icosahedral packing of the subunits greatly aided the image analysis. More recently, the technique has been successfully applied to large asymmetric complexes such as the ribosome (Frank, 2001) and the U1 SNRnp (Stark et al., 2001). There are now numerous examples where frozen-hydrated specimens have been examined by 2D and 3D image analysis, and several cogent reviews are available (Auer, 2000; Baumeister and Steven, 2000; Chiu et al., 1999, 2002; Henderson, 1995, 1998; Kühlbrandt and Williams, 1999; Orlova and Saibil, 2004; Ruprecht and Nield, 2001; Saibil, 2000; Subramaniam and Milne, 2004; Tao and Zhang, 2000). A benchmark in single-particle image analysis of a small membrane protein complex was the study of the 290-kDa transferrin receptor (Cheng et al., 2004). Although cryo-EM and image analysis were performed on a soluble ectodomain of the receptor in the absence of detergent, the resolution of 7.5 Å was impressive and demonstrated the potential for single-particle analysis to achieve higher resolution.

For cryo-EM, the sample on the grid is blotted to near dryness with filter paper and then immediately plunged into an ethane slush using a weighted drop rod (see Fig. 8 in Yeager et al., 1999). The freezing process occurs so rapidly that the buffer becomes vitrified, which is an excellent method to preserve native biological structure to high resolution. The specific steps for preparing frozen-hydrated specimens and transferring them to the electron microscope are detailed in Yeager et al. (1999) and will not be repeated here.

The blotting time is a critical variable. Insufficient blotting will result in a layer of ice that is so thick that it is opaque to the electron beam. If the blotting time is too long, the grid will have dry patches, the ice thickness will be too thin to support the sample, or the entire grid will be dry. Automated freezing devices (e.g., the Vitrobot, FEI) allow more control of the blotting time with the goal of attaining a more reproducible ice thickness.

After freezing, the grid is transferred to specimen boxes for long-term storage in liquid nitrogen, or the grid is transferred to a cryo-stage, which maintains the specimen at −180° after insertion into the cryo-microscope. The first step in cryo-EM is to assess the ice thickness to determine whether it is optimal for recording high-resolution images. It is not uncommon for contaminating areas of crystalline ice to form during freezing or transfer of the grid (Fig. 15.3, circles). Crystalline ice is usually easy to recognize in the images and can be distinguished from sample precipitates by observing the characteristic diffraction patterns in the diffraction mode of the microscope (Chiu, 1986; Dubochet et al., 1988; Unwin, 1986). The optimal ice thickness will be just sufficient to envelop the particles without exposing them at the air–water interface during blotting and freezing. If the ice is too thick, the actual magnification of the particles may vary in the micrographs due to the different depths of the particles in the ice. In some instances, such as virus particles that are several hundred Å in diameter, it is easy to assess the ice thickness because overlapping particles can be visualized. As the ice layer thins in the center of the hole, the particles tend to partition to the edges of the hole. This usually does not happen with integrins due to their small size compared with the thickness of the ice.

With reference to membrane protein samples such as full-length integrins, the presence of detergents greatly complicates the generation of cryo-grids with optimal ice thickness. The diminished surface tension in detergent solutions increases the possibility of blotting the grid to complete dryness. In attempting to compensate for this, we frequently encounter the opposite problem of ice that is too thick and opaque to the electron beam. Even when the ice is thin enough to allow imaging, it is often still too thick, and the diffuse scattering from the frozen aqueous buffer severely degrades the signal-to-noise ratio in the images. This is a particular problem with integrins, which are relatively small and require images with optimal contrast to be visualized. The majority of grids we prepare, even when using the controlled environment of the Vitrobot (FEI), are either completely inadequate or contain extensive regions that cannot be imaged. In addition, only a fraction of the images contains sufficient contrast to enable reliable visualization of integrins.

5. Low-Dose EM

Negative-stain EM is performed with particles on a continuous carbon layer (Fig. 15.2). Regions are identified with a uniform spread and minimal precipitation of stain and minimal sample aggregation. Rather than using a

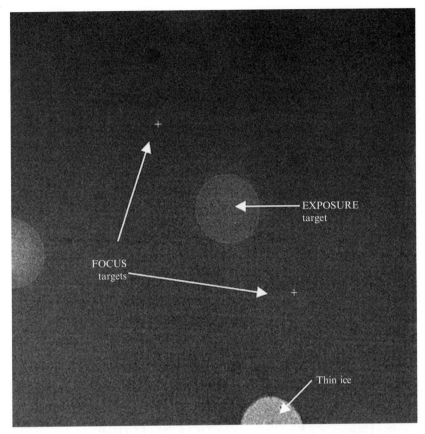

Figure 15.5 Low–dose EM. Almost the entire dose applied to the target occurs during the recording of the image in EXPOSURE mode at a typical magnification of 30,000 to 100,000× and a dose of 5 to 30 e$^-$/Å^2. In the center of the image is a hole with optimal ice thickness that was identified in SEARCH mode at low magnification using a very dim beam. The center of the hole is substantially darker than a nearby hole (thin ice). Having centered the target area, the beam is shifted about one beam diameter in either direction, and the magnification is typically increased to more than 100,000×. In this FOCUS mode, corrections are made for astigmatism, the level of defocus is set, and an assessment of drift of the stage is made. The biological specimen would be immediately incinerated if illuminated with the conditions used in FOCUS mode. The holes are 1.0 μm in diameter.

continuous carbon layer, cryo–EM of single particles is performed using a carbon film that is fenestrated with holes (Fig. 15.5). Carbon films with holes of defined size and location can now be generated by photolithography (Quantifoil substrates) (Ermantraut et al., 1998), which has certainly facilitated automated image acquisition (Carragher et al., 2000; Potter et al., 1999; Suloway et al., 2005).

Although the maximum magnification on current electron microscopes (~300,000×) allows visualization of individual atoms in hard materials, the exquisite beam sensitivity of biological materials limits the magnification to roughly 30,000 to 100,000×, so that the dose delivered to the specimen during imaging is ~5 to 10 $e^-/Å^2$. The approach used for visualizing biological materials is referred to as low-dose imaging (Fig. 15.5), in which almost all of the dose is delivered during the recording of the image (and not during identification of the field to be imaged or during focusing, corrections for astigmatism, and monitoring for stage drift). This approach is essential for high-resolution cryo-EM structure analysis so that the native structure of the macromolecular complex is preserved.

Low-dose operations involve the following three sequential imaging conditions: (1) SEARCH mode, where a target is identified at very low magnification with a very dim beam, so that the dose applied to the specimen is minimal; (2) FOCUS mode, in which a region adjacent to the target is imaged at high resolution to allow focusing and astigmatism adjustments, and an assessment of drift or movement of the stage is made; and (3) EXPOSURE mode, where the final image is recorded (Fig. 15.5). The specific steps for performing low-dose EM on Philips/FEI electron microscopes are described in Yeager et al. (1999) and will not be repeated here. Recently, there have been major advances in automating low-dose electron microscopy, which has had a major impact on increasing the efficiency and reliability of recording high-resolution images (Carragher et al., 2000; Potter et al., 1999; Suloway et al., 2005).

6. IMAGE APPRAISAL, PARTICLE SELECTION, AND PREPROCESSING

Particle selection is perhaps the most difficult procedure for single-particle studies of integrin complexes. Integrins are quite small for EM and thus inherently difficult to recognize in electron micrographs (Figs. 15.2 and 15.3). In addition, the molecules are highly asymmetric, and occur at all possible orientations in the micrographs. The combination of asymmetric shape and multiple viewing angles makes recognition of distinctive outlines difficult, due to the large number of very different 2D projection views of the complex. Integrins are also not very compact, and the distribution of mass makes recognition of domains extremely difficult. In cryo-micrographs, we find that the head domains (β-propeller and βA domains [Xiong et al., 2001]) are primarily visible as distinct regions of optical density (Figs. 15.3 and 15.6). The situation is only slightly improved in negative stain (Fig. 15.2).

For these reasons, our studies rely on the fact that our samples are more than 90% pure as estimated by electrophoresis. Thus, any projection in the micrograph is likely to be produced by integrins, and our principal strategy

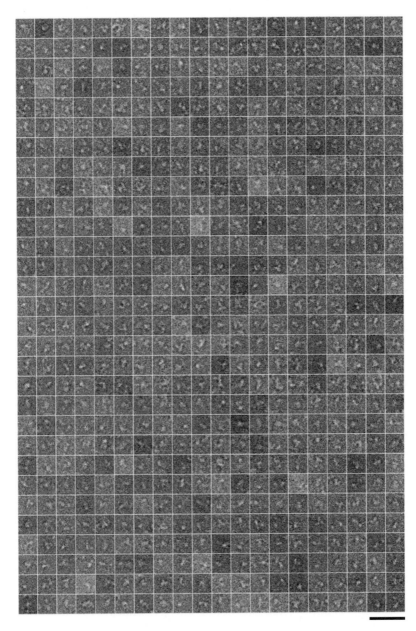

Figure 15.6 Gallery of frozen–hydrated particle images of $\alpha_{IIb}\beta_3$ automatically selected from several cryo-micrographs by the EMAN program *boxer*. Scale bar is 500 Å.

is to initially select any identifiable particle, regardless of its shape. The only caveat is to avoid selecting two or more integrins near one another, overlapping particles, or obvious aggregates (Figs 15.3, squares). This requires a sample that is sufficiently dispersed that the possibility for overlap is unlikely. Visual inspection of the boxed particles (Fig. 15.6) is necessary to ensure that protein density does not extend beyond the boundaries of the box. Only a single head domain should occur in any particle projection.

An alternative method that we have occasionally used is to manually select particle images that have a distinctive appearance and then aligning and averaging only those images. This analysis may be used to answer only very specific questions concerning the structure of a particular projection. Averaging a number of projections provides a much less noisy version of the projection, but interpretation of the projection is problematic. In contrast, a 3D analysis assigns every particle to a projection of a single 3D map. There are cases, however, where the analysis of a specific projection is informative. For instance, in our study of the $\alpha_V\beta_3$ ectodomain bound to fibronectin domains FN9–EDB–10 (Adair et al., 2005), we found that the FN fragment was conformationally flexible. To verify that the full FN fragment was bound to a bent conformation of $\alpha_V\beta_3$, side views of $\alpha_V\beta_3$ that clearly displayed an extended FN molecule were selected. The analysis of this subset clearly indicated that FN was binding to the bent form of the integrin. The difficulty with this analysis is that one cannot get a good feel for the relative prevalence of this particular conformation, as only a subset of particles was processed. To assess the major conformation, an accurate 3D model is required to provide projections that can identify all of the particles, rather than a hand-selected subset.

The production of contrast in electron micrographs is modeled by the contrast transfer function (CTF), which determines how the specimen image is modulated by the optics of the electron microscope (Fig. 15.7). Contrast in electron micrographs is enhanced by acquiring the image in an underfocused condition. This has the desirable effect of increasing the contrast, but at the expense of creating a phase reversal, where the phases of the amplitudes in the Fourier transform of the image are inverted by 180 degrees at certain spatial frequencies, which will depend on the discrepancy between true focus and the underfocus setting. EM studies must account for this, either by performing a low-pass Fourier filter on the image to exclude all inverted frequencies or by preprocessing the image to computationally restore the phases to the correct sign. Additional factors include a systematic drop-off in amplitude of the structure factors at high frequencies (the envelope function). If determined, the envelope function may be corrected during subsequent processing.

1. Examine electron micrographs recorded on Kodak SO163 film on a light box with a hand magnifier to determine if they contain recognizable particles. The integrin particles will be randomly oriented, which, when combined with their asymmetric shape, will make recognizing distinctive projections

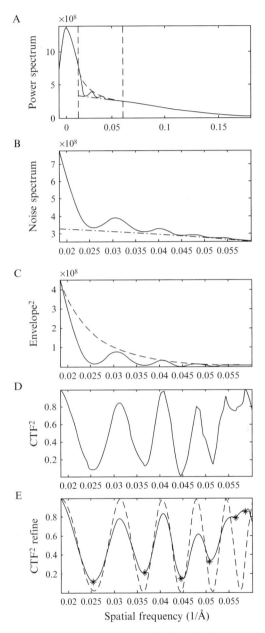

Figure 15.7 CTF estimation and correction. (A) Circularly averaged power spectrum from the micrograph. The vertical dashed lines indicate the region displayed in the following panels. (B) Estimation of background (dashed line). (C) Power spectrum after background subtraction showing the estimation of the envelope function (dashed line). (D) Power spectrum after background subtraction and correction for envelope function. (E) Estimation of defocus value (dashed line). Profiles were generated using the program ACE (from Mallick *et al.*, 2005).

difficult at this stage. Check the negatives for obvious drift or astigmatism. For negative-stain images, an optical diffraction bench (see Fig. 14 in Yeager *et al.*, 1999) is very useful to check for both conditions (see Fig. 12 in Yeager *et al.*, 1999), as the diffraction pattern from the carbon background will be greatly impacted by either condition. Film images should be digitized on a high-resolution scanner at a step size several times greater than the ultimate resolution that one hopes to achieve (see Fig. 15 in Yeager *et al.*, 1999). For integrins, we use steps between 3.0 and 4.7 Å/pixel, and we currently digitize micrographs using a Zeiss Axiophot scanner or record digital images directly on a CCD camera attached to the electron microscope.

2. We perform particle selection with the program *boxer*, which is distributed with the EMAN software package (Ludtke *et al.*, 1999) (http://blake.bcm. edu/EMAN/). Since integrins are so difficult to identify in micrographs, *boxer* contains median filters at various strengths, of which the 7 × 7 and 9 × 9 filters are most useful for enhancing the contrast so that integrin particles can be identified. To manually select particles, choose a large box size (~300 Å) and center an identifiable region of density within the box. Because the box is much larger than the maximum dimension of an integrin heterodimer, there should be no protein density near the edges of the box. By default, the output of *boxer* is a stack file in IMAGIC format (van Heel *et al.*, 1996). Two files are created that share the same prefix and have .img and .hdr suffixes. This format is used by subsequent EMAN routines.

3. An alternative to manual particle selection is to use an automated routine. We have used the routine in *boxer* that searches the micrograph with a set of user-provided references. The references may be either class averages produced by reference-free alignment and classification of hand-selected images (see below) or generated from a 3D map. In addition, the EMAN routines *pdb2mrc* followed with *makeboxref.py* may be used to generate a set of reference images from a PDB file. The *boxer* interface allows a number of parameters to be set interactively by the user. The program runs the references against a section of the micrograph selecting particles within it. It then presents the user with a set of sliders that control various parameters, updating the selected particles on the fly. When the operator is satisfied with the particles selected in the subregion, the parameters are used to select particles in the entire micrograph (Fig. 15.6). The parameters often have to be adjusted from micrograph to micrograph due to variability in the thickness of the stain and ice.

4. The final preprocessing stage is the determination of the CTF parameters for each micrograph (Fig. 15.7). This may be performed interactively with the EMAN program *ctfit*. Particle data sets for each micrograph are read into the program, which has controls to input parameters for defocus value, the envelope function and the background profile. The signal levels in cryomicrographs may be so low that it is difficult to accurately determine the defocus levels. In this case, regions of carbon

surrounding the hole may be selected. The use of carbon, free of protein, has advantages in the determination of the envelope function, since the measured diffraction pattern from the particle set includes the protein structure factors as well as the CTF, whereas the structure factors for carbon are constant. Once the CTF parameters have been determined, the particles are processed to restore phases that were flipped by the CTF. Additional CTF parameters are written into the file header.

7. Initial Particle Analysis and Generation of Reference-Free Aligned Class Averages

Whether you are examining negatively stained or unstained, frozen-hydrated specimens, the majority of the raw integrin particles are usually too noisy to interpret. Our initial analysis of selected particles begins with a reference-free classification and averaging stage, in which individual particle images are aligned and grouped into equivalent views and averaged (Fig. 15.8). The procedure has been implemented in both the SPIDER (Frank *et al.*, 1996) and EMAN (Ludtke *et al.*, 1999) image processing packages. The first step is alignment of the entire particle set to a common center and rotational angle. An initial particle is randomly selected from the data set as a reference, and each subsequent particle image is compared statistically. The image is shifted in x and y and rotated in the plane of the image (Euler angle ω) to maximize the value of the cross-correlation function. To avoid bias, the process is iterated, with the resulting average used as a reference in the next iteration. The iterations halt when the averages converge. If the particles are in random orientations or a large number of different orientations, the resulting global average will not be particularly meaningful. An additional stage to separate particles into classes with equivalent views is then necessary. This stage performs a multivariate statistical analysis in the global variance map to identify covariance factors, which may be used to classify individual particles. When the aligned particles are averaged, an average is generated for each pixel (displayed together as the averaged image), and the resulting variance is produced for each pixel. Eigenvectors are extracted from this variance map representing pixels with correlated variance. The resulting vectors represent the covariance in the image, or relative correlation between pixel values in the original data set. Particles are classified based on the relative strength of contribution of each of these eigenvectors in the particle image. The procedure outlined below uses routines in the SPIDER package. We have generated reference-free classes both with EMAN (via the routine *startnrclasses*) and with SPIDER. The SPIDER routines allow greater control over the procedure, and are outlined below. Boxed particles in IMAGIC format (van Heel *et al.*, 1996) may

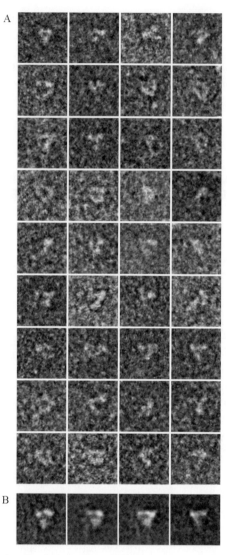

Figure 15.8 Example of reference–free alignment and averaging. (A) Raw particle images of the $\alpha_V\beta_3$ ectodomain in a buffer containing Mn^{2+}. The grid has been stained with uranyl acetate. Particles were selected based on the appearance of being side views of the bent conformation. (B) Class averages created from a data set containing the particles in (A). The boxes are 280 Å on a side.

be converted to SPIDER format with the EMAN command *proc2d*. For Intel–based workstations, the correct flag is *spiderswap*, which creates a single SPIDER stack. Individual images in a SPIDER stack are accessed from within SPIDER with the file@particle notation.

1. Perform a reference-free alignment of the boxed particle images with the SPIDER command AP SR. Use a randomly selected image as the reference. This procedure performs rotational as well as translational alignment, and is necessary to center the particles.

2. Rotate and center the particles using the x,y translational and ω rotational parameters determined by AP SR. It will be necessary to loop through the particle set to (a) read the x,y translations and rotation from the AP SR results file and (b) use RT SQ to rotate the particle image.

3. The SPIDER routine AP CA is then used to rotate the particle images and simultaneously perform the K-means classification. The most important parameters to set are the outer ring to be used in the alignment and the number of classes to be generated. As the outer ring is enlarged, more background pixels will be included, which will decrease the accuracy in determining the translational and rotational parameters. The outer ring should be set to a radius that will just include the boundary of the particle. For highly asymmetric particles such as integrins, this can be difficult to judge. The radius will depend on the shape of the complex as well as the types of views that have been selected. If the number of classes to be generated is too large, then each class will not have a sufficient number of particle images to have a reliable averaged image for that class. Conversely, if the number is too small, disparate views will be averaged. The number of classes is usually selected to have 50 to 100 particle images.

4. AP CA generates a list of rotations for each particle, as well as separate documents listing the particles in each class. To generate the class averages, it is necessary to loop through the particle images again and use RT SQ to rotate each (with 0,0 given for the translation). Finally, the SPIDER routine AS DC is used to generate averages from the class documents for each class.

8. THREE-DIMENSIONAL STRUCTURE REFINEMENT

The most common method for generating a 3D density map uses an iterative, projection-matching approach. In this approach (shown schematically in Fig. 15.9), a gallery of 2D back projections of a starting 3D model are generated at specific angular increments—the smaller the increment, the higher the potential resolution of the final 3D map. By this approach the Euler angles (θ, φ) for each back projection are known. In the most time-consuming stage of the algorithm, each particle is classified into one projection class by an exhaustive cross-correlation search against the gallery of all 2D projections of the 3D model. The cross-correlation search includes variation of the in-plane rotational angle ω and x,y translation of the particle image for each of the projections. Following classification, particles grouped

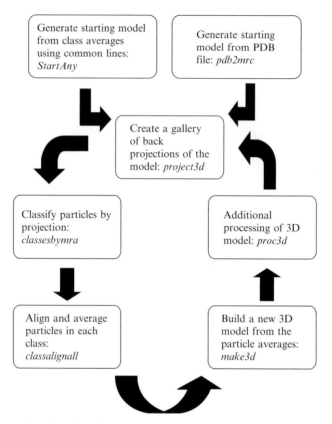

Figure 15.9   Data flow for refinement of a 3D structure using EMAN software (Ludtke *et al.*, 1999). The subprograms run by the *refine* command are italicized.

into a common projection of the model are averaged. At this stage, envelope corrections are commonly applied. By knowing the Euler angle relationships of the averaged classes, a new 3D map can be generated. For the next cycle of refinement, a new gallery of back projections is generated from the 3D map, and the process is repeated.

1. Generate an initial model to use in the refinement. There are several methods to do this, two of which we commonly use in deriving 3D maps of integrins.

   a. The first method is to employ the class averages generated above. The EMAN program *StartAny* may be used to generate an initial model from a set of class averages using the method of common lines. The common lines method generates a series of one-dimensional (1D) projections of the averages at various Euler angles. Any two projections from the same 3D object will have one of these 1D projections in common, and the angles used to generate the common 1D projection provide the orientation. The principal input variable is the selection of class averages to be used, typically 10 to 20. The method does not produce accurate starting maps with integrins due to the very low signal-to-noise ratio. Nevertheless, they are an adequate starting point for the refinement to proceed.

b. The second method is to calculate a molecular envelope from a pdb coordinate file by use of the EMAN program *pdb2mrc*. Input parameters are the pdb file and flags specifying the size of the box (box=), the voxel size (apix=), and a resolution for the final map (res=). The resolution is given in EM units, an integer N, where the resolution is N ÷ (size of box). The flag center is frequently necessary when the X-ray coordinates from the pdb file are not centered in the unit cell. It is important that the box size exactly equals the box size for the raw particle images, and the spacing should of course be the same as well.

2. The basic refinement loop is run with the EMAN command *refine*. The *refine* command generates a large number of files and requires starting files to have specific names, so it is best to run this routine in a unique subdirectory for each refinement. *Refine* requires that the raw particle images (which must be named start.hed/start.img) and the 3D model (which must be named threed.0a.mrc) are located in the same directory from which the routine is run. The main options to refine are an integer, to indicate the total number of iterations, and mask= to provide the radius (in pixels) that will be applied at various stages in the refinement. *Refine* automates the refinement loop, providing a wrapper for other programs. The following options are supplied on the command line, which apply at various locations in the loop:

project3d creates a set of projections from the final 3D map from the previous refinement cycle (threed.0a.mrc for the first round). The most important option here is ang=, which gives the angular spacing for the model. The number of projections produced is not easily determined from the angle, so it is best to run *project3d* on its own to see how many classes will be produced. For integrins, we prefer to have at least 50 particles per class on average (# of particles ÷ # of classes). Smaller angular spacings are better, if the number of particles is sufficiently large.

classesbymra performs the cross-correlation search between the reference projection set and the raw particles. It produces files that list the particles for each projection. The option slow changes the target function used by the search, which is sometimes useful.

classalignall aligns each of the classes to produce a class average. Corrections for the contrast transfer function (CTF) and the envelope function are applied at this stage (ctfc=). Due to the small size of the integrin particles, the 3D maps are typically at low resolution (20 to 30 Å), and we often omit this (using the option median instead). An important option is classiter=, an integer that specifies the number of iterations used to generate the class averages. The iteration proceeds by aligning all the particles to the initial projection and averaging them. To avoid bias from the initial projection, each new average is used to align the particles in successive rounds. We find that there is little bias using values of 8 to 12. An important parameter at this stage is classkeep=, which is a sigma multiplier for retaining images in the average. Images with sigma values relative to the average that are outside this parameter are excluded from the average. The images that are included/excluded may vary during the iteration as the average image changes. A value of 0.2 is stringent, whereas 0.8 is less stringent.

make3d generates the next 3D map from the class averages. The pad= parameter pads the projections in Fourier space to avoid Fourier series truncation artifacts. We often use a value 50% larger than the particle sizes. The parameter hard= is used to reject projections from the reconstruction when the phase residual exceeds the parameter. We typically use a value of 25.

proc3d is applied at the end to perform some specified processing of the model. We sometimes use filt3d=, which applies a low-pass Fourier truncation filter to the map. This may be useful with noisy maps, such as produced by cryo-EM data. The integer parameter in filt3d= is in EM units. Convert to real-world units by dividing the EM unit value by the length of the side of the box in Å. Another post-processing procedure that we have found useful is the amask= automasking feature. This requires three comma-separated numbers: the first is a radius that will contact some "good" density (protein, not noise), the second is a map cut-off or threshold value, and the third is a pad integer to extend the map a few voxels beyond the threshold. Basically, this procedure is analogous to "solvent flattening" for refinement in X-ray crystallography and attempts to set regions of the background to a uniform gray level to reduce contributions from noise. The threshold is the critical parameter, which is used to specify a contiguous isosurface beyond which the data are masked out. The xfiles= parameter must be used with *automask* to set the isosurface value of 1 to correspond to the protein volume. Xfiles= also takes three comma-separated values, the Å/pixel value, the mass of the protein in kDa, and a resolution to align to. When applied, an isosurface of 1 should enclose 100% of the protein volume. We often give amask= a threshold value of 0.5, which cuts out much of the background but includes all of the protein. Caution should be exercised with this parameter, as one might mask out weak extended density, such as thin transmembrane and cytoplasmic domains.

3. Refinement ceases when the resulting maps have converged to a common solution. We find that maps will converge to a rough solution within six to eight rounds, with noisy data taking longer. When starting from a known pdb file, we always iterate far longer than is required (e.g., ~20 rounds) to eliminate model bias. We have found that 12 rounds are sufficient with negative-stain data to completely eliminate model bias. Resulting models will only resemble starting models if the underlying data have this conformation.

4. The convergence of refinement is often checked by comparing the Fourier shell correlation (FSC) between preceding and subsequent rounds (e.g., see Fig. 4 in Ludtke *et al.*, 1999). This may be run manually with *proc3d* or from the graphical EMAN program.

5. The overall resolution of a final map is typically estimated by calculating the FSC, comparing two maps that are generated using half the data (Fig. 15.10). Raw particles are split into two stacks depending on even or odd numbering, and separate class averages and final maps are calculated. The two maps are compared in resolution shells in Fourier space. At very low resolution, the two maps are nearly identical and will have correlation coefficients close to 1. In moving to higher-resolution shells, the correlation coefficient will fall, and we use an arbitrary cut-off value of 0.5 to indicate the resolution of the 3D map. (Of course, the final map will include twice as many particles, so the 50% cut-off is a conservative estimate of the resolution.) The EMAN program *eotest* (using the same options as *refine*)

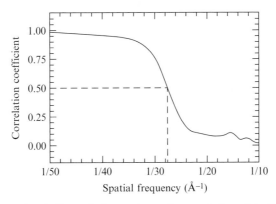

Figure 15.10 Fourier shell correlation to determine resolution of the 3D map of $\alpha_V\beta_3$. The total data set of particle images was divided randomly into two equal groups. The images in each data set were independently aligned with projections of the final 3D model, and two independent maps were correlated in resolution shells in Fourier space. An arbitrary correlation coefficient value of 0.5 is used to assess the resolution of the map (~27 Å).

calculates the two maps. The program *proc3d* is then used to calculate the FSC. Note that the output shells are in EM units (N/box side).

9. EVALUATION OF THE REFINEMENT

With unknown structures, it is always necessary to evaluate a final map to establish that it represents the bona fide 3D structure of the macromolecular complex. The initial quality of the refinement is given by the FSC (Fig. 15.10). The FSC calculation, while rapid and convenient, should not be relied on as the sole measure of the quality of the 3D reconstruction. The evaluation is biased in that while the data are randomly split into two sets, they have already been classified by the same set of map projections.

With a sufficiently large data set, two completely independent sub-data sets can be treated independently from the beginning of the analysis in order to generate two independent 3D maps. The FSC of these independently determined maps can then be calculated. We employed this method in our study of the unliganded $\alpha_V\beta_3$ ectodomain (Adair *et al.*, 2005). An extended refinement conducted on each half of the data yielded final maps which, when compared by FSC, gave the same resolution estimation as the refinement using the full data set. This approach of solving the 3D structure twice using completely independent, large data sets is more tractable now that software such as the LEGINON package is available for automated recording of EM images (Carragher *et al.*, 2000; Potter *et al.*, 1999).

The FSC is a good measure of the internal consistency of the data, and is thus a good measure of the Fourier domain frequencies that may be relied on in the data, but might give misleading results if the map is incorrect. This problem is more likely to occur if the refinement has been set up with a large number of particles and a relatively coarse projection sampling angle, such that a large number of particles are assigned to each projection class. In this case, a situation where multiple conformations exist in solution may nevertheless provide a reasonable evaluation by FSC since the number of particles being averaged in each half of the data set will still be very large, so that the averages for each of the halves will still be very close to one another. If there are, for instance, 500 particles on average in each class in the refinement, each half will have an average of 250 per class. If there are projections from multiple conformations in each class, the averages from each half will tend to resemble one another since the probability is good that equivalent numbers of both conformations are in each half, so that both averages are equally incorrect. A finer angular increment should be chosen to avoid this.

The correctness of the results can also be evaluated by reviewing the classified raw particle images (Fig. 15.11). The graphical EMAN program permits

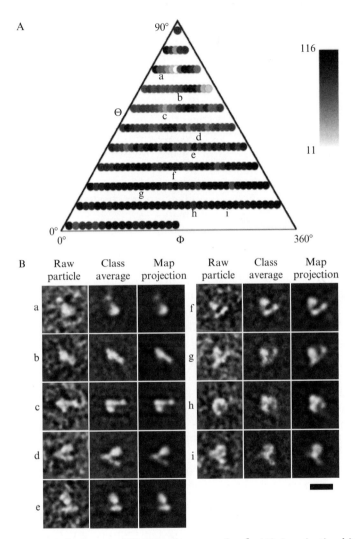

Figure 15.11 Results from an EMAN refinement of $\alpha_V\beta_3$. (A) A projection histogram displaying the number of particles at particular θ,φ orientation angles for the final round of refinement. Each circle represents a specific projection, and the grayscale is proportional to the number of particles belonging to that class, ranging from 11 to 116. (B) Examples of particles from the final round of refinement. Raw particle images are displayed in the left column, class averages in the middle column, and back projections of the final 3D map in the right column. The lower-case letters mark the positions of the class in the histogram in (A). Scale bar is 100 Å. (Adapted from Adair *et al.*, 2005.)

convenient viewing of particle classes with the attendant projections. EMAN generates a histogram showing the distribution of particles with different θ,φ orientation angles (Fig. 15.11A). One can easily assess whether there is an even distribution of particle orientations that have been included in the analysis.

For instance, the particles used for the analysis of $\alpha_V\beta_3$ were selected automatically, and there is an even distribution of orientations (Fig. 15.11A). In our previous analysis of $\alpha_{IIb}\beta_3$, the particles were selected manually, and the histogram showed that some orientations are not as well represented in the histogram, presumably because these orientations were difficult to discern in the micrographs (see Fig. 15.1C in Adair and Yeager, 2002).

For particular θ,φ Euler angles, the raw particle images can also be compared with the class averages and the back projections of the 3D map in the same view (Fig. 15.11B). Due to the low signal-to-noise ratios in micrographs of integrins, it will be difficult to easily observe motifs in every particle. Nevertheless, there should be concordance in the features and an increase in clarity in moving from single images, to the 2D class average, to the back projection of the final 3D map.

Another option for additional evaluation of the final map is to perform the refinement with multiple starting maps. Different maps may be generated by use of (1) the common lines method, using a different set of class averages to generate the 3D map, or (2) a pdb file, using a program such as O (Kleywegt and Jones, 1994) or chimera (Pettersen *et al.*, 2004) to manually modify the pdb file or to choose a related, but conformationally distinct protein. With integrins, two such models for the integrin heads are provided by the $\alpha_V\beta_3$ (Xiong *et al.*, 2001) and $\alpha_{IIb}\beta_3$ (Xiao *et al.*, 2004) X-ray structures, although the complete ectodomain in the latter case must be generated by docking the $\alpha_V\beta_3$ X-ray structure into the EM map. We employed a variation of this technique in evaluating the maps generated in our study of the $\alpha_V\beta_3$ ectodomain (Adair *et al.*, 2005). For the soluble ectodomain, we observed a bent conformation in buffers containing Mn^{2+} for both the free ectodomain and in a complex with the FN fragment. It was reassuring that when an unbent, extended structure was used as a starting model, the final map converged to a size and shape equivalent to that used when we started with a bent model.

A modification of this approach was used to evaluate the thin extension of density in the 3D map of $\alpha_{IIb}\beta_3$ that we ascribed to the transmembrane and cytoplasmic domains (Fig. 15.12). This density was computationally deleted, and the truncated 3D map was used as a refinement model. The reappearance of this density in the refined 3D map suggested that this was a bona fide feature of the structure.

A possible reason for the failure of a refinement to converge is that the macromolecular complex may have multiple conformations in solution. Multiple conformations have been reported from integrin ectodomain studies (Adair *et al.*, 2005; Takagi *et al.*, 2002), although in each case a single conformation seems to be favored. A sample with roughly equal distributions of quite different conformations will not converge on a single final map. In this situation it will be necessary to classify the raw particles based on projections from two or more maps. The EMAN command *multirefine*, which has the

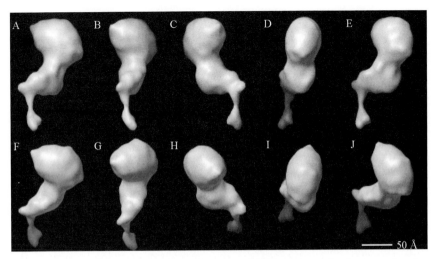

Figure 15.12 Surface-shaded 3D density map of full-length $\alpha_{IIb}\beta_3$. The isosurface has been set to enclose 100% of the expected volume of the complex. (A to E) Series of views sequentially rotated by 45 degrees. (F to G) The same orientations as in A to E, but rotated toward the viewer by 45 degrees.

same parameters as *refine*, may be used for this purpose. The only additional parameter on the command line is the number of maps to be used. A directory to run the refinement is set up as before, but with multiple subdirectories, named sequentially 0, 1, and so on, to contain different threed.0a.mrc maps. A refinement result may be used to permanently split the data. The particle classes are stored in each subdirectory. The *proc2d* command may be looped over the class files to create a new start.hed file for each subdirectory. Care must be taken to give the *proc2d* command the first=1 parameter to ensure that the map projection is not included in the new file. The results of the split data analysis may be evaluated by refining the data set classified as being projections from a single map, using as a starting map the other map that was used to split the data. The result should again converge to the same solution.

10. PSEUDOATOMIC MODELING OF EM DENSITY MAPS

A powerful strategy in the structural characterization of macromolecular complexes that may not be amenable to 3D crystallization is the wedding of low-resolution 3D structures provided by cryo-EM and high-resolution structures provided by X-ray crystallography and NMR spectroscopy (Baker and Johnson, 1996). By this approach pseudoatomic models can be constructed for macromolecular complexes that may have too much conformational flexibility for crystallization, that may be too large for NMR spectroscopy, or that may be too labile for crystallization, which may require many days to weeks. Rossman (2000) estimated that a proper fitting of a protein atomic model into EM density can reveal interesting interactions between residues and bound molecules. In addition, larger complexes that are not easily fitted into crystal lattices, unstable multiprotein complexes that become degraded in the time-frame of crystallization trials, and other complexes that may undergo structural changes in response to functional stimuli are all ideal specimens for analysis by cryo-EM combined with pseudoatomic modeling (Rossman, 2000). A single template atomic structure may be available for one conformational state that can be used for deriving pseudoatomic models for multiple intermediate conformational states. In turn, such intermediate states may be amenable to cryo-EM but may not be available in sufficient quantities or be sufficiently stable for 3D crystallization. This approach is exemplified by our analysis of the native structure of the tetravirus NωV, for which crystal structures were available for the individual subunits from the crystal structure of the mature capsid particle. However, the procapsid assembly intermediate could not be crystallized. Nevertheless, a pseudoatomic model for the procapsid was derived by cryo-EM and icosahedral image analysis combined with molecular modeling (Canady et al., 2000).

There are a number of quantitative approaches for docking known protein structures into EM maps, which can be based on local (Jiang et al., 2001; Roseman, 2000) or global density correlation (Volkmann and Hanein, 1999; Wriggers et al., 1999), or a combination of both (Grimes et al., 1997; Kikkawa et al., 2000). The large conformational changes seen in some of our integrin maps, and the low variation in internal density seen with negative stain, have led us to concentrate instead on interactive manual fitting. A number of software packages exist for interactive fitting of pdb files within EM density envelopes. We have used O (Kleywegt and Jones, 1994), AVS (Sheehan et al., 1996), and chimera (Pettersen et al., 2004). In all of these programs, a transparent or semitransparent representation of the map at a given isosurface level is displayed simultaneously with individual domains from the high-resolution structures (Fig. 15.13). X-ray structures have been determined for two related integrins: the ectodomain of $\alpha_V\beta_3$, where both chains have been terminated immediately prior to the putative transmembrane domain (Xiong et al., 2001), and a severely truncated $\alpha_{IIb}\beta_3$ ectodomain containing only the β-propeller domain of α_{IIb} and the A-like, hybrid, and PSI domains of β_3 (Xiao et al., 2004). The α and β domains of the ectodomain of $\alpha_V\beta_3$ contain 12 independently folded subdomains, which can be manipulated within the EM density envelope. The constraints on fitting include the boundary of the EM density map, the known connectivity of the subdomains, and the known location of the C-terminal domains of the α and β chains that are close to the membrane. A stereo display is advantageous, but not necessary, for manipulating the high-resolution structures to

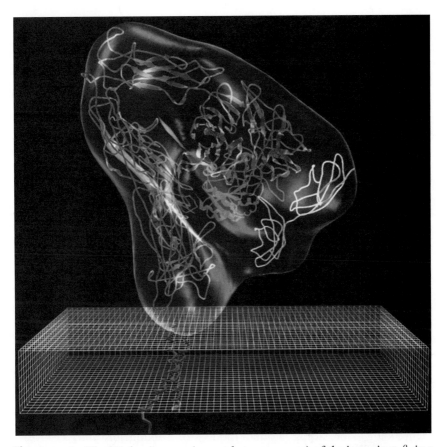

Figure 15.13 The 3D density map (grayscale transparency) of the integrin $\alpha_V\beta_3$ in a complex with fibronectin determined by EM and single-particle image analysis. Fibronectin domains 9 and 10 were localized by examining a difference map between the integrin with and without a bound fragment of fibronectin. The X-ray structures of the α_V (blue) and β_3 (red) chains (Xiong *et al.*, 2001) have been docked into the EM density envelope. Additional density (lower right) can accommodate fibronectin domain 10 adjacent to the ligand binding site (green) as well as domain 9 at the synergy site (yellow). The complex is shown adjacent to the white grid box, which represents the 30-Å–thick hydrophobic portion of the cellular membrane across which signals are transmitted. An R-handed α-helical coiled-coil (purple) is shown spanning the lipid bilayer, as we have proposed for the inactive conformation of $\alpha_{IIb}\beta_3$ (Adair and Yeager, 2002). (Modified from Adair and Yeager, 2005.) Graphics by Michael E. Pique and Mark Yeager. (See color insert.)

evaluate goodness of fit. Two conditions need to be fulfilled before this docking procedure can be started: (1) the EM map needs to be converted into a format which the program can read, and (2) the correct isosurface contour needs to be determined. We employ the program *em2em* (supplied with the Imagic program suite, van Heel *et al.* 1996) to convert between map formats, particularly between mrc and CCP4 formats, which are related but distinct. We subsequently use the program *mapman* (Kleywegt and Jones, 1993), supplied with O, to convert between CCP4 format and the DSN6 (brix) format required by O. The isosurface is typically set to a contour that encloses 100% of the protein mass, where the density is set to some typical value appropriate for the complex (e.g., a partial specific volume of $\sim 0.74 \text{ cm}^3/\text{g}$ for proteins). The program *VOLUME* supplied with EMAN will calculate a threshold for this value, based on a supplied mass and voxel spacing, and the assumed protein density.

Several approaches have been developed to fit atomic models into EM maps. These vary from manual/interactive domain fitting and modeling of available crystal structures, such as our analysis of $\alpha_{IIb}\beta_3$ (Adair and Yeager, 2002) and $\alpha_V\beta_3$ (Adair *et al.*, 2005) to the application of homology modeling to experimental EM maps to solve the entire 80S ribosome from *Saccharomyces cerevisiae* and predict subunit–subunit interactions (Spahn *et al.*, 2001). Technically, all these fitting methods involve either time-consuming, human manual docking (Cheng *et al.*, 1995; Sali, 2002) or quantitative docking, based on both local (Jiang *et al.*, 2001; Roseman, 2000) and global density correlation (Volkmann and Hanein, 1999; Wriggers *et al.*, 1999) or a combination of both (Grimes *et al.*, 1997; Kikkawa *et al.*, 2000). However, as noted by Roseman (2000), these methods are successful for highly constrained symmetrical systems or cases in which a specific component can be isolated from the map of the complex. Therefore, alternative, robust methods are essential to analyze low-resolution cryo-EM maps of asymmetric structures with smooth surface topography. For this purpose, it is desirable to combine automated global docking techniques with protein–protein patch prediction methodology for fitting atomic models of subdomains into cryo-EM maps. We are currently exploring the applicability of the computational biology package ICM (Internal Coordinates Mechanics) (Abagyan *et al.*, 1994), to model the protein sequences of the integrin $\alpha_{IIb}\beta_3$ and predict interaction patches between different domains. ICM is appealing because the calculations include energy contributions from van der Waals interactions, electrostatic interactions, hydrogen bonding, torsional, and solvation. We are encouraged by an analysis that used ICM for modeling α-helices within the membrane proteins glycophorinA, KcsA, and MscL (Kovacs *et al.*, 2007). In this approach, a simulated cryo-EM map at 6 Å in-plane resolution was used as a penalty term (acting on the backbone atoms only) during the energy minimization stage in order to keep the helices close to their experimental positions as given by the density map. This approach greatly reduced the number of conformations that had to be searched for each helix. The predicted conformations for GpA, KcsA, and MscL were close to the experimental structures, with rmsds between 0.9 and 1.9 Å for the backbone atoms.

11. FUTURE PROSPECTS FOR CRYO-EM

In spite of substantial recent progress in the solution of high resolution structures of membrane proteins (e.g., see the websites of Hartmut Michel, http://www.mpibp-frankfurt.mpg.de/michel/public/memprotstruct.html, and Stephen White, (http://blanco.biomol.uci.edu/Membrane_Proteins_xtal.html), it is expected that there will continue to be numerous membrane protein complexes that will not be amenable to high-resolution structure analysis by crystallographic or spectroscopic methods. Determination of the X-ray structure of the ectodomain of $\alpha_V\beta_3$ was certainly a pioneering milestone in the integrin field (Xiong et al., 2001). However, to date, there is no available crystal structure of a full-length integrin. We believe that cryo-EM and 3D image reconstruction of single particles combined with pseudoatomic modeling will continue to provide a valuable ancillary strategy to explore the structure and function of membrane proteins such as integrins. This may be especially true for integrins because it is likely that there may be multiple conformational states in the presence and absence of small molecule and physiologic ligands. Validation of the models requires the use of orthogonal techniques such as assessment of steric accessibility for antibody labeling and mutagenesis combined with activity measurements.

ACKNOWLEDGMENTS

We thank Barbie Ganser-Pornillos and Kevin Koehntop for helpful comments. Support for writing this review was provided by the National Institutes of Health (grant RO1 HL48908 [MY]) and a Wayne Green Postdoctoral Fellowship (BDA).

REFERENCES

Abagyan, R., Totrov, M., and Kuznetsov, D. (1994). ICM: A new method for structure modeling and design: Applications to docking and structure prediction from the distorted native conformation. J. Comput. Chem. **15**, 488–506.

Adair, B. D., Xiong, J.-P., Maddock, C., Goodman, S. L., Arnaout, M. A., and Yeager, M. (2005). Three-dimensional EM structure of the ectodomain of integrin $\alpha V\beta 3$ in a complex with fibronectin. J. Cell Biol. **168**, 1109–1118.

Adair, B. D., and Yeager, M. (2002). Three-dimensional model of the human platelet integrin $\alpha_{IIb}\beta_3$ based on electron cryomicroscopy and x-ray crystallography. Proc. Natl. Acad. Sci. USA **99**, 14059–14064.

Auer, M. (2000). Three-dimensional electron cryo-microscopy as a powerful structural tool in molecular medicine. J. Mol. Med. **78**, 191–202.

Baker, T. S., and Johnson, J. E. (1996). Low resolution meets high: Towards a resolution continuum from cells to atoms. Curr. Opin. Struct. Biol. **6**, 585–594.

Baumeister, W., and Steven, A. C. (2000). Macromolecular electron microscopy in the era of structural genomics. *Trends Biochem. Sci.* **25,** 624–631.

Böttcher, B., Wynne, S. A., and Crowther, R. A. (1997). Determination of the fold of the core protein of hepatitis B virus by electron cryomicroscopy. *Nature* **386,** 88–91.

Brooks, P. C., Clark, R. A. F., and Cheresh, D. A. (1994). Requirement of vascular integrin $\alpha v \beta 3$ for angiogenesis. *Science* **264,** 569–571.

Calvete, J. J. (1994). Clues for understanding the structure and function of a prototypic human integrin: The platelet glycoprotein IIb/IIIa complex. *Thromb. Haemost.* **72,** 1–15.

Canady, M. A., Tihova, M., Hanzlik, T. N., Johnson, J. E., and Yeager, M. (2000). Large conformational changes in the maturation of a simple RNA virus, Nudaurelia capensis ω virus (NωV). *J. Mol. Biol.* **299,** 573–584.

Carragher, B., Kisseberth, N., Kriegman, D., Milligan, R. A., Potter, C. S., Pulokas, J., and Reilein, A. (2000). Leginon: An automated system for acquistion of images from vitreous ice specimens. *J. Struct. Biol.* **132,** 33–45.

Cheng, R., Kuhn, R. J., Olson, N. H., Rossman, M. G., Choi, H.-K., Smith, T. J., and Baker, T. S. (1995). Nucleocapsid and glycoprotein in an enveloped virus. *Cell* **80,** 621–630.

Cheng, Y., Zak, O., Aisen, P., Harrison, S. C., and Walz, T. (2004). Structure of the human transferrin receptor-transferrin complex. *Cell* **116,** 565–576.

Chiu, W. (1986). Electron microscopy of frozen, hydrated biological specimens. *Ann. Rev. Biophys. Biophys. Chem.* **15,** 237–257.

Chiu, W. (1993). What does electron cryomicroscopy provide that X-ray crystallography and NMR spectroscopy cannot? *Ann. Rev. Biophys. Biomol. Struct.* **22,** 233–255.

Chiu, W., Baker, M. L., Jiang, W., and Zhou, Z. H. (2002). Deriving folds of macromolecular complexes through electron cryomicroscopy and bioinformatics approaches. *Curr. Opin. Struct. Biol.* **12,** 263–269.

Chiu, W., McGough, A., Sherman, M. B., and Schmid, M. F. (1999). High-resolution electron cryomicroscopy of macromolecular assemblies. *Trends Cell Biol.* **9,** 154–159.

Conway, J. F., Cheng, N., Zlotnick, A., Wingfield, P. T., Stahl, S. J., and Steven, A. C. (1997). Visualization of a 4-helix bundle in the hepatitis B virus capsid by cryo-electron microscopy. *Nature* **386,** 91–94.

Craig, R., and Woodhead, J. L. (2006). Structure and function of myosin filaments. *Curr. Opin. Struct. Biol.* **16,** 204–212.

DeSimone, D. W. (1994). Adhesion and matrix in vertebrate development. *Curr. Opin. Cell Biol.* **6,** 747–751.

Dubochet, J., Adrian, M., Chang, J. J., Homo, J. C., Lepault, J., McDowall, A. W., and Schultz, P. (1988). Cryo-electron microscopy of vitrified specimens. *Q. Rev. Biophys.* **21,** 129–228.

Ermantraut, E., Wohlfart, K., and Tichelaar, W. (1998). Perforated support foils with pre-defined hole size, shape and arrangement. *Ultramicroscopy* **74,** 75–81.

Faull, R. J., Du, X., and Ginsberg, M. H. (1994). Receptors on platelets. *Methods Enzymol.* **245,** 183–194.

Frank, J. (2001). Cryo-electron microscopy as an investigative tool: The ribosome as an example. *BioEssays* **23,** 725–732.

Frank, J., Radermacher, M., Penczek, P., Zhu, J., Li, Y., Ladjadj, M., and Leith, A. (1996). SPIDER and WEB: Processing and visualization of images in 3D electron microscopy and related fields. *J. Struct. Biol.* **116,** 190–199.

Grimes, J. M., Jakana, J., Ghosh, M., Basak, A. K., Roy, P., Chiu, W., Stuart, D. I., and Prasad, B. V. V. (1997). An atomic model of the outer layer of the bluetongue virus core derived from X-ray crystallography and electron cryomicroscopy. *Structure* **5,** 885–893.

Hayat, M. A., and Miller, S. E. (1990). "Negative Staining." McGraw-Hill, New York.

Henderson, R. (1995). The potential and limitations of neutrons, electrons, and X-rays for atomic resolution microscopy of unstained biological macromolecules. *Q. Rev. Biophys.* **28,** 171–193.

Henderson, R. (1998). Macromolecular structure and self-assembly. *Novartis Found. Symp.* **213,** 36–52.

Howe, A., Aplin, A. E., Alahari, S. K., and Juliano, R. L. (1998). Integrin signaling and cell growth control. *Curr. Opin. Cell Biol.* **10,** 220–231.

Hynes, R. O. (1992). Integrins: Versatility, modulation, and signaling in cell adhesion. *Cell* **69,** 11–25.

Hynes, R. O. (1999). Integrins: Bidirectional, allosteric signaling machines. *Cell* **110,** 673–687.

Jaffe, J. S., and Glaeser, R. M. (1984). Preparation of frozen-hydrated specimens for high resolution electron microscopy. *Ultramicroscopy* **13,** 373–377.

Jiang, W., Baker, M., Ludtke, S. J., and Chiu, W. (2001). Bridging the information gap: Computational tools for intermediate resolution structure interpretation. *J. Mol. Biol.* **308,** 1033–1044.

Kikkawa, M., Okada, Y., and Hirokawa, N. (2000). 15 Å resolution model of the monomeric kinesin motor, KIF1A. *Cell* **100,** 241–252.

Kleywegt, G. J., and Jones, T. A. (1993). "MAPMAN—the Manual." Uppsala, Sweden, Uppsala Software Factory.

Kleywegt, G. J., and Jones, T. A. (1994). "O2D—the Manual." Uppsala, Sweden, Uppsala Software Factory.

Kovacs, J. A., Yeager, M., and Abagyan, R. (2007). Computational prediction of atomic structures of helical membrane proteins aided by EM maps. *Biophys. J.* **93,** 1–10.

Kühlbrandt, W., and Williams, K. A. (1999). Analysis of macromolecular structure and dynamics by electron cryo-microscopy. *Curr. Opin. Chem Biol.* **3,** 537–543.

Ludtke, S. J., Baldwin, P. R., and Chiu, W. (1999). EMAN: Semiautomated software for high-resolution single-particle reconstructions. *J. Struct. Biol.* **128,** 82–97.

Mallick, S. P., Carragher, B., Potter, C. S., and Kriegman, D. J. (2005). ACE: Automated CTF estimation. *Ultramicroscopy* **104,** 8–29.

Mehta, R. J., Diefenbach, B., Brown, A., Cullen, E., Jonczyk, A., Güssow, D., Luckenbach, G. A., and Goodman, S. L. (1998). Transmembrane-truncated $\alpha v \beta 3$ integrin retains high affinity for ligand binding: Evidence for an "inside-out" suppressor? *Biochem. J.* **330,** 861–869.

Milligan, R. A., Brisson, A., and Unwin, P. N. (1984). Molecular structure determination of crystalline specimens in frozen aqueous solutions. *Ultramicroscopy* **13,** 1–9.

Orlova, E. V., and Saibil, H. R. (2004). Structure determination of macromolecular assemblies by single-particle analysis of cryo-electron micrographs. *Curr. Opin. Struct. Biol.* **14,** 584–590.

Ostermeier, C., and Michel, H. (1997). Crystallization of membrane proteins. *Curr. Opin. Struct. Biol.* **7,** 697–701.

Pettersen, E. F., Goddard, T. D., Huang, C. C., Couch, G. S., Greenblatt, D. M., Meng, E. C., and Ferrin, T. E. (2004). UCSF Chimera—A visualization system for exploratory research and analysis. *J. Comput. Chem.* **25,** 1605–1612.

Phillips, D. R., Charo, I. F., Parise, L. V., and Fitzgerald, L. A. (1988). The platelet membrane glycoprotein IIb-IIIa complex. *Blood* **71,** 831–843.

Phillips, D. R., Charo, I. F., and Scarborough, R. M. (1991). GPIIb-IIIa: The responsive integrin. *Cell* **65,** 359–362.

Potter, C. S., Chu, H., Frey, B., Green, C., Kisseberth, N., Madden, T. J., Miller, K. L., Nahrstedt, K., Pulokas, J., Reilein, A., Tchang, D., Weber, D., *et al.* (1999). Leginon: A system for fully automated acquisition of 1000 micrographs a day. *Ultramicroscopy* **77,** 153–161.

Price, D. T., and Loscalzo, J. (1999). Cellular adhesion molecules and atherogenesis. *Am. J. Med.* **107,** 85–97.

Pytela, R., Pierschbacher, M. D., Ginsberg, M. H., Plow, E. F., and Ruoslahti, E. (1986). Platelet membrane glycoprotein IIb/IIIa: Member of a family of Arg-Gly-Asp—specific adhesion receptors. *Science* **231,** 1559–1562.

Roseman, A. (2000). Docking structures of domains into maps from cryo-electron microscopy using local correlation. *Acta Cryst.* **D56,** 1332–1340.

Rossman, M. G. (2000). Fitting atomic models into electron-microscopy maps. *Acta Cryst.* **D56,** 1341–1349.

Ruprecht, J., and Nield, J. (2001). Determining the structure of biological macromolecules by transmission electron microscopy, single particle analysis and 3D reconstruction. *Prog. Biophys. Mol. Biol.* **75,** 121–164.

Saibil, H. R. (2000). Macromolecular structure determination by cryo-electron microscopy. *Acta Cryst.* **D56,** 1215–1222.

Sali, A. (2002). Modeling of molecular assemblies by satisfaction of spatial restraints. *In* "Biophysical Society Discussions" (A. Brünger, D. DeRosier, S. Harrison, and E. Nogales, eds.). Biophysical Society, Asilomar, California.

Shattil, S. J. (1993). Regulation of platelet anchorage and signaling by integrin $\alpha_{IIb}\beta_3$. *Thromb. Haemost.* **70,** 224–228.

Sheehan, B., Fuller, S. D., Pique, M. E., and Yeager, M. (1996). AVS software for visualization in molecular microscopy. *J. Struct. Biol.* **116,** 99–106.

Spahn, C. M. T., Kieft, J. S., Grassucci, R. A., Penczek, P. A., Zhou, K., Doudna, J. A., and Frank, J. (2001). Hepatitis C virus IRES RNA-induced changes in the conformation of the 40s ribosomal subunit. *Science* **291,** 1959–1962.

Stark, H., Dube, P., Lührmann, R., and Kastner, B. (2001). Arrangement of RNA and proteins in the spliceosomal U1 small nuclear ribonucleoprotein particle. *Nature* **409,** 539–542.

Subramaniam, S., and Milne, J. L. S. (2004). Three-dimensional electron microscopy at molecular resolution. *Annu. Rev. Biophys. Biomol. Struct.* **33,** 141–155.

Suloway, C., Pulokas, J., Fellmann, D., Cheng, A., Guerra, F., Quispe, J., Stagg, S., Potter, C. S., and Carragher, B. (2005). Automated molecular microscopy: the new Leginon system. *J. Struct. Biol.* **151,** 41–60.

Takagi, J., Petre, B. M., Walz, J., and Springer, T. A. (2002). Global conformational rearrangements in integrin extracellular domains in outside-in and inside-out signaling. *Cell* **110,** 599–611.

Takagi, J., Strokovich, K., Springer, T. A., and Walz, T. (2003). Structure of integrin $\alpha_5\beta_1$ in complex with fibronectin. *EMBO J.* **22,** 4607–4615.

Tao, Y., and Zhang, W. (2000). Recent developments in cryo-electron microscopy reconstruction of single particles. *Curr. Opin, Struct. Biol.* **10,** 616–622.

Taylor, K. A., and Glaeser, R. M. (1976). Electron microscopy of frozen hydrated biological specimens. *J. Ultrastruct Res.* **55,** 448–456.

Trus, B. L., Roden, R. B. S., Greenstone, H. L., Vrhel, M., Schiller, J. T., and Booy, F. P. (1997). Novel structural features of bovine papillomavirus capsid revealed by a three-dimensional reconstruction to 9 Å resolution. *Nature Struct. Biol.* **4,** 413–420.

Unwin, N. (1986). The use of cryoelectron microscopy in elucidating molecular design and mechanisms. *Ann. N. Y. Acad. Sci.* **483,** 1–4.

van Heel, M. V., Harauz, G., Orlova, E. V., Schmidt, R., and Schatz, M. (1996). A new generation of the IMAGIC image processing system. *J. Struct. Biol.* **116,** 17–24.

Volkmann, N., and Hanein, D. (1999). Quantitative fitting of atomic models into observed densities derived by electron microscopy. *J. Struct. Biol.* **125,** 176–184.

Wriggers, W., Milligan, R., and McCammon, J. A. (1999). Situs: A package for docking crystal structures into low-resolution maps from electron microscopy. *J. Struct. Biol.* **125,** 185–195.

Xiao, T., Takagi, J., Coller, B. S., Wang, J.-H., and Springer, T. A. (2004). Structural basis for allostery in integrins and binding to fibrinogen-mimetic therapeutics. *Nature* **432,** 59–67.

Xiong, J.-P., Stehle, T., Diefenbach, B., Zhang, R., Dunker, R., Scott, D. L., Joachimiak, A., Goodman, S. L., and Arnaout, M. A. (2001). Crystal structure of the extracellular segment of integrin $\alpha_V\beta_3$. *Science* **294,** 339–345.

Yeager, M., Unger, V. M., and Mitra, A. K. (1999). Three-dimensional structure of membrane proteins determined by two-dimensional crystallization, electron cryomicroscopy, and image analysis. *Methods Enzymol.* **294,** 135–180.

Zhao, F.-Q., and Craig, R. (2003a). Ca^{2+} causes release of myosin heads from the thick filament surface on the milliseconds time scale. *J. Mol. Biol.* **327,** 145–158.

Zhao, F.-Q., and Craig, R. (2003b). Capturing time-resolved changes in molecular structure by negative staining. *J. Struct. Biol.* **141,** 43–52.

CHAPTER SIXTEEN

INTRAVITAL IMAGING AND CELL INVASION

Milan Makale

Contents

Abstract

The main cause of cancer treatment failure is the invasion of normal tissues by cancer cells that have migrated from a primary tumor. An important obstacle to understanding cancer invasion has been the inability to acquire detailed, direct observations of the process over time in a living system. Intravital imaging, and the rodent dorsal skinfold window chamber in particular, were developed several decades ago to address this need. However, it is just recently, with the advent of sophisticated new imaging systems such as confocal and multiphoton microscopy together with the development of a wide range of fluorescent cellular and intracellular markers, that intravital methods and the window chamber have acquired powerful new potential for the study of cancer cell invasion. Moreover, the interaction of various cell signaling pathways with the integrin class of cell surface receptors has increasingly been shown to play a key role in cancer invasion. The window chamber in combination with integrin-knockout rodent models, integrin-deficient tumor cell lines, and integrin antagonists, allows real-time observation of integrin-mediated cancer invasion and angiogenesis. The present review outlines the history, uses, and recent methods

Moores UCSD Cancer Center, University of California, San Diego, La Jolla, California

Methods in Enzymology, Volume 426

ISSN 0076-6879, DOI: 10.1016/S0076-6879(07)26016-1

© 2007 Elsevier Inc.

All rights reserved.

of the rodent dorsal skinfold window chamber. The introduction of labeled tumor cells into the chamber is described, and imaging of tumors and angiogenic vessels within chambers using standard brightfield, confocal, and multiphoton microscopy is discussed in detail, along with the presentation of sample images.

1. INTRODUCTION

1.1. Overview

Invasion of surrounding normal tissues by cells from a primary tumor is one of the primary causes of cancer treatment failure. In pancreatic cancer and brain cancers, for example, the primary cause of death is the invasion of normal tissues and organ structures surrounding the primary tumor (Condeelis *et al.*, 2005). Accordingly, a goal of current research efforts is to understand and find effective methods by which to contain the spread of cancer cells from a primary tumor (Brooks *et al.*, 1997; Coomber *et al.*, 2003). However, cancer invasion is incompletely understood, and an important obstacle has been the inability to directly observe the process of local invasion developing in an animal model over time. New technologies involving the use of intravital imaging in live animals hold promise in terms of facilitating the nondestructive, sequential chronicling of cancer cell invasion in a living system (Condeelis *et al.*, 2005; Dewhirst *et al.*, 2002). Tissue-window–based intravital methods allow high-resolution imaging of implanted labeled tumor cells, extracellular matrix (ECM) components, and local blood vessels in the living animal over days and weeks.

One of the most widely used and versatile intravital model systems is the rodent dorsal skinfold window chamber. This elegant preparation, first described by Algire in 1943, consists of a lengthwise fold of dorsal skin with an implanted clear glass window that permits direct observation of the tumor implant, surrounding vessels, and the host's subcutaneous tissues (Fig. 16.1A and B). This preparation is presently experiencing a resurgence, primarily due to the introduction of sophisticated new imaging methods. These include laser-scanning confocal microscopy and scanning multiphoton microscopy, along with the availability of a variety of epitope binding and intracellularly expressed fluorescent dyes and markers, and various nanoparticles (Liu *et al.*, 2005; Sahai *et al.*, 2005). These methods allow tracking of the movement of tumor margins and individual cells, the growth of blood vessels, and the interaction between the tumor cells and local blood vessels.

The cell-surface integrin receptors have been shown in many studies to play a critical role in tumor cell migration and survival (Klemke *et al.*, 1994), leading to the creation of specific integrin-receptor knockout mice (Robinson *et al.*, 2006), and integrin-lacking tumor cell types (Felding-Habermann, 2003). Furthermore, intracellular pathways relating to integrin function have

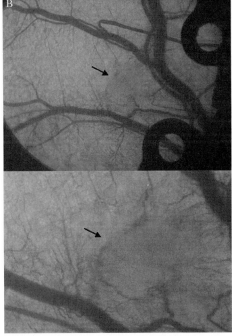

Figure 16.1 (A) Dorsal skinfold window chamber is shown implanted on a hairless athymic (*nu/nu*) mouse. The subcutaneous vasculature is visible. (B) Human pancreatic FG tumor (800-micron diameter) growing within the chamber. The top panel is a general view at about 1.5×, and the bottom image is a closer view of the same tumor at 2.5×. (See color insert.)

been elucidated, and knockout mice and inhibitors of the primary moieties in these pathways have been developed. All of these new biological tools can be used with the window chamber to directly observe the effects on tumor cell invasion by various molecular pathways, cell surface events, and cell–ECM interactions. Moreover, animals implanted with a dorsal skinfold window chamber can also be prepared with primary tumors at other anatomical locations to determine the humoral effects of a primary tumor on metastatic growth (Gohongi *et al.*, 2004), or to examine the effects of the local environment and tissue specific integrin–ECM interactions on particular tumor cell types.

This review will focus on the technical aspects of the dorsal skinfold window chamber and the application of this method for the study of cancer cell survival and invasion.

1.2. Background and applications

Various types of tissue-viewing chambers have been used extensively for the study of the microvasculature (Kerger *et al.*, 1995), tumors (Jain *et al.*, 2002; Pahernik *et al.*, 2002), cellular behavior (Algire, 1953), wound healing (Devoisselle *et al.*, 2001; Vranckx, 2002), and imaging of internal organs (Bertera *et al.*, 2003). Window chambers have been used as an observational tool in a variety of animal tissues including rabbit ear tissues (Arfors *et al.*, 1970; Sandison, 1924); mouse, rat, and hamster dorsal skinfold (Kerger *et al.*, 1995; Menger *et al.*, 2002); the body wall of the mouse (Bertera *et al.*, 2003); superficial tissues of the human arm (Branemark, 1971); and tissues of the human leg (Vranckx *et al.*, 2002). In 1934, Williams applied the viewing chamber method to the body skin (Williams, 1934). In 1943, Algire further modified this approach, using the dorsal skin of the mouse pinched into a fold, to allow direct observation of the subcutaneous tissues and the microcirculation (Algire, 1943).

Later the chamber became widely used for tumor studies in mice and was subsequently adapted to hamsters (Endrich *et al.*, 1980) and rats. The method comprised two metal plates supporting a longitudinal fold of dorsal skin, within which a 1-cm circular area of skin had been removed and the exposed tissue covered by a thin circular layer of glass. Initially the two parallel plates were machined from stainless steel. However, later improvements included a switch to titanium due to its light weight, biocompatibility, and low coefficient of thermal conductivity. Eventually the chamber was adapted for studies of the subcutaneous and tumor microcirculation (Endrich *et al.*, 1979) with indirect methods being developed, such as the use of phosphorescent dyes to measure tissue oxygen. More recently it was adapted for investigations involving tumor microcirculation (Dewhirst *et al.*, 2000; Kerger *et al.*, 1995). Jain and coworkers pioneered the use of the dorsal skinfold window chamber for the study of interstitial tumor pressure, fluid transport in tumors, penetration of chemotherapeutic agents into solid tumors, and tumor angiogenesis (Jain *et al.*, 2002; Leunig *et al.*, 1992; Yuan *et al.*, 1993).

Various laboratories have initiated the use of scanning multiphoton microscopy of tumors (Condeelis *et al.*, 2005), and Jain has used multiphoton microscopy with the dorsal skinfold window chamber to track tumor cells, angiogenesis, and tumor vascular perfusion (Abdul-Karim *et al.*, 2003;

Kashiwagi et al., 2005). Li and colleagues (2000) injected 20 to 30 fluorescently labeled tumor cells into the window chamber to examine the very beginning stages of tumor growth and angiogenesis, as well as tumor cell migration toward and around existing host microvessels.

Later, Makale, Gough, and coworkers (Makale et al., 2005) developed the two-sided chamber, in which a transparent window is on one side of the fold and a planar oxygen or glucose sensor is on the other. This arrangement allows the microcirculation and the oxygen levels in the sandwiched subcutaneous tissue to be measured. The window chamber is also used for transplantation studies of organs and tissues (Chen et al., 2006), for studies of the tissue compatibility of biomaterials (Menger et al., 1990), for studies of microsurgical techniques (Mordon et al., 1997), and for the study of the tumor penetration of drug carriers (Dreher et al., 2006).

Much remains to be learned about cancer cell invasion, but it is clear that the adhesive interactions between cells and the ECM mediated by integrins are crucial to cell motility and to the survival of migratory cells (Deryugina et al., 2000; Hegerfelt et al., 2002; Ingber, 1992; Stupack and Cheresh, 2002). The integrins are heterodimeric transmembrane glycoproteins comprised of α and β subunits, and bind mainly via RGD sites to an array of ECM substrates including collagens, laminins, vitronectin, and thrombin (Felding-Habermann, 2003). Integrins transmit signals directly to the focal adhesion kinase (Fak) and Src families, eliciting the activation of several major cell signaling pathways (Felding-Habermann, 2003; Sieg et al., 2000; Stupack and Cheresh, 2002). Tumor cells widely express the integrin avβ3 (Deryugina et al., 2000); invasive melanoma cells, for example, have been shown to depend on β-3 and β-1 integrins for their aggressive properties (Hegerfeldt et al., 2002; Petitclerc et al., 1999). Accordingly, integrin-knockout rodent models (Robinson et al., 2006) and integrin-deficient cell lines have been developed to facilitate understanding the role of integrins in tumor invasion, normal and pathological angiogenesis, and normal tissue remodeling (Deryugina et al., 2000; Eliceiri et al., 2002; Felding-Habermann, 2003; Larsen et al., 2006). Using the dorsal skinfold window chamber, one can view how tumors implanted into integrin knockout animals respond and the effects of integrin antagonists on tumor growth. In addition, normal tissue remodeling can be characterized using real-time or time-lapse imaging. The behavior of integrin-positive or -deficient cell lines, such an orthotopic model composed of the human melanoma M21 (avβ3+) and M21L (avβ3−) lines implanted into the murine skinfold chamber can be directly assessed in terms of invasiveness. This can also be done with other kinds of knockout mice (e.g., Src deficient) and/or inhibitors of Raf, Ras, Mek, PI-3-kinase, Fak, and Src.

The dorsal skinfold window chamber can be implanted on virtually any strain of mouse, as well as the golden hamster. The chamber can be installed onto rats, but the installation is more difficult and the chamber is not as clear owing to the relatively thick dorsal skin retractor muscle of the rat. The chamber can be modified with a stainless-steel cannula passing from the chamber exterior to its interior, so that substances may be injected into the chamber proper without removing the coverslip. A variety of mouse tumors as well as the hamster melanoma can be implanted into the window chamber tissue as either fragments or a dense suspension of cells. Window chambers placed on the athymic hairless mouse (*nu/nu*) and the SCID mouse can be used to grow a wide range of human tumors, and the growth of the tumor may be chronicled along with the development of blood vessels and the movement of tumors cells into the surrounding normal tissue. The chamber tissues are thin, approximately 500 microns, and the tumors are also compressed, so that scanning-confocal and -multiphoton microscopes can be used to image deep into the tumor and surrounding tissue. Individual cells can be clearly imaged, as can vascular growing tips and associated protrusions. Our laboratory has applied the window chamber to the study of a variety of tumors using fluorescent light microscopy, laser-scanning confocal microscopy, and laser-scanning multiphoton microscopy.

The window chamber, when properly prepared, should provide excellent images for 2 weeks. In some cases, 3- to 4-week-old chambers have produced clear images (Menger *et al.*, 2002). The chamber offers the primary advantage of allowing tumors and tissues to be imaged serially over days and weeks, with excellent resolution and clarity and without tissue damage. However, the chamber is limited by the fact that tumors grow in a more compressed configuration than they normally would. Additionally, after several weeks the chamber tissues may become cloudy and heal over, resulting in reduced image quality. Furthermore, the chamber is for many tumors not orthotopic, although melanomas are in their natural environment, and pancreatic, lung, colon, and a wide variety of other tumor types do grow vigorously within the *nu/nu* mouse dorsal skinfold chamber.

Our laboratory at the UCSD Moores Cancer Center has adapted, developed, and implemented methods for fabricating, assembling, and installing the window chambers, and for preparing and injecting tumor cells into the chambers. We have designed specific equipment and methods for standard brightfield, fluorescence, confocal, and multiphoton microscopy of the chamber and its contents. These methods have been tested extensively in our laboratory, they work well, and they comprise the primary scope of this article.

2. Methods

2.1. Basic configuration and fabrication of the window chamber

The window chamber apparatus, shown in Fig. 16.2, consists of two complementary titanium alloy frames, each with a 1.2-cm diameter opening, and on one of the frames the opening is fitted with a titanium ring for attachment of the glass coverslip. The ring extends 0.025 in (0.6 mm) to the tissue to

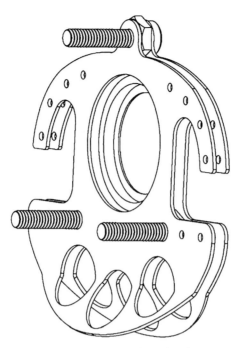

Figure 16.2 The mouse dorsal skinfold window chamber apparatus. The schematic drawing depicts two complementary titanium plates. The plates are held together with screws, nuts, and sutures.

provide a seal, and it is machined so that the coverslip is held as close to the exposed tissue as possible. The glass coverslip is secured in place by a copper-beryllium retaining ring. The internal lip, on the tissue side of the ring, is only 0.007 in (0.3 mm) thick. The contralateral frame has a circular opening only, without a ring or coverslip. The two frames are held together by three 10-mm screws that are separated by 4-mm hexagonal nuts (Figs. 16.1A and 16.2). The frames are secured together with another set of hexagonal nuts. The frames themselves are 0.015 in (0.4 mm) thick, and the entire assembly, including two frames, sealing ring, three screws, six nuts, glass coverslip, and retaining ring, weighs approximately 2.7 g. Each frame has a lateral angled flange to permit the chamber apparatus to rest comfortably on the subject's back, and each flange has three circular areas of metal removed to reduce weight and allow air circulation. The top margin of each chamber plate has small holes drilled into it to allow placement of supporting suture material.

2.2. Preparation of the window chamber hardware

The entire window chamber apparatus, whether it has been used previously or not, is prepared in a specific and careful manner. Proper cleaning is crucial to the prevention of infection and an inflammatory tissue reaction. Each chamber is disassembled and placed in a solution of 0.23% peracetic acid and 7.35% hydrogen peroxide (Sporgon, Decon Laboratories) for about 24 h. The glass coverslip is discarded, and the chamber is carefully scrubbed with a bristle-type scrub brush in distilled water and laboratory-grade soap. Adherent tissue is gently scraped away with fine tweezers and a scalpel blade, and the chamber plates are vigorously rubbed with various grades of Scotch-brite polishing pads (coarse, fine, and extra-fine; 3M Company). The chamber plates and associated hardware are rinsed in double-distilled water, and then placed in an ultrasonic cleaner with double-distilled water and instrument soap. After ultrasonic cleaning for 45 min, all the hardware is extensively rinsed in double-distilled water, placed on a clean, dry surface, gently patted dry with cotton gauze, and then placed in a sterile biological hood under ultraviolet light for approximately 24 h. Glass coverslips are soaked in 0.23% peracetic acid and 7.35% hydrogen peroxide for 2 h, rinsed in 70% ethanol, and patted dry with cotton. The chambers are assembled without coverslips and retaining rings, and then placed separately in autoclave pouches. The coverslips and retaining rings are wrapped in cotton gauze and placed in autoclave pouches. All the chamber materials are then autoclaved.

2.3. Specific variations in the window chamber for tumor studies

The dorsal-skinfold window chamber can be installed in various ways to emphasize particular tissue layers and to implant tumors or organs onto those layers. The usual configuration is to place the window over the A-0 feeding vessels of the dorsal skin, and to remove from one side of the fold a 1.2-cm circular area of skin, subcutaneous fat and fascia, and retractor muscle. The contralateral retractor muscle and skin remain intact. The tumor material is placed on or into the retractor muscle. This allows the margins of the cut retractor muscle to be reflected over the margins of the wound to help seal the chamber and to block the seepage of blood from the margins of the skin into the chamber. In other variants, the subcutaneous fat and fascia may remain intact, both retractor muscles may be left intact, or both retractor muscle layers may be removed and the tumor material placed on the remaining subcutaneous layer. These variations encompass different degrees of tissue trauma, and they do allow the effects of different tissue types, vascular densities, and patterns of circulation to be assessed in terms of tumor cell growth. The dorsal skinfold chamber may also be installed as a two-sided preparation in which one side of the double intact retractor muscle is covered by a glass coverslip, and the contralateral side is covered by a planar oxygen or glucose array. We have found that this sort of arrangement, because it adds weight and involves more extensive exposure of the retractor muscles, is more amenable with larger animals such as the adult hamster, the desert sand rat, or the typical laboratory rat.

2.4. Surgical installation of the dorsal skinfold window chamber

2.4.1. General presurgical procedures

Table 16.1 summarizes the supplies needed for window chamber implantation. Aseptic procedures are strictly observed, and all procedures performed in our laboratory receive prior approval from the University of California San Diego IACUC committee. The window chamber preparations are vulnerable to infection, especially in *nu/nu* mice. All instruments are cleaned in an ultrasonic cleaner, rinsed in distilled water, and autoclaved. All suture material and surgical sponges are either obtained in sterile form or are autoclaved before use. Surgeons and assistants wear hair bouffants, masks, gowns, and sterile gloves. Hairless athymic (nu/nu) mice receive the antibiotic SMZ in their drinking water for at least 2 days prior to surgery and for the entire postoperative period.

2.4.2. Preparation of mice/hamsters

The animals are checked for overall health status and then are injected intraperitoneally with 100 mg/kg of ketamine and 250 to 500 mcg/kg of medetomidine. The absence of corneal reflex and foot withdrawal to toe pinch indicates that the animals have attained a surgical plane of anesthesia. Animals with fur have the entire dorsum shaved with electric clippers and a depilatory is applied to the dorsum to remove fine hairs. The dorsum is then washed with normal saline. On all animals Betadine antiseptic is generously applied and allowed to remain for at least 60 sec. The back is then thoroughly cleaned with 70% ethanol swabs.

2.4.3. Surgical implantation of window chamber

The animal is placed on a temperature-controlled heated pad covered by a clean surgical drape. The heating pad and mouse are placed on a Plexiglas surgery stand (Fig. 16.3), and the dorsal skin is drawn up into a longitudinal fold using a straight suture needle (Roboz RS-7987) and sterile 4-0 silk. The skinfold is transilluminated so that the symmetrical pattern of blood vessels on either side of the dorsal midline can be matched so that the skinfold will be even. The chamber plate which holds the attachment screws is positioned so that the circular opening is over the feeding A-0 vessels. The top of the chamber plate is sutured to the top of the skinfold, and the two lower attachment screws are forced through the entire skin thickness via small incisions made by an ultrasharp stab knife (Fine Science Tools 10315-12). A 1.4-cm diameter circle is cut into the skin using fine, curved scissors (Roboz RS-5481). Care is taken to not cut any deeper than the outer layer of skin. Then the edge of the circle is grasped with Dumont forceps (Fine Science Tools) and microspring scissors (Roboz RS-5603) are used to snip away the fascia holding the round area of skin to the underlying

Table 16.1 Supplies needed for window chamber installation

Item	Supplier(s), Item Number	Rationale/Use
Fine scissors (curved)	Roboz RS-5481	Used for initial skin incision
Stab knife	Fine Science Tools 10315–12	Extremely sharp and fine, used to make very small incision to allow chamber screws to pass through skin
Fine forceps (curved)	Roboz RS-5079	For holding tissue layers
Microspring scissors (curved)	Roboz RS-5603	For cutting subcutaneous fat, fascia, and the retractor muscle
Ultrafine microscissors	Fine Science Tools 15000–00	Used to incise the mouse jugular vein for catheterization
Dumont fine forceps (curved)	Fine Science Tools #7, T1	For holding the skin while cutting
Hemostat clamp	Fine Science Tools 13009–12	For holding the skinfold suspending silk thread in place, and for clamping silk suture to apply traction to vessel for cannulation
Fine vascular clamp	Fine Science Tools 18055–04	For clamping off vessel blood flow during cannulation
Surgical needle (straight, thin eyelet)	Roboz RS-7987	For suturing chamber plates to dorsal skinfold
Surgical silk (4-0)	Harvard Apparatus 517615	For suturing skinfold and chamber plates together and for suspending skinfold while installing chamber
Window chamber surgery stand	Fabricated from Plexiglas®	For window chamber and catheterization procedures
Ring pliers	American Ring and Tool Co.	To place or remove the retaining ring with respect to the chamber-frame sealing ring that contains the glass coverslip

Item	Source	Use
Fiber optic light source	Fisher, VWR, Harvard Apparatus, and others	Transillumination of skinfold and tissue window
Bead sterilizer	Fisher, VWR, Harvard Apparatus, and others	Instruments can be sterilized during the course of the surgery
Antibiotic solution	Tobramycin 40 mg/ml USP (American Pharmaceutical Partners, Inc.)	Filling window chamber with fluid, washing chamber tissues
Plasmalyte A or normal saline	Plasma–Lyte A® injection pH 7.4 (multiple electrolytes injection Type 1 USP, Baxter, Deerfield, IL) 0.9% NaCl, preservative-free (Hospira)	For hydrating the subject with subcutaneous injection following surgery
Betadine, 1% Betadine solution	Purdue-Frederick	Swabbed on dorsum and neck for disinfection of skin
Ethanol swabs, 70%	Becton Dickenson	For cleaning Betadine
Cotton sponges (4 × 4)	Fisher	For swabbing Betadine
Sponges, spear type	Ultracell, Inc.	Extremely useful for washing chamber tissues and removing excess saline/antibiotic
Anesthetic cocktail	Ketamine (Fort Dodge Pharmaceuticals) and medetomidine (Fort Dodge Phamaceuticals)	Medetomidine diluted 10-fold (01 mg/ml) and mixed with medetomidine (100 mg/ml) to a ratio of 2.5:1
Antisedan	Fort Dodge Pharmaceuticals	Medetomidine reversing agent

(continued)

Table 16.1 (*continued*)

Item	Supplier(s), Item Number	Rationale/Use
Catheter tubing	1-Fr or 2-Fr polyurethane tubing (Harvard Apparatus) may be coated with tobramycin and heparin for cannulating the jugular vein. Alternatively, polyethylene PE-10 tubing (Clay-Adams) may be stretched thin to one-third of its diameter and used	Vessel catheterization
Veterinary bonding cement	MWI Veterinary Supplies	For gluing reflected cut retractor muscle ends to external skin around tissue chamber opening
SMZ antibiotic in drinking water	Sulfamethoxazole 200 mg and trimethopium 40 mg, Hi-Tech Pharmacal	For *nu/nu* mice before and after surgery. Mix 5 ml with 250 ml of autoclaved drinking water
Topical antibiotic ointment	MWI, triple ophthalmic ointment	Applied around skin opening to prevent infection tracking into the chamber
Binocular dissecting microscope	Nikon, Zeiss, Olympus, Fisher, VWR, Harvard Apparatus, and others	Some practitioners prefer to use a binocular dissecting microscope for window chamber surgery; others perform the surgery by naked eye
Glycerol (100% USP)	Sigma	Catheter lock solution
Heparin (sodium salt)	Sigma	Catheter lock solution (50 U/ml)
Sporgon Instrument Sterilizing Solution	Fisher	For disinfecting and sterilizing window chambers and hardware after use

Note: Described here are the essential instruments and hardware along with their use in the installation of a dorsal skinfold tissue window chamber and for jugular vein catheterization. The specific instruments types that are indicated are typical examples of what may be used, and similar devices are usually available from more than one supplier. Some items have to be machined or otherwise assembled.

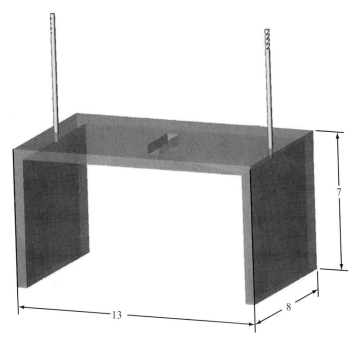

Figure 16.3 Surgery stand for the window chamber. The two notched vertical rods are used to secure strands of 4-0 silk that suspend the dorsal skin of the subject into a longitudinal fold. Note the slit in the horizontal section for the dorsal chamber used during catheterization. Dimensions are in inches.

tissue. Gentle traction is applied and the skin is entirely removed. This exposes the underlying fat and fascia, which are incised and removed using microspring scissors and fine forceps (Roboz RS 5079). The first retractor muscle layer now lies exposed, and this is incised and cut back using the microspring scissors and fine forceps. The muscle is cut back and then reflected over the external skin; the muscle is glued to the skin using small drops of veterinary bonding cement (MWI). Antibiotic ointment is applied to the margin of the skin opening and to the sites where screws and suture pass through the skin. The chamber plate is removed and placed on the cut side of the skinfold with the screws protruding through the previously established holes in the skin. The orientation of the chamber plates with screws may vary according to the microscope-stage mouse/hamster holder. In some cases, the plate with screws may remain on its original side of the skinfold and the complementary plate with the coverslip attached is placed over the cut side of the skinfold. After the first plate has been positioned, it is again sutured along its top margin to the top of the skinfold. Then the complementary plate is placed on the contralateral side of the skinfold, over the screws, secured by hexagonal nuts. The outer margins of both plates are sutured to each other and to the skinfold. Several drops of antibiotic solution (Tobramycin, APP Inc.) are instilled into the chamber open tissue area, and then a sterile glass coverslip is placed over the opening. The coverslip is secured with a retaining ring.

2.4.4. Catheterization of right external jugular vein

This is performed in some animals to facilitate repeated and frequent injections of solutions, fluorescent dyes, labeled tumor cells, and other agents. In animals receiving relatively few infusions, the tail vein method is used and the jugular is not catheterized. For jugular catheterization immediately following window chamber implantation, the animal is turned over to a supine position and the heating pad is moved somewhat so that the window chamber fits through a slit cut in to the surgery stand. The forelegs are taped to the surgery stand and the ventral aspect of the neck is swabbed with Betadine and washed with 70% ethanol. A 1-cm incision is made over the right external jugular region, and the vein is exposed using blunt dissection. Three lengths (3 to 4 cm) of sterile 4-0 silk are passed under the vein and loosely tied with two throw knots. Gentle traction is applied to the cranial end of the exposed vein, and the distal end is closed using a fine vascular clamp (Fine Science Tools 18055-04). The vein is carefully incised using microspring scissors (straight blade, ultrafine, Fine Science Tools 15000-00), and a 1- or 2-Fr polyurethane catheter or a stretched PE-10 tubing catheter filled with heparinized saline (150 U/ml) is advanced caudally until it reaches the distal clamp. The sutures are tied around the vein and catheter, and veterinary bonding cement is used to secure the sutures and vein. The saline in the catheter is slowly withdrawn using a syringe, and the catheter is then filled with 100% glycerol (USP) containing heparin (50 U/ml), and sealed. The free end of the catheter is drawn under the skin of the right shoulder and out between the shoulder blades. The catheter is secured to the window chamber frame. The incisions are closed with 4-0 silk and veterinary bonding cement.

2.4.5. Surgical recovery and postsurgical period

The animals will be placed on a heated pad and gently warmed; mice and hamsters are injected subcutaneously with 1 mg/kg of antisedan, a medetomidine-reversing agent. Mice and hamsters are injected subcutaneously with normal saline, 0.1 to 0.2 ml and 0.5 to 1.0 ml, respectively, to maintain hydration. Mice and hamsters with window chambers are individually housed in rat microisolator cages to prevent other animals from damaging window chambers and catheters. Hairless athymic mice (nu/nu) are placed in autoclaved microisolator cages and are given autoclaved water with the antibiotic SMZ added. The animals are monitored daily, and surgical sites are checked for inflammation and infection. The heparinized glycerol in catheters should be replaced every 3rd day.

2.4.6. Assessing the quality of the chamber and potential problems

The window chamber should remain very clear and be free of discoloration bleeding and opacity (Fig. 16.4). There must not be air bubbles in the chamber, and the subcutaneous tissue should appear to be very close to the glass coverslip. After a few days, the entire chamber apparatus will often tilt somewhat to one side or the other; this does not represent a problem regarding the chamber or the animal subject. When examined under 10× magnification, blood flow in the major and minor vessels should be readily apparent. We have observed that vigorous blood flow is the primary determinant of chamber installation success. Sluggish or absent blood flow indicates that the chamber will fail and should not be used for experimental tumor studies. Poor blood flow may result from excessive tightening of the chamber screw nuts and/or removal of too much fascia from the retractor muscle of interest. Bacterial infection of the chamber is evinced by a general absence of optical detail, and a red, inflamed appearance with or without regions of pink or pink-yellow discoloration. An occasional problem in immune-compromised mice (nu/nu, SCID) may be the development of a dark green-colored fungal infection (*Aspergillus* sp.) along the chamber tissue periphery. Strict adherence to sterile technique and the proper use of antibiotics help reduce the occurrence of infections. In the case of recurring fungal infections, the chamber plates and associated hardware may need to be soaked in concentrated HCl for 15 min to destroy fungal spores. In addition, our laboratory has found that the use of new copper-beryllium retaining rings to secure the glass coverslip is associated with a marked reduction in the incidence of fungal infection. The use of iron-steel rings is not recommended. Barring infection, and with the presence of robust blood flow, a carefully prepared chamber should remain very clear for at least 2 weeks.

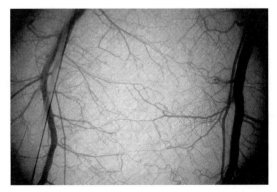

Figure 16.4   Typical window chamber. Note that the chamber is optically clear, free of bleeding and infection, and with good vascular perfusion.

2.5. Cell lines and preparation of cells for placement into window chamber tissue

2.5.1. Cell lines

A variety of human and mouse cell lines can be grown in the window chamber; human lines include pancreatic (FG and FGM), melanoma (M21 and M21L), lung, colon, brain, and prostate. Our laboratory typically cultures pancreatic (FG and FGM) and melanoma (M21 and M21L) lines in Minimal Eagles Medium (MEM) to which is added the following: 10% FBS, L-glutamine 1:100 (200 mM stock), antibiotics 1:10 (10,000 IU penicillin, 10,000 μg/ml streptomycin, 25 μg/ml amphotericin). The tumor cell lines are stably transfected to express green fluorescent protein (GFP) or DsRed. The cells are grown in 5% CO_2 at 37° in 100-mm tissue culture dishes. The cells are passaged at least once and when they are approximately 80% confluent they are ready to be placed into the window chambers. Generally it is best to wait 1 to 2 days after window chamber installation before instilling tumor cells; this allows time for inflammation and tissue irritation following the window chamber surgery to almost completely subside.

2.5.2. Preparation of tumor cells

The tumors may be placed within the window chamber as either a cell suspension or as small pieces. To prepare a suspension of cells, tumor cells grown on the tissue culture dishes are washed with sterile phosphate buffered saline (PBS), and then are treated with trypsin for approximately 2 min at 37°. The cells are washed off the dishes with medium, and then are centrifuged for about 5 min at 3000 rpm. The resultant pellet is resuspended in PBS, the cell density counted with a hemocytometer, and centrifuged and resuspended in PBS at a concentration of 200,000 cells per microliter in a microfuge tube placed on ice.

2.5.3. Preparation of solid tumors

To prepare solid pieces of the tumor, the tumor cells are grown in cell culture as described in section 2.5.1, and are then washed and suspended in PBS at a concentration of 1 million cells in approximately 40 μl of PBS. This volume (40 μl) is injected into the flank of a nude mouse and allowed to grow for 7 to 14 days. When the tumors have reached a palpable size, the mouse is sacrificed and the tumor is quickly dissected free and washed with cold PBS. The tumor is cut into 1-mm pieces with a sterile scalpel blade, and a piece from the periphery of the tumor is placed in PBS and on ice.

2.5.4. Injection of tumor cells into window chamber

The mice are comfortably secured in a clear Plexiglas holder and then moved into a laminar flow sterile hood. The glass coverslip is removed from the chamber and the chamber tissues are washed with a sterile (1:4) tobramycin–normal saline solution, and the excess fluid is gently blotted with a spear-tip sponge (Ultracell). An autoclaved 10-μl Hamilton syringe (Hamilton, Reno, NV) is used to inject 2 μl of the tumor cell suspension into the outer portion of the retractor muscle, usually between two major vessels and near the top of the window. It is important to ensure that the cells enter the outer muscle tissue and do not squirt back into the chamber, flooding the chamber with tumor cells. Injecting the cells near the top of the window allows the tumor to be visible as the skin of the dorsal fold sags over time. A few drops of Tobramycin solution are added to the open chamber and a fresh sterile coverslip is placed on the opening, care being taken to avoid and expel air bubbles. The animal can then be returned to its home cage. The tumor should grow vigorously in the chamber and have the general appearance depicted in Fig. 16.5A and B, and even single tumor cells in the chamber should be visible (Fig. 16.5C).

2.5.5. Placement of solid tumors into the window chamber

The mouse is placed in a holder as described in Section 2.5.4, the coverslip removed from the chamber, the chamber tissues washed with tobramycin solution, and a 1-mm (or less) piece of the tumor is placed near the top of the chamber. A few drops of tobramycin antibiotic solution are used to wet the chamber tissues, and a fresh, sterile coverslip is pressed down into the chamber opening. Pressing the coverslip down helps to flatten the tumor somewhat and prevent too large a gap between the chamber tissue and the coverslip. A smaller gap prevents air from entering, and decreases the microscope working distance that is required for imaging of the growing tumor and its vessels.

2.6. Imaging of the window chambers

2.6.1. Preparation and maintenance of subjects for microscopy

Animal movements must be minimized for successful imaging, so the mice or hamsters are injected intramuscularly with an anesthetic preparation. We have found the following subanesthetic dosages to be effective in preventing the animals from introducing movement artifact into the microscope images: 20 mg/kg Nembutal alone, 20 mg/kg Nembutal plus 125 mcg/kg medetomidine, or 50 mg/kg ketamine plus 125 mcg/kg medetomidine.

Animals sedated with Nembutal, and in particular those sedated with ketamine/medetomide, need to be kept warm either with a heated enclosure around the microscope or with a small thermal pad. Keeping the sedated subject warm prevents excessive hypotension, vasoconstriction, and a reduction of blood flow through the chamber tissues. Booster injections of 20 mg/kg of Nembutal given subcutaneously may be needed to keep the animals quiet for longer imaging procedures. The length of time that the animals can be imaged and the exact restraint and maintenance

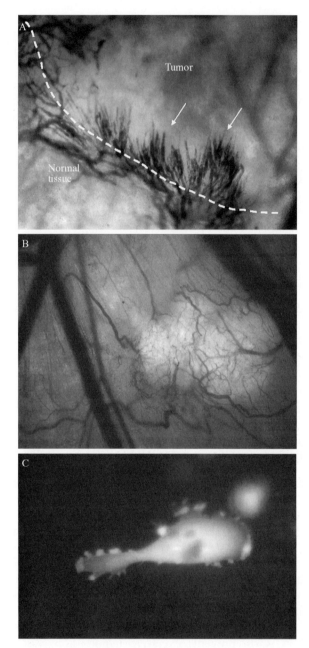

Figure 16.5   (A) Brightfield image of growing tips (10×, arrows) of human pancreatic tumor in the window chamber. (B) ~1.0-mm human pancreatic tumor imaged with fluorescence light microscopy. (C) Single migrating human melanoma cell, expressing the dSRed fluorophore, sandwiched between the chamber tissue and glass coverslip (20×).

procedures must be approved by the relevant institutional animal care committee.

Following sedation, the mice or hamsters are comfortably secured within a clear, perforated Plexiglas cylinder that has a slit to accommodate the window chamber assembly. The cylinder with the mouse is then placed within a specially designed flat Plexiglas holder that rests within the stage frame of a Nikon TE2000-E confocal microscope (Fig. 16.6A and B). The cylinder is also designed to be contained in a holder that is adapted for a Zeiss Axiovert100 inverted fluorescence microscope. This arrangement has also been adapted for imaging with an in-house constructed multiphoton microscope (D. Kleinfeld, UCSD Physics Department). The holders are generally similar in design, although the exact dimensions vary, and they are both made to hold the cover glass of the window chamber as close as possible to the microscope objective. The mouse cylinder rests in a trough within the holder, and the three 10-mm screws of the window chamber fit

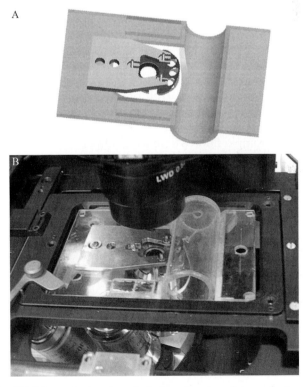

Figure 16.6   (A) Schematic drawing of window chamber plates and microscope holder (mouse not shown). (B) Photograph of mouse holder in Nikon confocal microscope stage. Note the machined stainless steel plate used to secure the mouse chamber and minimize movement.

through holes machined into a stainless steel plate that is bolted to the Plexiglas holder. The chamber screws are secured to the steel plate with hexagonal nuts. This configuration securely stabilizes the window and prevents breathing-induced movement of the window chamber tissues (Fig. 16.6A). The mouse cylinder is designed to fit into the confocal microscope stage holder and the inverted fluorescence microscope holder. Thus, the animal does not need to be removed from the cylinder when moving from one microscope to the other; instead the cylinder with the animal is simply transferred between the instruments.

2.6.2. Fluorescent dyes and labels for intravital microscopy

A variety of fluorescent markers are available for use with the brightfield fluorescence microscope and the scanning confocal microscope, as well as with scanning multiphoton instruments. These markers are injected via the tail vein or jugular catheter; some frequently used examples include FITC-lectin (fluorescein *Griffonia simplicifolia* lectin, FL-1102 Vector Laboratories), FITC-rhodamine (rhodamine *Griffonia simplicifolia* lectin, RL 1102 Vector Laboratories), and FITC-dextran (fluorescein isothiocyanate-dextran, 2×10^6 MW, Sigma). The labeled lectins are injected in 100 μl of normal saline containing 0.05 to 0.1 mg of lectin. The FITC-dextran is made up in sterile water to a concentration of 20 to 50 mg/ml, and 100 μl of this solution is injected (Fig. 16.7). Tumor cell lines may also be transfected so that they stably express GFP or dSRed, and mouse strains have been developed in which endothelial cells constitutively express GFP (e.g., tie-2-GFP mice).

When using multiple fluorophores, it is important in confocal microscopy to select those that are well separated in terms of emissions, to reduce

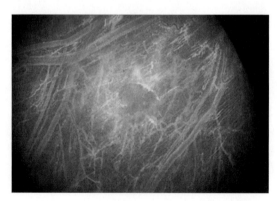

Figure 16.7   Fluorescence microscope image of window chamber with human pancreatic tumor after intravenous FITC-dextran (2 M mol. wt.) and FITC-lectin. Leakage around the tumor can be seen; tumor-associated vessels are permeable. The tumor cells are expressing dSRed and growing in a *nu/nu* mouse window chamber. (See color insert.)

color channel cross-talk. The maximum possible separation of emission spectra is rather critical for scanning multiphoton microscopy, as only one excitatory laser wavelength is used. In our experience, the combination of FITC and dSRed fluorophores yields good results in multiphoton and confocal microscopy. Some of the fluorophores recently developed for multiphoton microscopy, such as red fluorescent protein variants (Hc-RFP) (Tsai et al., 2006), emit in the longer wavelengths (600 nm or longer), which reduces the scattering of emitted light from deep within tissue samples and living tissues.

2.6.3. Confocal microscopy

The laser-scanning confocal microscope only allows the detection of light from a specific volume within the plane of focus, so that the resultant image is comparatively free of scattered light and attendant blurring. A pinhole aperture is placed at the conjugate focus of the objective so that sample-emitted light rays that are outside the focal plane of the objective lens, and therefore not in focus, are physically rejected and do not enter the detection system. This reduces the amount of light entering the system, but the advantage is a considerable improvement in image clarity and depth penetration within the sample. This permits sequential, thin, horizontal optical sections to be acquired axially through a sample (Semwogerere and Weeks, 2005). These individual images may be concatenated to create a three-dimensional rendering of the object of interest. The objective lenses should be long-working distance lenses that are adjusted for imaging through a cover glass.

A benefit of confocal microscopy is that it allows two or more excitatory laser sources of different wavelengths to be applied. Several fluorophores with widely divergent emission spectra may be precisely imaged using photomultiplier tube (PMT) detectors (Fig. 16.8). Combinations of lasers and detectors may be turned off and on so that only specific fluorophores may be fully excited and a relatively narrow band of each emission spectrum detected and imaged. Laser confocal excitation is relatively robust, so that the emitted light may be de-scanned through the scanning mirrors. De-scanning makes it possible to use a spectral detector, rather than PMTs alone, to cleanly detect the output of many fluorophores at once. Most commercially available confocal systems have, in addition to the PMTs or spectral detection system, another PMT positioned above or below the sample to detect the fraction of light transmitted through the tissue. Therefore, a transmitted light image may be acquired and combined with the florescent images.

2.6.4. Scanning multiphoton microscopy

The laser-scanning multiphoton microscope is designed to image features deep within scattering samples, such as the living rodent brain, thick organ and tumor sections, and small whole animal preparations. Imaging depths of

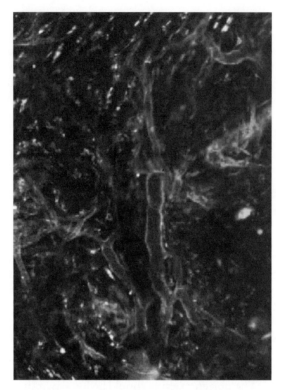

Figure 16.8   Confocal microscope image of a human M21 (av β3 integrin positive) melanoma tumor growing in the window chamber. Tumor cells (blue) are labeled with a fluorophore-linked nanoparticle that recognizes av β3 on the tumor cells. The typically chaotic tumor blood vessels are labeled with rhodamine-lectin, which appears red in the image. (See color insert.)

500 to 1000 microns can be realized by the application of a long-wavelength (infrared) 800- to 1200-nm laser, which can traverse tissue with less scattering and attenuation (Fig. 16.9A and B). Multiphoton imaging is based on the principle, first enunciated by Maria Geoppert-Mayer, that two temporally nearly coincident photons, each with half the energy and wavelength of a single photon that would normally excite a given fluorophore, will together be sufficient to excite the fluorophore and cause light to be emitted (Schenke-Layland *et al.*, 2006). This is true of virtually all fluorophores, and the fluorescence generated by multiphoton events is proportional to the incident laser intensity and the illuminated area (Tsai *et al.*, 2002). The amount of fluorescence generated from a given axial plane is seen to decrement as a quadratic function of the distance from the focal plane (Masters and So, 1999; Tsai *et al.*, 2002). Therefore, the photon flux and the probability of a two-photon event fall off precipitously as the distance from the focal plane is

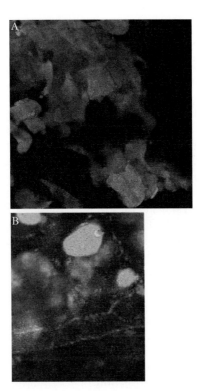

Figure 16.9 (A) Scanning multiphoton 3D rendering. The invading periphery of a human FGM tumor expressing dSRed was imaged at 40×. (B) Multiphoton image taken at 40×; the vessels, labeled with rhodamine–lectin show up in red, and the human melanoma (M21L, avβ3 integrin negative) cells in green (expressing GFP), surround the microvessels. (Images acquired in collaboration with Benjamin Migliori, and Philbert Tsai, Ph.D., and David Kleinfeld, Ph.D., Department of Physics, University of California, San Diego.) (See color insert.)

increased. This fact leads to four distinct advantages in terms of multiphoton imaging:

1. Fluorophore excitation is virtually nonexistent outside the focal plane; consequently, photobleaching is much reduced during multiphoton imaging (St. Croix *et al.*, 2006).
2. The energy of the incident laser is not effectively transferred to molecules outside the focal plane so that phototoxicity is greatly reduced (St. Croix *et al.*, 2006).
3. There is no interfering scattered light arising from above or below the sample because light is emitted only from the focal plane. This is an improvement over confocal microscopes, which have at least some degree of interfering scattered light.

4. No pinhole is needed because two-photon excitation events only occur in a sub-femtoliter volume within the focal plane. Without a pinhole, all the emitted light from the focal plane can be detected. Thus, more light is collected compared with confocal microscopy in which a pinhole is used.

In laser-scanning multiphoton microscopy, the lasers are pulsed on the femtosecond time scale to maximize intensity within a very short time frame, but the overall energy deposition to the sample is reduced. However, since a single long-wavelength laser is used, all fluorophores in the sample will be excited at once; thus, if two or more fluorophores are imaged they must be widely separated in terms of their emission spectra to minimize cross-talk between color channels. In addition, the objective lenses must be of high numerical aperture (>1.00) and they must have robust IR and UV transmission characteristics. The PMT should be situated just beyond the objective, that is, as close to the sample as possible, to maximize signal detection.

De-scanning of the multiphoton emission signal should be avoided, as this results in loss of signal amplitude due to distance, divergence of light, and scattering caused by imperfections in the scanning mirrors and other optical components (Majewska et al., 2000; Masters and So, 1999; Tsai et al., 2002). The incident laser should be introduced to the scanning mirrors as directly as is feasible, avoiding fiberoptics and correcting optics, as these tend to introduce significant laser pulse broadening and a reduction in multiphoton efficiency.

A properly configured multiphoton system should be able to provide optically sectioned images from depths of 500 microns or more below the surface of highly scattering tissues (Tsai et al., 2002). Multiphoton microscopy can acquire images with submicron lateral resolution and micron axial resolution on the millisecond time scale (Tsai et al., 2002). We have obtained good imaging even with window chambers of long duration that had occluding material adherent to the coverslip. The image clarity with deep-tissue microscopy using the multiphoton microscope generally exceeds what can be acquired with confocal microscopy.

ACKNOWLEDGMENTS

Rick Calou prepared the photograph of the microscope stage and the schematic drawings of the window chamber plates and the window-chamber microscope stage apparatus. Eric Murphy, PhD, assisted with the confocal imaging. Benjamin Migliori, Phil Tsai, PhD, and David Kleinfeld, PhD, collaborated with respect to the scanning multiphoton images. Virginia J. Makale, PhD, and Donita Serban, PhD, proofread the manuscript and provided helpful criticism.

REFERENCES

Abdul-Karim, M. A., Al-Kofahi, K., Brown, E. B., Jain, R. K., and Roysam, B. (2003). Automated tracing and change analysis of angiogenic vasculature from in vivo multiphoton confocal image time series. Microvasc. Res. **66**(2), 113–125.

Algire, G. H. (1943). An adaptation of the transparent chamber technique to the mouse. *J. Natl. Cancer Inst.* **4**, 1–11.

Algire, G. H. (1953). Visualizing cellular behavior *in vivo*. *J. Int. Chirg.* **13**, 381–384.

Arfors, K. E., Jonsson, J. A., and McKenzie, F. N. (1970). A titanium rabbit ear chamber: Assembly, insertion and results. *Microvasc. Res.* **2**, 516–518.

Bertera, S., Geng, X., Tawdrous, Z., Bottino, R., Balamurugan, A. N., Rudert, W. A., Drain, P., Watkins, S. C., and Trucco, M. (2003). Body window-enabled *in vivo* multicolor imaging of transplanted mouse islets expressing an insulin-timer fusion protein. *Biotechniques* **35**, 718–722.

Branemark, P—I (1971). *Intravascular Anatomy of Blood Cells in Man.* Basel: Karger.

Brooks, P. C., Klemke, R. L., Schon, S., Lewis, J. M., Schwartz, M. A., and Cheresh, D. (1997). Insulin-like growth factor receptor cooperates with integrin avβ5 to promote tumor cell dissemination *in vivo*. *J. Clin. Invest.* **99**(6), 1390–1398.

Chen, D. C., Agopian, V. G., Avansino, J. R., Lee, J. K., Farley, S. M., and Stelzner, M. (2006). Optical tissue window: A novel method for optimizing engraftment of intestinal stem cell organoids. *J. Surg. Res.* **134**(1), 52–60.

Condeelis, J., Singer, R. H., and Seagall, J. E. (2005). The great escape: When cancer cells hijack the genes for chemotaxis and motility. *Annu. Rev. Cell. Dev. Biol.* **21**, 695–718.

Coomber, B. L., Yu, J. L., Father, K. E., Plumb, C., and Rak, J. W. (2003). Angiogensis and the role of epigenetics in metastasis. *Clin. Exp Metastasis* **20**, 215–227.

Deryugina, E. I., Bourdon, M. A., Jungwirth, K., Smith, J. W., and Strongin, A. Y. (2000). Functional activation of integrin avβ3 in tumor cells expressing membrane–type 1 matrix metalloproteinase. *Int. J. Cancer* **86**, 15–23.

Devoisselle, J. M., Begu, S., Tourne-Peteilh, C., Desmettre, T., and Mordon, S. (2001). *In vivo* behaviours of long circulating liposomes in blood vessels in hamster inflammation and septic shock models—use of intravital fluorescence microscopy. *Luminescence* **16**, 73–78.

Dewhirst, M. W., Klitzman, B., Braun, R. D., Brizewl, D. M., Haroon, Z. A., and Secomb, T. W. (2000). Review of methods used to study oxygen transport at the microcirculatory level. *Int. J. Cancer* **90**, 237–255.

Dewhirst, M. W., Shan, S., Cao, Y., Moeller, B., Yuan, F., and Li, C. Y. (2002). Intravital fluorescence facilitates measurement of multiple physiologic functions and gene expression in tumors of live animals. *Dis. Markers* **18**(5–6), 293–311.

Dreher, M. R., Liu, W., Michelich, C. R., Dewhirst, M. W., Yuan, F., and Chilkoti, A. (2006). Tumor vascular permeability, accumulation, and penetration of macromolecular drug carriers. *J. Natl. Cancer Inst.* **98**(5), 335–344.

Eliceiri, B. P., Puente, X. S., Hood, J. D., Stupack, D. G., Schlaepfer, D. D., Huang, X. Z., Sheppard, D., and Cheresh, D. A. (2002). Src-mediated coupling of focal adhesion kinase to integrin avβ5 in vascular endothelial growth factor signaling. *J. Cell Biol.* **157**(1), 149–159.

Endrich, B., Intaglietta, M., Reinhold, H. S., and Gross, J. F. (1979). Hemodynamic characteristics in microcirculatory blood channels during early tumor growth. *Cancer Res.* **39**(1), 17–23.

Endrich, B., Asaishi, K., Gotz, A., and Messmer, K. (1980). Technical report—a new chamber technique for microvascular studies in unanesthetized hamsters. *Res. Exp. Med. (Berl.)* **177**, 125–134.

Felding-Habermann, B. (2003). Integrin adhesion receptors in tumor metastasis. *Clin. Exp. Metastasis* **20**, 203–213.

Gohongi, T., Todoroki, T., Fukumura, D., and Jain, R. K. (2004). Influence of the site of human gallbladder xenograft (Mz-ChA-1) on angiogenesis at the distant site. *Oncol. Rep.* **11**(4), 803–807.

Hegerfelt, Y., Tusch, M., Brocker, E., and Friedl, P. (2002). Collective cell movement in primary melanoma explants: plasticity of cell-cell interaction, β1-integrin function, and migration strategies. *Cancer Res.* **62**, 2125–2130.

Ingber, D. E. (1992). Extracellular matrix as a solid-state regulator in angiogenesis: Identification of new targets for anti-cancer therapy. *Sem. Cancer Biol.* **3**, 57–63.

Jain, R. K., Munn, L. L., and Fukumara, D. D. (2002). Dissecting tumor pathophysiology using intravital microscopy. *Nat. Rev. Cancer* **2**, 266–276.

Kashiwagi, S., Izumi, Y., Gohongi, T., Demzou, Z. N., Xu, L., Huang, P. L., Buerk, D. G., Munn, L. L., Jain, R. K., and Fukumura, D. (2005). NO mediates mural cell recruitment and vessel morphogenesis in murine melanomas and tissue-engineered blood vessels. *J. Clin. Invest.* **115**(7), 1816–1827.

Kerger, H., Torres Filho, I. P., Rivas, M., Winslow, R. M., and Intaglietta, M. (1995). Systemic and subcutaneous microvascular oxygen tension in conscious Syrian golden hamsters. *Am. J. Physiol. Heart Circ. Physiol.* **268**, H802–H810.

Klemke, R. L., Yebra, M., Bayna, E. M., and Cheresh, D. A. (1994). Receptor tyrosine kinase signaling required for integrin $av\beta5$-directed cell motility but not adhesion on vitronectin. *J. Cell Biol.* **127**(3), 859–866.

Larsen, M., Artym, V., Green, J. A., and Yamada, K. M. (2006). The matrix reorganized: Extracellular matrix remodeling and integrin signaling. *Curr. Opin. Cell Biol.* **18**(5), 463–471.

Leunig, M., Yuan, F., Menger, M. D., Boucher, Y., Goetz, A. E., Messmer, K., and Jain, R. K. (1992). Angiogenesis, microvascular architecture, microhemodynamics, and interstitial fluid pressure during early growth of human adenocarcinoma LS174T in SCID mice. *Cancer Res.* **52**, 6553–6560.

Li, C. Y., Shan, S., Cao, Y., and Dewhirst, M. W. (2000). Role of incipient angiogenesis in cancer metastasis. *Cancer Metastasis Rev.* **19**, 7–11.

Liu, P., Zhang, A., Xu, Y., and Xu, L. X. (2005). Study of non-uniform nanoparticle in liposome extravasation in tumor. *Int. J. Hyperthermia* **21**(3), 259–270.

Majewska, A., Yiu, G., and Yuste, R. (2000). A custom-made two-photon microscope and deconvolution system. *Eur. J. Physiol.* **441**, 398–408.

Makale, M., Lin, J., Calou, R., Tsai, A., Chen, P., and Gough, D. A. (2003). Tissue window chamber system for validation of implanted oxygen sensors. *Am. J. Physiol. Heart Circ. Physiol.* **284**, H2288–H2294.

Masters, B. R., and So, P. T. C. (1999). Multi-photon excitation microscopy and confocal microscopy imaging of *in vivo* human skin: A comparison. *Microsc. Microanal.* **5**, 282–289.

Makale, M. T., Chen, P. C., and Gough, D. A. (2005). Variants of the tissue-sensor array window chamber. *Am. J. Physiol. Heart Circ. Physiol.* **289**, 57–65.

Menger, M. D., Laschke, M. W., and Vollmer, B. (2002). Viewing the microcirculation through the window: Some twenty years experience with the hamster dorsal skinfold chamber. *Eur. Surg. Res.* **34**, 83–91.

Menger, M. D., Walter, P., Hammersen, F., and Messmer, K. (1990). Quantitative analysis of neovascularization of different PTFE-implants. *Eur. J. Cardiothorac. Surg.* **4**(4), 191–196.

Mordon, S., Desmettre, T., Devoisselle, J. J., and Mitchell, V. (1997). Selective laser photocoagulation of blood vessels in a hamster skin flap model using a specific ICG formulation. *Lasers Surg. Med.* **21**, 365–373.

Pahernik, S., Harris, A. G., Schmitt-Sody, M., Kransnici, S., Goetz, A. E., Dellian, M., and Messner, K. (2002). Orthogonal polarization spectral imaging as a new tool for the assessment of antivascular tumor treatment *in vivo*: A validation study. *Br. J. Cancer* **86**, 1622–1627.

Petitclerc, E., Stromblad, S., von Schalscha, T., Mitjans, F., Piulats, J., Montgomery, A. M. P., Cheresh, D. A., and Brooks, P. C. (1999). Integrin $av\beta3$ promotes M21 melanoma growth in human skin by regulating tumor cell survival. *Cancer Res.* **59**, 2724–2730.

Robinson, S. D., Wilson, S., and Hodivala-Dilke, K. M. (2006). Generation of genetically modified embryonic stem cells for the development of knockout mouse animal model systems. *Methods Mol. Med.* **120,** 465–477.

Sahai, E., Wyckoff, J., Philippar, U., Seagall, J. E., Gertler, F., and Condeelis, J. (2005). Simultaneous imaging of GFP, CFP and collagen in tumors *in vivo* using multiphoton microscopy. *BMC Biotechnol.* **23,** 5–14.

Sandison, J. C. (1924). A new method for the microscopic study of living growing tissues by the introduction of a transparent chamber in the rabbits ear. *Anat. Rec.* **28,** 281–287.

Schenke-Layland, K., Riemann, I., Damour, O., Stock, U. A., and Konig, K. (2006). Two-photon microscopes and *in vivo* multiphoton tomographs—powerful diagnostic tools for tissue engineering and drug delivery. *Adv. Drug Delivery Rev.* **58**(7), 878–896.

Semwogerere, D., and Weeks, E. R. (2005). Confocal microscopy. *In* "Encyclopedia of Biomaterials and Biomedical Engineering." Taylor and Francis, London.

Sieg, D. J., Hauck, C. R., Ilic, D., Klingbeil, C. K., Schaefer, E., Damsky, C. H., and Schlaepfer, D. D. (2000). FAK integrates growth-factor and integrin signals to promote cell migration. *Nat. Cell Biol.* **2,** 249–257.

St. Croix, C. M., Leelavanichkul, K., and Watkins, S. C. (2006). Intravital fluorescence microscopy in pulmonary research. *Adv. Drug Delivery Rev.* **58,** 834–840.

Stupack, D. G., and Cheresh, D. A. (2002). Get a ligand, get a life: Integrins, signaling and cell survival. *J. Cell Sci.* **115,** 3729–3738.

Tsai, P. S., Nishimura, N., Yoder, E., Dolnick, E., White, G. A., and Kleinfeld, D. (2002). Principles, design, and construction of a two-photon laser-scanning microscope for *in vitro* and *in vivo* brain imaging. *In* "*In Vivo* Optical Imaging of Brain Function" (Ron D. Frostig, ed.), pp.115–171. CRC Press, Boca Raton, FL.

Tsai, T. H., Lin, C. Y., Tsai, H. J., Chen, S. Y., Tai, S.-P., Lin, K. H., and Sun, C. K. (2006). Biomolecular imaging based on far-red fluorescent protein with a high two-photon excitation action cross section. *Optics Lett.* **31**(7), 930–932.

Vranckx, J. J., Slama, J., Preuss, S., Perrez, N., Svensjo, T. S., Breuing, K., Bartlett, R., Pribaz, J., Weiss, D., and Eriksson, E. (2002). Wet wound healing. *Plast. Reconstr. Surg.* **110,** 1680–1687.

Williams, R. G. (1934). An adaptation of the transparent-chamber technique to the skin of the body. *Anat. Rec.* **60,** 487–491.

Yuan, F., Leunig, M., Berk, D. A., and Jain, R. K. (1993). Microvascular permeability of albumin, vascular surface area, and vascular volume measured in human adenocarcinoma LS174T using dorsal chamber in SCID mice. *Microvasc. Res.* **45,** 269–289.

CHAPTER SEVENTEEN

USING *XENOPUS* EMBRYOS TO INVESTIGATE INTEGRIN FUNCTION

Douglas W. DeSimone,* Bette Dzamba,† *and* Lance A. Davidson‡

Contents

Abstract

Xenopus embryos are a useful and important system for cell biological studies of integrin adhesion and signaling. Explants prepared from gastrulating embryos undergo normal morphogenetic movements when cultured in simple salt solutions. These preparations are accessible to a variety of experimental perturbations and time-lapse imaging at high resolution, making it possible to elucidate mechanisms of integrin function in intact tissues and whole embryos. Methods used for the visualization of integrins, cadherins, extracellular matrix, and cytoskeletal linkages in both fixed and live tissues are described. We also discuss the use of a novel explant preparation suitable for following the normal deposition and assembly of fibronectin fibrils by ectoderm and mesoderm at gastrulation.

* Department of Cell Biology, and Morphogenesis and Regenerative Medicine Institute, School of Medicine, University of Virginia, Charlottesville, Virginia
† Department of Cell Biology, School of Medicine, University of Virginia, Charlottesville, Virginia
‡ Department of Bioengineering, University of Pittsburgh, Pittsburgh, Pennsylvania

Methods in Enzymology, Volume 426
ISSN 0076-6879, DOI: 10.1016/S0076-6879(07)26017-3
© 2007 Elsevier Inc.
All rights reserved.

1. Introduction

Amphibian eggs and embryos have long been a preferred system for experimental embryology owing to their ready availability, ease of culture, and rapid and robust development. Most amphibian embryos are relatively large and thus amenable to microinjection and microsurgical techniques. Cells and tissues can be removed from specific embryonic regions and cultured in simple salt solutions without added CO_2, serum, or other supplements. In many cases, these explanted cells and tissues will recapitulate the normal behaviors and movements that they undergo in the intact embryo. These particular properties make amphibian embryos ideal for addressing cell biological questions, particularly those involving mechanisms of adhesion mediated by cadherins and integrins. This chapter focuses on techniques and preparations that have been used successfully to study integrins and their ligands in a biologically relevant context—amphibian gastrulation and the cellular movements and rearrangements that drive this process.

The South African clawed toad *Xenopus laevis* is currently the most widely used amphibian species for embryologic investigation (Kay and Peng, 1991; Sive *et al.*, 2000). It has been a particularly useful model for studies of cell adhesion and migration and for elucidating the functions of integrins and extracellular matrix (ECM) in these cellular processes (Dzamba *et al.*, 2001). Prior to the initiation of gastrulation movements, fibronectin (FN) fibrils are deposited along the blastocoel roof (Lee *et al.*, 1984) and serve as an adhesive substrate for migrating mesendodermal cells (Davidson *et al.*, 2002). Disruptions of either FN or integrin $\alpha5\beta1$ perturb not only *Xenopus* mesendoderm migration, but also convergent extension movements and the spreading of the superficial ectoderm by epiboly (Davidson *et al.*, 2006; Marsden and DeSimone, 2001, 2003). At least two major integrin-related questions arise from studies of this type. First, how are FN matrices assembled *in vivo*? Second, what roles do integrins play in coordinating the adhesive and migratory behaviors of cells and tissues. The majority of other studies devoted to similar questions have utilized avian and mammalian cell lines cultured on planar surfaces. The methods described here illustrate the utility of *Xenopus* embryos as an *in vivo* system suitable for detailed investigations of integrin functions.

The purpose of this chapter is to provide examples of preparations that are useful for visualizing integrins and the associations of these transmembrane receptors with extracellular ligands and cytoskeletal elements in tissues undergoing morphogenesis at gastrulation. First, we discuss the isolation, fixation, and immunostaining of tissue explants. Second, we describe the use of GFP-tagged integrin subunits for real-time visualization of integrins in protrusive cells. Finally, we discuss a novel preparation that enables the real-time monitoring of integrin-dependent FN fibrillogenesis in a living tissue. A number of useful resources are available that serve as a general guide for the preparation of *Xenopus* embryos and the most common methodologies used to investigate function (Kay and Peng, 1991; Sive *et al.*, 2000). The latter include knockdowns of protein expression by antisense morpholino, expression of transcripts synthesized *in vitro*, and the production of transgenic frogs. We do not discuss these methods in detail because they are standard techniques in the *Xenopus* field, and are described in numerous primary papers and methods reviews (e.g., Sive *et al.*, 2000). Instead, we focus here on a selection of specialized techniques used to visualize integrin and ECM interactions in tissues; these methods are applicable to a range of experimental strategies used to address function. In addition, a number of investigators have used *Xenopus* explants and individual cells from dissociated tissues to study cell migration on FN *in vitro*. Detailed methods for several of these preparations are described in DeSimone *et al.* (2005).

It is important to mention that a lot of interest in recent years has been directed towards *Xenopus tropicalis* as an alternative and/or complementary model system to *X. laevis*. *X. tropicalis* shares the primary advantages of *X. laevis* but adds rapid development, fast generation times, a diploid genome (*X. laevis* is tetraploid), and the potential for genetic analyses. Substantial resources are available for *X. tropicalis*, including 8X sequence coverage of the genome by the U.S. Department of Energy Joint Genome Institute (JGI) (see http://genome.jgi-psf.org/Xentr4/Xentr4.info.html), a database of more than 900,000 expressed sequence tags (ESTs) along with arrayed embryonic cDNA libraries (Gilchrist *et al.*, 2004), DNA microarrays, and extensive databases of gene markers (Amaya, 2005). These features, coupled with the fact that key signaling processes and morphogenetic behaviors in amphibia are conserved with mammals and other vertebrates, make *X. tropicalis* a particularly promising system for studies of early development. The methods described in this chapter focus on *X. laevis* but are adaptable to *X. tropicalis*.

2. Visualization of Integrins, Extracellular Matrix, and Cytoskeleton in Embryo Explants

2.1. Embryo fixation and microsurgery

We routinely fix both whole embryos and explants in 4% formaldehyde in PBS for either 2 h at room temperature (RT) or overnight at 4°. The formaldehyde must be free of methanol for phalloidin staining. Round-bottom microfuge tubes (2.0 ml) placed on a Nutator or similar device are convenient containers for fixation and subsequent washing and incubation steps (described below). Explants may be prepared either before or after fixation. Fixation of whole embryos before microdissection has the advantage that all of the samples can be taken at the same time point, assuming that the embryos selected for dissection are from the same clutch of eggs fertilized at the same time. This is particularly important when investigating dynamic changes occurring over a short developmental time period, such as FN fibril assembly and the cytoskeletal rearrangements that accompany rapid morphogenetic movements at gastrulation. A second advantage to fixing embryos prior to preparing explants is that it avoids the stretching of tissues and wound responses that are inevitably generated when dissecting live embryos. These perturbations can alter the organization of the matrix or cytoskeleton, particularly along wound margins. One of the primary disadvantages of fixation prior to dissection is that the tissue becomes brittle and thus subject to cracking during dissection, although this is of little consequence at high magnifications when the goal is to image single cells within the tissue. In addition, some tissues may be more difficult to separate from the rest of the embryo following fixation.

For gastrula-stage embryos, the blastocoel roof tends to collapse during fixation if the vitelline envelope is left on. This can be avoided by removing the vitelline envelope before fixation. The vitelline envelope or membrane, as it is often called, is an extracellular network of glycoproteins that becomes readily visible following fertilization due to swelling of the perivitelline space (i.e., between egg membrane and vitelline layer). With practice and a deft hand, the vitelline envelope can be removed rapidly with watchmaker's forceps. However, if timing is critical and a large number of samples is needed, it may not be practical to remove the vitelline envelopes before fixation. In this case, embryos are fixed for a brief time period (30 min to 1 h) before removing the vitelline membrane. Embryos can then be placed back in the fix for an additional 1 to 1.5 h at RT or overnight at 4°. An alternative is to incubate the fixed embryos in TBS containing 0.1% Tween before dissection, which often has the effect of reversing the collapse of the blastocoel roof.

A range of microdissection "tools" may be used for explant making, depending on both the type of explant being produced and the personal preferences of the investigator. Watchmaker's forceps, tungsten or glass needles, and "eyebrow knives" are suitable for this type of work (Sive *et al.*, 2000). Dissections are typically done in a plastic (nontissue culture) Petri dish that has been "roughed up" with a razor blade to make it less slippery or in a dish lined with modeling clay (i.e., plasticine) where the surface can be shaped to accommodate and hold embryos in place for dissection. One important thing to bear in mind when working with live devitellinized embryos and tissue explants is that surface tension at the air–water interface will cause "impressively instantaneous destruction" of amphibian cells and tissues—as most beginners learn firsthand following their initial attempts at microsurgery. This is easily avoided by transferring devitellinized embryos and dissected tissues by pipette with care to avoid contact of the biological material with the interface.

2.2. Methods for immunostaining of *Xenopus* explants for confocal microscopy

We find the following general protocols suitable for immunostaining of ECM molecules such as FN, membrane glycoproteins such as integrins, and the actin cytoskeleton in thick tissue explants.

1. Incubate explants for 2 h at RT (or overnight at 4°) with primary antibody (concentration determined empirically) diluted in TBS containing 0.1% Tween-20 (TBST). For example, we routinely use the anti-FN mAb 4H2 (Ramos and DeSimone, 1996) at 1 μg/ml in order to visualize FN fibrils in animal cap tissues (Fig. 17.1).
2. Wash three times for 5 min each with TBST.
3. Incubate 2 h at RT (or overnight at 4°) with suitable secondary antibody. We typically use Alexa Fluor conjugated antibodies obtained from Molecular Probes (Eugene, OR) at a dilution of 1:2000.
4. Wash three times for 5 min each with TBST.
5. Mount on glass slides in 50% glycerol using silicone grease to suspend a coverslip over the explant with gentle compression. The latter may require some trial and error to identify a level of compression sufficient to obtain a relatively flat surface for imaging without distorting the tissue or crushing cells. The coverslip is then sealed to the microscope slide with clear nail polish.

2.3. Visualization of actin in *Xenopus* explants

If colocalization with the actin cytoskeleton is required, we have determined that the following additional steps yield satisfactory results for simultaneous imaging of both ECM and integrins or cadherins (Fig. 17.1).

1. Following fixation as described above, explants are permeabilized with 0.5% saponin in TBS for 5 min. If both actin and ECM are to be visualized simultaneously, the permeabilization should be carried out before antibody incubation.
2. Wash three times at 5 min each with TBST as above.
3. Incubate 2 h at RT with two units per milliliter of Alexa Fluor 488- or 546-conjugated Phalloidin (Molecular Probes, Eugene, OR) in TBS. Note that the phalloidin can be added with either the primary or secondary antibody if both actin and other molecules (e.g., FN, integrins, cadherins) are also to be visualized.

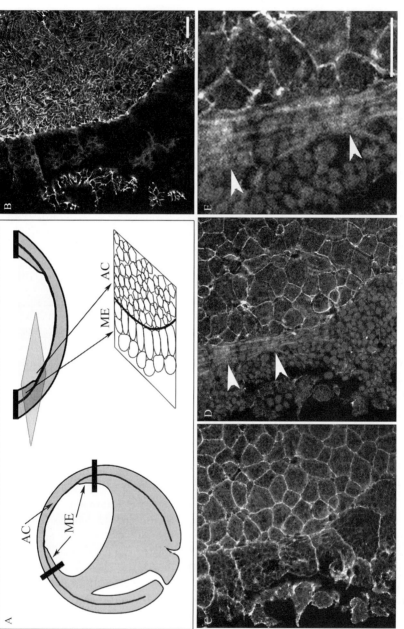

Figure 17.1 Single confocal section of a triple-labeled gastrula stage animal cap. (A) Schematic of explant preparation and area imaged (B through E). Animal caps (A and C) and the leading edge mesendoderm (ME) were microdissected from embryos fixed at late gastrula (stage 11.5). The area of the animal cap in each image (B through E) contains large mesendoderm cells that are migrating as a cohesive sheet across the blastocoel roof, which is made up of smaller animal cap cells. (B through E) A single confocal section of the animal cap at the level of the

4. Wash three times for 5 min each in TBST.
5. Mount as described above and image using confocal microscopy.

2.4. Microscopy: general considerations

One of the biggest challenges involved in imaging both live and fixed tissues obtained from frog embryos is the presence of yolk platelets within the cytoplasm. Aside from autofluorescence, yolk platelets cause significant scattering of light and can also mask other subcellular structures. This is particularly problematic using widefield microscopes and standard epifluorescence. Confocal microscopy and/or digital deconvolution of both widefield and confocal images can alleviate much of this problem. Confocal microscopy is particularly helpful for visualization of actin filaments in *Xenopus* tissues. Fluorescent dextrans injected into the cytoplasm of egg or early cleavage–stage blastomeres can be very useful as contrasting agents that help reveal cell borders in *Xenopus* by virtue of the fact that the dextrans are effectively "excluded" from the yolky cytoplasm but visible in the yolk-free subcortical region (Davidson and Keller, 1999; Marsden and DeSimone, 2001). In addition, membrane-bound GFP (Davidson *et al.*, 2002) or other GFP-tagged molecules such as integrins (Fig. 17.2) can be expressed in embryos where they outline cells and reveal dynamic changes in membrane protrusions using low-light imaging methods and time-lapse microscopy (Davidson *et al.*, 2006).

Typical preparations of both fixed and live explants from *Xenopus* gastrulae are shown in Figs. 17.1 and 17.2, respectively. The blastocoel roof of a fixed mid-gastrula stage embryo with migratory mesendoderm attached was removed (Fig. 17.1A) and triple-labeled to visualize FN fibrils (Fig. 17.1B), C-cadherin, and actin filaments (Fig. 17.1C and D). This preparation reveals the migratory mesendodermal cell sheet as it progresses across the blastocoel roof of the animal cap, which expresses a dense fibrillar FN matrix (fibrils are partially obscured by the overlying mesendoderm in Fig. 17.1B). C-cadherin staining reveals the polygonal shapes of the animal cap cells of the blastocoel roof and outlines the large, yolk-filled mesendodermal cells (Fig. 17.1C). The organization of the actin cytoskeleton differs dramatically in the mesendoderm and animal cap cells. Actin becomes

fibronectin matrix at the surface of the animal cap cells. Scale bars = 25 μm. (B) Fibronectin visualized using mAb 4H2 and Alexa 546 anti-mouse IgG. Fibronectin forms a dense fibrillar matrix on the AC, which is partially obscured at left by the mesendodermal cell sheet. (C) C-cadherin localized with polyconal antisera and Alexa 647 anti-rabbit IgG. Animal cap cells are packed in a polygonal array. C-cadherin also stains and reveals the outlines of the much larger cells of the mesendodermal cell sheet at left. (D) Actin cytoskeleton labeled with Alexa 488 conjugated phalloidin. Actin is arranged at the periphery of animal cap cells and forms long bundles along the edge of the mesendoderm (arrowheads). (E) Higher magnification view of the leading-edge mesendoderm from D.

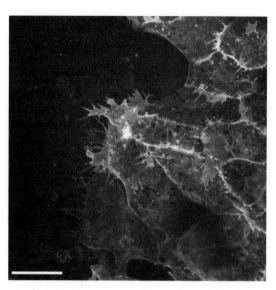

Figure 17.2   Projection of confocal sections of an animal cap expressing integrin α5–GFP. The animal cap was dissected from an early gastrula (stage 10) embryo and cultured on a fibronectin-coated coverslip for live imaging using a Nikon C1 confocal microscope and a 100× objective. A total of 135 Z-sections were collected at 0.1-μm intervals. Scale bar = 25 μm.

localized cortically in animal cap cells as FN fibrillogenesis proceeds (Dzamba and DeSimone, in preparation), while in the migratory mesendoderm, actin filaments form long filaments along the leading edge of the tissue (Fig. 17.1D and E, arrowheads).

Of course, the dynamic events hinted at in fixed preparations such as that shown in Fig. 17.1, are ideally studied in living tissues, and we have developed methods for real-time visualization of integrins (Fig. 17.2B), FN (Fig. 17.3), and actin (Davidson, Keller, and DeSimone, unpublished observations) at gastrulation. Figure 17.2 is a z–series projection of confocal sections of animal cap cells that are expressing a *Xenopus* integrin α5 subunit with GFP attached to the C–terminal cytoplasmic tail. This α5–GFP construct dimerizes with endogenous integrin β1, and is expressed at the plasma membrane (Dzamba, Davidson, and DeSimone, unpublished observations) where it can be followed in filopodial and lamellipodial protrusions.

3. Live Imaging of Fibronectin Fibrils

We have developed a novel preparation that enables the dynamic assembly and remodeling of FN fibrils by ectodermal and mesodermal tissues to be visualized (Fig. 17.3). One particular advantage of this

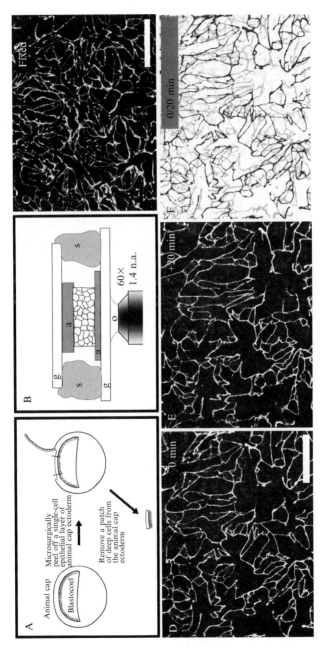

Figure 17.3 (A) Explants with epithelial cell layers were prepared from late blastula or early gastrula stage embryos. (B) Explants were sandwiched between a thin (lower) and thick (upper) sheet of agarose, and a glass coverslip bridge held in place by silicone grease. The thin sheet of agarose (typically less than 25 μm) allowed high resolution of Cy3-tagged mAb 4H2 bound to fibronectin fibrils. (C) Fibrils organization in equivalent stage animal caps fixed with 5% TCA and immunofluorescently stained with mAb 4H2. (D and E) Frames from a time-lapse sequence of an animal cap explant separated by 20 minutes. (F) The two frames (D and E) overlaid with the later frame shown in grey. Scalebars in (C) and (D) indicate 50 μm. (D), (E), and (F) are at the same scale.

preparation is that FN fibrils are free to "move" with the tissues that assemble them at their surfaces. This is possible because the tissue is sandwiched between thin sheets of agarose, which are optically transparent, hold the tissue flat for observation, and do not restrict the movements of nascent FN fibrils as they are assembled at cell surfaces. A Cy3-labeled mAb directed against *Xenopus* FN is included in the culture chamber in order to decorate the FN fibril, which is subsequently imaged by confocal time-lapse microscopy. The dynamic nature of these fibrils is demonstrated in Fig. 17.3, which also illustrates the explant preparation and agarose viewing chamber used for imaging.

3.1. Preparation of thin agarose sheets

A heat block is preheated to 60° and filled with water. One-percent agarose in double-distilled water is prepared and kept molten in the heat block. A single large coverslip is then placed on the heat block and a single drop of agarose pipetted onto the warmed coverslip, followed immediately by a smaller coverslip fragment. The thickness of the resulting agarose gel can be controlled by the amount of molten agarose sandwiched between the two glass coverslips. The coverslip with the agarose is then immediately placed into a humid chamber to prevent evaporation from the edges of the agarose gel. In 15 min, the coverslip with the agarose is removed from the humid chamber and immersed in $1/3 \times$ MBS (Modified Barths Saline), and the top coverslip fragment is removed with forceps.

3.2. Explant preparation

Explants consisting of deep ectoderm are isolated using hair loops and eyebrow knives (Sive *et al.*, 2000). Late-blastula or early-gastrula stage embryos are manually removed from their vitelline envelopes with forceps. The outer, single cell–layered epithelium is then peeled back and a small explant consisting entirely of deep animal cap cells removed (Fig. 17.3A). The explant is then transferred to a confocal viewing chamber containing DFA (Sive *et al.*, 2000) supplemented with 0.1 mg/ml bovine plasma FN (Roche Molecular Biochemicals), antibotic/antimycotic (Sigma), and 5 μg/ml Cy3-4H2 (see discussion of antibody preparation below). Confocal viewing chambers were made from rectangular-shaped custom pieces of acrylic milled to provide small volume for culture media. Acrylic chambers are glued to large coverslips using silicone grease (Dow-Corning) and sealed with small, square glass coverslips. Explants are transferred to the well along with two thin rectangular sheets of agarose. The explant is then sandwiched between the agarose and held in place by a small coverslip fragment and silicone grease (Fig. 17.3B).

3.3. Live imaging of FN fibrils

For live imaging of FN fibrils, the mAb 4H2 (a non-function–blocking antibody raised against the central cell binding domain of frog fibronectin (Ramos and DeSimone, 1996) was conjugated to the Cy3 fluorophore using an NHS-ester form of Cy3 (FluoroLink, Amersham Pharmacia Biotech). Cy3-4H2 was dialyzed against DFA (without BSA) and concentrated with cutoff filters (Centricom YM-30 centrifuge filters, Amicon).

Confocal time-lapse sequences were collected using a confocal laser scan head PCM2000 (Nikon, Melville, NY) mounted on an inverted compound microscope (Nikon). High-resolution images were acquired with a 1.40 numerical aperature 60 × oil-immersion plan apochromat objective. The Cy3 signal was excited using a 543-nm Green HeNe laser and collected with the 565-nm emission filter. Independent neutral-density filters were used to reduce the excitation intensities to minimal levels to reduce photo damage to the samples. Furthermore, the confocal pinhole was opened to the "middle" position to collect from a larger z-slab and to increase brightness. Time-lapse sequences were collected and stored as uncompressed animation stacks on a computer running image acquisition software (Compix Inc., Cranberry Township, PA). These stacks were transferred to a computer and processed using ImageJ (ImageJ, Wayne Rasband, NIMH; see http://rsb.info.nih.gov/ij/). Processing was limited to linear adjustments to the lookup table and to one round of smoothening and one round of sharpening. Figure 17.3D and E show the initial and subsequent position of fibrils at 0 min and 20 min, respectively. These two images are superimposed in Fig. 17.3E as described in the legend.

4. Conclusions

Xenopus embryos provide a powerful system for the analysis of integrin interactions with both the ECM and actin cytoskeleton. Using explant preparations and confocal imaging techniques, it is possible to attain an unprecedented level of detail regarding dynamic changes in integrin–adhesive interactions on both sides of the membrane in tissues undergoing biologically relevant morphogenetic movements. Future work in our laboratory will seek to establish transgenic lines of *X. tropicalis* expressing integrins and related interacting molecules tagged with fluorescent proteins in order to more easily facilitate live cell imaging in embryos and explants.

ACKNOWLEDGMENTS

This work was supported by the U.S. Public Health Service (grants HD26402 and DE14365 to D.W.D., HD44750 to L.A.D.).

REFERENCES

Amaya, E. (2005). Xenomics. *Genome Res.* **15,** 1683–1691.

Davidson, L. A., Hoffstrom, B. G., Keller, R., and DeSimone, D. W. (2002). Mesendoderm extension and mantle closure in *Xenopus laevis* gastrulation: Combined roles for integrin alpha(5)beta(1), fibronectin, and tissue geometry. *Dev. Biol.* **242,** 109–129.

Davidson, L. A., and Keller, R. E. (1999). Neural tube closure in *Xenopus laevis* involves medial migration, directed protrusive activity, cell intercalation and convergent extension. *Development* **126,** 4547–4556.

Davidson, L. A., Marsden, M., Keller, R., and Desimone, D. W. (2006). Integrin alpha5-beta1 and fibronectin regulate polarized cell protrusions required for *Xenopus* convergence and extension. *Curr. Biol.* **16,** 833–844.

DeSimone, D. W., Davidson, L., Marsden, M., and Alfandari, D. (2005). The *Xenopus* embryo as a model system for studies of cell migration. *Methods Mol. Biol.* **294,** 235–245.

Dzamba, B. D., Bolton, M. A., and DeSimone, D. W. (2001). The integrin family of cell adhesion molecules. *In* "Cell Adhesion" (M. C. Beckerle, ed.), Vol. 39. Oxford University Press, Oxford.

Gilchrist, M. J., Zorn, A. M., Voigt, J., Smith, J. C., Papalopulu, N., and Amaya, E. (2004). Defining a large set of full-length clones from a *Xenopus tropicalis* EST project. *Dev. Biol.* **271,** 498–516.

Kay, B. K., and Peng, H. B. (1991). "*Xenopus laevis*: Practical Uses in Cell and Molecular Biology." New York: Academic Press, New York.

Lee, G., Hynes, R., and Kirschner, M. (1984). Temporal and spatial regulation of fibronectin in early *Xenopus* development. *Cell* **36,** 729–740.

Marsden, M., and DeSimone, D. W. (2001). Regulation of cell polarity, radial intercalation and epiboly in *Xenopus*: Novel roles for integrin and fibronectin. *Development* **128,** 3635–3647.

Marsden, M., and DeSimone, D. W. (2003). Integrin-ECM interactions regulate cadherin-dependent cell adhesion and are required for convergent extension in *Xenopus. Curr. Biol.* **13,** 1182–1191.

Ramos, J. W., and DeSimone, D. W. (1996). Xenopus embryonic cell adhesion to fibronectin: position-specific activation of RGD/synergy site-dependent migratory behavior at gastrulation. *J. Cell Biol.* **134,** 227–240.

Sive, H. L., Grainger, R. M., and Harland, R. M. (2000). "Early development of *Xenopus laevis*: A Laboratory Manual." Cold Spring Harbor Laboratory Press, Cold Spring Harbor, NY.

CHAPTER EIGHTEEN

METHODS TO STUDY LYMPHATIC VESSEL INTEGRINS

Barbara Garmy-Susini,* Milan Makale,* Mark Fuster,[†] *and* Judith A. Varner*

Contents

Abstract

The lymphatic system plays a key role in the drainage of fluids and proteins from tissues and in the trafficking of immune cells throughout the body. Comprised of a network of capillaries, collecting vessels, and lymph nodes, the lymphatic system plays a role in the metastasis of tumor cells to distant parts of the body. Tumors induce lymphangiogenesis, the growth of new lymphatic vessels, in the peritumoral space and also within tumors and lymph

* Moores UCSD Cancer Center, University of California, San Diego, La Jolla, California
† Department of Medicine, University of California, San Diego, California

Methods in Enzymology, Volume 426
ISSN 0076-6879, DOI: 10.1016/S0076-6879(07)26018-5

© 2007 Elsevier Inc.
All rights reserved.

nodes. Tumor lymphangiogenesis has been shown to play a role in promoting tumor metastasis. As mediators of lymphatic endothelial cell adhesion, migration, and survival, integrins play key roles in the regulation of lymphangiogenesis. Recent studies indicate that select integrins promote lymphangiogenesis during development and disease and that inhibitors or loss of expression of these integrins can block lymphangiogenesis. In this report, we describe methods to isolate and culture murine and human lymphatic endothelial cells as well as methods to analyze the expression of integrins on these cells. We also show how to assess integrin-mediated adhesion, migration, and tube formation *in vitro*. We demonstrate how to evaluate integrin function during lymphangiogenesis in a variety of animal models *in vivo*. Additionally, we show how to study lymphangiogenesis using intravital microscopy.

1. INTRODUCTION

Lymph nodes are the initial or frequent sites of metastasis for many tumors, including human pancreatic, gastric, breast, and prostate carcinomas, melanomas, and other tumors. Lymphangiogenesis, the growth of new lymphatic vessels, in tumors has been linked to the formation of lymph node metastases (Bando *et al.*, 2006; Dadras *et al.*, 2005; Karkkainen *et al.*, 2004; Massi *et al.*, 2006; Roma *et al.*, 2006). Lymphatic capillaries, unlike typical blood capillaries, lack pericytes and a continuous basal lamina. Due to their greater permeability, lymphatic capillaries are thought to be more effective than blood capillaries in allowing passage of tumor cells into and out of vessels (Fig. 18.1). Tumor-secreted factors such as VEGF-C (Karkkainen *et al.*, 2004; Skobe *et al.*, 2001) and VEGF-A (Hirakawa *et al.*, 2005; Nagy *et al.*, 2002) have been shown to promote lymphangiogenesis within tumors. These factors activate VEGFR3, a tyrosine-kinase VEGF family receptor that is expressed primarily on lymphatic endothelium (He *et al.*, 2005). Expression of VEGF-C is correlated with increased lymph node metastasis and poor clinical outcome in a variety of tumors (Bando *et al.*, 2006; Karkkainen *et al.*, 2004; Massi *et al.*, 2006; Roma *et al.* 2006; Skobe *et al.*, 2001). Indeed, in animal models of metastasis, inhibitors of VEGF-C (soluble VEGFR3) inhibited tumor lymphangiogenesis and tumor metastasis to lymph nodes (He *et al.*, 2005). Lymphatic endothelial cell (LEC) integrin, integrin $\alpha 4\beta 1$, regulates lymphangiogenesis as well as tumor metastasis to the lymph nodes.

While growth factors and their receptors play critical roles in angiogenesis and lymphangiogenesis, the integrin family of cell adhesion proteins controls cell attachment to the extracellular matrix and promotes the survival, proliferation, and motility of many cell types. Angiogenesis, the development of new blood vessels, depends not only on soluble growth factors such as VEGF-A, but also on survival and migratory signals transduced by the

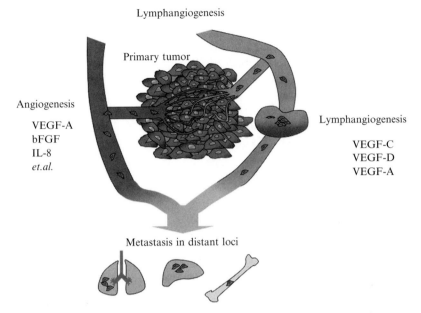

Figure 18.1 Model of tumor lymphangiogenesis. Current theory holds that tumors spread via both hematogenous (vascular) and lymphatic routes. Tumors secrete a number of factors including VEGF-A, VEGF-C, and others that induce both lymphangiogenesis and angiogenesis.

integrins $\alpha v\beta 3$ (Brooks *et al.*, 1994), $\alpha v\beta 5$ (Friedlander *et al.*, 1995), $\alpha 5\beta 1$ (Kim *et al.*, 2000), and/or $\alpha 4\beta 1$ (Garmy-Susini *et al.*, 2005). Integrin $\alpha 9$ has been shown to play a key role in lymphatic vessel development as animals lacking this integrin develop chylothorax (Huang *et al.*, 2000). Recent studies in our lab indicate that tumors and select growth factors induce lymphangiogenesis and lymphatic vessel expression of a fibronectin-binding integrin, integrin $\alpha 4\beta 1$. This single integrin promotes lymphangiogenesis and subsequent metastasis by regulating migration and survival of LECs during lymphangiogenesis. The recent identification of selective markers of lymphatic versus vascular endothelial cells has allowed identification of the mechanisms that regulate lymphangiogenesis. Lymphatic endothelia selectively express Lyve-1, a member of the CD44 hyaluronic acid receptor family (Banerji *et al.*, 1999), Prox-1, a lymphatic vessel–specific homeobox transcription factor (Wigle and Oliver, 1999), and VEGFR3, a receptor for VEGF-C and VEGF-D (Kaipainen *et al.*, 1995).

As our lab has investigated the roles of integrins in angiogenesis and lymphangiogenesis (Garmy-Susini *et al.*, 2005; Kim *et al.*, 2000), we describe here a number of techniques that can be used to study the roles of integrins and other receptors during lymphangiogenesis. We show how to isolate and culture murine and LECs. We demonstrate how to assess integrin-mediated adhesion, migration, and tube formation *in vitro*. We also show how to study lymphangiogenesis *in vivo* in a variety of animal models. Finally, we show how to study lymphangiogenesis using intravital microscopy.

2. ISOLATION OF HUMAN LYMPHATIC ENDOTHELIAL CELLS

To study human lymphatic endothelial cells (LEC) in culture, we purified LECs from dermal human microvascular endothelial cells (HMVEC), which are a mixture of blood and lymphatic endothelial cells. HMVEC are comprised of equal proportions of CD34-Lyve-1+ lymphatic endothelial cells and CD34+Lyve-1− blood endothelial cells. To separate these two cell populations, HMVEC were grown in endothelial growth medium (EGM-2) containing 2% fetal bovine serum, bFGF, and VEGF (Cambrex, Inc., East Rutherford, NJ). To sort out lymphatic endothelial cells, 5 × 10⁶ HMVEC cells were aseptically incubated in 1 ml of PBS containing 100 μg/ml of PE-conjugated anti-CD34 for 1 h at 4°. Cells were washed three times in cold, sterile phosphate buffered saline (PBS) containing 5% RIA-grade bovine serum albumin (BSA). Fluorescence activated cell sorting was then performed on these dermal cells; CD34-negative cells were collected and further cultured. To confirm that these cells were lymphatic endothelial cells, they were incubated in rabbit anti–human Lyve-1 antibodies (Research Diagnostics, Inc.), followed by Alexa-488 conjugated secondary antibodies. Immunostained cells were fixed in 3.7% paraformaldehyde and analyzed by flow cytometry. More than 98% of cells were CD34-Lyve-1+ (Fig. 18.2). Alternatively, human dermal lymphatic endothelial cells (LEC) purified by immunoselection for expression of podoplanin can be purchased from Cambrex. These cells have similar phenotypes as those isolated using the methods described above.

3. ISOLATION OF MURINE LYMPHATIC ENDOTHELIAL CELLS

A variety of methods have been employed for isolation of LEC from various species. The general approaches developed for harvesting LECs from lymphatic vessels or whole tissues include the following: (1) release of LEC directly into culture from the luminal surfaces of large lymphatic vessels by digestion from bovine thoracic duct (Weber *et al.*, 1996) or by direct contact

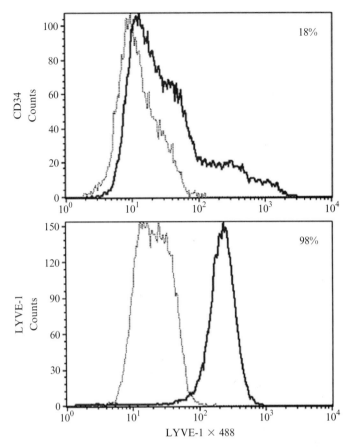

Figure 18.2 Purification of lymphatic endothelial cells. Expression of CD34 and LYVE-1 in purified lymphatic endothelial cells was evaluated by FACS analysis. Nonspecific antibody binding was analyzed by incubating cells in isotype matched IgG. The number in the upper right corner of each profile indicates the percentage of positive cells.

of individual inverted rat mesenteric lymphatics with culture-plate surfaces (Hayes *et al.*, 2003); (2) purification of LECs from whole-tissue digests using LEC-specific markers; and (3) induction of LECs to high abundance *in vivo* (by generation of lymphangiomas) with subsequent harvesting of the abundant lymphatic endothelia into culture. Techniques using lymphatic-specific antibodies to purify such endothelia from tissue digests include use of anti-podoplanin antibodies (Kriehuber *et al.*, 2001; Petrova *et al.*, 2002) or anti-LYVE-1 antibodies (Podgrabinska *et al.*, 2002) with subsequent selection for the labeled LECs by flow cytometry–mediated cell sorting or magnetic-bead purification and passage into culture. Isolation of lymphatic endothelia from mice is especially appealing for studying the *in vitro* behavior of genetically altered LECs isolated from knockout animals. A few publications have applied the technique of Mancardi and coworkers (1999) with minor modifications (Ando *et al.*, 2005; Nakamura *et al.*, 2004; Sironi *et al.*, 2006), wherein primary LECs are isolated from experimental abdominal lymphangiomas in mice. Herein, we provide protocols that we have developed based on these approaches.

Murine LECs were isolated from the murine experimental lymphangioma protocol. A 1:1 emulsion of PBS and Incomplete Freund's Adjuvant is prepared by rapid mixing of both ingredients using two syringes connected by a stopcock adapter. The white emulsion can be injected into C57Bl/6 mice (200 μl intraperitoneally), with repeat injection at day 10, and harvesting of abdominal lymphangioma material at day 21. The harvest is performed immediately after sacrificing the mouse, using sterile scissors/forceps to dissect lymphangioma plaques from the abdominal side of the diaphragm and surface of the liver. Lymphangioma material is then minced into fragments smaller than 1 mm in a 3-cm tissue culture dish followed by addition of DMEM containing 0.1% collagenase-A (Sigma-Aldrich, St. Louis, MO), and 0.1% FCS, supplemented with penicillin/ streptomycin. After incubation for 1 h at 37° in a humidified 5% CO_2 incubator, fragments are further disrupted by up/down passage through a 1-ml pipette (or a few times through an 18G needle/syringe if difficult to pass). Cells are washed twice in serum-free DMEM, and the contents are passed through a 100-μm cell strainer to clear any undigested material. After one extra DMEM wash, cells are resuspended in either of the following two LEC-complete media for propagation/ subculture on gelatin-coated culture plates:

Complete Medium 1: DMEM/ 20% FCS/ 100 μg/ml heparin/ 30 μg/ml endothelial cell growth supplement (ECGS) (Sigma-Aldrich, St. Louis, MO), and supplementation with Pen/Strep and nonessential amino acids.
Complete Medium 2: EBM-2 medium (Cambrex) with EGM-2 MV growth factor supplements (CambrexJ). An initial seeding density of 2.5 × 10^5 cells/cm^2 is adequate.

LECs isolated using this method express LYVE-1 (as assessed by FACS) for one or two passages, after which they tend to lose LYVE-1 expression while remaining strongly podoplanin positive for several passages. Others (Sironi *et al.*, 2006) have also noted lack of LYVE-1 expression (with continued Prox-1 expression) using a similar isolation method.

We have also used an antibody-based method to isolate a highly pure podoplanin-positive cell population from primary microvascular endothelial cultures derived from whole mouse lungs. The initial isolation of a

CD31+ population of microvascular endothelia from murine lungs is detailed in Wang *et al.* (2005). Endothelia plated into culture using this method can be harvested using Accutase (Innovative Cell Tech., San Diego, CA), with resuspension in DMEM containing 1% FCS (running medium). Cells are then resuspended in the same medium containing 5 μg/ml podoplanin antibodies (Syrian hamster anti-mouse podoplanin; Research Diagnostics) at a cell concentration of $\sim$2000 cells/μl, and incubated at room temperature for 15 min followed by two washes in running medium. Cells are then resuspended in running medium containing 1:200 rabbit anti-Syrian hamster FITC labeled antibody (Jackson Immunoresearch, West Grove, PA) and incubated for 1 h at 4° while stirring. After washing in PBS containing 0.5% BSA/2mM EDTA, cells are labeled with goat anti-rabbit magnetic beads (Miltenyi Biotech, Auburn, CA), and purified using magnetic columns according to manufacturer (Miltenyi Biotech, Auburn, CA) instructions. Eluted cells should be cultured at a density of greater than 1.2×10^4/cm^2, and can be grown in either of the complete media (A or B) specified above. At the time of plating, an aliquot can be assessed for purity by FACS.

4. CHARACTERIZATION OF LEC INTEGRIN EXPRESSION

Integrin expression on purified LECs was analyzed by flow cytometry by incubating cells for 1 h on ice in mouse anti-human a4β1 (HP1/2) from Biogen-Idec, mouse anti-human a5β1 (JBS5), from Chemicon, mouse anti-human avβ5 (P1F6), or mouse anti-human avβ3 (LM609) from David Cheresh, Moores UCSD Cancer Center, La Jolla, CA. Isotype control antibodies (IgG2a and IgG2b) were purchased from BD Bioscience. Cells were washed three times with PBS and incubated for 1 h in fluorochrome conjugated goat anti-mouse IgG on ice. Alternatively, fluorochrome conjugated primary antibodies for cell sorting and cytometry (anti-CD34, anti-β2 integrin, anti-a4β1, anti-a5β1, anti-avβ3, and anti-avβ5) were purchased from BD Pharmingen. Cells were well washed, fixed with 3.7% paraformaldehyde, and analyzed by flow cytometry. Exemplary patterns of integrin expression are shown in Fig. 18.3.

5. *IN VITRO* CELL ADHESION ASSAYS

To determine which integrins mediate lymphatic endothelial cell attachment to various extracellular matrix proteins, cell adhesion assays were performed in 48-well plates coated with 5 μg/ml of the extracellular

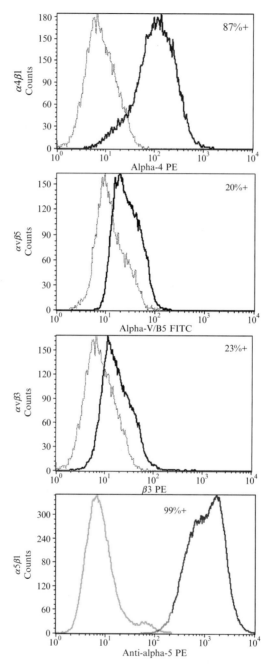

Figure 18.3   Analysis of lymphatic endothelial cell integrin expression. Expression of αvβ3, a4β1, a5β1, and avβ5 (black line) and isotype matched IgG (gray line) on purified LECs was evaluated by FACS analysis. The percentage of positive cells is indicated by the number in the upper right corner of each profile.

matrix proteins (ECM) vitronectin, fibronectin, and CS-1 fibronectin diluted in PBS for 12 h at 4°. Plates that were not coated with ECM protein were used as negative controls. Plates were blocked by incubation with 2% heat-denatured bovine serum albumin in PBS for 1 h at room temperature. Lymphatic endothelial cells were resuspended in adhesion buffer (Hanks balanced salt solution, 10 mM HEPES, pH 7.4, 2 mM MgCl$_2$, 2 mM CaCl$_2$, 0.2 mM MnCl$_2$, 1% BSA). Blocking antibodies—anti-avβ3 (LM609, Chemicon), anti-avβ5 (P1F6, Chemicon), anti-a4β1 (HP1/2, Chemicon), and anti-a5β1 (JBS5, Chemicon)—were used as competitive inhibitors of cell adhesion to ECM proteins at 25μg/ml final concentration. Cells were then incubated without agitation at 37°, 5% CO$_2$ for 10 to 25 min. Plates were washed three times with warm adhesion buffer, and nonadherent cells were removed by aspiration. Cells remaining in plates were fixed by incubation in 3.7% paraformaldehyde for 1 h and were stained by incubation in 1% crystal violet in buffer for 1 h. Plates were well washed with water to remove excess crystal violet, air dried overnight, and extracted by incubation in 200 μl of acetic acid. The absorbance of 100 μl of each of the extracts was measured at 560 nm using a tunable plate reader. All *in vitro* assays were performed three times with triplicate samples per group. The mean number of cells adhering (plus/minus standard error of measurement [SEM]) for the entire treatment group in each experiment was determined.

6. MIGRATION ASSAYS

To determine whether specific integrins regulate LEC migration, cell migration assays were performed using Costar Transwells. The undersides of 8-μm transwell inserts were coated with fibronectin, CS-1 fibronectin, or vitronectin (5 μg/ml) for 12 h at 4°. Nonspecific binding sites were blocked by incubation with 3% BSA in PBS for 1 h at 37°. Cells were resuspended in migration buffer (Hepes-buffered M199 medium containing 1% BSA, 1.8 mM CaCl$_2$, 1.8 mM MgCl$_2$, and 0.2 mM MnCl$_2$, pH 7.4), and 50,000 cells were added to the upper chamber and incubated at 37°, 5% CO$_2$. Cells were allowed to migrate from the upper to the lower chamber for 4 h at 37°. Nonmigratory cells were removed from the upper chamber by wiping the upper surface with an absorbent tip. Cells that had migrated to the lower side of the transwell insert were then fixed for 15 min with 3.7% paraformaldehyde for 15 min and with a 2% crystal violet solution. After extensive water washing to remove excess crystal violet, the number of cells that had migrated (to the bottom of the insert) was counted in five random 200x fields per replicate. All *in vitro* assays were performed three times with triplicate samples per group. The mean number of cells migrating (±SEM) for the entire treatment group in each experiment was determined (Fig. 18.4).

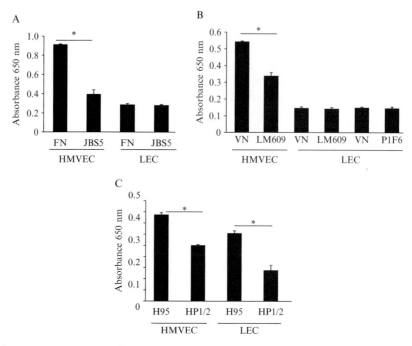

Figure 18.4   Integrin-mediated migration of lymphatic endothelial cells (LECs). (A) Migration of human microvascular endothelial cells (HMVEC) or LECs to the H95 fragment of CS-1 fibronectin in the absence (Cntl) or presence of anti-a4β1 function-blocking antibodies. (B) Migration of HMVEC or LEC to plasma fibronectin in the absence (Cntl) or presence of anti-a5β1 function-blocking antibodies. (C) Migration of HMVEC or LEC to vitronectin in the absence (Cntl) or presence of anti-avβ3 and anti-avβ5 (P1F6) function-blocking antibodies. Asterisk indicates statistical significance.

7. MATRIGEL TUBE FORMATION

To assess the ability of cultured LECs to form vessel-like structures in culture (tube formation), Matrigel (BD Biosciences) was added to the wells of an eight-well chamber slide in a volume of 150 μl and allowed to solidify at 37° for 30 min. After the Matrigel solidified, human LECs (5×10^4 cells) were plated in 300 μl of media EGM-2 without serum but containing 50 ng/ml of VEGFC. To test the roles of integrins $\alpha4\beta1$ and $\alpha5\beta1$ in *in vitro* tube formation, 25 μg/ml final concentration of anti-a4β1 (HP1/2) and anti-a5β1 (JBS5) were added to the cells before plating. Cultures were incubated at 37° with humidified 95% air/5% CO_2 for 24 h. Cultures were photographed and the mean number of tubes ($\pm$SEM) per microscopic field was determined. Studies were performed in triplicate with triplicate samples per group (Fig. 18.5).

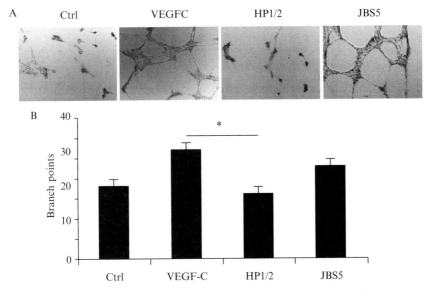

Figure 18.5 Lymphatic endothelial cell–mediated tube formation. (A) Micrographs of tube formation by LECs in Matrigel containing saline or VEGF-C in the presence or absence of anti-a4β1 or anti-a5β1 antibodies. (B) Quantification of the number of branch points formed. Asterisk indicates statistical significance.

8. FROZEN SECTION IMMUNOFLUORESCENCE MICROSCOPY

To identify lymphatic vessels in tissues, cryosections were immunostained to detect Lyve-1, a member of the CD44 hyaluronic acid receptor family (Banerji et al., 1999), Prox-1, a lymphatic vessel specific homeobox transcription factor (Wigle and Oliver, 1999), and VEGFR3, a receptor for VEGF-C and VEGF-D (Kaipainen et al., 1995), each of which is a marker of lymphatic endothelium. Tissue sections were fixed with ice-cold acetone for 2 min and then air dried for 30 min. Five-micrometer-thick cryosections were rehydrated by incubating for 5 min three times in PBS baths. Tissues were permeabilized by incubation for 60 sec in PBS containing 0.1% Triton X-100. Nonspecific binding sites were saturated by incubation of the slides for 1 h in PBS containing 5% BSA. Cryosections were incubated for 1 h at room temperature with primary antibodies. To detect lymphatic vessels in murine tissues, cryosections were incubated for 1 h at room temperature in a humidified chamber in anti Lyve-1 (2 μg/ml RDI-103PA50, Research & Diagnostics), anti-podoplanin (5 μg/ml, clone 103-M40 from Fitzgerald) or anti-prox-1 (5 μg/ml, RDI-102PA30,

Research & Diagnostics). Slides were washed three times in PBS before incubating in 1 μg/ml Alexa 488 or 647-conjugated cross-absorbed goat anti-rat immunoglobulin (Invitrogen) for 1 h at room temperature in a humidified chamber. Nuclei were detected by incubating slides with 200 nM Dapi for 5 min. Slides were briefly rinsed in PBS and thin coverslips were mounted in Dako Cytomation fluorescent mounting medium (Dako Corporation). Lymphatic vessels in human tissues were detected by incubating sections in anti-Lyve-1 (2 μg/ml, RDI-102PA50X Research & Diagnostics). Murine blood vessels were detected by incubating slides in 1 μg/ml of rat anti-mouse CD31 (clone MEC13.3, BD Pharmingen). Blood vessels in human tissues were detected by incubating slides in 1 μg/ml mouse anti-human CD31 (clone P2B1 from Chemicon International).

To detect the expression of integrins on lymphatic vessels in murine tissues, cryosections were incubated in 2 μg/ml rat anti-mouse integrin a5β1 (clone 5H10), rat anti-mouse integrin beta 5 (AB1926 from Chemicon International), or rat anti-mouse integrin anti-beta 3 (clone AP3 from BD Bioscience). To detect expression of integrins on human lymphatic vessels, cryosections were incubated in 1 μg/ml goat anti-integrin alpha 4 (sc-6590) (Santa Cruz Biotechnology), 1 mg/ml mouse anti-α5β1 (JBS5) (Chemicon International), 1 μg/ml mouse anti-human αvβ3 (LM609) or 1 μg/ml mouse anti-human αvβ5 (P1F6) (David Cheresh, University of California, San Diego), or rabbit anti-beta 5 (AB1926) (Chemicon International). Slides were washed three times in PBS before incubating in 1 μg/ml Alexa 488-, 594- or 647-conjugated cross-absorbed goat anti-mouse or rat immunoglobulin (Invitrogen) for 1 h at room temperature in a humidified chamber. Nuclei were detected by incubating slides with 200 nM Dapi for 5 min. Slides were briefly rinsed in PBS and thin coverslips were mounted in Dako Cytomation fluorescent mounting medium (Dako Corporation). Integrin α4β1 expression was detected on lymphatic vessels in human breast tumors as shown in Fig. 18.6.

To detect metastases, cryosections of murine lymph nodes were immunostained with mouse anti-human cytokeratin (5μg/ml CBL 272) (Chemicon International). Alternatively, metastases were detected in some animal studies by imaging of red fluorescent tumor cells within lymph nodes. Slides were washed three times in PBS before incubating in 1 μg/ml Alexa 488-, 594-, or 647-conjugated cross-absorbed goat anti-mouse or rat immunoglobulin (Invitrogen) for 1 h at room temperature in a humidified chamber. Nuclei were detected by incubating slides with 200 nM Dapi for 5 min. Slides were briefly rinsed in PBS and thin coverslips were mounted in Dako Cytomation fluorescent mounting medium (Dako Corporation). All thin sections were photographed at $\times$200 to 900 magnification. Experiments were performed 5 to 10 times. For quantification of vessels within tissues, Lyve-1, prox-1, or podoplanin-positive vessels in 5 to 10 microscopic fields

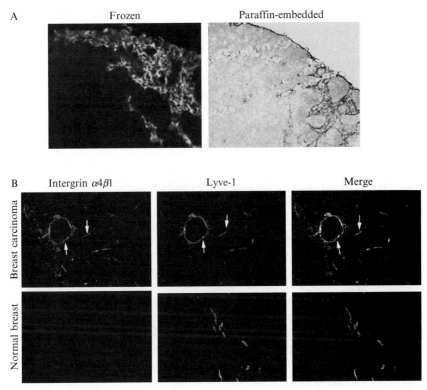

Figure 18.6 Immunostaining to detect lymphatic vessels in tissues. (A) Cryosections (left) or paraffin-embedded sections (right) of murine lymph nodes were immunostained using anti-Lyve-1 antibody. Nuclei were respectively stained with DAPI (blue) or Nuclear Fast Red (pink). Magnification x200. (B) Integrin $\alpha4\beta1$ expression (red) on lymphatic endothelium (green) was detected in human breast carcinomas. Magnification $\times200$.

per cryosection (per animal) were quantified, and the mean number of vessels (plus/minus the standard error of the mean) for the entire treatment group determined.

9. PARAFFIN-EMBEDDED SECTION IMMUNOHISTOCHEMISTRY

To detect lymphatic vessels in archived human or murine tissues, paraffin-embedded sections were de-paraffined in xylene and rehydrated according to standard protocols before fixation in methanol. Antigen retrieval was performed in murine or human tissue sections by incubation with proteinase K (0.1 mg/ml) for 10 min before blocking in PBS, 0.05% Tween-20, and 3% BSA. Murine cryosections were incubated with anti-Lyve1 antibody (RDI-103PA50, Research & Diagnostics), diluted at 1–5 μg/ml, and human cryosections were incubated in 4 μg/ml rabbit anti-human Lyve-1 (Reliatech). Sections were then incubated in 1 μg/ml cross-absorbed goat anti-rabbit secondary antibodies coupled to horseradish peroxidase and DAB color substrate (Vector Laboratories). Comparison of immunostaining of murine lymphatic vessels in frozen and paraffin embedded tissues are shown in Fig. 18.6.

10. GROWTH-FACTOR–INDUCED LYMPH NODE LYMPHANGIOGENESIS

To evaluate lymphatic vessel changes in lymph nodes of growth factor stimulated mice, 6- to 8-week-old C57Bl/6 female mice were subcutaneously injected with 400 μl of cold Matrigel containing 400ng/ml of VEGF-A, bFGF, or VEGF-C (R&D Systems) ($n = 8$). After 7 to 15 days, mice were euthanized. Inguinal, brachial, and mesenteric lymph nodes and Matrigel plugs were collected and frozen in OCT. Lymphatic vessel density was quantified by immunostaining tissue sections with anti-Lyve-1 or anti-podoplanin. Detection of integrins expressed on resting or growing lymphatic vessels was performed by immunostaining cryosections with anti-integrin antibodies as described below.

To evaluate the functional roles of integrins in lymphangiogenesis, VEGF-C implanted tumor bearing mice were injected intravenously at days 1, 3, 6, and 9 with 200 μg/mouse of low-endotoxin, sterile, function-blocking rat anti-mouse integrin $\alpha5\beta1$, $\alpha4\beta1$, or isotype-matched control antibodies. Alternatively, 400 μl of growth factor–depleted Matrigel containing 400 ng VEGF-C and 25 μg rsVCAM or saline were implanted subcutaneously in mice. As a positive control for inhibition of lymphangiogenesis, mice were also treated with anti-VEGFR-3 (AFL4, eBioscience, San Diego, CA) or isotype-matched control antibodies ($n = 7$). After 14 days, lymph nodes were removed, embedded in OCT, frozen, and sectioned. Thin sections (5 μm) were immunostained with anti–Lyve-1 antibodies. At least eight microscopic fields per tissue section were analyzed for quantification studies (Fig. 18.7).

11. MURINE LYMPHANGIOMA MODEL

Lymphangiogenesis in the context of inflammation was modeled in mice by injecting C57Bl/6 mice, 6 weeks of age, twice at 15-day intervals in the intraperitoneal cavity with 200 μl of incomplete Freund's adjuvant

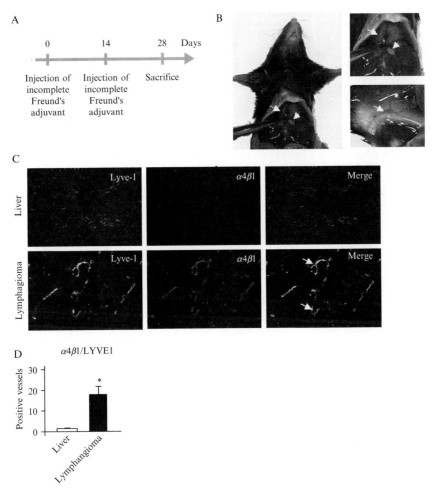

Figure 18.7 Lymphangioma mouse model. (A) Murine lymphangioma was induced by injection of incomplete Freund's adjuvant in the peritoneal cavity at 2-week intervals. (B) Mice developed inflammatory foci on the peritoneal side of the diaphragm and upper surface of the liver, as assessed after 28 days. (C) Lyve-1, integrin $\alpha 4\beta 1$-positive vessels in lymphangiomas and normal liver. Magnification x200. (D) Quantification of lymphatic vessels in lymphangiomas and normal liver.

diluted 1:1 in PBS (Mancardi *et al.*, 1999). Mice were sacrificed after 4 weeks. The lymphangiomas were then removed from the diaphragm and upper surface of both lobes of the liver as were proximal lymph nodes. Tissues were embedded in OCT, frozen, and sectioned for immunohistological analysis of lymphatic vessel density by Lyve-1 immunostaining (Fig. 18.8), blood vessel density (by CD31 immunodetection), and integrin expression.

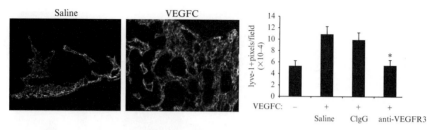

Figure 18.8 Matrigel-induced lymph node lymphangiogenesis. (Left) Cryosections of lymph nodes from VEGF-C or saline-stimulated mice were immunostained to detect Lyve-1. Magnification ×200. (Right) Quantification of mean number of Lyve-1-positive vessels/field (± standard error of measurement) in animals treated by intravenous injection with function-blocking anti-VEGFR3 or isotype-matched antibodies.

 ## 12. Tumor Models

To study the roles of integrins in lymphatic vessel growth within tumors and lymph nodes of tumor-bearing animals, four murine animal models were used. Lymphatic vessel development in the MMTV-PyMT model of spontaneous breast tumor development was first explored. In this model, female mice in the FVB/N background uniformly develop breast adenocarcinomas by 2 months of age and ultimately exhibit lung metastases in 90% of animals (Guy *et al.*, 1992). In our studies, 3-month-old PyMT mice in the FVB/N background with breast tumors in multiple fat pads were sacrificed, and primary tumors as well as inguinal, brachial, and mesenteric lymph nodes were removed and cryopreserved in OCT. Metastases in lymph nodes were detected by anti-cytokeratin immunostaining as described below. Lymphatic vessels were detected by anti-Lyve-1 or anti-podoplanin immunostaining as described above.

One syngeneic and one xenograft model of subcutaneous tumor development were also studied. Lewis Lung Carcinoma cells (LLC) or human colon carcinoma cells (HT29) were cultured in DMEM medium supplemented with l-glutamine and 10% fetal bovine serum. Tumor cells (5 × 10^6) in 200 μl PBS were injected subcutaneously into 6-week-old nude mice (HT29) or syngeneic C57Bl6 mice (LLC). A time course of tumor and lymph node lymphangiogenesis, metastasis and integrin expression was performed from 1 to 28 days (LLC, $n = 4$) or 1 to 6 weeks after tumor cell inoculation (HT29, $n = 6$) (Fig. 18.9). In some studies, animals bearing tumors were treated with saline, 200 μg/mouse anti-VEGFR3, anti-$\alpha4\beta1$, or isotype-matched control anti-CD1 antibodies three times per week for 2 weeks, beginning either 1 day (LLC) or 3 days (HT29) after tumor cell inoculation ($n = 8$, LLC; $n = 10$, HT29). LLC tumors were treated for

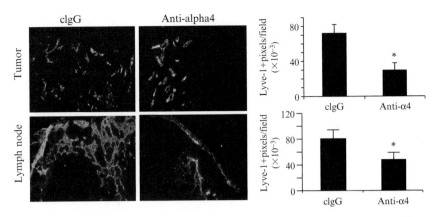

Figure 18.9 Lymphatics in LLC model. (A) Animals bearing subcutaneous Lewis lung carcinoma tumors were treated with anti-integrin $\alpha 4\beta 1$ or isotype-matched control antibodies and were immunostained to detect Lyve-1 (red) in primary tumors and inguinal lymph nodes. Nuclei are counterstained with Dapi (blue). (B) Quantification of mean ($\pm$ standard error of measurement [SEM]) Lyve-1+ vessels/field in primary tumors. (C) Quantification of mean ($\pm$SEM) Lyve-1+Pixels/field in proximal (inguinal) lymph nodes.

14 days and HT29 tumors were treated for 28 days. The tumors and draining inguinal lymph nodes, brachial, and mesenteric lymph nodes from LLC tumor-bearing mice were then removed, embedded in OCT, frozen, and sectioned for immunohistological analysis of lymphatic vessel density (by Lyve-1 immunostaining) and for metastases by cytokeratin immunodetection. At least five microscopic fields per tissue section were analyzed for quantification studies. Mean lymphatic vessel density or cytokeratin expression was determined after quantifying positive staining on 5 to 10 microscopic fields per animal. Each study was performed two to three times.

To study the role of lymph node lymphangiogenesis and integrin a4β1, in pancreatic cancer metastasis, together with Dr. Michael Bouvet, University of California, San Diego, we established an orthotopic model of pancreatic cancer. This model of pancreatic cancer can be imaged in real time and exhibits many of the features of human pancreatic disease, including lymph node and local peritoneal metastasis (Bouvet *et al.*, 2002, 2005; Katz *et al.*, 2003; Yamauchi *et al.*, 2005). Red fluorescent protein (RFP) expressing pancreatic tumors growing subcutaneously in nude mice at the exponential growth phase were resected aseptically. Necrotic tissues were cut away, and the remaining healthy tumor tissues were minced with scissors into approximately 3 × 3 × 3-mm pieces in Hanks' balanced salt solution containing 100 units/ml penicillin and 100 μg/ml streptomycin. Each piece was weighed and adjusted with scissors to 50 mg. The abdomen of a recipient nude mouse was sterilized with alcohol. An incision was then made through the left upper abdominal pararectal line and peritoneum.

The pancreas was carefully exposed and three tumor pieces were transplanted on the middle of the pancreas with a 6-0 Dexon (Davis-Geck, Inc., Manati, Puerto Rico) surgical suture. The pancreas was then returned to the peritoneal cavity, and the abdominal wall and skin were closed with 6-0 Dexon sutures. Animals were kept in a sterile environment. Measurable tumors developed within 14 days and intra–abdominal and lymph node metastases were observed within four weeks. RFP tumors were imaged on a weekly basis using whole animal imaging systems. An Olympus OV100 Whole Mouse Imaging System (Olympus Corp., Tokyo, Japan), containing an MT–20 light source (Olympus Biosystems, Planegg, Germany) and DP70 CCD camera (Olympus Corp., Tokyo, Japan), was used for whole-body imaging of fluorescent tumor growth and metastasis. High–resolution images were captured directly on a PC (Fujitsu Siemens, Munich, Germany). Images were processed for contrast and brightness and analyzed with the use of Paint Shop Pro 8 and Image ProPlus 3.1 software. Tumors and lymph nodes from these mice were cryopreserved and immunostained with anti-Lyve to quantify lymphatic vessel density.

13. Endothelial Cell–Specific Integrin α4 Deletion Mutant

To evaluate the role of integrin α4β1 in lymphangiogenesis, C57BL/6 × 129 mice homozygous for the floxed α4 allele (Scott *et al.*, 2003) were crossed to C57BL/6 mice expressing Cre under the control of the Tie2 promoter (B6. Cg-Tg Tek-cre 12Flv/J from Jackson Labs mice) to generate Tie2Cre+ α4 flox/+ mice. Alternatively, C57BL/6 × 129 mice homozygous for the floxed α4 allele were crossed to C57BL/6 mice expressing Cre under the control of the VE-Cadherin promoter (Alva *et al.*, 2006) to generate Tie2Cre+ α4 flox/ + mice. Mating of these Cre+ α4 flox/+ mice with α4 flox/flox mice yielded Cre+ α4 flox/flox, Cre+ α4 flox/+, Cre- α4 flox/flox mice, and Cre- α4 flox/+ mice. Genotyping was performed by PCR on mouse tail DNA using the primers (F) 5'-CGGGATCAGAAAGAATCCAAA-3' and (R) 5'-CTGGCATGGGGTTAAAA TTG-3' to yield a 180-bp product in wild-type mice and a 250-bp product in floxed mice. Primers (F) 5'-CCACCTGGTGTATGAAAGC-3' and (R) 5'-CTGGCATGGGGT-TAAAATTG-3' were used to identify the excised α4 flox allele. Expression of Cre in both sets of mice was evaluated by PCR amplification of tail DNA with the primers (F) 5'GCGGTCTGGCAGT AAAAACTATC3' and (R) 5'GTGAAACAGCATTGCTGTCACTT3' to yield a 100-bp product.

Lymphangiogenesis assays were performed on these mice by subcutaneously injecting 400 μl of cold Matrigel containing 400 ng/ml VEGF-C or

VEGF-A into Tie2Cre+α4flox/flox, Tie2Cre+α4flox/+ and Tie2Cre-α 4fl/fl ($n = 5$) mice, or VE-CadherinCre+ α4flox/flox, VE-CadherinCre-α 4 flox/flox, and VECadherinCre+α4 flox/+ mice ($n = 5$). After 7 to 21 days, Matrigel plugs were removed, embedded in OCT, and frozen, and sectioned. Thin sections (5 μm) were immunostained with anti-CD31 and anti-Lyve-1 antibodies. At least five microscopic fields per tissue section were analyzed for quantification studies.

14. INTRAVITAL MICROSCOPY OF LYMPH NODES

To image lymphatic vasculature, lymphangiogenesis, and metastasis within lymph nodes, dorsal skinfold window chambers can be implanted into mice of varying backgrounds over lymph nodes. Mice are anesthetized with ketamine (100 mg/kg) and medetomidine (250 mcg/kg) injected intraperitoneally. The entire dorsum is swabbed with Betadine, and then 70% ethanol is used to swab the dorsum clean. All surgical instruments and gauze are sterilized prior to use, and a bead sterilizer used to re-sterilize instruments during surgery. The mice are placed on a temperature-controlled pad, and covered with a clean drape, for surgery. The level of anesthesia is assessed throughout the surgery and booster injections of anesthetic cocktail are administered when deemed necessary to maintain the anesthetic plane. Skin chamber implantation requires 15 to 30 min and catheterization an additional 15 min. The skin of the dorsum is gently drawn up by sutures at the edge of the skinfold. A round area of skin and underlying tissue of 1.5-cm diameter is removed from one side of the symmetrical fold. This exposed tissue is then covered by a micro-cover glass incorporated into one-half of the frame while the other side remains in its natural environment. A few drops of sterile buffer solution containing 0.05 ml of tobramycin (40 mg/ml of tobramycin, broad-spectrum antibiotic) per 10 ml is applied to the chamber, and Neosporin antibiotic cream is applied to the skin surrounding the chamber incisions.

Animals are injected subcutaneously with 0.2 ml of buffer solution (Plasma-Lyte A to help maintain hydration and electrolytes) containing 75 μg Antisedan (atipmezole HCL) to antagonize the anesthetic effects of medetomidine and hasten recovery. The mice are returned to clean cages, kept warm, and monitored continuously until regaining consciousness.

15. ABDOMINAL WINDOW IMPLANTATION

This procedure is an adaptation of the abdominal wall method first developed by Tsuzuki and coworkers (2001) for studying pancreatic tumors. The body wall is incised over the inguinal region, the inguinal lymph node gently exposed, and an autoclaved titanium circular mount with eight holes on the edge (15-mm outer diameter, 11-mm inner diameter, and 1-mm thick, custom made in the campus research machine shop) is inserted between the skin and the abdominal wall. This ring is sutured in place with 4-0 prolene. Then an autoclaved circular glass coverslip (11.7-mm diameter, Fisher Scientific) is placed on top and fixed by a snap ring. The inguinal surface is kept moist throughout the procedure with prewarmed buffered saline solution.

15.1. Imaging inguinal window chamber mice

The mice are injected via the tail vein with 100 μl of rhodamine–lectin. They are then sedated with 20 mg/kg of Nembutal given intraperitoneally and placed in a perforated plastic cylinder with an opening in the plastic wall oriented over the mouse's abdominal window. The cylinder is secured in a specially constructed holder designed to fit on the stage of an inverted fluorescent microscope, and the vasculature of the inguinal lymph node is imaged.

15.2. Imaging of vascular and lymphatic vessel perfusion

To observe vascular perfusion of the lymph nodes in the chamber, animals can be tail-vein injected with FITC- or rhodamine-conjugated tomato-lectin (Fig. 18.10). To observe lymphatic vessels in the chamber, animals can be imaged by injection of FITC- or rhodamine-conjugated tomato-lectin or FITC-dextran (150,000 Da) in the footpad (Fig. 18.10). Animals can be implanted with fluorescent tumor cells to envision tumor metastasis to the lymph nodes. Twice per week animals with window chambers can be imaged to document tumor cell number and growth. In all cases, the animals are comfortably restrained in a clear, perforated Plexiglas holding tube and secured to the stage of a confocal scanning microscope or an inverted fluorescent microscope. Images of the node, vascular, and lymphatic connections, as well as tumor cell number and growth, can be obtained for about 1 h. The animals are lightly sedated with 20 mg/kg of Nembutal intramuscularly to minimize movement. The confocal microscope can detect fluorescent labels deep within the lymph node (diameter approximately 100 microns). The animals are returned to their home cages and observed for 1 h until fully recovered. If body temperature drops, they are gently warmed with an isothermal wrap.

For repeated injections of imaging agents into the vascular circulation, catheters can be inserted. To install a catheter, the mouse can be placed supine, and a 1-cm longitudinal incision made to the right of the midline. The right external jugular is exposed by blunt dissection a 1-Fr cannula made from polyurethane tubing with a polished tip, coated with heparin and

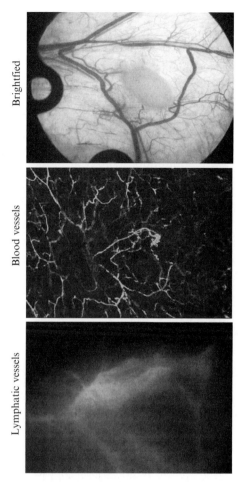

Figure 18.10 Lymphatics in intravital microscopy. Blood vessel (A) and lymphatic vessel (B) imaging in lymph nodes under dorsal skin-fold windows after intravenous and footpad injection, respectively, of FITC-dextran.

tobramycin, and containing 100% glycerol USP with 50 U/ml of heparin, will be introduced into the vein, secured, and the free end will be drawn beneath the skin of the right flank to exit between the shoulder blades. The catheter is sealed and coiled, and secured with tape to the window chamber frames. The surgical incisions are closed with veterinary bonding cement and 4-0 silk. The catheters and incision sites are checked daily for integrity and for any signs of inflammation or infection. The catheters are flushed every 3 days by withdrawing the glycerol and replenishing the catheter glycerol with 0.05 ml of fresh glycerol/heparin. In our experience, jugular catheters with glycerol do not require daily flushing, and minimal flushing is preferred as flushing introduces heparin into the general circulation of the subject.

16. Intradermal and Footpad Injections

In some studies intradermal injections of up to 50 μL of saline or 200 ng of VEGF-C were performed every day for 7 days. Intradermal injections of 50 μl of saline, 50 μg anti-$\alpha 4\beta 1$ (PS2), anti-VEGFR3 or isotype control antibodies were performed every 3 days for 21 days ($n = 5$ mice). Mice were sacrificed after 21 days and tumors as well as inguinal, brachial, and mesenteric lymph nodes were removed and embedded in OCT for cryosection and immunohistology analysis.

In other studies, to introduce tracers directly into the lymphatic vessels, mice were anesthetized with ketamine and then 100 μl of FITC-labeled dextran (FITC-Dextran, 30 mg/ml, Sigma Aldrich), Evans Blue (5 mg/ml) or rhodamine-conjugated tomato lectin from *Lycopersicon esculentum* were injected in the footpad. Footpads were gently massaged to encourage lymphatic flow. After 1 h, lymphatic vessels could be observed in dorsal skinfold chambers or in resected inguinal, brachial, and mesenteric lymph nodes.

REFERENCES

Alva, J. A., Zovein, A. C., Monvoisin, A., Murphy, T., Salazar, A., Harvey, N. L., Carmeliet, P., and Iruela-Arispe, M. L. (2006). VE-cadherin-Cre-recombinase transgenic mouse: A tool for lineage analysis and gene deletion in endothelial cells. *Dev. Dyn.* **235**, 759–767.

Ando, T., Jordan, P., Wang, Y., Harper, M. H., Houghton, J., Elrod, J., and Alexander, J. S. (2005). Homogeneity of mesothelial cells with lymphatic endothelium: Expression of lymphatic endothelial markers by mesothelial cells. *Lymphat. Res. Biol.* **3**, 117–125.

Bando, H., Weich, H. A., Horiguchi, S., Funata, N., Ogawa, T., and Toi, M. (2006). The association between vascular endothelial growth factor-C, its corresponding receptor, VEGFR-3, and prognosis in primary breast cancer: A study with 193 cases. *Oncol. Rep.* **15**, 653–659.

Banerji, S., Ni, J., Wang, S. X., Clasper, S., Su, J., Tammi, R., Jones, M., and Jackson, D. G. (1999). LYVE-1, a new homologue of the CD44 glycoprotein, is a lymph-specific receptor for hyaluronan. *J. Cell Biol.* **144**, 789–801.

Bouvet, M., Wang, J., Nardin, S. R., Nassirpour, R., Yang, M., Baranov, E., Jiang, P., Moossa, A. R., and Hoffman, R. M. (2002). Real-time optical imaging of primary tumor growth and multiple metastatic events in a pancreatic cancer orthotopic model. *Cancer Res.* **62**, 1534–1540.

Bouvet, M., Spernyak, J., Katz, M. H., Mazurchuk, R. V., Takimoto, S., Bernacki, R., Rustum, Y. M., Moossa, A. R., and Hoffman, R. M. (2005). High correlation of whole-body red fluorescent protein imaging and MRI on an orthotopic model of pancreatic cancer. *Cancer Res.* **65**, 9829–9833.

Brooks, P., Clark, R. A., and Cheresh, D. A. (1994). Requirement of vascular integrin alpha v beta 3 for angiogenesis. *Science* **264**, 569–571.

Dadras, S. S., Lange-Asschenfeldt, B., Velasco, P., Nguyen, L., Vora, A., Muzikansky, A., Jahnke, K., Hauschild, A., Hirakawa, S., Mihm, M. C., and Detmar, M. (2005). Tumor lymphangiogenesis predicts melanoma metastasis to sentinel lymph nodes. *Mod. Pathol.* **18**, 1232–1242.

Friedlander, M., Brooks, P. C., Shaffer, R. W., Kincaid, C. M., Varner, J. A., and Cheresh, D. A. (1995). Definition of two angiogenic pathways by distinct alpha v integrins. *Science* **270**, 1500–1502.

Garmy-Susini, B., Jin, H., Zhu, Y., Sung, R. J., Hwang, R., and Varner, J. (2005). Integrin alpha4beta1-VCAM-1-mediated adhesion between endothelial and mural cells is required for blood vessel maturation. *J. Clin. Invest.* **115**, 1542–1551.

Guy, C., Cardiff, R., and Muller, W. (1992). Induction of mammary tumors by expression of polyomavirus middle T oncogene: A transgenic mouse model for metastatic disease. *Mol. Cell. Biol.* **12**, 954–961.

Hayes, H., Kossmann, E., Wilson, E., Meininger, C., and Zawieja, D. (2003). Development and characterization of endothelial cells from rat microlymphatics. *Lymphat. Res. Biol.* **1**, 101–119.

He, Y., Rajantie, I., Pajusola, K., Jeltsch, M., Holopainen, T., Yla-Herttuala, S., Harding, T., Jooss, K., Takahashi, T., and Alitalo, K. (2005). Vascular endothelial cell growth factor receptor-3-mediated activation of lymphatic endothelium is crucial for tumor cell entry and spread via lymphatic endothelium. *Cancer Res.* **65**, 4739–4746.

Hirakawa, S., Kodama, S., Kunstfeld, R., Kajiya, K., Brown, L. F., and Detmar, M. (2005). VEGF-A induces tumor and sentinel lymph node lymphangiogenesis and promotes lymphatic metastasis. *J. Exp. Med.* **201**, 1089–1099.

Huang, X. Z., Wu, J. F., Ferrando, R., Lee, J. H., Wang, Y. L., Farese, R. V., and Sheppard, D. (2000). Fatal bilateral chylothorax in mice lacking the integrin alpha9beta1. *Mol. Cell. Biol.* **20**, 5208–5215.

Kaipainen, A., Korhonen, J., Mustonen, T., van Hinsbergh, V. W., Fang, G. H., Dumont, D., Breitman, M., and Alitalo, K. (1995). Expression of the fms-like tyrosine kinase 4 gene becomes restricted to lymphatic endothelium during development. *Proc. Natl. Acad. Sci. USA* **92**, 3566–3570.

Karkkainen, M. J., Haiko, P., Sainio, K., Partanen, J., Taipale, J., Petrova, T. V., Jeltsch, M., Jackson, D. G., Talikka, M., Rauvala, H., Betsholtz, C., and Alitalo, K. (2004). Vascular endothelial growth factor C is required for sprouting of the first lymphatic vessels from embryonic veins. *Nat. Immunol.* **5**, 74–80.

Katz, M. H., Takimoto, S., Spivack, D., Moossa, A. R., Hoffman, R. M., and Bouvet, M. (2003). A novel red fluorescent protein orthotopic pancreatic cancer model for the preclinical evaluation of chemotherapeutics. *J. Surg. Res.* **113**, 151–160.

Kim, S., Harris, M., and Varner, J. A. (2000). Regulation of angiogenesis *in vivo* by ligation of integrin alpha5beta1 with the central cell-binding domain of fibronectin. *Am. J. Pathol.* **156**, 1345–1362.

Kriehuber, E., Breiteneder-Geleff, S., Groeger, M., Soleiman, A., Schoppmann, S. F., Stingl, G., Kerjaschki, D., and Maurer, D. (2001). Isolation and characterization of dermal lymphatic and blood endothelial cells reveal stable and functionally specialized cell lineages. *J. Exp. Med.* **194**, 797–808.

Mancardi, S., Stanta, G., Dusetti, N., Bestagno, M., Jussila, L., Zweyer, M., Lunazzi, G., Dumont, D., Alitalo, K., and Burrone, O. R. (1999). Lymphatic endothelial tumors induced by intraperitoneal injection of incomplete Freund's adjuvant. *Exp. Cell. Res.* **246**, 368–375.

Massi, D., Puig, S., Franchi, A., Malvehy, J., Vidal-Sicart, S., Gonzalez-Cao, M., Baroni, G., Ketabchi, S., Palou, J., and Santucci, M. (2006). Tumour lymphangiogenesis is a possible predictor of sentinel lymph node status in cutaneous melanoma: A case–control study. *J. Clin. Pathol.* **59**, 166–173.

Nagy, J., Vasile, E., Feng, D., Sundberg, C., Brown, L. F., Detmar, M. J., Lawitts, J. A., Benjamin, L., Tan, X., Manseau, E. J., Dvorak, A. M., and Dvorak, H. F. (2002). Vascular permeability factor/vascular endothelial factor induces lymphangiogenesis as well as angiogenesis. *J. Exp. Med.* **196,** 1497–1506.

Nakamura, E. S., Koizumi, K., Kobayashi, M., and Saiki, I. (2004). Inhibition of lymphangiogenesis-related properties of murine lymphatic endothelial cells and lymph node metastasis of lung cancer by the matrix metalloproteinase inhibitor MMI270. *Cancer Sci.* **95,** 25–31.

Petrova, T. V., Makinen, T., Makela, T. P., Saarela, J., Virtanen, I., Ferrell, R. E., Finegold, D. N., Kerjaschki, D., Yla-Herttuala, S., and Alitalo, K. (2002). Lymphatic endothelial reprogramming of vascular endothelial cells by the Prox-1 homeobox transcription factor. *EMBO J.* **21,** 4593–4599.

Podgrabinska, S., Braun, P., Velasco, P., Kloos, B., Pepper, M. S., and Skobe, M. (2002). Molecular characterization of lymphatic endothelial cells. *Proc. Natl. Acad. Sci. USA* **99,** 16069–16074.

Roma, A. A., Magi-Galluzz, C., Kral, M. A., Jin, T. T., Klein, E. A., and Zhou, M. (2006). Peritumoral lymphatic invasion is associated with regional lymph node metastases in prostate adenocarcinoma. *Mod. Pathol.* **19,** 392–398.

Scott, L. M., Priestley, G. V., and Papayannopoulou, T. (2003). Deletion of alpha4 integrins from adult hematopoietic cells reveals roles in homeostasis, regeneration, and homing. *Mol. Cell. Biol.* **23,** 9349–9360.

Sironi, M., Conti, A., Bernasconi, S., Fra, A. M., Pasqualini, F., Nebuloni, M., Lauri, E., De Bortoli, M., Mantovani, A., Dejana, E., and Vecchi, A. (2006). Generation and characterization of a mouse lymphatic endothelial cell line. *Cell Tissue Res.* **325,** 91–100.

Skobe, M., Hawighorst, T., Jackson, D. G., Prevo, R., Janes, L., Velasco, P., Riccardi, L., Alitalo, K., Claffey, K., and Detmar, M. (2001). Induction of tumor lymphangiogenesis by VEGF-C promotes breast cancer metastasis. *Nat. Med.* **7,** 192–198.

Tsuzuki, Y., Mouta-Carreira, C., Bockhorn, M., Xu, L., Jain, R. K., and Fukumura, D. (2001). Pancreas microenvironment promotes VEGF expression and tumor growth: Novel window models for pancreatic tumor angiogenesis and microcirculation. *Lab Invest.* **81,** 1439–1451.

Wang, L., Fuster, M., Sriramarao, P., and Esko, J. D. (2005). Endothelial heparan sulfate deficiency impairs L-selectin– and chemokine-mediated neutrophil trafficking during inflammatory responses. *Nat. Immunol.* **6,** 902–910.

Weber, E., Lorenzoni, P., Raffaelli, N., Cavina, N., and Sacchi, G. (1996). Glioma cells induce gamma-glutamyl transpeptidase activity in cultured blood but not lymphatic endothelial cells. *Biochem. Biophys. Res. Commun.* **225,** 1040–1044.

Wigle, J. T., and Oliver, G. (1999). Prox1 function is required for the development of the murine lymphatic system. *Cell* **98,** 769–778.

Yamauchi, K., Yang, M., Jiang, P., Yamamoto, N., Xu, M., Amoh, Y., Tsuji, K., Bouvet, M., Tsuchiya, H., Tomita, K., Moossa, A. R., and Hoffman, R. M. (2005). Real-time *in vivo* dual-color imaging of intracapillary cancer cell and nucleus deformation and migration. *Cancer Res.* **65,** 4246–4252.

CHAPTER NINETEEN

ANALYSIS OF INTEGRIN SIGNALING IN GENETICALLY ENGINEERED MOUSE MODELS OF MAMMARY TUMOR PROGRESSION

Yuliya Pylayeva,* Wenjun Guo,† *and* Filippo G. Giancotti‡

Contents

Abstract

Cancer progression—the evolution of malignant tumors towards metastatic dissemination—is a complex, multistep process orchestrated by neoplastic

* Cell Biology Program, Memorial Sloan-Kettering Cancer Center, and Sloan-Kettering Division, Weill Graduate School of Medical Sciences, Cornell University, New York, New York
† Whitehead Institute for Biomedical Research, Cambridge, Massachusetts
‡ Sloan-Kettering Division, Weill Graduate School of Medical Sciences, Cornell University, New York, New York

Methods in Enzymology, Volume 426
ISSN 0076-6879, DOI: 10.1016/S0076-6879(07)26019-7

© 2007 Elsevier Inc.
All rights reserved.

cells but aided by elements of the tumor microenvironment such as macrophages, activated fibroblasts, and endothelial cells. During tumor progression, cancer cells acquire a number of traits, such as the ability to undergo unrestrained proliferation, to resist pro-apoptotic insults, and to invade through tissue boundaries. Genetic and epigenetic changes conspire to drive the emergence of these traits against the backdrop of host selection. It is becoming increasingly clear that certain integrins and integrin-signaling components amplify oncogenic signaling to promote tumor progression. Mouse models of cancer provide useful, if not necessary, experimental systems to study tumor initiation and progression *in vivo* and to test novel therapeutic approaches. We have utilized mouse models of mammary tumorigenesis to examine the role of integrin $\alpha6\beta4$ signaling in tumor progression *in vivo*. In this chapter, we describe a collection of cell biological and genetic methods that may aid in characterizing the roles of integrin signals in mammary tumorigenesis.

1. INTRODUCTION

Although it represents the major cause of mortality in cancer patients, metastatic dissemination remains poorly understood, in large part because of the complexity of molecular and cellular changes that it requires (Gupta and Massague, 2006). In order to become overtly metastatic, cancer cells must acquire multiple genetic and epigenetic lesions that endow them with the ability to invade surrounding tissues, to elicit an angiogenic response, to penetrate into the vasculature, to survive shear stresses and the lack of a supporting matrix in the blood stream, and to extravasate and give rise to metastatic outgrowths in the target organ. Alterations in cell adhesion and signaling are key to proper execution of all these steps in the metastatic cascade (Guo and Giancotti, 2004; Hood and Cheresh, 2002).

Integrins are a family of cell adhesion receptors consisting of an α- and a β-subunit. In mammals, 18 α-subunits and 8 β-subunits variously combine to form 24 receptors characterized by distinct, but often overlapping, ligand-binding specificities. Cells usually express several distinct integrins at a time, allowing attachment to a wide variety of matrix proteins and leading to intracellular signaling events that regulate cell proliferation, migration, and survival (Hynes, 2002; Miranti and Brugge, 2002). Normal cells rely on specific integrin signals as well as other adhesive signals to thrive within the confines of the organs and tissues to which they belong (Giancotti and Ruoslahti, 1999). A major hallmark of neoplastic cells is their decreased reliance on attachment to the extracellular matrix for proliferation and survival. However, it is becoming apparent that various oncogenic mutations hijack the adhesion signaling machinery to promote proliferation and survival in spite of the absence of a supporting matrix. The hostile environment of surrounding host tissues puts forth tremendous pressures that the newly transformed cells have to cope with. The repertoire of integrin subunits expressed by cancer cells can be dynamically altered in response to these pressures. Although the underlying mechanisms are not known, cancer cells often display elevated levels of those integrins that upon cooperation with various oncogenic pathways can promote their survival, migration, and proliferation, whereas they tend to lose expression of those integrins that may exert the opposite effects (Guo and Giancotti, 2004; Hood and Cheresh, 2002).

The transition from carcinoma *in situ* to invasive carcinoma is marked by a series of stereotypical changes in cell adhesion and signaling. Cancer cells disassemble intercellular junctions and penetrate the basement membrane, thus gaining entry into the interstitial matrix. In order to spread to distant organs, cancer cells then need to gain access to the circulatory system. They achieve this goal by penetrating newly formed blood vessels or lymphatics. Because tumor vessels have a discontinuous basement membrane, incomplete pericyte coverage, and frayed endothelial junctions, cancer cells do not necessarily need to disrupt the vessel wall to gain access to the blood or lymphatic stream. While in the circulation, cancer cells often form platelet-containing emboli, which arrest in the microcirculation of target organs. Upon extravasating into the parenchyma of target organs, cancer cells initially form micrometastases. Outgrowth of these lesions is thought to require both an ample blood supply and a supportive stroma. It is becoming increasingly clear that integrin signals play key roles during various steps of this metastatic cascade (Guo and Giancotti, 2004; Hood and Cheresh, 2002).

The a6β4 integrin is a laminin receptor that binds with the highest affinity to laminin-5 and mediates assembly of hemidesmosomes in normal stratified epithelia (Sonnenberg, 1993). Several observations suggest that α6β4 integrin plays a preeminent role during tumor progression. A significant fraction of carcinomas of the skin, head and neck, gastrointestinal tract, and mammary glands express elevated levels of α6β4, often portending a poor prognosis (Mercurio *et al.*, 2001). In addition, cell biological studies have indicated that the ErbB2, EGF-R, and Met receptor tyrosine kinases, which are often overexpressed or constitutively activated by mutation during tumor progression, combine with α6β4 to promote pro-invasive growth (Falcioni *et al.*, 1997; Mariotti *et al.*, 2001; Trusolino *et al.*, 2001).

We have hypothesized that a6β4 promotes carcinoma progression by amplifying oncogenic signaling. Most previous studies on the role of integrins in cancer progression have used *in vitro* models. Tumor initiation and progression to metastasis are complex biological processes, encompassing disruption of growth control, invasion into the interstitial matrix, angiogenesis, entry into the bloodstream, and extravasation and outgrowth in target organs. Traditional *in vitro* assays do not recapitulate most of the salient features of these varied processes, in particular the evolving organ/

tissue microenvironments, as well as of the various anatomical restrictions that are imposed during metastatic progression *in vivo*. It is therefore imperative to utilize models that can recapitulate all of the steps involved in the evolution of cancer. In addition, the majority of these *in vitro* assays often rely on established cancer cell lines that have been indefinitely propagated *in vitro*. Although these lines possess complex genomes as natural tumors do, it is unclear how similar they are to the tumors from which they were derived. Furthermore, it is difficult to define the specific step of tumor initiation or progression that they represent.

Our laboratory has recently examined the role of integrin $\beta 4$ signaling in ErbB2-induced mammary tumorigenesis (Guo *et al.*, 2006). The cytoplasmic portion of $\beta 4$, which measures over 100 kD (mol. wt.) and contains no homology to other known integrin β subunits, mediates association with hemidesmosomal cytoskeleton and directs recruitment of the signaling adaptor protein Shc and of PI-3K (Mainiero *et al.*, 1995, 1997; Shaw *et al.*, 1997; Spinardi *et al.*, 1995). Mice carrying a targeted deletion of the entire cytoplasmic domain of $\beta 4$ die at birth, because their epidermis lacks hemidesmosomes and thereby detaches from the underlying basement membrane (Murgia *et al.*, 1998). Thus, we were unable to analyze the role of $\beta 4$ signaling in postnatal development and adult life using these mice. To circumvent this problem, we have recently generated mice carrying a selective deletion of the $\beta 4$ signaling domain, which includes all major tyrosine phosphorylation sites but not the plectin-binding site required for assembly of hemidesmosomes ($\beta 4$ 1355T mice). The $\beta 4$ 1355T homozygous mice are viable and fertile, and the protein encoded by the mutant allele retains cytoskeletal function but is unable to signal through Shc and PI-3K (Nikolopoulos *et al.*, 2004).

2. Experimental Approach

To study the role of integrin signaling in mammary tumorigenesis, our laboratory utilizes transgenic mice that express defined oncogenes in their mammary glands, and thereby develop mammary tumors with relatively short latency and almost complete penetrance. Specifically, we have used mice that express an oncogenic form of Neu (YD) (MMTV-Neu (YD)) or the polyoma virus middle-T protein from the MMTV promoter (MMTV-PyMT), because these mouse models exhibit traits characteristic of human cancer progression, develop mammary gland tumors with short latency, and progress to lung metastasis (Guy *et al.*, 1992; Muller *et al.*, 1988). Employment of mouse models with genetically defined lesions enables dissection of the signaling pathways that are instrumental in cancer progression. In addition, mouse models that produce spontaneous metastases allow examination of all the limiting steps of metastasis, such as selection for metastasis-prone cells in primary tumors, intravasation, escape from stress in the circulation, and homing and outgrowth in the target organ.

Introduction of deletions or even point mutations in a gene of interest can lead to embryonic or perinatal lethality (as was the case with complete deletion of the cytoplasmic domain of $\beta4$). If haplo-insufficiency is anticipated, it may be informative to study heterozygous mutant mice. In many cases, however, it is necessary to generate mice carrying a conditional mutation of the gene, using strategies that allow either temporal or spatial (tissue-specific) control of the genetic perturbation (Glaser *et al.*, 2005). In addition, it is important to consider that loss of function of the gene of interest can produce unwanted effects during development of the organ to be studied. This can create difficulties in the analysis of tumor initiation, since developmental defects can result in a reduction in the number of stem and committed progenitor cells that are thought to be the target of oncogenic mutations, thus giving rise to an artifactual impairment of tumorigenesis in the mutant animals. Combining spatial and temporal control of the deletion event allows for avoidance of this problem (Glaser *et al.*, 2005).

If mutation of the gene of interest alters primary tumor initiation and growth, determining the effects on metastatic seeding becomes more complicated. To alleviate this problem, it is possible to analyze rates of metastatic dissemination in mice carrying primary tumors of comparable total weight. In addition, mouse allograft transplant models represent an appropriate system to study the metastasis of cells in a living organism. Such models entail the injection of transformed cells into immunocompromised mice and have the benefit of producing organ-specific metastases in a reasonable time frame and with reproducible results.

3. MEASUREMENT OF PRIMARY TUMOR GROWTH

To assess tumor initiation and growth, we use both genetically engineered and allograft mouse models. To avoid the confounding effect of potential modifier loci, we introduce the integrin or integrin-signaling mutation in the same background of the mouse model of cancer to be used, through at least five back-crosses. We then generate mice heterozygous at this locus, but homozygous at the transgenic oncogene locus. These mice are finally interbred to generate cohorts of tumor-prone mice with or without the integrin or integrin-signaling mutation.

Cancerous lesions of the mammary glands are normally monitored by palpation, and the time of onset is recorded. When tumors reach a diameter of approximately 4 mm, their dimensions are measured using a caliper. Because tumors are ovaloid, it is sufficient to measure two dimensions (length and width) to calculate tumor volumes and plot tumor growth over time (volume $= 0.52 \times LW^2$, where L is the length and W is width of the tumor being measured). Once the experimental animal is euthanized, the tumors are dissected out, and their weight and appearance (normal, hemorrhagic, or cystic) are recorded for future reference. We utilize two different schedules for collection of tumor samples: in the first, we wait until the mice need to be sacrificed according to institutional guidelines; and in the second, we sacrifice the cohorts of mice at specific time points up until the age when the control cohort needs to be sacrificed according to institutional guidelines. The former approach requires long-term monitoring of mouse cohorts to document tumor onset and growth rates in the form of tumor-free survival curves and graphs depicting cumulative tumor burden over time. The latter approach allows for direct comparisons of tumor burden between control and experimental groups, at least until the mice with the largest tumor burden need to be sacrificed.

Allograft methods enable the assessment of *in vivo* tumorigenic properties of cancer cells that have been explanted from normal mammary glands or mammary tumors and genetically modified *in vitro*. For these studies, we inject 2×10^5 to 1×10^6 cells either subcutaneously or surgically into mouse mammary fat pad #4 of 4- to 6-week-old *nu/nu* or *NOD/SCID* immunocompromised female mice. The cells are briefly trypsinized, washed with serum-containing media, and counted with Trypan blue (Sigma) to confirm cell viability. Next, 2×10^5 to 1×10^6 cells are then resuspended in 25 μl of PBS. Immediately prior to the injection, the cell suspension is mixed on ice with an equal volume of undiluted Matrigel (BD Biosciences) and a total volume of 50 μl is injected into the mammary fat pad.

4. Spontaneous and Allograft Lung Metastasis Assays

The mouse models we utilize in the laboratory (MMTV–Neu (YD) and MMTV–PyMT) develop mammary carcinomas that frequently metastasize to the lung, allowing an examination of the effect of integrin signaling on the multiple steps of the metastatic cascade over the natural time course of disease progression *in vivo*. We have compared the extent of spontaneous pulmonary metastases in Neu(YD)/β4-WT and Neu(YD)/β4-1355T mice carrying primary tumors of comparable size (Fig. 19.1). Analysis of hematoxylin and eosin (H&E)-stained lung sections indicated a lower number of metastatic foci in Neu(YD)/β4-1355T mice as compared to their wild-type counterparts.

An alternative approach is to inject mammary tumor cells orthotopically into the mammary fat pad of immunocompromised or syngeneic mice and

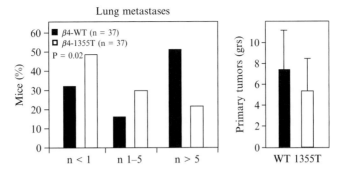

Figure 19.1 Mice were sacrificed approximately 7.5 weeks after detection of their first palpable tumor. Sagittal lung sections (two to four per mouse). were stained with hematoxylin and eosin (H&E) and examined microscopically. The graph on the left shows the percentage of mice of the indicated genotype with less than 1, 1 to 5, or more than 5 metastases per lung section. The graph on the right shows the mean cumulative tumor burden ($\pm$ standard deviation) in each cohort of mice at the time of euthanasia. (With permission, from Guo, W., Pylayeva, Y., Pepe, A., Yoshioka, T., Muller, W. J., Inghirami, G., and Giancotti, F. G. (2006). Beta 4 integrin amplifies ErbB2 signaling to promote mammary tumorigenesis. *Cell* 126, 489–502.)

monitor tumor progression to metastatic disease. This method enables monitoring of synchronized cohorts of tumor-bearing animals for metastatic progression. However, in order to study integrin signaling, it is necessary to isolate mammary cancer cells from integrin- or integrin signaling–deficient mouse models of cancer and reinject them into immunocompatible mice or to isolate primary mammary epithelial cells from the integrin mutant mice and transform them *in vitro* before orthotopic re-injection in mice (Elenbaas *et al.*, 2001; Welm *et al.*, 2005).

The transition of primary lesions from mammary intraepithelial neoplasia (MIN) to invasive carcinoma and the outgrowth of metastases in the lung can be assessed by histological criteria and immunohistochemistry with antibodies to proteins that are either lost or newly expressed during these processes. The study of intravasation, however, requires quantification of cancer cells circulating in the blood stream. To improve the sensitivity of such assays, cancer cells can be separated from blood by using FACS sorting with antibodies to epithelial cell-specific (e.g. epithelial-specific antigen [ESA]) or tumor-specific (e.g., Neu) antigens and then quantified by performing nested PCR with primers that amplify cancer cell-specific genomic sequences (Kim *et al.*, 1998), such as those of the transgenic oncogene, or by using clonogenic assays (Yang *et al.*, 2004).

The subsequent steps in metastatic colonization, comprising survival in the circulation, extravasation, and outgrowth in the target organ, can be studied using allograft methods where tumor cells are introduced directly into the circulation of the recipient mice. For example, to assay metastasis to the lungs, we routinely introduce genetically modified mammary cancer cells via intravenous injection. This method allows for efficient delivery of the cells to the vascular bed of the lung, which results in reproducible rates of lung metastasis.

We have used this method to compare the ability of either Neu-β4-WT or Neu-β4-1355T cells to initiate lung metastatic growth. Cells of the indicated genotype were injected intravenously into nude mice and lungs were extracted and processed for histology 30 days later (Fig. 19.2B). The percentage of lung area occupied by metastatic foci was calculated by subjecting H&E-stained lung sections to image analysis (NIH ImageJ). Whereas wild-type cells gave rise to numerous lung metastases, Neu-β4-1355T cells were only capable of producing a few micrometastatic lesions (Fig. 19.2B).

In preparation for tail vein injection, the cells are trypsinized, washed first with serum-containing medium and then with PBS, counted with Trypan Blue, and then resuspended at a density of 5×10^5 or 1×10^6 cells per 100 μl of PBS. Cells are kept in suspension on ice until they are ready to be injected. Mice are injected with 100 μl of the cell suspension, via the lateral tail vein, using a 27-gauge needle. The number and size of metastatic lesions depend on the type of cells that are being injected, their number, and the length of time from injection to euthanasia. The animals need to be monitored on a regular basis, since advanced lung metastases may cause weight loss and breathing impairment.

Until recently it was technically difficult to image metastatic growth in live animals. Therefore, the tail vein injection method was used primarily as an end-point assay. In addition, although the number of cells effectively introduced into the circulation via tail vein injection is somewhat variable, there was no way to correct for this error. These obstacles, however, can now be avoided by injecting cancer cells that have been genetically modified to allow bioluminescent imaging *in vivo* (Levin, 2005). This method, which measures the visible light produced by the enzyme luciferase upon reaction with its substrate (D-luciferin in case of firefly luciferase), provides for a quantitative measurement of metastatic growth over time. Furthermore, it helps to identify whether a specific integrin or integrin-signaling mutation affects survival in the bloodstream, homing to the lung, extravasation, or outgrowth, especially when used in conjunction with immunofluorescent analysis of tissue sections. In order to label the cells, we resort to the use of a retroviral vector encoding thymidine kinase, GFP, and firefly luciferase (TGL vector) (Ponomarev *et al.*, 2004). The cells are infected with concentrated virus *in vitro* prior to injecting them into the circulation of mice as described above. Bioluminescence can be visualized and quantitatively recorded by imaging mice in an IVIS chamber (Xenogen). After introducing the cancer cells into the tail vein, we immediately image them by injecting 100 μl of 15 mg/ml D-luciferin (D-luciferin

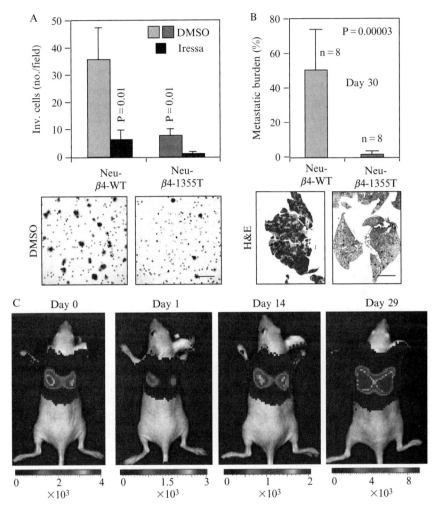

Figure 19.2 (A) Cells treated with Iressa (10 μM) or vehicle alone (DMSO) were subjected to matrigel invasion assay in response to FBS. The graph shows the mean number of invaded cells (± standard deviation [SD]) per microscopic field from triplicate samples. The bottom panel shows representative fields. Scale bar = 50 μm. (B) Cells were injected in the tail vein of nude mice. Percentages of lung section areas occupied by metastases (±SD) were quantified 30 days later by image analysis. Bottom panels show representative images of sagittal lung sections. Scale bar = 2 mm. (C) Primary MMTV-polyoma MT–derived tumor cells (1 × 10⁶ cells/mouse) were infected with TGL vector and injected in the tail vein of nude mice. Representative images taken at four time points (day 0, day 1, day 14, and day 29) are shown. Color scale represents number of photons per second (photon flux). (With permission, panels A and B from Guo, W., Pylayeva, Y., Pepe, A., Yoshioka, T., Muller, W. J., Inghirami, G., and Giancotti, F. G. (2006). Beta 4 integrin amplifies ErbB2 signaling to promote mammary tumorigenesis. *Cell* 126, 489–502.) (See color insert.)

potassium salt, Xenogen) retro-orbitally. This day-0 data point ensures that the injection was successful. The animals are then imaged, and data are recorded systematically over time (depending on the general well-being of animals, or when the imaged signal is strong enough to ensure the presence of metastases). We have performed an experiment where primary tumor cells derived from mammary carcinoma arising in PyMT mice were infected with TGL vector and injected intravenously (Fig. 19.2C). A strong, lung-specific signal was observed 20 min post-injection (day 0). The signal declined during the 1st week of the experiment, presumably because not all of the cells that had arrested in the lung penetrated into the parenchyma. Significant signal was detected starting at week 2 and continued to increase over time.

The normalized photon flux can be plotted as a function of time and provides a quantitative measure of metastatic growth over time. In addition, expression of GFP in the injected cells allows for easy visualization of micrometastatic lesions on tissue sections. Immunofluorescent labeling of blood vessel components and subsequent confocal imaging at early time points helps to define whether the integrin or integrin-signaling mutation affects homing, extravasation, or initial outgrowth.

5. ANALYSIS OF TUMOR SECTIONS

Histological, immunochemical, and molecular cytology studies can provide mechanistic insight into tumor progression and metastasis in mouse models. Parameters such as histological grade, proliferative and apoptotic rates, disruption of epithelial adhesion and polarity, expression of genes associated with epithelial-to-mesenchymal transition (EMT), and extent of neoangiogenesis can be investigated through analysis of tissue sections.

Processing fixed samples for paraffin embedding enables optimal preservation of tissue architecture. Because paraffin-embedded sections are stained by using immunohistochemistry (IHC) and counterstained with hematoxylin, it is often quite easy to identify the cell type expressing a given antigen, and determine its extracellular or subcellular localization. The tissues are serially dehydrated and embedded in paraffin. The blocks can then be cut into 5-μm-thick sections. Some sections can be stained with H&E to assess pathological appearance of the cancerous lesion (see, e.g., Fig. 19.2B), whereas the others are subjected to IHC (see, e.g., Fig. 19.3). We use the mouse-on-mouse kit (MOM) for mouse monoclonal antibodies and ABC kit for other antibodies (both kits are available from Vector). Staining using antibodies against Ki-67 (Novocastra), cleaved caspase 3 (Cell Signaling), PECAM1 (BD Biosciences), and E-cadherin (BD Biosciences) provides an estimate of cell proliferation, apoptosis, neovascularization, and loss of epithelial adhesion in the tumor or

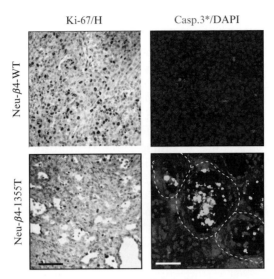

Figure 19.3 Orthotopic tumors were stained as indicated. Dotted lines were drawn along the basement membrane of pseudoglandular structures. Scale bars = 50 μm. (With permission, from Guo, W., Pylayeva, Y., Pepe, A., Yoshioka, T., Muller, W. J., Inghirami, G., and Giancotti, F. G. (2006). Beta 4 integrin amplifies ErbB2 signaling to promote mammary tumorigenesis. *Cell* 126, 489–502.)

metastatic lesion. We have compared the extent of proliferation in the tumor samples from Neu(YD)/β4-WT and Neu(YD)/β4-1355T mice (Fig. 19.3, left panels). Paraformaldehyde-fixed (PFA), paraffin-embedded sections were stained with antibody against Ki-67 and counterstained with hematoxylin. Significantly fewer Ki-67 positive cells were detected in tumors derived from Neu(YD)/β4-1355T mice, suggesting that β4 signaling promotes tumor cell proliferation.

Tumor tissues can also be stained with antibodies to smooth muscle α-actin or the proteoglycan NG2 to visualize pericyte coverage of the vasculature and with antibodies to the hypoxia induced transcription factor HIF1a to assess the extent of hypoxic stress endured by tumor cells. Due to the inherent heterogeneity in the appearance of cancerous lesions, multiple samples should be analyzed and the extent of positive staining systematically quantified using imaging software (i.e., NIH ImageJ).

Paraffin embedding can sometimes be detrimental to the integrity of antigenic epitopes. When working with antibodies to these epitopes, it is advisable to embed the tissue samples in OCT compound (VWR) and then freeze them. In our experience, immunofluorescence is an excellent technique for staining frozen samples (Fig. 19.3) and is not as time consuming as IHC. Double- and triple-labeling methods allow simultaneous staining of two or three proteins, respectively, whereas counterstaining with DAPI labels all nuclei. Antibodies to $\beta4$ integrin (BD Pharmingen), E-cadherin (BD Biosciences), β-catenin (Zymed), PECAM 1(BD Biosciences), and cleaved caspase 3 (Fig. 19.3) have been successfully used in our laboratory.

Some antigens are destroyed by PFA fixation. For example, the epitope recognized by the antibody to the tight junction protein ZO-1 (Zymed) is lost after PFA fixation. To stain with this antibody, we omit the fixation step and embed the tissue directly into OCT after dissecting it out. The slides are later fixed in cold acetone, according to our IF protocol. Freezing in OCT is not recommended if antigen retrieval is to be used, since boiling frozen tissue samples damages their integrity.

We have compared the extent of apoptosis in primary tumor samples derived from Neu(YD)/$\beta4$-WT and Neu(YD)/$\beta4$-1355T mice by using immunofluorescent staining (Fig. 19.3, right panel). Paraformaldehyde-fixed, frozen sections were stained with antibodies against the cleaved form of caspase3 and counterstained with DAPI to visualize nuclei. The basement membrane was identified through staining an adjacent section with antibodies to laminin-5 (data not shown, basement membrane is delineated by dotted lines). Tumor samples derived from Neu(YD)/$\beta4$-1355T mice exhibited significantly more caspase3-positive cells inside the lumens of pseudoglandular structures, suggesting that $\beta4$ signaling protects tumor cells from anoikis.

Whereas primary mammary tumors can be dissected, fixed, or left unfixed, and then frozen in OCT or embedded in paraffin, certain precautions have to be applied before processing metastatic lungs, because these organs contain an extensive microcirculatory system and, if distended, a significant amount of air. After dissecting the lungs, we perfuse them with PBS to flush out the red blood cells, which can contribute to nonspecific background after staining. To achieve this, we cut the splenic artery and inject approximately 5 ml of cold PBS into the right atrium of the heart. The lung is then carefully dissected out and inflated by introducing ice-cold 4% PFA into the trachea. The tissue is finally fixed overnight in 4% PFA and stored in 70% ethanol at 4° until sectioning.

5.1. Methods

5.1.1. Fixation and embedding in OCT

1. Fix the tissue samples overnight in 4% PFA at 4° and wash them on the next day in ice-cold PBS.
2. Incubate the tissues in sterile PBS-30% sucrose at 4° with gentle shaking overnight or until they sink to the bottom of the container.
3. Incubate the tissues once in PBS-30% sucrose diluted 1:1 with OCT compound for 2 h on ice, twice in fresh OCT compound (each incubation

is 30 min on ice), and then embed them in OCT by flash-freezing the blocks.

5.1.2. Immunohistochemistry

1. Place the slides containing paraffin on a 56° hot plate overnight. Cool the slides to room temperature the following morning, and de-wax (only for paraffin sections) them through serial immersions in Histoclear (National Diagnostics) and ethanol solutions.
2. Fix frozen and/or de-waxed paraffin sections in ice-cold acetone for 10 min, air dry, and immerse three times in PBS containing 0.1% BSA. To quench endogenous peroxidase activity, treat the slides with 0.1% H_2O_2 in PBS for 15 min at room temperature and then wash three times with PBS-0.1% BSA.
3. When antigen retrieval is necessary (e.g., anti-Ki67, anti-cleaved caspase3), immerse the slides in a boiling solution of 0.01 M citric acid (pH 6.0) three times for 5 min and then cool them to room temperature.
4. Since IHC protocols utilize antibodies that are Biotin- and Avidin-linked, it is necessary to incubate the slides in avidin and biotin blocking solutions in order to reduce nonspecific background (avidin and biotin blocking kit, Vector). Place the slides in a humid chamber and incubate with a few drops of blocking solution for 15 min, wash with PBS-0.1% BSA, reincubate with blocking solution for 15 min, and finally wash with PBS-0.1% BSA.
5. In order to reduce nonspecific background produced by secondary antibodies, it is useful to incubate the slides in PBS-2% BSA containing 10% normal serum for 30 min at room temperature in a humidified chamber. Use serum from the same species in which the secondary antibodies were produced. The slides are then incubated in a solution of primary antibody in PBS-2% BSA overnight at 4°.
6. On the next morning, wash the slides with PBS-0.1% BSA and incubate in a solution of biotinylated secondary antibody (Vector) diluted in PBS-2% BSA for 1 h at room temperature. Wash the slides and incubate them with a tertiary reagent (peroxidase-linked avidin, A&B reagents, Vector) dissolved in PBS-2% BSA for 30 min at room temperature.
7. 3,3′-diaminobenzidine (DAB) solution is prepared from tablets according to manufacturer's instructions (Sigma) and is combined with hydrogen peroxide (final concentration of 0.02%) to create active substrate for HRP. Treat the slides with 0.5% Triton X-100 (Sigma) for 4 min and then immerse them in the DAB/H_2O_2 solution. The incubation time can range from seconds to minutes, depending on the affinity and concentration of both primary and secondary antibodies used. In order to compare staining intensity between slides in the same experiment,

however, it is crucial to develop slides that are stained with the same primary antibody for the same amount of time.

8. After the chromogenic reaction is complete, the slides can be counter-stained with hematoxylin (Gills Hematoxylin, Fisher Scientific), dehydrated in a series of alcohol solutions, and mounted with Permount (Fisher Scientific).

5.1.3. Immunofluorescent staining

1. Dry frozen slides in the fume hood for 1.5 to 2 h, and fix them in ice-cold acetone for 10 min.
2. In order to reduce nonspecific background that can be produced by secondary antibody binding, incubate the slides in a solution of 10% normal serum in PBS-2% BSA for 30 min at room temperature in a humidified chamber. Use serum of the same species in which the secondary antibody was produced. After blocking, incubate the slides with primary antibody diluted in PBS-2% BSA, overnight at 4°.
3. On the next morning, wash the slides three times with PBS-0.1% BSA, and incubate them with fluorescently conjugated secondary antibodies diluted in PBS-2% BSA for 1 h at room temperature in the dark.
4. To visualize nuclei, wash the slides with PBS and incubate them with 50 ng/ml solution of DAPI in PBS for 10 min at room temperature. Mount the slides using an aqueous mounting medium.

6. Detergent Extraction of Tumor Samples

In addition to histological techniques, tumor tissues can be lysed in detergent and subjected to biochemical analysis. In this case, tumor tissues are flash-frozen in liquid nitrogen, crushed into fine powder using mortar and pestle, and then suspended in an appropriate lysis buffer. We routinely use RIPA buffer that contains 20 mM Tris-HCl pH 7.5, 1% NP40, 0.5% Na deoxycholate, 0.1% SDS, 10% glycerol, 137 mM NaCl, 1 mM CaCl$_2$, 1 mM MgCl$_2$, 50 mM NaF, 1 mM Na vanadate, and protease inhibitor cocktail (Roche). Due to the abundance of DNA in tissue samples prepared using this method, we routinely sonicate the lysates and clarify them by centrifugation at 13,000 rpm for 15 min at 4°. Lysates prepared in RIPA buffer are suitable for a number of biochemical analyses, such as immunoblotting, immunoprecipitation, and ELISA.

7. *Ex Vivo* Culture of Mammary Tumor Cells

In order to apply cell biological methods to the study of tumor progression and metastasis in the mouse, it is necessary to explant and propagate in culture neoplastic cells from primary tumors and metastatic lesions. These cells can be profiled for various genetic lesions, aberrant signaling properties, and proliferative and migratory status. Additional assays include assessment of epithelial junction integrity, EMT, matrix adhesion (Hughes *et al.*, 2001), and cell migration (Keely, 2001). The cells can also be genetically altered via infection with retroviral or lentiviral constructs encoding short hairpin RNAs or wild-type or mutant versions of various genes. In addition, the cells can be used for allograft and *ex-vivo* pharmacological studies.

Our laboratory has experience with the isolation and culture of epithelial cells derived from the mouse mammary gland. We have dissected tumors from Neu(YD)/β4-WT and Neu(YD)/β4-1355T mice and performed a series of enzymatic digestions in order to dissociate the tumor cells from the surrounding host tissue and extracellular matrix and to eliminate the viscous genomic DNA released by dying cells. Western blotting was used to assess the levels of expression of integrin β4 and ErbB-2 in the primary tumor cells (Fig. 19.5A). Nontransformed mouse mammary epithelial cells can be isolated using a similar protocol (Welm *et al.*, 2005).

7.1. Method

1. Dissect the tumors, clear them of adjacent tissue, and wash them in cold PBS. Use a clean razor blade to mince the tissue and resuspend the fragments in 10× volumes of DME HG:F12 (1:1) supplemented with nonessential amino acids (NEAA) containing 1.25 mg/ml collagenase III and 1 mg/ml hyalorunidase (both from Worthington Biochemicals). Digest the tissue fragments at 37° for 2 to 3 h on a rotating shaker.

2. Spin down the cells and aggregates at 1000 rpm for 5 min and resuspend them in 5 ml of 0.25% trypsin/1 mM EDTA solution for 10 min at 37°. Spin down again and resuspend the cells in 5 ml of 4.5 μg/ml DNAse I (Worthington Biochemicals) in DME HG:F12 media for 10 min at 37°.

3. After DNAse treatment, wash the cells and aggregates in PBS–5% FBS and pipette the samples up and down several times to obtain single-cell suspensions. Resuspend the cells in complete medium (DME HG:F12 with NEAA, 10% FBS, 1 μg/ml hydrocortisone, 10 μg/ml insulin, 1 nM cholera toxin, and 10 ng/ml EGF) and plate them onto dishes that have been coated with 20 μg/ml of collagen I (Upstate Biotechnologies).

8. GENETIC MODIFICATION OF PRIMARY MAMMARY TUMOR CELLS

The derivation of primary cell cultures from tumors provides invaluable opportunities to study oncogenic signaling pathways. Furthermore, the cells can be genetically modified by infection with viral vectors encoding interfering RNA sequences or various cDNAs. However, since the MMTV promoter is regulated by hormones (Ucker *et al.*, 1981), its activity tends to decline once primary cells have been explanted in culture (even if the cultures are supplemented with hydrocortisone). It is therefore often necessary to re-introduce the oncogene into the primary cells by using a vector driven by a general promoter, such as CMV. Primary mammary gland epithelial cells can be difficult to transfect, so we usually utilize retroviral and/or lentiviral infection protocols. We have achieved successful knockdowns using the lentiviral vectors pLKO.1 (Open Biosystems) and pLVTH, which allow stable expression of shRNAs (Chatterjee *et al.*, 2004) (Fig. 19.4). We typically use at least two different sequences for hairpin constructs in order to exclude that the phenotype is due to off-target effects. To study the consequences of abrogating β4 signaling in otherwise isogenic neoplastic cells, we have also used a combined knockdown/reconstitution approach. To this end, we have used a retroviral LTRH1 vector (Barton and Medzhitov, 2002), expressing both shRNA targeting the mRNA encoding endogenous mouse β4 and an expression cassette for human integrin β4 (also β4-1355T).

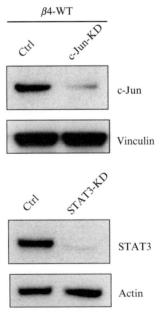

Figure 19.4 Neu-β4-Wt cells infected with lentiviral vectors encoding a control shRNA or shRNAs targeting c-Jun or STAT3 were subjected with immunoblotting with the indicated antibodies. (With permission, from Guo, W., Pylayeva, Y., Pepe, A., Yoshioka, T., Muller, W. J., Inghirami, G., and Giancotti, F. G. (2006). Beta 4 integrin amplifies ErbB2 signaling to promote mammary tumorigenesis. *Cell* 126, 489–502.)

8.1. Production of retroviral/lentiviral stocks and cell infection *in vitro*

1. In order to generate retroviral stocks, transfect the expression vector into packaging cells (such as Phoenix; (Grignani *et al.*, 1998)), which express the necessary viral structural proteins. Lentiviral vectors require transfection of a three-plasmid system: the vector of interest (such as pLVTH), the packaging construct (pCMV-dR8.74), and the envelope plasmid (pMD2.G-VSVG). In this case, transfect the three plasmids into 293FT cells (Invitrogen), which are typically cultured in the presence of standard DMEM with 10% FBS.

2. Cells should be at least 80% confluent at the time of transfection. Mix the DNA with Lipofectamine 2000 at a ratio of 1 μg of DNA per 3 μl of Lipofectamine 2000 (Invitrogen) in Optimem (Gibco) (add 4 ml of transfection mix in Optimem to 12 ml of regular DMEM with 10% FBS for each 150-mm dish). The ratio of vector plasmid:packaging construct: VSVG should be 1:1:1, and 8.75 μg of each construct should be transfected when using a 150-mm dish. Incubate the transfection mix overnight at 37°.

3. On the next morning, replace the medium with regular DMEM containing 10% FBS. The media can be collected at 48 h and 72 h post-transfection and concentrated viral supernatants (see below) can be stored at–80°. Pretreat all the plasticware and media with 10% bleach before discarding.

4. Prior to infecting cells, filter the viral supernatants to eliminate cellular contaminants. Centrifuge viral supernatants at 1400 rpm for 5 min and then filter them through a 0.45-μm filter (Millipore). Use of 0.22-μm diameter filters is not recommended, since passage through the pores may shear the viral envelope.

5. In order to concentrate the virus, centrifuge the filtered supernatant at 18,000 rpm for 2 h at 4°, aspirate most of the medium, and resuspend the visible pellet in 1 to 2 ml of medium. Infect target cells by adding the concentrated viral stock in combination with 5 μg/ml polybrene directly to the plate. Incubate the cells overnight at 37°, and then replace the medium with standard growth medium. Allow the cells to recover for 24 h, and then split them in drug selection medium. Effective multiplicity of infection (MOI) can be determined through viral stock titration. The cells are plated into multiwell plates and infected with serial 10-fold dilutions of viral stock. Productive infection is assessed through drug selection or FACS analysis for GFP positive cells.

Note: It is imperative to perform a horizontal transfer test every time a fresh viral stock is produced. The test is needed to ensure that an accidental generation of replication-competent virus has not occurred. The medium is collected from the virus-infected target cells, filtered, combined with polybrene, and added to a fresh plate of previously unmanipulated target cells. Cells are allowed to recover and subjected to drug selection together with control cells. Both sets of cells should die with the same kinetics. If the test for horizontal transfer is positive, all the stocks, cells, and media generated during the experiment should be destroyed by bleach treatment and autoclaving.

9. Disruption of Epithelial Adhesion and Growth Control

Integrin signaling disrupts both epithelial adhesion and growth control, contributing to invasive growth (Avizienyte *et al.*, 2002; Guo *et al.*, 2006; Weaver *et al.*, 1997). We describe in the following sections *in vitro* assays that can be used to examine the contribution of integrin signaling to tumor cell hyperproliferation, resistance to apoptosis, disruption of epithelial adhesion and polarity, and invasion.

9.1. Cell proliferation assay

1. To estimate growth factor–independent proliferation, deprive cells of growth factors for 24 to 36 h. Detach the cells with 0.25% Trypsin-1 mM EDTA and replate them onto eight-well chamber slides (Nunc) coated with a matrix protein of choice (we typically use collagen I) at a density of 15,000 cells per well. Incubate the cells in serum-free medium containing BrdU (Roche, 5-Bromo-2'-deoxy-uridine Labeling and Detection Kit, BrdU at 1:1000 dilution) for 24 h. To estimate proliferation in the presence of growth factors, do not starve the cells, and perform BrdU incorporation in presence of serum.
2. Fix the chamber slides in 100% ice-cold methanol, wash with PBS, and incubate with anti-BrdU solution diluted 1:10 in incubation buffer (Roche Kit) for 30 min at 37° in a humidified chamber.
3. Wash the slides with PBS and incubate them in a solution of 2 μg/ml of fluorescently conjugated secondary anti-mouse antibodies in PBS-0.1% BSA for 1 h at room temperature in the dark.
4. Wash the slides, counterstain them with DAPI (see above), and mount coverslips using aqueous mounting medium.

9.2. Apoptosis assay

1. To estimate the ability of integrin signaling to oppose apoptosis caused by growth factor deprivation, detach asynchronously growing cells in 0.05% Trypsin-1 mM EDTA and replate them at a density of 15,000

cells per well onto eight-well chamber slides (Nunc) coated with the extracellular matrix protein of choice. Incubate the cells in serum-free medium for 24 and 48 h and fix them with 4% PFA.

Next, to evaluate apoptosis caused by loss of matrix adhesion (anoikis), detach the cells using 3 mM EDTA, resuspend them in serum–containing medium supplemented with 1% BSA, and keep them in suspension at 37° with gentle rocking. Cells can be harvested at various points throughout the assay (typically 8, 16, and 24 h) and plated onto coverslips coated with 100 μg/ml polylysine-L (Sigma) for 10 min (both live and dead cells attach readily to polylysine). Fix the cells with 4% PFA.

2. Visualize apoptotic cells by TUNEL assay (*In Situ* Cell Death Detection Kit, Roche; follow manufacturer's instructions).

9.3. Matrigel invasion assay

1. Briefly trypsinize the cells, wash them once with serum-containing medium, and resuspend them in serum-free medium at a density of 1×10^5 per 300 μl of medium.

2. Place the transwell inserts (8-μm pore size, BD Falcon) into the wells of a 24-well plate and coat them with 2 μg of Matrigel. For this purpose, dissolve 2 μg of Matrigel in 50 μl of sterile water and apply 50 μl of solution to each well. Allow the mixture to solidify by keeping wells uncovered in the tissue culture hood for 2 to 3 h.

3. Apply 300 μl of cells (at the density described in Step 1) in serum-free medium into the Transwell chamber. Add 800 μl of serum-containing medium to the bottom of the well and incubate the cells at 37° for various times (we typically use 6-, 12-, and 24-h time points).

4. Thoroughly scrape the top of the Transwell membrane with a cotton swab to remove any cells that have not entered the Matrigel. Fix the cells that have reached the bottom of the Transwell with 4% PFA for 10 min at room temperature. Wash the filters with PBS and stain them with 0.5% crystal violet (make a solution of 0.5% crystal violet in 20% methanol) for 5 min. Thoroughly wash the filters with water and allow them to dry overnight. The efficiency of invasion can be quantified by counting the number of invaded cells per microscopic field. A picture of a representative field can be taken using a brightfield microscope (Fig. 19.2A).

9.4. Culture in three-dimensional matrigel

1. Detach the cells by trypsinization, and pipette gently to generate a single cell suspension. Wash the cells once with serum-free three-dimensional (3D) culture medium (DME HG:F12 (1:1) + NEAA, 1 μg/ml

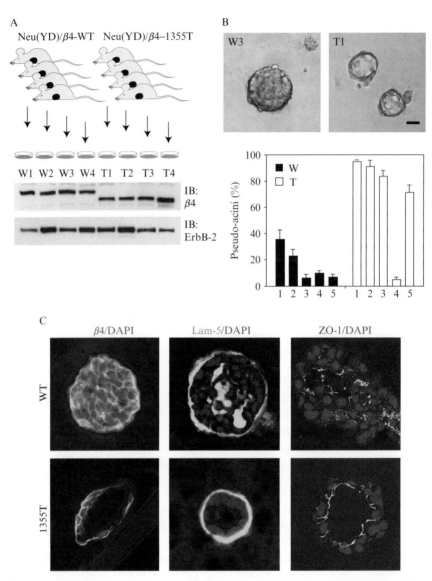

Figure 19.5 (A) Four populations of primary Neu(YD)/β4-WT (W1-W4) and Neu (YD)/β4–1355T (T1-T4) tumor cells were isolated and subjected to immunoblotting with anti-β4 and anti-ErbB2. (B) W1-W5 and T1-T5 cells were cultured in 3D Matrigel in serum-free medium for 13 days. The upper panel shows phase-contrast images of W3 and T1 cells. Scale bar = 50 μm. The graph shows the percentages (± standard deviation) of pseudoacini formed by each primary tumor cell population in 3D Matrigel. (C) 3D Matrigel cultures (*n*=3 per genotype) were fixed at day 13, sectioned, and stained as indicated. Scale bar = 50 μm. (With permission, from Guo, W., Pylayeva, Y., Pepe, A., Yoshioka, T., Muller, W. J., Inghirami, G., and Giancotti, F. G. (2006). Beta 4 integrin amplifies ErbB2 signaling to promote mammary tumorigenesis. *Cell* 126, 489–502.)

hydrocortisone, 1 nM cholera toxin, 1xITS (Sigma)) and resuspend at the density of 2×10^5cells/ml in the same medium.

2. Mix the cell suspension with Matrigel at 1:1 ratio and plate into a 96-well plate (50 μl per well) to examine tumor cell growth in 3D by phase contrast microscopy (see Fig. 19.5B) or into Transwell chambers (100 μl per well) for immunostaining (see Fig. 19.5 C).

3. Add 3D culture medium to a solidified bed of Matrigel at a volume of 50 μl per well in a 96-well plate. Add 3D culture medium at a volume of 100 μl into the top chamber and 600 μl into the bottom chamber of Transwells. Reapply the medium every 3 days.

4. To process the Transwell 3D cultures for immunostaining, remove the medium from the Transwells and fix the inserts with 4% PFA for 15 min at room temperature. If the antigen is sensitive to PFA fixation, fix the inserts in 1:1 methanol:acetone for 10 min at $-20°$.

5. After washing with PBS, cut the gel (together with the membrane) out of the Transwell, remove the insert, incubate once in OCT compound for 30 min on ice, and then freeze. Eight-micrometer sections can be cut and further processed for immunostaining (see Fig. 19.5C).

ACKNOWLEDGMENTS

We thank the Transgenic and Knock-out Mouse Facility and the Molecular Cytology Facility of MSKCC for technical assistance. These studies were supported by the National Institutes of Health (grants R37 CA58976 to F.G.G., RO1 CA113996 to F.G.G., and P30 CA08748 to Memorial Sloan-Kettering Cancer Center).

REFERENCES

Avizienyte, E., Wyke, A. W., Jones, R. J., McLean, G. W., Westhoff, M. A., Brunton, V. G., and Frame, M. C. (2002). Src-induced de-regulation of E-cadherin in colon cancer cells requires integrin signalling. *Nat. Cell. Biol.* **4,** 632–638.

Barton, G. M., and Medzhitov, R. (2002). Retroviral delivery of small interfering RNA into primary cells. *Proc. Natl. Acad. Sci. USA* **99,** 14943–14945.

Chatterjee, M., Stuhmer, T., Herrmann, P., Bommert, K., Dorken, B., and Bargou, R. C. (2004). Combined disruption of both the MEK/ERK and the IL-6R/STAT3 pathways is required to induce apoptosis of multiple myeloma cells in the presence of bone marrow stromal cells. *Blood* **104,** 3712–3721.

Elenbaas, B., Spirio, L., Koerner, F., Fleming, M. D., Zimonjic, D. B., Donaher, J. L., Popescu, N. C., Hahn, W. C., and Weinberg, R. A. (2001). Human breast cancer cells generated by oncogenic transformation of primary mammary epithelial cells. *Genes Dev.* **15,** 50–65.

Falcioni, R., Antonini, A., Nistico, P., Di Stefano, S., Crescenzi, M., Natali, P. G., and Sacchi, A. (1997). Alpha 6 beta 4 and alpha 6 beta 1 integrins associate with ErbB-2 in human carcinoma cell lines. *Exp. Cell Res.* **236,** 76–85.

Giancotti, F. G., and Ruoslahti, E. (1999). Integrin signaling. *Science* **285,** 1028–1032.

Glaser, S., Anastassiadis, K., and Stewart, A. F. (2005). Current issues in mouse genome engineering. *Nat. Genet.* **37,** 1187–1193.

Grignani, F., Kinsella, T., Mencarelli, A., Valtieri, M., Riganelli, D., Grignani, F., Lanfrancone, L., Peschle, C., Nolan, G. P., and Pelicci, P. G. (1998). High-efficiency gene transfer and selection of human hematopoietic progenitor cells with a hybrid EBV/ retroviral vector expressing the green fluorescence protein. *Cancer Res.* **58,** 14–19.

Guo, W., and Giancotti, F. G. (2004). Integrin signalling during tumour progression. *Nat. Rev. Mol. Cell Biol.* **5,** 816–826.

Guo, W., Pylayeva, Y., Pepe, A., Yoshioka, T., Muller, W. J., Inghirami, G., and Giancotti, F. G. (2006). Beta 4 integrin amplifies ErbB2 signaling to promote mammary tumorigenesis. *Cell* **126,** 489–502.

Gupta, G. P., and Massague, J. (2006). Cancer metastasis: building a framework. *Cell* **127,** 679–695.

Guy, C. T., Cardiff, R. D., and Muller, W. J. (1992). Induction of mammary tumors by expression of polyomavirus middle T oncogene: A transgenic mouse model for metastatic disease. *Mol. Cell. Biol.* **12,** 954–961.

Hood, J. D., and Cheresh, D. A. (2002). Role of integrins in cell invasion and migration. *Nat. Rev. Cancer* **2,** 91–100.

Hughes, P. E., Oertli, B., Han, J., and Ginsberg, M. H. (2001). R-Ras regulation of integrin function. *Methods Enzymol.* **333,** 163–171.

Hynes, R. O. (2002). Integrins: Bidirectional, allosteric signaling machines. *Cell* **110,** 673–687.

Keely, P. J. (2001). Ras and Rho protein induction of motility and invasion in T47D breast adenocarcinoma cells. *Methods Enzymol.* **333,** 256–266.

Kim, J., Yu, W., Kovalski, K., and Ossowski, L. (1998). Requirement for specific proteases in cancer cell intravasation as revealed by a novel semiquantitative PCR-based assay. *Cell* **94,** 353–362.

Levin, C. S. (2005). Primer on molecular imaging technology. *Eur. J. Nucl. Med. Mol. Imaging.* **32**(Suppl. 2), S325–S345.

Mainiero, F., Murgia, C., Wary, K. K., Curatola, A. M., Pepe, A., Blumemberg, M., Westwick, J. K., Der, C. J., and Giancotti, F. G. (1997). The coupling of alpha6beta4 integrin to Ras-MAP kinase pathways mediated by Shc controls keratinocyte proliferation. *EMBO J.* **16,** 2365–2375.

Mainiero, F., Pepe, A., Wary, K. K., Spinardi, L., Mohammadi, M., Schlessinger, J., and Giancotti, F. G. (1995). Signal transduction by the alpha 6 beta 4 integrin: Distinct beta 4 subunit sites mediate recruitment of Shc/Grb2 and association with the cytoskeleton of hemidesmosomes. *EMBO J.* **14,** 4470–4481.

Mariotti, A., Kedeshian, P. A., Dans, M., Curatola, A. M., Gagnoux-Palacios, L., and Giancotti, F. G. (2001). EGF-R signaling through Fyn kinase disrupts the function of integrin alpha6beta4 at hemidesmosomes: Role in epithelial cell migration and carcinoma invasion. *J. Cell Biol.* **155,** 447–458.

Mercurio, A. M., Bachelder, R. E., Chung, J., O'Connor, K. L., Rabinovitz, I., Shaw, L. M., and Tani, T. (2001). Integrin laminin receptors and breast carcinoma progression. *J. Mammary Gland Biol. Neoplasia* **6,** 299–309.

Miranti, C. K., and Brugge, J. S. (2002). Sensing the environment: A historical perspective on integrin signal transduction. *Nat. Cell. Biol.* **4,** E83–E90.

Muller, W. J., Sinn, E., Pattengale, P. K., Wallace, R., and Leder, P. (1988). Single-step induction of mammary adenocarcinoma in transgenic mice bearing the activated c-neu oncogene. *Cell* **54,** 105–115.

Murgia, C., Blaikie, P., Kim, N., Dans, M., Petrie, H. T., and Giancotti, F. G. (1998). Cell cycle and adhesion defects in mice carrying a targeted deletion of the integrin beta 4 cytoplasmic domain. *EMBO J.* **17,** 3940–3951.

Nikolopoulos, S. N., Blaikie, P., Yoshioka, T., Guo, W., and Giancotti, F. G. (2004). Integrin beta4 signaling promotes tumor angiogenesis. *Cancer Cell* **6,** 471–483.

Ponomarev, V., Doubrovin, M., Serganova, I., Vider, J., Shavrin, A., Beresten, T., Ivanova, A., Ageyeva, L., Tourkova, V., Balatoni, J., Bornmann, W., Blasberg, R., and Gelovani Tjuvajev, J. (2004). A novel triple-modality reporter gene for whole-body fluorescent, bioluminescent, and nuclear noninvasive imaging. *Eur. J. Nucl. Med. Mol. Imaging* **31,** 740–751.

Shaw, L. M., Rabinovitz, I., Wang, H. H., Toker, A., and Mercurio, A. M. (1997). Activation of phosphoinositide 3-OH kinase by the alpha6beta4 integrin promotes carcinoma invasion. *Cell* **91,** 949–960.

Sonnenberg, A. (1993). Integrins and their ligands. *Curr. Top. Microbiol. Immunol.* **184,** 7–35.

Spinardi, L., Einheber, S., Cullen, T., Milner, T. A., and Giancotti, F. G. (1995). A recombinant tail-less integrin beta 4 subunit disrupts hemidesmosomes, but does not suppress alpha 6 beta 4-mediated cell adhesion to laminins. *J. Cell Biol.* **129,** 473–487.

Trusolino, L., Bertotti, A., and Comoglio, P. M. (2001). A signaling adapter function for alpha6beta4 integrin in the control of HGF-dependent invasive growth. *Cell* **107,** 643–654.

Ucker, D. S., Ross, S. R., and Yamamoto, K. R. (1981). Mammary tumor virus DNA contains sequences required for its hormone-regulated transcription. *Cell* **27,** 257–266.

Weaver, V. M., Petersen, O. W., Wang, F., Larabell, C. A., Briand, P., Damsky, C., and Bissell, M. J. (1997). Reversion of the malignant phenotype of human breast cells in three-dimensional culture and *in vivo* by integrin blocking antibodies. *J. Cell Biol.* **137,** 231–245.

Welm, A. L., Kim, S., Welm, B. E., and Bishop, J. M. (2005). MET and MYC cooperate in mammary tumorigenesis. *Proc. Natl. Acad. Sci. USA* **102,** 4324–4329.

Yang, J., Mani, S. A., Donaher, J. L., Ramaswamy, S., Itzykson, R. A., Come, C., Savagner, P., Gitelman, I., Richardson, A., and Weinberg, R. A. (2004). Twist, a master regulator of morphogenesis, plays an essential role in tumor metastasis. *Cell* **117,** 927–939.

CHAPTER TWENTY

DESIGN AND CHEMICAL SYNTHESIS OF INTEGRIN LIGANDS

Dominik Heckmann *and* Horst Kessler

Contents

Abstract

The design and synthesis of peptidic and nonpeptidic integrin ligands derived from the most abundant natural tripeptide sequence, RGD, are described in this article. Special emphasis is placed on the activity and selectivity of the ligands to integrin subtypes. Two approaches are described—ligand- and structure-oriented design. When no structure of the complex or the target is known, one may derive suitable starting points from natural peptide sequences, which often require conformational restriction for a further optimization. A "spatial screening" procedure was used to identify highly active and selective ligands for the integrin subtypes $\alpha v\beta 3$ and $\alpha IIb\beta 3$. Structure-based methods require knowledge of the binding domain of the target. Hence, the first structure of the $\alpha v\beta 3$ integrin with bound cilengitide was a landmark for the structure-based approach. Meanwhile, a design using homology models of other integrin subtypes has also been successfully applied. To improve the ADME profile, nonpeptidic ligands have been developed using the information of the spatial distances and orientations of the most important pharmacophoric groups (especially the

Department of Chemistry, Technical University München, Garching, Germany

Methods in Enzymology, Volume 426
ISSN 0076-6879, DOI: 10.1016/S0076-6879(07)26020-3

© 2007 Elsevier Inc.
All rights reserved.

carboxyl group and the basic moiety at the other end of the molecule). Applications of the $\alpha v\beta 3$ ligands as drugs in antiangiogenic tumor therapy for molecular imaging of metastases and for improvement of biocompatibility of grafts are briefly described.

1. INTRODUCTION

The discovery of fibronectin as a major extracellular matrix (ECM) protein and the observation that it interacts with cell surfaces stimulated great interest in the nature of ECM—cell interactions (Eble and Kühn, 1997), years before the term "integrin" was introduced by Tamkun and Hynes (Tamkun et al., 1986). The fact that in vertebrates most cells are associated with supportive structures, such as collagen or fibronectin, and the observation that attachment of cells to a solid substratum is essential for normal cell physiology, highlighted the importance of such interactions for biological and medicinal research. Pioneering studies by Ruoslahti and Pierschbacher in the early 1980s revealed that binding of integrins to CAMs occurs not only in the tripeptide sequence RGD in fibronectin, but also in small peptide sequences such as Arg-Gly-Asp-Ser-Pro (Pierschbacher et al., 1981, 1982). Due to its appearance in various ECM proteins such as fibrinogen and vitronectin, the RGD sequence was initially—and prematurely—named "universal recognition motif." Although several different recognition motifs have been identified (Isacke and Horton, 2000; Plow et al., 2000), the RGD motif represents the most prominent recognition sequence involved in cell adhesion. The corresponding ECM-binding receptors, the integrins, have been described and classified as a family of homologue $\alpha\beta$ heterodimers that recognize several motifs such as RGD or LDT in their natural ligands and are involved in both cell adhesion and outside-in/inside-out signaling pathways. The findings that integrins are involved in a variety of pathological processes such as inflammation, vascular homeostasis, thrombosis, restenosis, osteoporosis, cardiovascular disorders, cancer invasion, metastasis, and tumor angiogenesis, made them valuable targets for pharmaceutical/medicinal chemistry, especially for cancer research (Arndt et al., 2005; Clemetson and Clemetson, 1998; Jin and Varner, 2004; Rojas and Ahmed, 1999; Yusuf-Makagiansar et al., 2002).

Peptide ligands with various sequences were synthesized to achieve selectivity for the different receptor subtypes that recognize the RGD-sequence, especially to discriminate $\alpha IIb\beta 3$ from $\alpha v\beta 3$ integrin. Furthermore, early attempts were made to synthesize selective peptidomimetics for the platelet integrin $\alpha IIb\beta 3$, which has been identified as a target for treatment of thromboses. In 1991, we showed that head-to-tail cyclization of pentapeptides containing both the RGD sequence and at the same time distinct positions of D–amino acids increased binding activity to $\alpha v \beta 3$ by three orders of magnitude, whereas in the same compounds the binding to $\alpha IIb \beta 3$ is drastically reduced (Aumailley *et al.*, 1991). In contrast, suitable hexapeptides show reversed selectivity (Müller *et al.*, 1994).

These findings together with extensive conformational analysis pointed out that integrin receptors vary according to the ligand conformations they recognize. In subsequent years, a large number of modified integrin ligands were published, whereas the main focus shifted from peptides toward nonpeptidic mimetics. The general aim of all these studies was to identify highly potent, orally available, metabolically stable, nontoxic analogues that specifically address distinct integrin subtypes. We will concentrate here on the development of ligands for $\alpha v \beta 3$, $\alpha v \beta 5$, and $\alpha 5 \beta 1$, which are involved in angiogenesis and metastatic cancers but also in retinopathy (Friedlander *et al.*, 1996; Hammes *et al.*, 1996) or acute renal injury (Goligorsky *et al.*, 1997). Apart from their function as anchors for cell adhesion, integrins are involved in bidirectional signaling pathways. A ligand for a particular integrin could, on the one hand, serve as an agonist that triggers a signaling pathway; on the other, it could act as an antagonist by competitive suppression of binding to its natural ligand. Due to the complexity of integrin-signaling pathways, the agonist/antagonist nature of artificial ligands is still a matter of debate (Humphries *et al.*, 2003; Hynes, 2002). The intention of this article is to provide a brief overview of the strategies used in the design of peptidic and nonpeptidic integrin ligands and the rational and combinatorial methods used in this process.

2. DESIGN OF INTEGRIN LIGANDS: OVERVIEW

As long as the structure of an integrin's extracellular head group was not identified, the development of drug candidates concentrated on optimizing RGD peptides or screening methods to identify a suitable ligand. This approach is called "ligand-oriented drug design." The first complete structure of an integrin's extracellular head group was published for $\alpha v \beta 3$ in 2001 (Xiong *et al.*, 2001), and a year later (Xiong *et al.*, 2002) the structure for the same integrin in complex with the cyclic peptide "cilengitid" (Dechantsreiter *et al.*, 1999) was published. The first integrin ligands were proteins (e.g., Fn-fragments, antibodies) or smaller peptides containing the RGD recognition motif (Pierschbacher and Ruoslahti, 1984a,b). The preparation of peptidic ligands is facilitated by relatively easy solid-phase synthesis. Linear peptides are usually highly flexible and are found in numerous conformations, but cyclization can be used to reduce the conformational space (Kessler, 1982). In addition, smaller head-to-tail cyclic peptides (penta- or hexa-peptides) are stable against enzymatic degradation.

If a conformationally constrained molecule is biologically active, it can serve to localize the spatial orientation of the pharmacophoric groups. This allowed the synthesis of biased libraries (a compound library, whose diversity is limited by structural considerations or a general pharmacological profile) that reduce the effort involved and facilitate understanding structure—function relationships. However, screening of compounds to identify the scaffolds carrying the pharmacophors is still needed to find suitable ligands. Due to their therapeutic relevance, a vast number of highly active compounds have been published.

Although other non–RGD-binding integrins such as $\alpha 4\beta 1$ or $\alpha 4\beta 7$ (both recognizing the LDT sequence) have been the subject of pharmacological research, we will concentrate here on RGD as the most prominent recognition motif. Due to space constraints, we will concentrate on a few examples that document the process of development from a lead structure to highly active and selective compounds. After the structure of the integrin head group that includes the binding site is known, switching to structure-based optimization ("rational drug design")—streamlining the whole process of drug development—becomes possible.

2.1. Ligand-based design

2.1.1. Peptidic ligands

Although many integrins recognize a common motif (e.g., RGD), and the receptors show a high degree of homology, they are nonetheless able to distinguish among different natural ligands that contain the same recognition sequence. It is known that residues adjacent to RGD contribute to receptor selectivity, but the data cannot sufficiently explain the specificity of

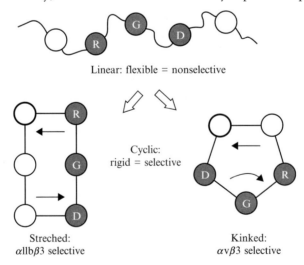

Figure 20.1   Induction of selectivity to the RGD-sequence by head-to-tail-cyclization.

RGD recognition by distinct integrins or the fact that RGD in a number of proteins is nonfunctional. This leads to the assumption that integrin receptors also recognize a distinct conformation of the RGD sequence that in natural ligands is maintained by the secondary and tertiary structure of the protein. The integration of a recognition motif into a cyclic peptide is a feasible way to restrict the conformational space of the amino acid sequence, and was demonstrated to show an impact on binding affinity and receptor specificity (Kessler, 1982). Restriction in conformational freedom may increase binding affinity to a receptor, but only if the biologically active conformation is included in the allowed conformational space (*matched case*). The resulting activity gain is owed to the decrease in conformational entropy that is lost upon binding, and to a preinduced strain toward adoption of the binding conformation. In the *mismatched case*, where the peptide is not able to adopt a biologically active conformation, the affinity toward the target receptor is lost. In particular, backbone-cyclized penta- and hexa-peptides are known to stabilize distinct conformations by adaptation of turn-like structures that can be examined using NMR spectroscopy and MD simulations (Aumailley *et al.*, 1991; Gurrath *et al.*, 1992; Kessler, 1982). A "spatial screening" (Haubner, 1997; Weide *et al.*, 2007) of various conformations (keeping the chemical nature of the side chains unchanged) of a cyclized peptide sequence can be achieved by varying the chirality of selected amino acids or altering ring size or sequence reversion (retro-inverso concept) (Shemyakin *et al.*, 1969). A library of both active and inactive peptides with assigned conformation allows detailed structure–function relationships. In our studies on the design of highly active cyclic peptides, we have chosen the linear RGDFV as the lead sequence based on findings that in RGDX, peptide activity decreases in the following series: $X = F > V > C > Q > S$ (Aumailley *et al.*, 1991; Haubner *et al.*, 1996a; Tranqui *et al.*, 1989). Conformational control was introduced by a D-residue and/or prolin, which induces characteristic turn motifs within the cyclic peptides. (Gurrath *et al.*, 1992; Kessler *et al.*, 1995). For the design of cyclic hexapeptides, a spacer residue (Ala or Gly) was added flanking the RGD sequence.

Figure 20.2 shows the schematic representation of turn motifs in cyclic peptides with one D-amino acid. The systematic permutation of the one D-amino acid residue of the sequence leads to a shift of the sequence around the template in the same conformation, which results in a different three-dimensional (3D) representation of the recognition sequence (pharmacophoric groups) with identical chemical constitution (e.g., identical functional side chain and sequence) in each peptide (Kessler and Kutscher, 1986).

The nature of the turn motifs was examined by NMR spectroscopy (Table 20.1), together with the inhibitory capacity toward cell adhesion on vitronectin (Gurrath *et al.*, 1992). Although a large variety of turn motifs could be generated by a D-amino acid scan and the introduction of proline,

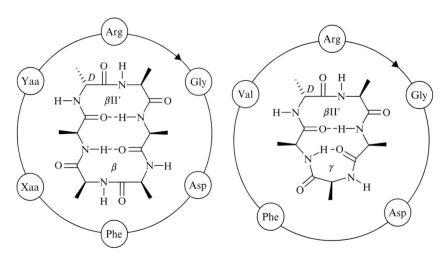

Figure 20.2 Schematic representation of the structural templates for the design of cyclic RGD-containing hexapeptides (left) and pentapeptides (right). The D-amino acid is set in the upper left corner at the *i+1* position of the upper βII' turn. The arrows indicate that the amino acid sequence is shifted systematically around the template by using different amino acids of the sequence in D-configuration.

Table 20.1 Sequence, sructure, and inhibitory capacity (IC50) of cyclic hexa- and penta-peptides

Entry	Sequence	Turn (i+1 Amino Acid)	IC50 [μM]
1	c(–rGDFVG–)	II'(r1), βI(F4)	>120
2	c(–RGDFPG–)	*cis:* βVIa(F4), βII(R1)	>100
		trans: βII(P5), βII'(G2)	
3	c(–RGDFpG–)	βII'(p5), βII'(G2)	>100
4	c(–RGDfLA–)	βII'(f4), βII(R1)	28
5	c(–RGDfLG–)	βII'(f4), βII(R1)	>120
6	c(–rGDFV–)	βII'(r1),γ(F4)	>120
7	c(–RGdFV–)	n.d.	>120
8	c(–RGDfV–)	βII'(f4),γ(G2)	0.1
9	c(–RADfV–)	βII'(f4),γ(A2)	41
10	c(–RGDFv–)	βII'(v5),γ(G2)	30
11	RGDFv	—	>170
12	GRGDS	—	18

Notes: IC50 values are in μM and represent 50% inhibition of HBL-100 cell (high in αv integrins) adhesion on immobilized vitronectin. Amino acids are presented in the one letter code as capital letters (L-amino acids) and lower-case letters (D-amino acids).
n.d., no data.

in general, cyclic hexapeptides show no or little activity toward αvβ3 (entries 1 to 5). The D-amino acid scan of cyclic pentapeptides (entries 6, 7, 8, 10) afforded the peptide *cyclo*(-RGDfV-) (entry 8) (Aumailley *et al.*, 1991), which displays a 100-fold improved inhibitory capacity on the αvβ3-vitronectine integrin interaction than the reference peptide GRGDS. The dramatic loss of activity on substitution of glycine by alanine (entry 9) points out the importance of glycine; the activity gain between entries 10 and 11 is owed to cyclization.

From the structural point of view, the favored conformation of the RGD sequence forms a kink around glycine. At the same time, the ability to inhibit binding of the platelet integrin αIIbβ3 to fibrinogen is strongly reduced (Pfaff *et al.*, 1994). On the other hand, among the hexapeptides, which showed no significant inhibitory capacity on the cell-vitronectin-adhesion mediated mainly by the integrin αvβ3, some proved to be highly active inhibitors of platelet aggregation, which is mediated by integrin αIIbβ3. By comparing the calculated structures of several ligands with different specificities for either αvβ3 or αIIbβ3, a strong dependence of selectivity from the 3D arrangement of the RGD sequence could be observed (Lohof *et al.*, 2000). In *cyclo*(-RGDfV-), as a representative αvβ3 ligand, the RGD sequence is kinked, resulting in a shorter distance between the Cα of Arg and Glu than in the cyclic αIIbβ3 ligands (Bach II *et al.*, 1994; Müller *et al.*, 1994), which arrange RGD in a straightened form.

A number of modifications of the peptide structure have been performed (Haubner, 1997; Kessler *et al.*, 1995). Examples included the systematical substitution of peptide bonds with thioamides (Geyer *et al.*, 1993) or reduced amide bonds (Geyer *et al.*, 1994) and the incorporation of turn mimetics (Haubner *et al.*, 1996b) or sugar amino acids (Lohof *et al.*, 2000). Those structural modifications mostly afforded less active peptides.

To determine how the peptide bonds are involved in the binding of *cyclo* (-RGDfV-), we synthesized all possible stereoisomers of this structure as well as its *retro*-sequence (Wermuth *et al.*, 1997). An important result was the significantly reduced activity of the *retro-inverso* (Shemyakin *et al.*, 1969) peptide with the following stereochemistry cyclo(-VfdGr-) (Wermuth *et al.*, 1997). The NMR-MD–based structure exhibited an almost identical orientation of all side chains but a significantly reduced activity. This pointed out the involvement of at least one peptide bond in the interactions with the receptor, as in the retro-sequences the orientation of all peptide bonds is switched. A simple removal of one specific peptide bond via a mimetic, such as a reduced peptide bond or a thioamide, gives no conclusive answer, because such a modification typically induces strong conformational changes and the origin of the changed biological activity cannot be traced. The most important structural modification turned out to be the incorporation of N-methylated amino acids into the peptide sequence. N-methylation has been shown to be a valuable tool in structure-function-relationship studies

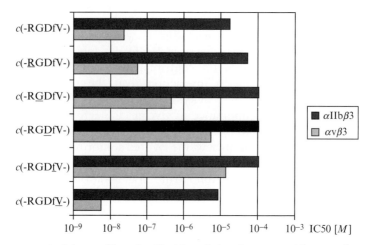

Figure 20.3   Activity profiles of cyclic, *N*-methylated pentapeptides on αvβ3 integrin and the platelet receptor αIIbβ3. The underlined residue resembles the *N*-methylated amino acid, lower case letters resemble D–amino acids.

(Gilon *et al.*, 2003). This modification often induces a *cis N*-methylated peptide bond, a change in the lipophilicity profile, and sterical hindrance. This also results in increased proteolytic stability for *N*-methylated peptides (Manavalan and Momany, 1980; Mazur *et al.*, 1980; Tonelli, 1976; Turker *et al.*, 1972). The concept of *N*-methylation has been applied on the cyclic pentapeptides binding αvβ3 integrin in order to identify new compounds with an improved affinity and selectivity profile. In case of the pentapeptide *cyclo*(-RGDfV-), an *N*-methyl scan afforded a highly active αvβ3 ligand with 1.5-fold selectivity against αIIbβ3 (Fig. 20.3) (Dechantsreiter *et al.*, 1999).

The peptide *cyclo*(-RGDfNMeVal-), which is now developed by MERCK KGaA, Darmstadt under the name cilengitide, is in phase II of clinical investigation for patients with glioblastoma multiforme, metastatic prostate cancer, and lymphoma (http://www.clinicaltrials.gov/ct/search?term=Cilengitide).

2.1.2. From peptides to peptidomimetics

A peptidomimetic is defined as a substance having a secondary structure as well as other structural feature analogues to that of the original peptide, which allows it to displace the original peptide from receptors or enzymes (Gante, 1994). They may offer advantages over physiologically active linear peptides by improving oral bioavailability and better stability against enzymatic degradation within the organism. Small cyclic peptides, however, are enzymatically stable but in general not orally active. One strategy for isosteric substitution is the use of *aza*-peptides (Dutta and Morley, 1975; Gante *et al.*, 2003). Based on the observation that glycine can be

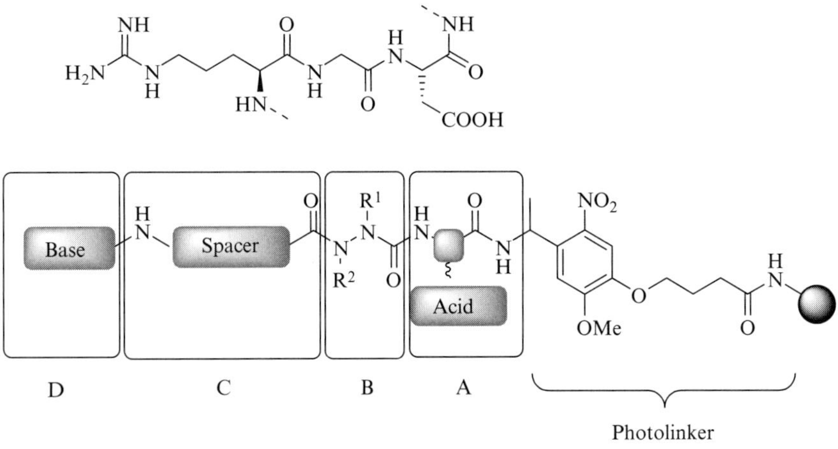

Figure 20.4 Transformation of the RGD-sequence into a modulary assembled library.

replaced by *aza*-glycine in the peptide *cyclo*(-RGDfV-) with preservation of biological activity and selectivity, we synthesized a noncoded library of linear *aza*-glycine peptides using combinatorial split-mix synthesis (Gibson *et al.*, 2001a) (Fig. 20.4).

The use of *aza*-glycine was stimulated by the observation that the N—N bond in those analogues was twisted by 90 degrees, leading to a kink in the structure that was previously determined as crucial for αv selectivity (Schmitt, 1998; Wermuth, 1996). The lead structure Arg-Gly-Asp was transformed into an RGD mimetic that could be assembled step by step on solid support according to the Fmoc strategy with the building blocks A through D to yield 660 compounds for a biological assay.

In the case of chiral building blocks, the racemic compound was synthesized to allow the detection of activity for both enantiomers. A photolinker was used as the anchor for photolytic cleavage of the compounds after on-bead testing using soluble αvβ3 integrin. After selection of the beads showing affinity toward αvβ3 (identified by an ELISA-type experiment) and cleavage from the resin, the active compounds were investigated by mass spectroscopy. Most compounds could be identified by their exact mass, while isomeric compounds were assigned according to their fragmentation patterns.

All RGD mimetics showing activity on αvβ3 were resynthesized—in case of chiral compounds in both enantiomers—and tested in an assay on isolated receptors (αvβ3, αvβ5, and αIIbβ3). Five potential lead structures could be identified, all showing an asparagin as building block A. In all cases, the natural L-enantiomer was more active than the D-enantiomer. The most active hit (Fig. 20.5) displayed a higher activity (IC50 = 150 n*M*) than the reference peptide GRGDSPK (400 n*M*), but still exhibited a high

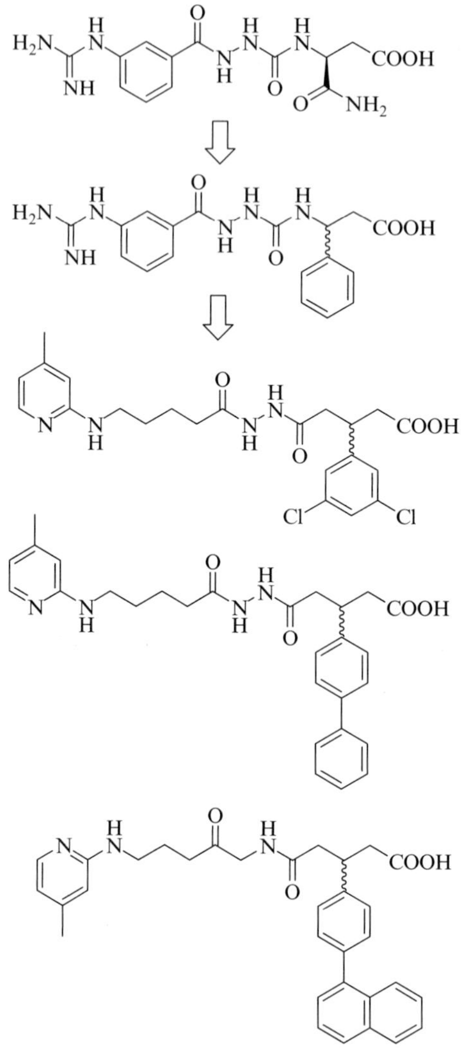

1 $IC_{50}(\alpha v\beta 3) = 150\ nM$

2 $\mathbf{IC_{50}(\alpha v\beta 3) = 0.8\ nM}$; $IC_{50}(\alpha v\beta 5) = 281\ nM$

3 $\mathbf{IC_{50}(\alpha v\beta 3) = 6\ nM}$; $IC_{50}(\alpha v\beta 5) = 5700\ nM$; $IC_{50}(\alpha v\beta 6) = 44\ nM$

4 $\mathbf{IC_{50}(\alpha v\beta 3) = 4\ nM}$; $IC_{50}(\alpha v\beta 5) = 8700\ nM$; $IC_{50}(\alpha v\beta 6) = 53\ nM$

5 $IC_{50}(\alpha v\beta 3) = 8.0\ nM$; $IC_{50}(\alpha v\beta 5) = 5170\ nM$; $\mathbf{IC_{50}(\alpha v\beta 6) = 0.1\ nM}$

Figure 20.5 From RGD-peptide to small molecules. Development of RGD mimetics with variable selectivity profile towards different αv integrins.

degree of polarity—an indicator of an unfavorable pharmacokinetic profile. Further modifications improved activity and the bioavailability profile. The process of lead optimization and the activities for the different αv integrin receptor subtypes is outlined in Fig. 20.5.

It has been shown in the past that hydrophobic residues in the β-position of the carboxylic group are well tolerated by the αvβ3 receptor. For this reason, the terminal carboxamide function was replaced by a phenyl residue, which resulted in the highly active, more lipophilic compound **2**. The solid-phase synthesis of *aza*-glycine compound **2** is described in the experimental section (p. 486). After the crystal structure of $\alpha v \beta 3$ in complex with cilengitide was solved (Xiong *et al.*, 2002), we used it for the elucidation of the binding mode into the receptor after removing cilengitide from the complex (Marinelli *et al.*, 2003). All well-established features of ligand–receptor interaction derived from ligand-oriented design in the integrin field are observed. In addition, the aromatic spacer unit is optimally oriented for parallel π-stacking interaction to $(\alpha v)Tyr^{178}$.

Hydrogen bonds between the $(\beta 3)Arg^{216}$ backbone carbonyl and two NH donors of the diacylhydrazine core can also be observed. However, the docking experiments resulted in two different binding modes of the peptidomimetic, which differ in the position of the phenyl ring. In one of them, the phenyl ring occupies a well-defined pocket adjacent to the MIDAS region, while in the other the phenyl ring sticks out of the receptor. It is reasonable to assume that the position of the aromatic ring is determined by sterical demand: a small aromatic moiety will fit in the pocket while larger groups will stick out of the binding pocket without a reduction of binding affinity. This is supported by the observation that attachment of large spacer units to the phenyl group for immobilization purposes did not interfere with the binding potency (Dahmen *et al.*, 2004). As the guanidinium group is usually considered a handicap to bioavailability, it can often be substituted by less basic, more lipophilic heterocyclic groups, such as an aminopyridin. Replacement of the spacer in compound **3** (Fig. 20.5) by an alkyl chain and a 2-aminopyridin group resulted in a slight decrease of affinity, which may result from the loss of the aromatic interaction of the spacer and the loss of one hydrogen donor, which allows only contact to one aspartic acid residue of the α-subunit (Sulyok *et al.*, 2001).

Further optimizations, especially concerning the aromatic moiety near the carboxylic function, led to compounds with different selectivity profiles among the integrins $\alpha v \beta 3$, $\alpha v \beta 5$, and $\alpha v \beta 6$. While $\alpha v \beta 3$ and $\alpha v \beta 5$ were investigated as targets for osteoporosis therapy (Horton *et al.*, 1991) and antiangiogenetic cancer therapy (Brooks *et al.*, 1994; Smith *et al.*, 1990), $\alpha v \beta 6$ is an ephithelia-specific integrin that is upregulated during inflammation (Huang *et al.*, 1996), tumor proliferation (Agrez *et al.*, 1994), and wound healing (Clark *et al.*, 1996).

The platelet receptor $\alpha IIb \beta 3$ was the first integrin target in medicinal research (Feuerstein *et al.*, 1996; Mousa, 2000; Rose *et al.*, 2000). Many nonpeptidic lead structures mimicking the RGD sequence of its natural ligand, fibrinogen, have since been identified. The optimal distance between the most important pharmacophors, the basic moiety (guanidine or mimetic), and the carboxyl group was determined from cyclic peptides (Müller *et al.*, 1994), and allowed the incorporation of various spacer groups between both pharmacophors. For a structure-based design using a tyrosine

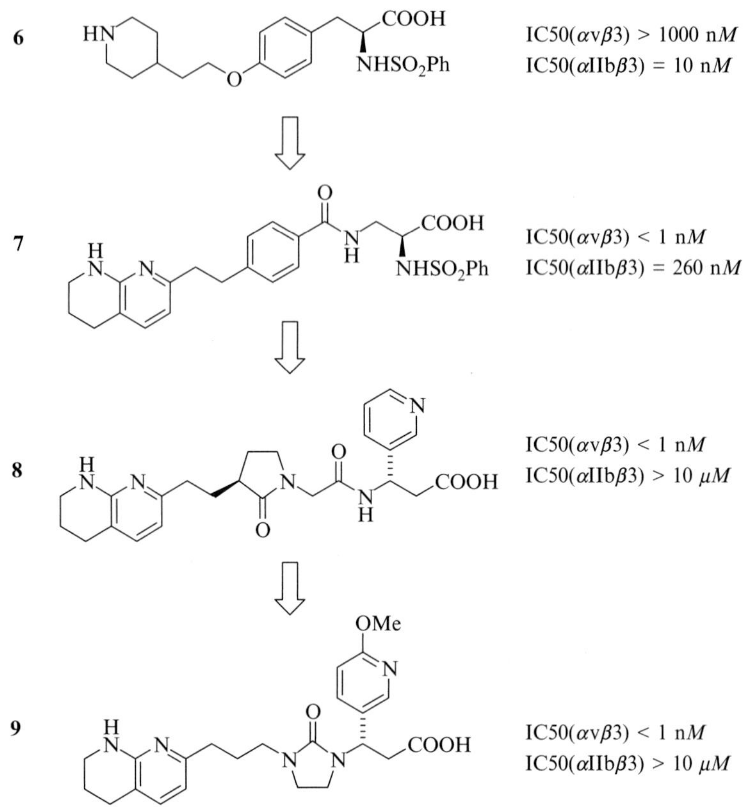

Figure 20.6 Inverting selectivity: From αIIbβ3 ligands to highly active αvβ3 ligands.

template (Fig. 20.6) (compound **6**), which was established by Merck Laboratories (Egbertson *et al.*, 1994; Hartman *et al.*, 1992), screening for different basic moieties resulted in a reversed selectivity toward αvβ3 instead of αIIbβ3 (Duggan *et al.*, 2000).

Although providing only one hydrogen donor, the tetrahydronaphthyridine (compound **7**) acts as a good guanidine mimetic in a series of compounds. Compound **8** represents a more RGD-like structure, which reveals subnanomolar activity on αvβ3 and low activity on αIIbβ3. While the stereochemistry on both chiral carbons was of minor importance, the substitution of the pyridine moiety by other annulated systems slightly affected affinity and could furthermore be used to optimize the pharmacokinetic properties (Coleman *et al.*, 2002). The development ended with compound **9**, a chain-shortened, constrained ligand displaying good *in vivo* potency and an improved pharmacokinetic profile that was selected for clinical trials (Coleman *et al.*, 2004). The proposed binding mode (Feuston *et al.*, 2002) for these compounds completely differs from the crystal structure of the $\alpha v \beta 3$-cilengitide complex (Feuston *et al.*, 2002; Xiong *et al.*, 2002). Whether this is a possible alternative binding mode or just a calculation artifact caused by using the unligated crystal structure with the unoccupied MIDAS region for these docking studies remains unknown.

Using the apo-$\alpha v \beta 3$ integrin structure, we could provide evidence that considerable conformational rearrangement is involved in the ligand binding (Gottschalk *et al.*, 2002), which was confirmed by the complex structure 1 year later (Xiong *et al.*, 2002). Important for the selectivity between $\alpha v \beta 3$ and $\alpha IIb \beta 3$ are the positions of the hydrogen donors at the guanidine mimetic. The high number of basic moieties evaluated by various groups suggested an *end-on* binding of the guanidine group for $(\alpha IIb)Asp^{224}$, which can be directly addressed by, for example, a 4-piperidyl group, while αv integrins show a *side-on* binding by $(\alpha v)Asp^{150}$ and/or $(\alpha v)Asp^{218}$. Therefore, the guanidine group is potentially biselective while a tetrahydronaphthyridine is only capable of *side-on* binding and thus promotes selectivity toward $\alpha v \beta 3$. Furthermore, a distance of 16 Å between the guanidine function and the carboxylate is favorable for $\alpha IIb3$ binding, while shorter distances are optimal for $\alpha v \beta 3$, a fact that has already been shown in peptide libraries (Müller *et al.*, 1994).

2.2. Rational structure-based design

The premise of the rational design of ligands is the knowledge about the structure of the receptor-binding pocket, preferably in complex with the natural ligand. For proteins, the best way to determine structure with high accuracy is X-ray diffraction of protein crystals. The first high-resolution crystallographic structure of an integrin was published by the Arnaout group for the extracellular segment of the integrin $\alpha v \beta 3$ in the unligated state (Xiong *et al.*, 2001) and later cocrystallized with the cilengitide (*cyclo*(-RGDfNMeVal-)) (Xiong *et al.*, 2002). The second integrin structure solved by X-ray spectroscopy was $\alpha IIb \beta 3$, in complex with the high-affinity ligand tirofiban (Xiao *et al.*, 2004). The binding mode of the $\alpha IIb \beta 3$ ligand tirofiban (Hartman *et al.*, 1992), compared to the binding mode of cilengitide in $\alpha v \beta 3$, is shown in Fig. 20.7.

The guanidinium group of *cyclo*(-RGDfNMeVal-) is fixed inside a narrow groove formed by the D3-A3 and D4-A4 loops of the β-propeller of αv by a bidentate salt bridge to (αv)-Asp^{218} at the bottom of the groove and by an additional salt bridge with (αv)-Asp^{150} at the rear. Contacts between the Asp of the ligand and the βI domain of $\alpha v \beta 3$ primarily involve the Asp carboxylate group, which protrudes into the MIDAS region of the $\beta 3$ subunit. The carboxylate function coordinates a Mn^{2+} ion at the MIDAS, and is also involved in a hydrogen bond with the backbone amide proton of $(\beta 3)$-Asn^{215} (not shown).

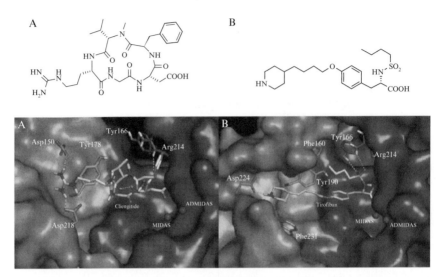

Figure 20.7 Structure of two ligand–integrin complexes based on X-ray structures. (A) Cilengitide binding αvβ3. (B) Tirofiban binding αIIbβ3. The receptors are shown as Connolly-surface representation (α-subuntits in blue, β-subunits in red). Interacting residues and metal ions are highlighted, hydrogen bonds shown as dot lines. (See color insert.)

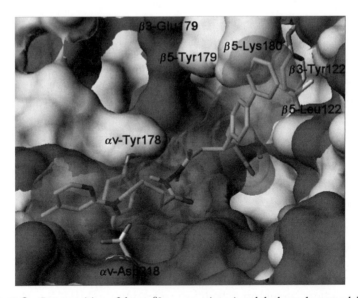

Figure 20.8 Superposition of the αvβ3 receptor (grey) and the homology model of the αvβ5 receptor (white) both represented as Connolly surface. The αvβ3-selective ligand (4) is docked into the αvβ3 receptor (green transparent), the MIDAS cation is shown as golden sphere. (See color insert.)

Further hydrogen bonds are formed with the backbone carbonyl of (β3)-Arg^{216} and the side-chain of (β3)-Arg^{214}. The D-Phe residue contributes to the binding by a weak $\pi-\pi$ interaction with (β3)-Tyr^{122}. The glycine allows the formation of a kinked conformation in the cyclic peptide but does not interact with the receptor itself. The binding mode of Tirofiban to αIIbβ3 strongly corresponds to this structure.

For decades, the main pharmacophors, the carboxyl group, and the amine or guanidine group, have been known to play an essential role in binding. Structural details are now established. The carboxylic function of the tyrosine scaffold coordinates the metal ion at the MIDAS, while the basic piperidine moiety is engaged in a salt bridge with the (αIIb)Asp^{224}. In contrast to the αv subunit, there is only one aspartic acid residue present in the α subunit to interact with the piperidine. This aspartate is more immersed in the receptor leading to a longer groove in the β-propeller. An αIIb ligand therefore requires an elongation to reach both anchoring points and the right orientation of the basic moiety toward the (αIIb)Asp^{224}. As previously identified from the study of restricted cyclic peptides (Müller *et al.*, 1994), the optimal length for a αIIbβ3 ligand is ~16 Å, an observation that has been extensively utilized in the design of selective compounds. The (αv)-Asp^{218} is replaced by (αIIb)-Phe^{231}, which, together with (αIIb)-Phe^{160} and (αIIb)-Tyr^{190}, results in a significantly more hydrophobic environment compared to αv. This hydrophobic cleft is occupied by the *n*-butyl-side chain of the sulfonamide of tirofiban, which is positioned by two hydrogen bonds with the (β3)-Tyr^{166}-hydroxyl function and the guanidine group of (β3)-Arg^{214}.

As the substrate-based design has already shown, these structural differences between the integrins αvβ3 and αIIbβ3 can be utilized to trigger selectivity in both directions. Since the role of integrin αvβ3 in tumor angiogenesis has been discovered, this selectivity against the platelet integrin αIIbβ3 is an important factor for the design of drug candidates to reduce side effects. Although over the years most attention has been paid to the role of αvβ3 as a target for antiangiogenetic therapy, there are other integrins involved in this process, such as the closely related αvβ5 integrin (Brooks *et al.*, 1994; Smith *et al.*, 1990) and α5β1 (Hynes, 2002).

Using integrin-selective antibodies, it was demonstrated that neovascular disease induces two distinct integrin-mediated pathways. *In vivo* angiogenesis induced by basic fibroblast growth factor (bFGF) or transforming growth factor-α (TGF-α) is dependent on αvβ3, whereas angiogenesis initiated by vascular endothelial growth factor (VEGF) depends on integrin αvβ5 (Friedlander *et al.*, 1995). The biological relevance of these pathways is still unknown, but it can be assumed that antiangiogenic ligands such as cilengitide—which displays dual selectivity to β3 and β5—interfere with both pathways to block angiogenesis. Due to the high degree of homology between the β3- and β5-subunits, homology modeling based on the αvβ3-crystal structure is a feasible way to gain insight into the receptor (Marinelli *et al.*, 2004).

Apart from some modifications in the specificity-determining-loop (SDL), the major parts of the binding pocket are highly conserved, especially regarding all residues surrounding the metals at MIDAS, ADMIDAS, and LIMBS. Thus, the supposed binding mode for bioactive ligands is similar to that of the integrin $\alpha v\beta3$. The major structural difference between both subunits are three rather bulky residues, $((\beta5)\text{Leu}^{122}, (\beta5)\text{Tyr}^{179}$, and $(\beta5)\text{Lys}^{180})$, adjacent to the MIDAS region, which are replaced by smaller residues in $\alpha v\beta3$. This limits the space that in most integrin ligands is occupied by the aromatic moiety and causes the decrease of activity of, for instance, compound **4** (Fig. 20.5) toward $\alpha v\beta5$ (Marinelli *et al.*, 2004). Hence, it is not surprising that all ligands that have high $\alpha v\beta5$ activity also exhibit high activity on $\alpha v\beta3$, and a selectivity for only $\alpha v\beta5$ seems hard to achieve.

To our knowledge, a $\alpha v\beta5$-selective ligand has not yet been identified. Although many observations using $\alpha v\beta3/\alpha v\beta5$ integrin ligands confirm their fundamental role in the process of angiogenesis, some experiments with genetically altered mice seriously questioned the idea that both integrins are unambiguously proangiogenetic or even necessary for angiogenesis (Bader *et al.*, 1998; Reynolds *et al.*, 2002; Taverna *et al.*, 2004; Yang, 2001).

While the discussion about the discrepancy between genetic ablation experiments and those carried out using integrin ligands is ongoing (Hynes, 2002), the proangiogenetic relevance of the integrin $\alpha5\beta1$ is clearly established via knockout experiments as well as from tests with inhibiting antibodies (Kim *et al.*, 2000). These findings drew $\alpha5\beta1$ into the focus of recent research on anticancer drugs and increased the demand for peptidic or nonpeptidic $\alpha5\beta1$ ligands. Although $\alpha5\beta1$ was identified very early on as a major fibronectin receptor, the structure of $\alpha5\beta1$ is not known in detail. Only rough models have been derived from X-ray scattering (Mould *et al.*, 2003) or electron microscopy (Takagi *et al.*, 2003), both with a resolution larger than 10 Å. The findings indicate that $\alpha5\beta1$ and $\alpha v\beta3$ are similar in shape and domain structure and show a similar binding mode for fibronectin. The fact that both integrins show affinity toward similar ligands, together with the high sequence homology between both integrins ($\alpha v:\alpha5 = 53\%$ and $\beta3:\beta1 = 55\%$), allowed the calculation of a homology model of the $\alpha5\beta1$ integrin on the basis of the crystal structure of the $\alpha v\beta3$ head group in the ligand-bound state (Marinelli *et al.*, 2005).

Only a few $\alpha5\beta1$ ligands have been published thus far, mainly due to a lack of testing systems, since it can be assumed that many of the ligands synthesized for $\alpha v\beta3$ show dual selectivity but have not been tested for $\alpha5\beta1$. Apart from some peptidic ligands with low micromolar activities on

$\alpha 5\beta 1$, a small nonpeptidic library based on a *spiro*-oxazoline scaffold (SJ749) is reported to show excellent activities (0.2 nM) on $\alpha 5\beta 1$, and at least 200-fold selectivity against $\alpha v\beta 3$ (Smallheer *et al.*, 2004). Using the ligand-bound $\alpha v\beta 3$ structure as a template, a homology model was set up by means of comparative protein-modeling methods yielding 10 different models. The models, which differ mostly in side-chain orientation, were used one at a time for docking studies with the ligand SJ749 using Autodock®. Docking results obtained for each simulation were carefully inspected to evaluate agreement with the experimental data for SJ749 (from structure–activity relationships and mutagenesis data), convergence, and free binding energy. The ligand was found to fit into one of the models preferably, which was energetically minimized and checked for stereochemical quality using the program PROCHECK®. The major interactions of the ligand SJ749 in $\alpha 5\beta 1$ are very similar to the corresponding interactions in $\alpha v\beta 3$ (Fig. 20.9): The carboxylic group coordinates the MIDAS, while the 2-aminopyridine interacts with the $(\alpha 5)Asp^{227}$ (corresponds to $(\alpha v)Asp^{218}$).

In contrast to αv, the $\alpha 5$-subunit contains only one aspartic acid residue to interact with the guanidine-mimetic, as the $(\alpha v)Asp^{150}$ is mutated to $(\alpha 5)$ Ala^{149}. In our model, both carbamate oxygens are interacting with residues

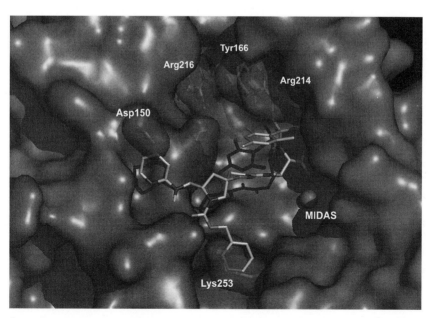

Figure 20.9 Connolly surface of $\alpha 5\beta 1$ with SJ749 modelled into. For comparison, mutated residues in $\alpha v\beta 3$ are pictured as sticks, labeled and shown with transparent Connolly surface. (See color insert.)

in $\alpha5\beta1$, an interaction that is not present in $\alpha v\beta3$. The $\alpha5\beta1$ receptor further shows a hydrophobic pocket in the vicinity of the MIDAS that is blocked by $(\beta3)Arg^{214}$ and $(\beta3)Tyr^{166}$ in $\alpha v\beta3$—a structural difference that could be exploited to gain selectivity against $\alpha v\beta3$ (Marinelli et al., 2005). Synthetic approaches toward rationally designed, selective $\alpha5\beta1$ ligands by means of the homology model have been successfully performed to develop specific ligands for $\alpha5\beta1$ with selectivity against $\alpha v\beta3$ and vice versa, both with low affinity towards $\alpha IIb\beta3$ (Heckman et al., 2007, Stragies et al., 2007).

3. Application of Integrin Ligands for Imaging and Surface Coating

3.1. Coating of material surfaces

Coating of biomaterials for enhanced biocompatibility of implant materials is an important issue in medicine (Castner and Ratner, 2002). ECM proteins certainly can be used for this purpose, but there are several disadvantages for such a procedure, such as degradablility of the protein, denaturation on the surface (Gray, 2004; Roach et al., 2005), contamination of the proteins or viruses that may elicit undesirable immune response and increase infection risks and, last but not least, proteins are expensive. On the other hand, the binding of peptidic or nonpeptidic integrin ligands to their targets can also be used to functionalize surfaces to promote attachment of cells. This opens up promising applications in implant surgery as well as cell culturing for tissue engineering and for the mechanical and microbiological examination of cell behavior. There are many materials used in this field that show different biocompatibilities, such as metals (titanium alloys or stainless steel), polymers (PMMA, polyethylene, polystyrene or silicones), ceramics, glass, and silicon or bone surrogates on the basis of potassium phosphate. Although those materials are nontoxic, their nonphysiological character often results in adverse effects resulting in inflammation and implant encapsulations to thrombosis (Thull, 2001). A successful biomimetic material should meet the following requirements:

- The *implant material* should be nontoxic, and proteolytically and mechanically stable. However, biodegradable materials are needed for special applications.
- The *anchor group* should allow attachment of the ligand on the surface by formation of covalent bonds (e.g., by chemoselective ligation) or by ionic interactions/self-assembly. This strongly depends on the nature of the implant material.
- *Spacer units* have to provide a minimal distance between the ligand and the surface. Examples are ε-aminohexanoic acid, 20-amino-3,6,9,12,15,

18-hexaoxaeicosanoic acid (HEGAS), lysine (used as branching point), and a variety of photolabile and photoswitchable units.
• The *integrin ligands* should be specific for the integrin subtype of the adhering cells to avoid side effects from binding other cells, such as platelets. They should be stable against proteolytic cleavage.

Thus far, there is limited knowledge about the specific integrin pattern on distinct cell types under specific biological conditions. Coating of various materials with integrin ligands enhances their biocompatibility, and must be used to address specific cells expressing a certain integrin subtype. At the beginning, linear or disulfide-cyclized peptides were successfully employed in surface coating. However, apart from the implant material, the nature of the immobilized ligands is crucial for the efficiency and stability of the coating. The use of the appropriate anchor group should lead to high immobilization densities, while the spacer unit has to provide the required minimum distance between the ligand and the surface. It seems that different spacer lengths are required for different materials. On polymethyl methacrylate (PMMA), we have found a minimal spacer length of 27 Å, while on titanium shorter spacers can be used (Kantlehner *et al.*, 2000). A schematic overview on integrin ligands with different space units and appropriate anchors is given in Fig. 20.10.

Linear peptides containing the RGD recognition motif were found to reach higher densities than proteins and to overcome the problem of immunogenicity. Still, linear peptides are prone to proteolytic cleavage and—in contrast to the natural ECM proteins—lack selectivity toward one integrin family. This factor is especially important to guarantee adhesion of specific cells to allow homogenic growth of tissue around the implant. The cyclic RGD peptides described in section 2.1.1 have been found to achieve high selectivities with improved proteolytic stability, but still are hampered by the synthetic effort (Aumailley *et al.*, 1991; Haubner *et al.*, 1996a). Peptidomimetics represent the most promising class of integrin ligands, as they are potentially highly selective, can be synthesized at low cost, and can reach high stability against proteases as well as in the sterilization process (Dahmen *et al.*, 2004).

On account of the large number of integrin ligands that have been attached to different surfaces, we will concentrate on a few selected examples (Hersel *et al.*, 2003). Immobilization of a certain integrin ligand requires functionality on the molecule, where the spacer-anchor units can be attached without loss of activity. This was, for example, achieved by immobilization of various linear RGDX peptides on a PE/AA copolymer that were found to promote cell attachment but totally lack selectivity toward a certain cell line (Hirano *et al.*, 1993). On the *cyclo*(-RGDfV-) peptide, the substitution of valine by lysine or glutamic acid derivatives retained full integrin-binding capacity providing an anchor point for surface attachment. This corresponds to the binding mode of cilengitde, where the

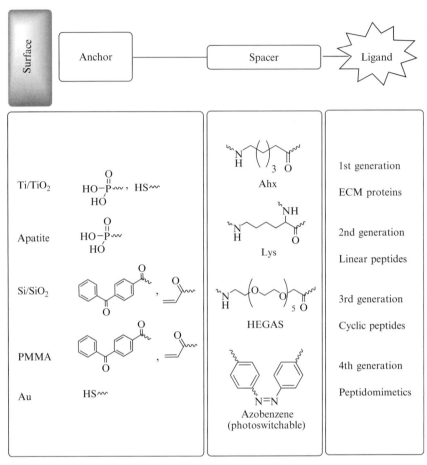

Figure 20.10 Selection of integrin ligands, spacer units, and anchor units that can be combined for the coating of different materials.

valine side chain is not important for binding to the receptor. The ε-amino function of the lysine could be further functionalized and turn *cyclo*(-RGDfK-) into a versatile building block for the immobilization of various surfaces (Haubner *et al.*, 1996a). The selectivity of the ligand toward $\alpha v\beta 3$ is used for the coating of implants (bone grafts). As the new bone matrix is formed by osteoblasts, which are known to express high levels of $\alpha v\beta 3$ integrin, the attachment of osteoblasts is stimulated by the corresponding ligands.

One of the most commonly used implant materials is titanium. It interacts with biologic fluids through an oxide layer, which is formed on contact with air and gives titanium its biocompatibility (Lacefield, 1999; Sykaras *et al.*, 2000; Yang, 2001). Such titanium implants can be improved with a calcium phosphate coating, which resembles the mineral phase of normal bone tissue (Lacefield, 1999; Sykaras *et al.*, 2000; Yang, 2001).

The peptidic ligand *cyclo*(-RGDfV-) was chosen because of its potency toward αvβ3 integrin and selectivity against αIIbβ3—an important feature to avoid thrombosis as a result of platelet activation on the implant surface. Phosphonate groups have been found to be very suitable anchors on titanium or titanium alloys over a wide pH range (pH 1 to 9) (Pechy *et al.*, 1995). An extremely stable connection to the Ti surface was obtained by tetrameric phosphonates using the multimer effect (Auernheimer *et al.*, 2005b; Mammen *et al.*, 1998). An anchor block was synthesized containing four phosphonic acids linked together by a branching unit that was made up of three lysine residues or one lysine residue and two bisphosphonomethyl-benzoic acids (Auernheimer and Kessler, 2006; Auernheimer *et al.*, 2005b) (Fig. 20.11).

An additional gain in proteolytic stability of the branching unit can be achieved by incorporation of D-lysine. The anchor-spacer units were prepared on solid phase using standard Fmoc chemistry (Carpino and Han, 1972; Carpino *et al.*, 1990; Fields and Noble, 1990), finally cleaved from the resin, and fragment-coupled to the partially deprotected cyclic peptide at the free lysine amino function. Comparison with the single thiol anchor gave a higher immobilization density for the multimeric phosphonate and a stable anchoring. Furthermore, the coated Ti surfaces could significantly enhance cell adhesion of mouse osteoblasts (62%) compared with uncoated titanium (16% cell attachment). Dentrimeric

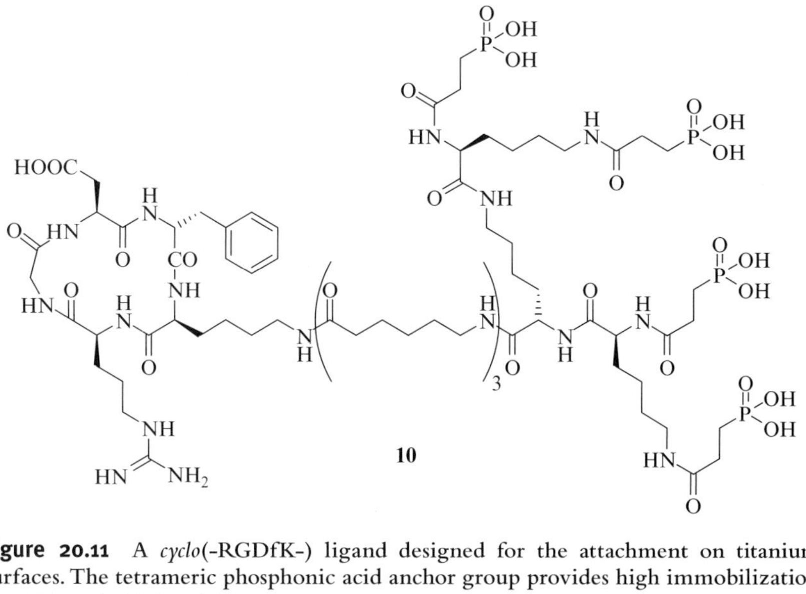

Figure 20.11 A *cyclo*(-RGDfK-) ligand designed for the attachment on titanium surfaces. The tetrameric phosphonic acid anchor group provides high immobilization densities and a high stability.

structures—for example, a lysine core—can also be utilized to synthesize multivalent RGDs to study the multimeric effect on cell adhesion. Cell attachment as a function of RGD concentration gives a sigmoidal increase, indicating that there is a critical minimum density for cellular response (Danilov and Juliano, 1989; Jeschke *et al.*, 2002; Kantlehner *et al.*, 2000). Experiments with highly structured surfaces revealed that a distance of less than 65 nm between ligand-coated gold dots that were applied on a nonadhesive surface using block-copolymer micelle nanolithography was needed for stable adhesion (Arnold *et al.*, 2004).

Recently, a photoswitchable attachment of cells to polymer surfaces has been achieved. An azobenzene unit has been used to change the distance between the binding peptide and the PMMA surface, which prevents binding by photoswitchable *trans-cis* isomerization (Auernheimer *et al.*, 2005a).

3.2. Tumor imaging and therapy

Integrins are expressed in different combinations of their subtypes in various cells and tissues. In some cases, specific integrins dominate on the surface of cells—for instance, $\alpha IIb\beta 3$ on activated platelets—that are, for example, assembled in a thrombus. The integrin $\alpha v\beta 3$ was found to be expressed in various tumors in which it mediates tumor survival (Montgomery *et al.*, 1994), proliferation (Felding-Habermann *et al.*, 1992), and metastasis (Nip *et al.*, 1995; Seftor *et al.*, 1992). The ability to distinguish among different tissues—especially those involved in pathogenic processes—can be utilized to target a certain tissue for imaging or drug delivery purposes. Furthermore, concerning tumors, the use of specific integrin ligands may provide information about the receptor status of the tumor and enable specific therapeutic planning. First attempts to directly address specific cells included the use of antibodies binding to certain surface structures on cells that were conjugated with a radionuclide (Gates *et al.*, 1998; Wiseman *et al.*, 1999, 2002). The disadvantages of antibodies were mainly their immunogenicity, which restricted multiple dosage, the limited uptake in tumor tissue, and slow clearance, which is due to high molecular weight.

Cyclic peptides and peptidomimetics represent an attractive alternative, given their improved pharmacokinetic properties. However, for the labeling of small molecules, attaching an additional functional group under retention of activity can be complicated. Among the large number of methods used in tumor imaging, we will concentrate on the single photon emission tomography (SPECT) and positron emission tomography (PET) (Weber *et al.*, 1999). While SPECT uses mostly γ-emitters like ^{123}I, ^{99m}Tc, and ^{111}In, the most frequently used positron emitters are ^{18}F, [124]I, and [11]C. It is important to note that SPECT is more often available in hospitals;

however, PET has a much higher resolution, and has the advantage of being applicable for direct quantification of receptor density *in vivo*. As the radio-nuclides have strictly limited half-lives, implementation of the nuclide should represent the final step of the synthesis and should allow quick purification before administration.

The two major methods of radiolabeling are *direct labeling* (Fichna and Janecka, 2003; Fischman *et al.*, 1993; Thakur, 1995), where the radionu-clide binds directly to active groups present in the molecule, and *chelating methods* (Eisenwiener *et al.*, 2000; Fichna and Janecka, 2003; Fischman *et al.*, 1993; Thakur, 1995), which use a bifunctional chelating group to be attached to the ligand followed by labeling of the conjugate. In general, the modification of a pre-existing integrin ligand toward application in tumor imaging is similar to the modifications for immobilization purposes as far as part of the molecule has to be identified, where a radionuclide can be attached without loss of binding affinity.

The attachment of either the chelating group or the radionuclide itself should be performed in a chemoselective way to avoid reaction of the introduced group with other functionalities that might be crucial for biological activity. A commonly applied chelating agent is DOTA (1,4,7,10-tetraazacy-clododecane-N,N',N'',N'''-tetraacetic acid). Because it forms very stable complexes with a variety of trivalent cations such as ^{66,67,68}Ga, ^{86,90}Y, and ^{111}In, chelators play an important role in medical applications (de Jong *et al.*, 1997; DeNardo *et al.*, 1995; McMurry *et al.*, 1992; Otte *et al.*, 1997).

The 2-hydrazinonicotic acid (HYNIC) is a good ligand especially for complexation of ^{99m}Tc (Abrams *et al.*, 1990). It only occupies one or two coordination sites of the metal, and thus requires a coligand to form stable complexes. This coligand can be used to modify the hydrophilicity and pharmacokinetic properties of the complex (Edwards *et al.*, 1997). In contrast to the chelating groups, which are mostly attached to the ligand before incorporation of the nuclide, the ^{18}F radiolabels are synthesized as fluorobenzaldehyde, fluorobenzoic acid, or 2-fluoropropionic acid.

All labeled agents can be synthesized in a short time with $K^{18}F$ as F^- donor followed by a oxime ligation (Thumshirn *et al.*, 2003) in case of the aldehyde (Fig. 20.12) or coupling of the *N*-Hydroxysuccinyl ester to an amino function in case of the acids (Wester *et al.*, 1996). This labeling method has been extensively used on our RGD peptides and peptidomimetics.

Figure 20.13 shows examples of integrin ligands that have been success-fully employed in tumor imaging (Meyer *et al.*). The αIIbβ3 selective compound **11** (DMP444), reported by Barrett *et al.* (1996, 1997), is used as a tool for imaging deep vein thrombosis (DVT) without altering platelet formation or hemodynamics of the coagulation cascade. The overexpres-sion of integrin αvβ3 in highly angiogenesis-dependent tumors allows the development of RGD-based radiolabeled compounds as promising tumor targets (Haubner and Wester, 2004). Compounds **12** (Poethko *et al.*, 2003)

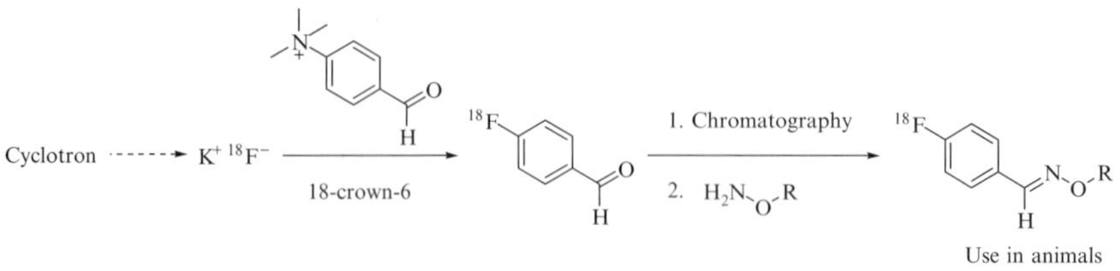

Figure 20.12 Preparation of 4-(^{18}Fluoro)benzaldehyde by nucleophilic aromatic substitution and labeling of hydroxylamine functionalized ligands by oxime ligation.

and **13** (Janssen *et al.*, 2002) are examples of homodimeric RGD ligands (selective for $\alpha v \beta 3$) that could be labeled with ^{18}F (**12**) or ^{111}In-DOTA (**13**).

For imaging purposes, the ligand should be highly hydrophilic to enhance solubility, facilitate renal clearance, and to avoid accumulation in liver and intestines. While the nonpeptidic compounds **16** (Haubner *et al.*, 2001a) and **17** benefit from the improved properties of a peptidomimetic and are highly hydrophilic, the cyclic peptide **15** is functionalized with a sugar amino acid to improve biokinetics. This compound **15** (Galacto-RGD) has been widely used in cancer patients to detect metastasis of tumors expressing high levels of αv-integrins (Haubner *et al.*, 2001c, 2005). The free amino group of the sugar amino acid was labeled using (4-Nitrophenyl)-2-^{18}F-propionate (Haubner *et al.*, 2001b).

A widely used method for direct labeling of ligands is the chemoselective iodination that is used to introduce ^{125}I into molecules for SPEC tomography. The reagent bis(pyridine)iodonium(I) tetrafluoroborate was used for selectively monoiodination of unprotected phenols at room temperature (Barluenga *et al.*, 1993). This method does not affect other amino acid residues, and thus can be applied on tyrosine residues in peptides and hydroxyl aromates in peptidomimetics. The corresponding peptide **14** (Haubner *et al.*, 1999) can be synthesized after substitution of phenyl alanine with tyrosine, which is possible without loss of activity.

Iodinated tyrosine often leads to reduced affinity, and therefore this method cannot be used without proving its effect. A promising way to increase binding affinities of ligands toward their receptors is the synthesis of multivalent compounds that expose several binding epitopes for receptor interaction (Mammen *et al.*, 1998). The following mechanisms lead to improved binding properties:

- A multivalent ligand increases the concentration of binding epitopes in the vicinity of the receptor pocket.
- It may furthermore interact with secondary binding sites, if present in the receptor.
- Oligomeric receptors are preferably bound by multimeric ligands (*chelate effect*).

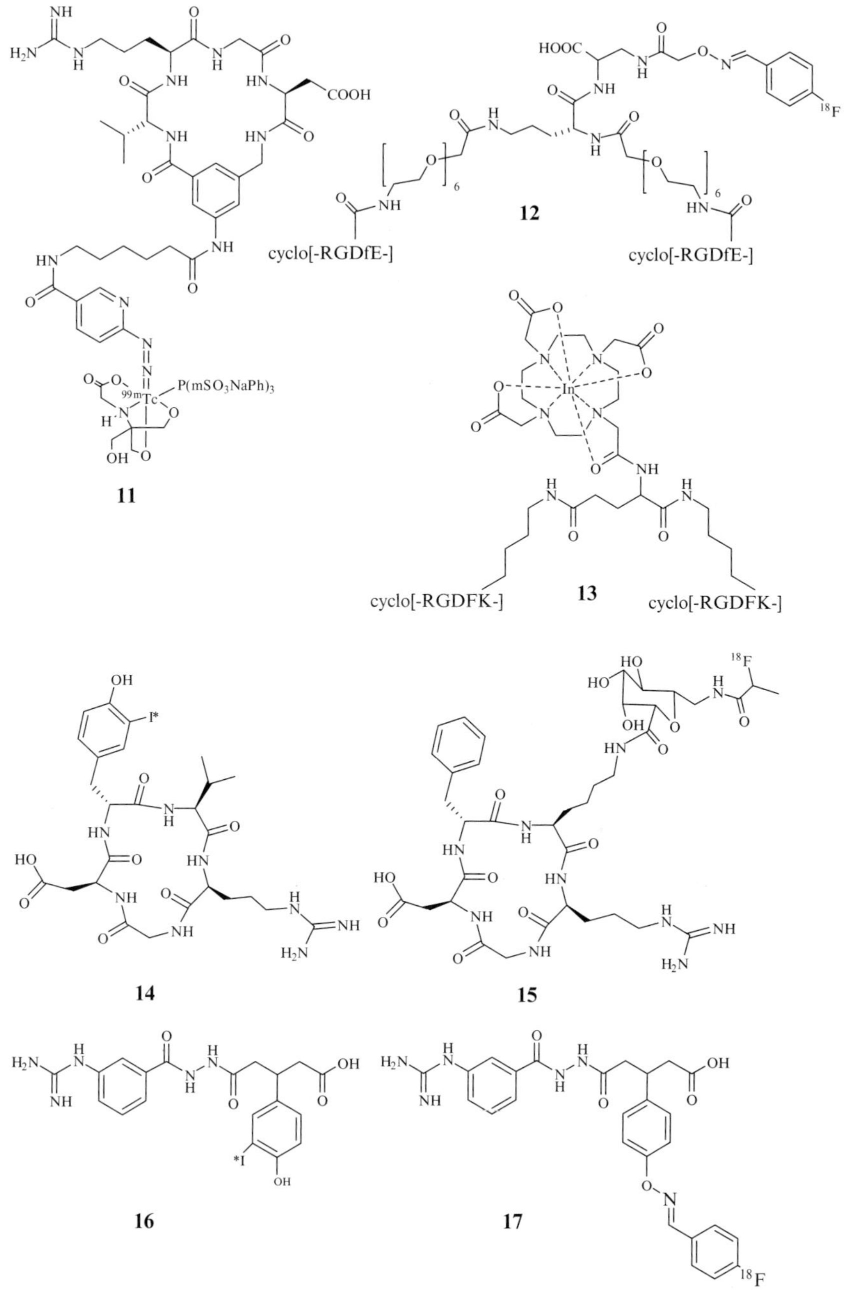

Figure 20.13 Examples of radiolabeled peptides and peptidomimetics targeting integrin $\alpha IIb\beta3$ (**11**) or $\alpha v\beta3$ (**12–17**).

- Binding of a multimeric ligand may cause oligomerization (*clustering*) of the receptors, thus improving binding affinity.

In recent years, evidence has increased for models of integrin activation that involve the clustering of integrin subunits (Gottschalk and Kessler, 2004; Li *et al.*, 2001, 2003). The data suggest that on binding of a (multi-meric) ligand, the integrin subunits cluster and enable various signaling pathways leading, for example, to formation of focal adhesion contacts. For the radiolabeling of tumors, this would mean that multivalent ligands show a dramatic increase in binding to tumor tissue.

4. Experimental Section

The synthesis of peptides on solid phase, established by Merrifield in the early 1960s (Merrifield, 1963, 1985), has dramatically facilitated the preparation of peptides. More and more organic reactions have been transferred on solid phase—with numbers steadily increasing. In this section, we will provide a detailed description of how solid–phase synthesis can be utilized for the preparation of a cyclic peptide (*cyclo*(-RGDfNMeVal-), cilengitide) and, on the other side, of an *aza*-glycine–based peptidomimetic. Both compounds have been proven to be highly active ligands for the $\alpha v \beta 3$ integrin with excellent selectivities against the platelet integrin $\alpha IIb\beta 3$.

The solid–phase syntheses were mainly performed on the scale of 100 mg to 1 g resin. The reaction vessel of choice was a syringe (e.g., 20 ml for a batch of 1 g resin) equipped with a polypropylene frit and a canula. The reagents used in solid–phase synthesis were dissolved in an appropriate medium (*N*-methylpyrrolidinone or DCM) and directly sucked into the syringe. Mixing of the reaction mixture was achieved by gentle rotation of the syringe stuck into a rubber stopper that was connected to a rotor. It should be emphasized that adequate swelling of the resin and a careful washing after each reaction step are crucial for efficient synthesis. In the following, the single reaction steps are given in greater detail as general procedures. For the solid–phase synthesis, the amounts of reactants needed are given in *equivalents* referring to the real loading of the resin. The loaded resin can be stored in dry form (washing with DCM, five times, 2 min each, and evaporation *in vacuo*), preferably with Fmoc-protected *N*-terminus.

4.1. General procedures

4.1.1. Procedure A: immobilization of Fmoc amino acid

A suitable reaction vessel was filled with Cl–TCP–resin (~1 mmol/g). The amino acid (1.2 eq.) was dissolved in dry DCM (8 ml/g resin), DIEA (3 eq.) was added, and the resulting mixture added to the dry resin. After 1 h of shaking, 0.5 ml DIEA and 3 ml methanol (per gram of resin) were added and shaken for additional 15 min. The solution was removed by filtration and the resin washed carefully with DCM (3 times), NMP, NMP/methanol (1:1), and pure methanol. The loaded resin was dried for several hours *in vaccuo* and weighed (Barlos *et al.*, 1989). The real loading of the resin can be calculated using following equation:

$$c[mol/g] = \frac{m_{total} - m_{resin}}{(MW - 36.461) \times m_{total}} \tag{1}$$

where m_{total} is mass of loaded resin, m_{resin} is mass of unloaded resin, and MW is molecular weight of the immobilized amino acid.

4.1.2. Procedure B: Fmoc de-protection

The swollen resin was treated with a solution of piperidine (20%) in NMP (two times, 15 min each). After filtration, the resin was washed carefully with NMP (five times, 1 min each) (Carpino and Han, 1972).

4.1.3. Procedure C1: standard coupling of Fmoc-amino acid according to HOBt/TBTU method

If the resin has been dried before, it was allowed to swell in NMP for 20 min. The Fmoc-amino acid (2 eq.) was dissolved in a 0.2 M solution of HOBt and TBTU (2 eq. each). After addition of DIEA (5.6 eq.), the mixture was added to the resin in a suitable reaction vessel and shaken for 1 h. The reaction was terminated by washing the resin with NMP (five times, 1 min each) (Knorr *et al.*, 1989).

4.1.4. Procedure C2: coupling of aromatic acids with HATU

The resin loaded with the free amine/hydrazine was washed in NMP. To a solution of 3-(Fmoc-amino)benzoic acid (2.5 eq.) and HATU (Carpino, 1993) (2.4 eq.) in NMP (10 ml/mmol), collidine (10 eq.) was added, and the resulting mixture was shaken with the resin for 12 h. The reaction was terminated by washing the resin with NMP (five times, 1 min each) (Gibson *et al.*, 2001b).

4.1.5. Procedure C3: coupling on *N*-methylated amino acids with HATU/HOAt

The resin loaded with the *N*-methylated amine was washed in NMP. To a solution of the amino acid (2 eq.), HATU (2 eq.), and HOAt (2 eq.) in NMP (10 ml/mmol), collidine (10 eq.) was added, and the resulting mixture was shaken with the resin for 12 h. The reaction was terminated by washing the resin with NMP (five times, 1 min each) (Gibson *et al.*, 2001b).

4.1.6. Procedure D1: cleavage from the TCP-resin with retention of permanent protecting groups

The resin was washed with DCM (three times, 2 min each) and then treated with a mixture of DCM/acetic acid/2,2,2-trifluoroethanol (6:3:1, 10 ml/g resin) for 1 h. The mixture was filtered and the filtrate collected. The procedure was repeated, the resin finally washed with the cleavage mixture, the collected filtrates diluted with toluene and the solvents removed *in vaccuo*. The coevaporation with toluene was repeated to remove all acetic acid. The side–chain protected peptide was obtained as acetate.

4.1.7. Procedure D2: cleavage from TCP resin with de-protection of permanent protecting groups

The resin was washed with DCM (three times, 1 min each) and then treated with a mixture of trifluoro acetic acid/DCM/triisopropylsilane (47.5:47.5:5, 8 ml/g resin) with shaking for 1 h. After filtration and collection of the filtrate, the procedure was repeated. The collected filtrates were diluted with toluene and evaporated to remove TFA.

4.1.8. Procedure E: backbone cyclization with DPPA

The side–chain protected peptide was dissolved in DMF (10^{-3} to 10^{-4} M). NaHCO$_3$ (5 eq.) and DPPA (diphenylphosphonic acid azide, 3 eq.) were added and stirred for ~24 h. The reaction was monitored by HPLC, until the starting material was consumed. The cyclic, less polar product could be detected with HPLC and ESI-MS. The spectrum showed a new, more unpolar compound that resembles the cyclized peptide. The DMF was evaporated *in vaccuo* at a temperature lower than 50°. The residue was taken up in DMF (1 to 2 ml) and precipitated on addition of water (~30 ml). The suspension was centrifuged (alternatively filtered), the solid washed with water, and centrifuged again. After a third washing step, the solid was dried *in vaccuo*.

4.1.9. Procedure F: cleavage of permanent protecting groups

In an appropriate reaction vessel (e.g., round bottom flask, Falcon tube), the crude, side–chain protected peptide was dissolved in a mixture of TFA/water/triisopropylsilane (95:2.5:2.5, ~3 ml). If the initial deep red color of the solution did not fade, additional drops on TIPS were added. The reaction was stirred/shaken for ~2 h, and eventually monitored by ESI-MS analysis. The deprotection was terminated by addition of diethyl ether (~30 ml), which caused precipitation of the peptide. The solid was collected by filtration (centrifugation), washed twice with diethyl ether, and dried *in vaccuo* to yield the crude peptide as TFA salt (colorless solid).

4.1.10. Procedure G: Fmoc protection of an amino acid

In a round flask equipped with dropping funnel and magnetic stirrer, the amino acid (10 mmol) was dissolved in a mixture of 10% Na_2CO_3 solution (20 ml) and dioxane (10 ml). Over 30 min, a solution of Fmoc-chloride (10.5 mmol) in dioxane (20 ml) was added dropwise at $0°$ under stirring. After 1 h, the cold bath was removed and stirring was continued for 24 h. The reaction mixture was then poured into 200 ml of ice water and extracted with diethyl ether (3× 100 ml). The aqueous phase was acidified with concentrated HCl to pH 1, and the resulting suspension extracted with ethyl acetate (2 × 100 ml). The combined organic phases were dyed with Na_2SO_4, filtered, and evaporated. The residue was recrystalized from ethyl acetate/hexane (Chang et al., 1980).

4.1.11. Procedure H: synthesis of aza-glycine building block

A two-phase suspension of Fmoc-hydrazine (synthesized according to Carpino and Han, 1972) in DCM/saturated aqueous $NaHCO_3$ solution (1 ml/mmol, respectively) was shaken vigorously for 5 min at $0°$. Stirring was paused and phosgene (1.9 M in toluene, 3 eq.) was added via a syringe to the organic (lower) phase. Stirring was resumed for 10 min, and then the phases were separated and the aqueous phase extracted two times with DCM. The combined organic phases were dried with Na_2SO_4 (Gibson et al., 1999). After filtration and evaporation of the solvent, the desired compound was obtained in ∼90% yield as a colorless solid, which was readily used in the following coupling procedure.

4.1.12. Procedure I: coupling of aza-glycine building block

The resin loaded with the de-protected amino acid was washed with dry DCM (twice, 2 min each). 5-(9H-Fluoren-9-ylmethoxy)-1,3,4-oxadiazol-2(3H)-on (3 eq.) was dissolved in dry DCM (7 ml/mmol) and added to the resin. After 90 min of shaking at ambient temperature, the reaction was terminated and the resin washed with DCM (six times, 1 min each) and NMP (three times, 1 min each) (Gibson et al., 1999).

4.1.13. Procedure J: guadinylation of immobilized amines

The resin containing the free amine function was washed three times with dry chloroform. The resin was then placed in an appropriate reaction vessel that was placed into a heated shaker. A 1-M solution of N,N'- Bis-Boc-guanylpyrazole (Wu et al., 1993) (10 eq.) was added and the mixture shaken for 20 h at $50°$. After cooling to room temperature, the mixture was filtered, and resin washed six times with DCM (Gibson et al., 1999). N,N'-Bis-Boc-guanylpyrazole could be recycled from the filtrate by evaporation of the solvent and recrystalization from hexane/ethyl acetate.

4.2. Preparation of *cyclo*(-RGDf*N*MeVal)-)

The Fmoc-protected amino acid building blocks were purchased from Iris, Bachem, or Novosys. Fmoc-*N*Me-valine can be prepared from Fmoc-valine in a two-step procedure (Freidinger *et al.*, 1983), or obtained from commercial sources. An optimized, efficient alternative for the synthesis of a large variety of *N*-methylated amino acid building blocks compatible with Fmoc-solid phase synthesis has been recently developed by us (Biron and Kessler, 2005; Biron *et al.*, 2005). For the synthesis of backbone cyclized peptides, the choice of the right C-terminal amino acid (first amino acid in the solid-phase synthesis) is of great importance. The pre-organization of the linear sequence in a loop facilitates the ring closure when, if the strain of the linear peptide against a cyclic conformation is too high, the yields in the cyclization step decrease dramatically. Usually, the D-amino acids or the glycine at the C-terminus yields the best results.

For SPPS of cilengitide, the Steps 1 to 13 were performed as described in Table 20.2. The final de-protection yields the crude peptide that was purified using reverse-phase HPLC techniques. For the synthesis of the peptide *cyclo* (-RGDfV-), the unmethylated Fmoc-Val-OH was used in Step 5, and the following amino acid coupled according to general procedure C1 (Step 7).

4.3. Synthesis of *aza*-glycine mimetic 2

The synthesis of the *aza*-glycine mimetic 2 was performed on solid phase (Merrifield, 1963, 1985) (TCP-resin) according to the Fmoc strategy (Carpino and Han, 1972), starting from the commercially available 3-amino-3-phenylpropionic acid (Fig. 20.16).

Table 20.2 Solid-phase synthesis of *cyclo*(-RGDf*N*MeVal-) on TCP resin

Step	Amino acid	Procedure
1. Loading of TCP resin with 1st amino acid	Fmoc-Gly-OH	A
2. Fmoc- deprotection		B
3. Coupling of 2nd amino acid	Fmoc-Arg(Pbf)-OH	C1
4. Fmoc- deprotection		B
5. Coupling of 3rd amino acid	Fmoc-NMe-Val-OH	C1
6. Fmoc deprotection		B
7. Coupling of 4th amino acid	Fmoc-D-Phe-OH	C3
8. Fmoc deprotection		B
9. Coupling of 5th amino acid	Fmoc-Asp(OtBu)-OH	C1
10. Fmoc deprotection		B
11. Cleavage from resin		D1
12. Cyclization		E
13. Deprotection		F

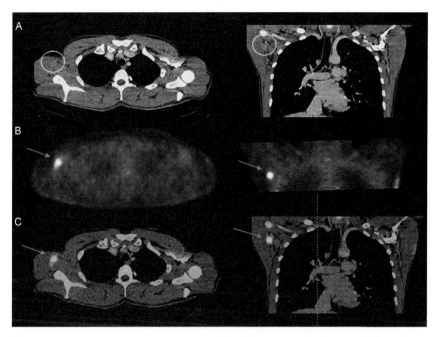

Figure 20.14   Lateral (left) and longitudinal (right) sections from a patient with axillary lymph node metastasis. (A) represents the CT-image, (B) the PET image after application of 220 MBq. Galacto-RGD (15), 2 h p.i., and (C) a superposition of both images. The site of the tumor is marked by a circle or arrows (Haubner *et al.*, 2005). (See color insert.)

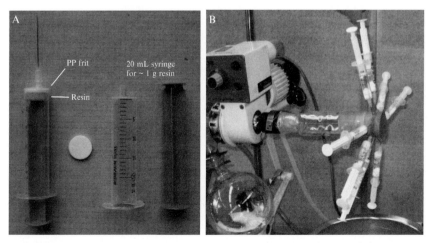

Figure 20.15   Syringes as reaction vessels for solid phase synthesis (A). Mixing of the reaction mixtures was achieved using a rubber stopper attached to a rotor (e.g., from a rotary evaporator, B).

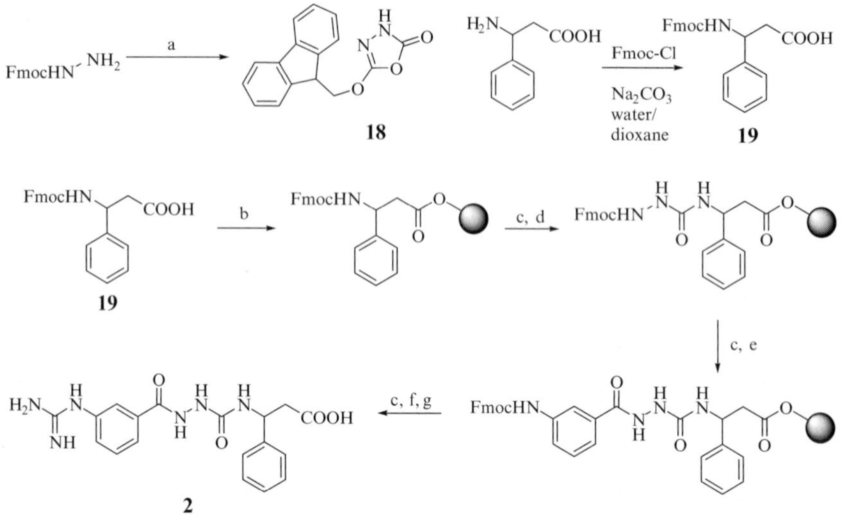

Figure 20.16 Synthesis of *aza*-glycine-mimetic 2. (a) Phosgene (1.9 M in toluene), sat. NaHCO₃, CH₂Cl₂; (b) TCP-resin, DIEA (2.5 eq.), CH₂Cl₂; (c) 20% piperidine in NMP; (d) 5-(9*H*-fluoren-9ylmethoxy)-1,3,4-oxydiazol-2-(3*H*)-one (18), NMP; (e) 3-(*N*-Fmoc-amino)benzoic acid (2 eq.), HATU (2 eq.), collidine (10 eq.); (f) *N,N'*-bis-Boc-1-guanylpyrazole (10 eq.), CHCl₃, 50°; (g) TFA, 5% TIPS.

Table 20.3 Combined solution and solid-phase synthesis of *aza*-glycine mimetic 2

Step	Reagent	Procedure
1. Loading of TCP resin	3-(Fmoc-amino)-3-phenylpropionic acid (19)	A
2. Fmoc deprotection		B
3. Introduction of *aza*-glycine	5-(9*H*-fluoren-9-ylmethoxy)-1,3,4-oxydiazol-2-(3*H*)-one (18)	I
4. Fmoc deprotection		B
5. Coupling of spacer unit	3-(*N*-Fmoc-amino)benzoic acid	C2
6. Fmoc deprotection		B
7. Guadinylation	*N,N'*-bis-Boc-1-guanylpyrazole	J
8. Cleavage from resin and deprotection		D2

Differently substituted 3-amino-3-phenylpropionic acids—for instance, for the introduction of spacer units—were prepared from the corresponding benzaldehydes according to accepted procedures (Cardillo *et al.*, 1998). The amino acid was first Fmoc protected following the procedure G.

The following synthetic steps were performed as described in Table 20.3. The *aza*-glycine building block **18** was prepared according to procedure H shortly before its utilization in Step 3 (Table 20.3) to avoid decomposition. The crude product was purified using reverse phase HPLC techniques and lyophilized to yield the desired compound (12 mg, 1 g resin, 4% overall yield) as TFA salt (colorless solid).

REFERENCES

Abrams, M. J., Juweid, M., Tenkate, C. I., Schwartz, D. A., Hauser, M. M., Gaul, F. E., Fuccello, A. J., Rubin, R. H., Strauss, H. W., and Fischman, A. J. (1990). Technetium-99m-human polyclonal IgG radiolabeled via the hydrazino nicotinamide derivative for imaging focal sites of infection in rats. *J. Nucl. Med.* **31,** 2022–2028.

Agrez, M., Chen, A., Cone, R. I., Pytela, R., and Sheppard, D. (1994). The alpha v beta 6 integrin promotes proliferation of colon carcinoma cells through a unique region of the beta 6 cytoplasmic domain. *J. Cell Biol.* **127,** 547–556.

Arndt, T., Arndt, U., Reuning, U., and Kessler, H. (2005). Integrins in angiogenesis: Implications for tumor therapy. *In* "Cancer Therapy: Molecular Targets in Tumor–Host Interactions" (G. F. Weber, ed.). Horizon Bioscience, Hethersett, Norwich, United Kingdom.

Arnold, M., Cavalcanti-Adam, E. A., Glass, R., Blümmel, J., Eck, W., Kantlehner, M., Kessler, H., and Spatz, J. P. (2004). Activation of integrin function by nanopatterned adhesive interfaces. *Chem. Phys. Chem.* **5,** 383–388.

Auernheimer, J., Dahmen, C., Hersel, U., Bausch, A., and Kessler, H. (2005a). Photoswitched cell adhesion on surfaces with RGD peptides. *J. Am. Chem. Soc.* **127,** 16107–16110.

Auernheimer, J., and Kessler, H. (2006). Benzylprotected aromatic phosphonic acids for anchoring peptides on titanium. *Bioorg. Med. Chem. Lett.* **16,** 271–273.

Auernheimer, J., Zukowski, D., Dahmen, C., Kantlehner, M., Enderle, A., Goodman, S. L., and Kessler, H. (2005b). Titanium implant materials with improved riocompatibility through coating with phosphonate-anchored cyclic RGD peptides. *Chem. Biol. Chem.* **6,** 2034–2040.

Aumailley, M., Gurrath, M., Müller, G., Calvete, J., Timpl, R., and Kessler, H. (1991). Arg-Gly-Asp constrained within cyclic peptides: Strong and selective inhibitors of cell adhesion to vitronectin and laminin fragment P1. *FEBS Lett.* **291,** 50–54.

Bach II, A. C., Eyermann, C. J., Gross, J. D., Bower, M. J., Harlow, R. L., Weber, P. C., and DeGrado, W. F. (1994). Structural studies of a family of high affinity ligands for GPIIb/IIIa. *J. Am. Chem. Soc.* **116,** 3207–3219.

Bader, B. L., Rayburn, H., Crowley, D., and Hynes, R. O. (1998). Extensive vasculogenesis, angiogenesis, and organogenesis precede lethality in mice lacking all alpha v integrins. *Cell* **95,** 507–519.

Barlos, K., Gatos, D., Kallitsis, J., Papaphotiu, G., Sotiriu, P., Yao, W. Q., and Schäfer, W. (1989). Synthesis of protected peptide-fragments using substituted triphenylmethyl resins. *Tetrahedron Lett.* **30,** 3943–3946.

Barluenga, J., Gonzalez, J. M., Garciamartin, M. A., Campos, P. J., and Asensio, G. (1993). Acid-mediated reaction of bis(pyridine)iodonium(i) tetrafluoroborate with aromatic-compounds—a selective and general iodination method. *J. Org. Chem.* **58,** 2058–2060.

Barrett, J. A., Crocker, A. C., Damphousse, D. J., Heminway, S. J., Liu, S., Edwards, D. S., Lazewatsky, J. L., Kagan, M., Mazaika, T. J., and Carroll, T. R. (1997). Biological evaluation of thrombus imaging agents utilizing water soluble phosphines and tricine as coligands when used to label a hydrazinonicotinamide-modified cyclic glycoprotein IIb/IIIa receptor antagonist with 99mTc. *Bioconjug. Chem.* **8,** 155–160.

Barrett, J. A., Damphousse, D. J., Heminway, S. J., Liu, S., Edwards, D. S., Looby, R. J., and Carroll, T. R. (1996). Biological evaluation of 99mTc-labeled cyclic glycoprotein IIb/IIIa receptor antagonists in the canine arteriovenous shunt and deep vein thrombosis models: effects of chelators on biological properties of [99mTc]chelator-peptide conjugates. *Bioconjug. Chem.* **7**, 203–208.

Biron, E., Chatterjee, J., and Kessler, H. (2005). Efficient synthesis of N-methyl amino acids compatible for Fmoc solid-phase synthesis. *Biopolymers* **80**, 522–523.

Biron, E., and Kessler, H. (2005). Convenient synthesis of N-methylamino acids compatible with Fmoc solid-phase peptide synthesis. *J. Org. Chem.* **70**, 5183–5189.

Brooks, P. C., Montgomery, A. M., Rosenfeld, M., Reisfeld, R. A., Hu, T., Klier, G., and Cheresh, D. A. (1994). Integrin alpha v beta 3 antagonists promote tumor regression by inducing apoptosis of angiogenic blood vessels. *Cell* **79**, 1157–1164.

Cardillo, G., Gentilucci, L., Tolomelli, A., and Tomasini, C. (1998). Stereoselective synthesis of (2R,3S)-N-benzoylphenylisoserine methyl ester. *J. Org. Chem.* **63**, 2351–2353.

Carpino, L. A. (1993). 1-Hydroxy-7-azabenzotriazole—an efficient peptide coupling additive. *J. Am. Chem. Soc.* **115**, 4397–4398.

Carpino, L. A., and Han, G. Y. (1972). 9-Fluorenylmethoxycarbonyl amino-protecting group. *J. Org. Chem.* **37**, 3404–3409.

Carpino, L. A., Sadat-Aalaee, D., Chao, A. G., and DeSelms, R. H. (1990). [(9-Fluorenyl-methyl)oxy]carbonyl (FMOC) amino acid fluorides. Convienient new peptide coupling reagents applicable to the FMOC/tert-butyl strategy for solution and solid-phase syntheses. *J. Am. Chem. Soc.* **112**, 9651–9652.

Castner, D. G., and Ratner, B. D. (2002). Biomedical surface science: Foundations of frontiers. *Surface Sci.* **500**, 28–60.

Chang, C. D., Felix, A. M., Jimenez, M. H., and Meienhofer, J. (1980). Solid-phase peptide-synthesis of somatostatin using mild base cleavage of N-alpha-9-fluorenylmethyl-oxycarbonylamino acids. *Int. J. Pept. Prot. Res.* **15**, 485–494.

Clark, R. A., Ashcroft, G. S., Spencer, M. J., Larjava, H., and Ferguson, M. W. (1996). Re-epithelialization of normal human excisional wounds is associated with a switch from alpha v beta 5 to alpha v beta 6 integrins. *Br. J. Dermatol.* **135**, 46–51.

Clemetson, K. J., and Clemetson, J. M. (1998). Integrins and cardiovascular disease. *Cell. Mol. Life Sci.* **54**, 502–513.

Coleman, P. J., Askew, B. C., Hutchinson, J. H., Whitman, D. B., Perkins, J. J., Hartman, G. D., Rodan, G. A., Leu, C. T., Prueksaritanont, T., Fernandez-Metzler, C., Merkle, K. M., Lynch, R., *et al.* (2002). Non-peptide alpha(v)beta(3) antagonists. Part 4: Potent and orally bioavailable chain-shortened RGD mimetics. *Bioorg. Med. Chem. Lett.* **12**, 2463–2465.

Coleman, P. J., Brashear, K. M., Askew, B. C., Hutchinson, J. H., McVean, C. A., Duong le, T., Feuston, B. P., Fernandez-Metzler, C., Gentile, M. A., Hartman, G. D., Kimmel, D. B., Leu, C. T., *et al.* (2004). Nonpeptide alphavbeta3 antagonists. Part 11: Discovery and preclinical evaluation of potent alphavbeta3 antagonists for the prevention and treatment of osteoporosis. *J. Med. Chem.* **47**, 4829–4837.

Dahmen, C., Auernheimer, J., Meyer, A., Enderle, A., Goodman, S. L., and Kessler, H. (2004). Improving implant materials by coating with nonpeptidic, highly specific integrin ligands. *Angew. Chem. Int. Ed.* **43**, 6649–6652.

Danilov, Y. N., and Juliano, R. L. (1989). (Arg-Gly-Asp)n-albumin conjugates as a model substratum for integrin-mediated cell adhesion. *Exp. Cell Res.* **182**, 186–196.

de Jong, M., Bakker, W. H., Krenning, E. P., Breeman, W. A., van der Pluijm, M. E., Bernard, B. F., Visser, T. J., Jermann, E., Behe, M., Powell, P., and Macke, H. R. (1997). Yttrium-90 and indium-111 labelling, receptor binding and biodistribution of [DOTA0,d-Phe1,Tyr3]octreotide, a promising somatostatin analogue for radionuclide therapy. *Eur. J. Nucl. Med.* **24**, 368–371.

Dechantsreiter, M. A., Planker, E., Marthä, B., Lohof, E., Hölzemann, G., Jonczyk, A., Goodman, S. L., and Kessler, H. (1999). N-methylated cyclic peptides as highly active and selective avb3 integrin antagonists. *J. Med. Chem.* **42,** 3033–3040.

DeNardo, S. J., Zhong, G. R., Salako, Q., Li, M., DeNardo, G. L., and Meares, C. F. (1995). Pharmacokinetics of chimeric L6 conjugated to indium-111- and yttrium-90-DOTA-peptide in tumor-bearing mice. *J. Nucl. Med.* **36,** 829–836.

Duggan, M. E., Duong, L. T., Fisher, J. E., Hamill, T. G., Hoffman, W. F., Huff, J. R., Ihle, N. C., Leu, C. T., Nagy, R. M., Perkins, J. J., Rodan, S. B., Wesolowski, G., *et al.* (2000). Nonpeptide alpha(v)beta(3). antagonists. 1. Transformation of a potent, integrin-selective alpha(IIb)beta(3). antagonist into a potent alpha(v)beta(3) antagonist. *J. Med. Chem.* **43,** 3736–3745.

Dutta, A. S., and Morley, J. S. (1975). Polypeptides. Part XIII. Preparation of alpha-aza-amino-acid (carbazic acid) derivatives and intermediates for the preparation of alpha-aza-peptides. *J. Chem. Soc. Perkin Trans.* **1,** 1712–1720.

Eble, J. A., and Kühn, K. (1997). "Integrin-Ligand Interaction." Springer-Verlag, Heidelberg.

Edwards, D. S., Liu, S., Barrett, J. A., Harris, A. R., Looby, R. J., Ziegler, M. C., Heminway, S. J., and Carroll, T. R. (1997). New and versatile ternary ligand system for technetium radiopharmaceuticals: water soluble phosphines and tricine as coligands in labeling a hydrazinonicotinamide-modified cyclic glycoprotein IIb/IIIa receptor antagonist with 99mTc. *Bioconjug. Chem.* **8,** 146–154.

Egbertson, M. S., Chang, C. T., Duggan, M. E., Gould, R. J., Halczenko, W., Hartman, G. D., Laswell, W. L., Lynch, J. J., Jr., Lynch, R. J., Manno, P. D., Naylor, A. M., Prugh, J. D., *et al.* (1994). Non-peptide fibrinogen receptor antagonists. 2. Optimization of a tyrosine template as a mimic for Arg-Gly-Asp. *J. Med. Chem.* **37,** 2537–2551.

Eisenwiener, K. P., Powell, P., and Macke, H. R. (2000). A convenient synthesis of novel bifunctional prochelators for coupling to bioactive peptides for radiometal labelling. *Bioorg. Med. Chem. Lett.* **10,** 2133–2135.

Felding-Habermann, B., Mueller, B. M., Romerdahl, C. A., and Cheresh, D. A. (1992). Involvement of integrin alpha V gene expression in human melanoma tumorigenicity. *J. Clin. Invest.* **89,** 2018–2022.

Feuerstein, G., Ruffolo, R., Jr., and Samanen, J. (1996). The integrin αIIbβ3 (GPIIb/IIIa). A target for novel anti-platelet drugs. *Pharm. Rev. Commun.* **8,** 257–265.

Feuston, B. P., Culberson, J. C., Duggan, M. E., Hartman, G. D., Leu, C. T., and Rodan, S. B. (2002). Binding model for nonpeptide antagonists of α(v)β(3) integrin. *J. Med. Chem.* **45,** 5640–5648.

Fichna, J., and Janecka, A. (2003). Synthesis of target-specific radiolabeled peptides for diagnostic imaging. *Bioconjug. Chem.* **14,** 3–17.

Fields, G. B., and Noble, R. L. (1990). Solid phase peptide synthesis utilizing 9-fluorenylmethoxycarbonyl amino acids. *Int. J. Pept. Protein Res.* **35,** 161–214.

Fischman, A. J., Babich, J. W., and Strauss, H. W. (1993). A ticket to ride: Peptide radio-pharmaceuticals. *J. Nucl. Med.* **34,** 2253–2263.

Freidinger, R. M., Hinkle, J. S., Perlow, D. S., and Arison, B. H. (1983). Synthesis of 9-fluorenylmethyloxycarbonyl-protected N-alkyl amino-acids by reduction of oxazoli-dinones. *J. Org. Chem.* **48,** 77–81.

Friedlander, M., Brooks, P. C., Shaffer, R. W., Kincaid, C. M., Varner, J. A., and Cheresh, D. A. (1995). Definition of two angiogenic pathways by distinct alpha v integrins. *Science* **270,** 1500–1502.

Friedlander, M., Theesfeld, C. L., Sugita, M., Fruttiger, M., Thomas, M. A., Chang, S., and Cheresh, D. A. (1996). Involvement of integrins alpha v beta 3 and alpha v beta 5 in ocular neovascular diseases. *Proc. Natl. Acad. Sci. USA* **93,** 9764–9769.

Gante, J. (1994). Peptidomimetics—tailored enzyme-inhibitors. *Angew. Chem. Int. Ed.* **33,** 1699–1720.

Gante, J., Kessler, H., and Gibson, C. (2003). Synthesis of azapeptides. *In* "Houben-Weyl: Methods of Organic Chemistry" (A. F.M Goodmann, L. Moroder, and C. Toniolo, eds.), Vol. E22c: Synthesis of Peptides and Peptidomimetics. Thieme Verlag, Stuttgart, New York.

Gates, V. L., Carey, J. E., Siegel, J. A., Kaminski, M. S., and Wahl, R. L. (1998). Nonmyeloablative iodine–131 anti-B1 radioimmunotherapy as outpatient therapy. *J. Nucl. Med.* **39,** 1230–1236.

Geyer, A., Mierke, D. F., Unverzagt, C., and Kessler, H. (1993). Synthesis of cyclic thiohexapeptides. *In* "Peptides 1992" (C. H. Schneider and A. N. Eberle, eds.). ESCOM Science Publishers, Leiden, Netherlands.

Geyer, A., Müller, G., and Kessler, H. (1994). Conformational-analysis of a cyclic Rgd peptide-containing a Psi[CH2-NH] bond—a positional shift in backbone structure caused by a single dipeptide mimetic. *J. Am. Chem. Soc.* **116,** 7735–7743.

Gibson, C., Goodman, S. L., Hahn, D., Hölzemann, G., and Kessler, H. (1999). Novel solid-phase synthesis of azapeptides and azapeptoides via Fmoc-strategy and its application in the synthesis of RGD-mimetics. *J. Org. Chem.* **64,** 7388–7394.

Gibson, C., Sulyok, G. A., Hahn, D., Goodman, S. L., Hölzemann, G., and Kessler, H. (2001a). Nonpeptidic $\alpha(v)\beta(3)$ integrin antagonist libraries: On-bead screening and mass spectrometric identification without tagging. *Angew. Chem. Int. Ed.* **40,** 165–169.

Gibson, C., Sulyok, G. A. G., Hahn, D., Goodman, S. L., Hölzemann, G., and Kessler, H. (2001b). Nonpeptidic $\alpha v \beta 3$ integrin antagonist libaries: on-bead screening and mass spectometric identification without tagging. *Angew. Chem. Int. Ed.* **40,** 165–169.

Gilon, C., Dechantsreiter, M. A., Burkhart, F., Friedler, A., and Kessler, H. (2003). Synthesis of N-methylated peptides. *In* "Houben-Weyl: Methods of Organic Chemistry" (A. F. M. Goodmann, L. Moroder, and C. Toniolo, eds.), Vol. E22c. Thieme Verlag, Stuttgart, New York.

Goligorsky, M. S., Noiri, E., Kessler, H., and Romanov, V. (1997). Therapeutic potential of RGD peptides in acute renal injury. *Kidney Int.* **51,** 1487–1492.

Gottschalk, K. E., Gunther, R., and Kessler, H. (2002). A three-state mechanism of integrin activation and signal transduction for integrin alpha(v)beta(3). *Chem. Biol. Chem.* **3,** 470–473.

Gottschalk, K. E., and Kessler, H. (2004). A computational model of transmembrane integrin clustering. *Structure* **12,** 1109–1116.

Gray, J. J. (2004). The interaction of proteins with solid surfaces. *Curr. Opin. Struct. Biol.* **14,** 110–115.

Gurrath, M., Müller, G., Kessler, H., Aumailley, M., and Timpl, R. (1992). Conformation/ activity studies of rationally designed potent anti-adhesive RGD peptides. *Eur. J. Biochem.* **210,** 911–921.

Hammes, H. P., Brownlee, M., Jonczyk, A., Sutter, A., and Preissner, K. T. (1996). Subcutaneous injection of a cyclic peptide antagonist of vitronectin receptor-type integrins inhibits retinal neovascularization. *Nat. Med.* **2,** 529–533.

Hartman, G. D., Egbertson, M. S., Halczenko, W., Laswell, W. L., Duggan, M. E., Smith, R. L., Naylor, A. M., Manno, P. D., Lynch, R. J., Zhang, G., Chang, C. T.-C., and Gould, R. J. (1992). Non-peptide fibrinogen receptor antagonists. 1. Discovery and design of exosite inhibitors. *J. Med. Chem.* **35,** 4640–4642.

Haubner, R., Fissinger, D., and Kessler, H. (1997). Stereoisomeric peptide libraries and peptidomimetics for designing selective inhibitors of the $\alpha v \beta 3$ integrin for a new cancer therapy. *Angew. Chem. Int. Ed.* **36,** 1374–1389.

Haubner, R., Gratias, R., Diefenbach, B., Goodman, S. L., Jonczyk, A., and Kessler, H. (1996a). Structural and functional aspects of RGD-containing cyclic pentapeptides as highly potent and selective integrin $\alpha v \beta 3$ antagonists. *J. Am. Chem. Soc.* **118,** 7461–7472.

Haubner, R., Schmitt, W., Hölzemann, G., Goodman, S. L., Jonczyk, A., and Kessler, H. (1996b). Cyclic RGD peptides containing beta-turn mimetics. *J. Am. Chem. Soc.* **118**, 7881–7891.

Haubner, R., Sulyok, G., Weber, W., Linke, W., Bodenstein, C., Wester, H. J., Kessler, H., and Schwaiger, M. (2001a). Synthesis and first *in vivo* evaluation of [I-123]Aza-RGD-1: A radiolabled RGD-mimetic for the noninvasive determination of alpha(v)beta3 integrin expression. *Eur. J. Nucl. Med.* **28**, 1175–1175 (P5389 Suppl. S, Aug. 2001).

Haubner, R., Weber, W. A., Beer, A. J., Vabuliene, E., Reim, D., Sarbia, M., Becker, K. F., Goebel, M., Hein, R., Wester, H. J., Kessler, H., and Schwaiger, M. (2005). Noninvasive visualization of the activated alphavbeta3 integrin in cancer patients by positron emission tomography and [18F]galacto-RGD. *PLoS Med.* **2**, 244–252.

Haubner, R., and Wester, H. J. (2004). Radiolabeled tracers for imaging of tumor angiogenesis and evaluation of anti-angiogenic therapies. *Curr. Pharm. Des.* **10**, 1439–1455.

Haubner, R., Wester, H. J., Burkhart, F., Senekowitsch-Schmidtke, R., Weber, W., Goodman, S. L., Kessler, H., and Schwaiger, M. (2001b). Glycosylated RGD-containing peptides: Tracer for tumor targeting and angiogenesis imaging with improved biokinetics. *J. Nucl. Med.* **42**, 326–336.

Haubner, R., Wester, H. J., Reuning, U., Senekowitsch-Schmidtke, R., Diefenbach, B., Kessler, H., Stocklin, G., and Schwaiger, M. (1999). Radiolabeled alpha(v)beta3 integrin antagonists: A new class of tracers for tumor targeting. *J. Nucl. Med.* **40**, 1061–1071.

Haubner, R., Wester, H. J., Weber, W. A., Mang, C., Ziegler, S. I., Goodman, S. L., Senekowitsch-Schmidtke, R., Kessler, H., and Schwaiger, M. (2001c). Noninvasive imaging of alpha(v)beta3 integrin expression using 18F-labeled RGD-containing glycopeptide and positron emission tomography. *Cancer Res.* **61**, 1781–1785.

Hersel, U., Dahmen, C., and Kessler, H. (2003). RGD modified polymers: Biomaterials for stimulated cell adhesion and beyond. *Biomaterials* **24**, 4385–4415.

Hirano, Y., Okundo, M., Hayashi, T., Goto, K., and Nakajima, A. (1993). Cell-attachment activities of surface immobilized oligopeptides RGD, RGDS, RGDV, RGDT and YIGSR towards five cell lines. *J. Biomater. Sci. Polym. Ed.* **4**, 235–243.

Horton, M. A., Taylor, M. L., Arnett, T. R., and Helfrich, M. H. (1991). Arg-Gly-Asp (RGD). peptides and the anti-vitronectin receptor antibody 23C6 inhibit dentine resorption and cell spreading by osteoclasts. *Exp. Cell Res.* **195**, 368–375.

Huang, X. Z., Wu, J. F., Cass, D., Erle, D. J., Corry, D., Young, S. G., Farese, R. V., Jr., and Sheppard, D. (1996). Inactivation of the integrin beta 6 subunit gene reveals a role of epithelial integrins in regulating inflammation in the lung and skin. *J. Cell Biol.* **133**, 921–928.

Humphries, M. J., McEwan, P. A., Barton, S. J., Buckley, P. A., Bella, J., and Mould, A. P. (2003). Integrin structure: Heady advances in ligand binding, but activation still makes the knees wobble. *Trends Biochem. Sci.* **28**, 313–320.

Hynes, R. O. (2002). A reevaluation of integrins as regulators of angiogenesis. *Nat. Med.* **8**, 918–921.

Isacke, C. M., and Horton, M. A. (2000). "The Adhesion Molecule Facts Book." Academic Press, London.

Janssen, M. L., Oyen, W. J., Dijkgraaf, I., Massuger, L. F., Frielink, C., Edwards, D. S., Rajopadhye, M., Boonstra, H., Corstens, F. H., and Boerman, O. C. (2002). Tumor targeting with radiolabeled alpha(v)beta(3) integrin binding peptides in a nude mouse model. *Cancer Res.* **62**, 6146–6151.

Jeschke, B., Meyer, J., Jonczyk, A., Kessler, H., Adamietz, P., Meenen, N. M., Kantlehner, M., Goepfert, C., and Nies, B. (2002). RGD-peptides for tissue engineering of articular cartilage. *Biomaterials* **23**, 3455–3463.

Jin, H., and Varner, J. (2004). Integrins: Roles in cancer developement and as treatment agents. *Brit. J. Cancer* **90**, 561–565.

Kantlehner, M., Schaffner, P., Finsinger, D., Meyer, J., Jonczyk, A., Diefenbach, B., Nies, B., Hölzemann, G., Goodman, S. L., and Kessler, H. (2000). Surface coating with cyclic RGD peptides stimulates osteoblast adhesion and proliferation as well as bone formation. *Chem. Biol. Chem.* **1,** 107–114.

Kessler, H. (1982). Conformation and biological activity of cyclic peptides. *Angew. Chem. Int. Ed.* **21,** 512–523.

Kessler, H., Diefenbach, B., Finsinger, D., Geyer, A., Gurrath, M., Goodman, S. L., Hölzemann, G., Haubner, R., Jonczyk, A., Müller, G., vonRoedern, E. G., and Wermuth, J. (1995). Design of superactive and selective integrin receptor antagonists containing the RGD sequence. *Lett. Pept. Sci.* **2,** 155–160.

Kessler, H., and Kutscher, B. (1986). Synthesis of cyclic pentapeptide analogs of thymopoietin—cyclization with carbodiimide and 4-(dimethylamino)pyridine. *Liebigs Ann. Chem.* 869–892.

Kim, S., Bell, K., Mousa, S. A., and Varner, J. A. (2000). Regulation of angiogenesis *in vivo* by ligation of integrin alpha5beta1 with the central cell-binding domain of fibronectin. *Am. J. Pathol.* **156,** 1345–1362.

Knorr, R., Trzeciak, A., Bannwarth, W., and Gillessen, D. (1989). New coupling reagents in peptide chemistry. *Tetrahedron Lett.* **30,** 1927–1930.

Lacefield, W. R. (1999). Materials characterics of uncoated/ceramic-coated implant materials. *Adv. Dent. Res.* **13,** 21–26.

Li, R., Babu, C. R., Lear, J. D., Wand, A. J., Bennett, J. S., and DeGrado, W. F. (2001). Oligomerization of the integrin alphaIIbbeta3: Roles of the transmembrane and cytoplasmic domains. *Proc. Natl. Acad. Sci. USA* **98,** 12462–12467.

Li, R., Mitra, N., Gratkowski, H., Vilaire, G., Litvinov, R., Nagasami, C., Weisel, J. W., Lear, J. D., DeGrado, W. F., and Bennett, J. S. (2003). Activation of integrin alpha IIb beta 3 by modulation of transmembrane helix associations. *Science* **300,** 795–798.

Lohof, E., Planker, E., Mang, C., Burkhart, F., Dechantsreiter, M. A., Haubner, R., Wester, H. J., Schwaiger, M., Hölzemann, G., Goodman, S. L., and Kessler, H. (2000). Carbohydrate derivatives for use in drug design: Cyclic alpha(v)-selective RGD peptides. *Angew. Chem. Int. Ed.* **39,** 2761–2764.

Mammen, M., Choi, S. K., and Whitesides, G. M. (1998). Polyvalent interactions in biological systems: Implications for design and use of multivalent ligands and inhibitors. *Angew. Chem. Int. Ed.* **37,** 2755–2794.

Manavalan, P., and Momany, F. A. (1980). Conformational energy studies on N-methylated analogs of thyrotropin releasing hormone, enkephalin, and luteinizing hormone-releasing hormone. *Biopolymers* **19,** 1943–1973.

Marinelli, L., Gottschalk, K. E., Meyer, A., Novellino, E., and Kessler, H. (2004). Human integrin alphavbeta5: Homology modeling and ligand binding. *J. Med. Chem.* **47,** 4166–4177.

Marinelli, L., Lavecchia, A., Gottschalk, K. E., Novellino, E., and Kessler, H. (2003). Docking studies on alphavbeta3 integrin ligands: pharmacophore refinement and implications for drug design. *J. Med. Chem.* **46,** 4393–4404.

Marinelli, L., Meyer, A., Heckmann, D., Laveccia, A., Novellino, E., and Kessler, H. (2005). Ligand binding analysis for human alpha5beta1 integrin: Strategies for designing new alpha5beta1 integrin antagonists. *J. Med. Chem.* **48,** 4204–4207.

Mazur, R. H., James, P. A., Tyner, D. A., Hallinan, E. A., Sanner, J. H., and Schulze, R. (1980). Bradykinin analogues containing N alpha-methyl amino acids. *J. Med. Chem.* **23,** 758–763.

McMurry, T. J., Brechbiel, M., Kumar, K., and Gansow, O. A. (1992). Convenient synthesis of bifunctional tetraaza macrocycles. *Bioconjug. Chem.* **3,** 108–117.

Merrifield, R. B. (1963). Solid phase peptide synthesis. 1. Synthesis of a tetrapeptide. *J. Am. Chem. Soc.* **85,** 2149–2154.

Merrifield, R. B. (1985). Solid-phase synthesis (Nobel Lecture). *Angew. Chem. Int. Ed.* **24**, 799–810.

Meyer, A., Auernheimer, J., Modlinger, A., and Kessler, H. (2006). Targeting RGD recognizing integrins: Drug development, biomaterial research, tumor imaging and targeting. *Curr. Pharm. Des.* **12**, 1723–2747.

Montgomery, A. M., Reisfeld, R. A., and Cheresh, D. A. (1994). Integrin alpha v beta 3 rescues melanoma cells from apoptosis in three-dimensional dermal collagen. *Proc. Natl. Acad. Sci. USA* **91**, 8856–8860.

Mould, A. P., Symonds, E. J., Buckley, P. A., Grossmann, J. G., McEwan, P. A., Barton, S. J., Askari, J. A., Craig, S. E., Bella, J., and Humphries, M. J. (2003). Structure of an integrin-ligand complex deduced from solution x-ray scattering and site-directed mutagenesis. *J. Biol. Chem.* **278**, 39993–39999.

Mousa, S. A. (2000). Integrins as novel drug discovery targets: Potential therapeutic and diagnostic implications. *Emerging Ther. Targets* **44**, 143–153.

Müller, G., Gurrath, M., and Kessler, H. (1994). Pharmacophore refinement of gpIIb/IIIa antagonists based on comparative studies of antiadhesive cyclic and acyclic RGD peptides. *J. ComputerAided Mol. Des.* **8**, 709–730.

Nip, J., Rabbani, S. A., Shibata, H. R., and Brodt, P. (1995). Coordinated expression of the vitronectin receptor and the urokinase-type plasminogen activator receptor in metastatic melanoma cells. *J. Clin. Invest.* **95**, 2096–2103.

Otte, A., Jermann, E., Behe, M., Götze, M., Bucher, H. C., Roser, H. W., Heppeler, A., Müller-Brand, J., and Mäcke, H. R. (1997). DOTATOC: A powerful new tool for receptor-mediated radionuclide therapy. *Eur. J. Nucl. Med.* **24**, 792–795.

Pechy, P., Rotzinger, F. P., Nazeeruddin, M. K., Kohle, O., Zakeeruddin, S. M., Humphrybaker, R., and Gratzel, M. (1995). Preparation of phosphonated polypyridyl ligands to anchor transition-metal complexes on oxide surfaces—application for the conversion of light to electricity with nanocrystalline Tio2 Films. *J. Chem. Soc. Chem. Commun.* **1**, 65–66.

Pfaff, M., Tangemann, K., Müller, B., Gurrath, M., Müller, G., Kessler, H., Timpl, R., and Engel, J. (1994). Selective recognition of cyclic RGD peptides of NMR defined conformation by alpha IIb beta 3, alpha V beta 3, and alpha 5 beta 1 integrins. *J. Biol. Chem.* **269**, 20233–20238.

Pierschbacher, M. D., Hayman, E. G., and Ruoslahti, E. (1981). Location of the cell-attachment site in fibronectin with monoclonal antibodies and proteolytic fragments of the molecule. *Cell* **26**, 259–267.

Pierschbacher, M. D., and Ruoslahti, E. (1984a). Cell attachment activity of fibronectin can be duplicated by small synthetic fragments of the molecule. *Nature* **309**, 30–33.

Pierschbacher, M. D., and Ruoslahti, E. (1984b). Variants of the cell recognition site of fibronectin that retain attachment-promoting activity. *Proc. Natl. Acad. Sci. USA* **81**, 5985–5988.

Pierschbacher, M. D., Ruoslahti, E., Sundelin, J., Lind, P., and Peterson, P. A. (1982). The cell attachment domain of fibronectin. Determination of the primary structure. *J. Biol. Chem.* **257**, 9593–9597.

Plow, E. F., Haas, T. A., Zhang, L., Loftus, J., and Smith, J. W. (2000). Ligand binding to integrins. *J. Biol. Chem.* **275**, 21785–21788.

Poethko, T., Thumshirn, G., Hersel, U., Rau, F., Haubner, R., Schwaiger, M., Kessler, H., and Wester, H. J. (2003). Improved tumor uptake, tumor retention and tumor/background ratios of pegylated RGD-multimers. *J. Nucl. Med.* **44**(Suppl.), 46.

Reynolds, L. E., Wyder, L., Lively, J. C., Taverna, D., Robinson, S. D., Huang, X., Sheppard, D., Hynes, R. O., and Hodivala-Dilke, K. M. (2002). Enhanced pathological angiogenesis in mice lacking beta3 integrin or beta3 and beta5 integrins. *Nat. Med.* **8**, 27–34.

Roach, P., Farrar, D., and Perry, C. C. (2005). Interpretation of protein adsorption: Surface-induced conformational changes. *J. Am. Chem. Soc.* **127,** 8168–8173.

Rojas, A. J., and Ahmed, A. R. (1999). Adhesion receptors in health and disease. *Crit. Rev. Oral Biol. Med.* **10,** 337–358.

Rose, D. M., Pozzi, A., and Zent, R. (2000). Integrins as therapeutic targets. *Emerging Ther. Targets* **4,** 397–411.

Schmitt, J. S. (1998). Lineare und zyklische Peptoide und Aza-Peptide als potentielle Tumortherapeutika und Isotopenmarkierung der humanen synovialen Phosphorlipase A2. Department of Organic Chemistry and Biochemistry, Technical University, Munich.

Seftor, R. E., Seftor, E. A., Gehlsen, K. R., Stetler-Stevenson, W. G., Brown, P. D., Ruoslahti, E., and Hendrix, M. J. (1992). Role of the alpha v beta 3 integrin in human melanoma cell invasion. *Proc. Natl. Acad. Sci. USA* **89,** 1557–1561.

Shemyakin, M. M., Ovchinnikov, Y. A., and Ivanov, V. T. (1969). Topochemical investigations of peptide systems. *Angew. Chem. Int. Ed.* **8,** 492–499.

Smallheer, J. M., Weigelt, C. A., Woerner, F. J., Wells, J. S., Daneker, W. F., Mousa, S. A., Wexler, R. R., and Jadhav, P. K. (2004). Synthesis and biological evaluation of non-peptide integrin antagonists containing spirocyclic scaffolds. *Bioorg. Med. Chem. Lett.* **14,** 383–387.

Smith, J. W., Vestal, D. J., Irwin, S. V., Burke, T. A., and Cheresh, D. A. (1990). Purification and functional characterization of integrin alpha v beta 5. An adhesion receptor for vitronectin. *J. Biol. Chem.* **265,** 11008–11013.

Stragies, R., Osterkamp, F., Zischinsky, G., Vossmeyer, D., Kalkhof, H., Reimer, U., Zahn, G. (2007). Design and sysnthesis of a new class of selective integrin $\alpha5\beta1$ antagonists. *J. Med. Chem.* (in press).

Sulyok, G. A., Gibson, C., Goodman, S. L., Hölzemann, G., Wiesner, M., and Kessler, H. (2001). Solid-phase synthesis of a nonpeptide RGD mimetic library: New selective alphavbeta3 integrin antagonists. *J. Med. Chem.* **44,** 1938–1950.

Sykaras, N., Iacopino, A. M., Marker, V. A., Triplett, R. G., and Woody, R. D. (2000). Implant materials, designs, and surface topographies: Their effect on osseointegration. A literature review. *Int. J. Oral Maxillofac. Implants* **15,** 675–690.

Takagi, J., Strokovich, K., Springer, T. A., and Walz, T. (2003). Structure of integrin alpha5beta1 in complex with fibronectin. *EMBO J.* **22,** 4607–4615.

Tamkun, J. W., DeSimone, D. W., Fonda, D., Patel, R. S., Buck, C., Horwitz, A. F., and Hynes, R. O. (1986). Structure of integrin, a glycoprotein involved in the transmembrane linkage between fibronectin and actin. *Cell* **46,** 271–282.

Taverna, D., Moher, H., Crowley, D., Borsig, L., Varki, A., and Hynes, R. O. (2004). Increased primary tumor growth in mice null for beta3- or beta3/beta5-integrins or selectins. *Proc. Natl. Acad. Sci. USA* **101,** 763–768.

Thakur, M. L. (1995). Radiolabelled peptides: Now and the future. *Nucl. Med. Commun.* **16,** 724–732.

Thull, R. (2001). Surface functionalization of materials to initiate auto-compatibilization *in vivo. Materialwiss. Werkst.* **32,** 949–952.

Thumshirn, G., Hersel, U., Goodman, S. L., and Kessler, H. (2003). Multimeric cyclic RGD peptides as potential tools for tumor targeting: Solid-phase peptide synthesis and chemoselective oxime ligation. *Chemistry* **9,** 2717–2725.

Tonelli, A. (1976). The effects of isolated N-methylated residues on the conformational characteristics of polypeptides. *Biopolymers* **15,** 1615–1622.

Tranqui, L., Andrieux, A., Hudry-Clergeon, G., Ryckewaert, J. J., Soyez, S., Chapel, A., Ginsberg, M. H., Plow, E. F., and Marguerie, G. (1989). Differential structural requirements for fibrinogen binding to platelets and to endothelial cells. *J. Cell Biol.* **108,** 2519–2527.

Turker, R. K., Hall, M. M., Yamamoto, M., Sweet, C. S., and Bumpus, F. M. (1972). A new, long-lasting competitive inhibitor of angiotensin. *Science* **177**, 1203–1205.

Weber, W. A., Avril, N., and Schwaiger, M. (1999). Relevance of positron emission tomography (PET) in oncology. *Strahlenther. Onkol.* **175**, 356–373.

Weide, T., Modlinger, A., and Kessler, H. (2007). Spatial screening for the identification of integrin ligands. *Top. Curr. Chem.* **272**, 1–50.

Wermuth, J. (1996). Alpha v beta 3 Inhibitoren durch räumliches Screening. Department of Organic Chemistry and Biochemistry, Technical University, Munich.

Wermuth, J., Goodman, S. L., Jonczyk, A., and Kessler, H. (1997). Stereoisomerism and biological activity of the selective and superactive alpha(v)beta(3) integrin inhibitor cyclo (-RGDfV-) and its retro-inverso peptide. *J. Am. Chem. Soc.* **119**, 1328–1335.

Wester, H. J., Hamacher, K., and Stocklin, G. (1996). A comparative study of N.C.A. fluorine-18 labeling of proteins via acylation and photochemical conjugation. *Nucl. Med. Biol.* **23**, 365–372.

Wiseman, G. A., Gordon, L. I., Multani, P. S., Witzig, T. E., Spies, S., Bartlett, N. L., Schilder, R. J., Murray, J. L., Saleh, M., Allen, R. S., Grillo-Lopez, A. J., and White, C. A. (2002). Ibritumomab tiuxetan radioimmunotherapy for patients with relapsed or refractory non-Hodgkin lymphoma and mild thrombocytopenia: A phase II multicenter trial. *Blood* **99**, 4336–4342.

Wiseman, G. A., White, C. A., Witzig, T. E., Gordon, L. I., Emmanouilides, C., Raubitschek, A., Janakiraman, N., Gutheil, J., Schilder, R. J., Spies, S., Silverman, D. H., and Grillo-Lopez, A. J. (1999). Radioimmunotherapy of relapsed non-Hodgkin's lymphoma with zevalin, a 90Y-labeled anti-CD20 monoclonal antibody. *Clin. Cancer Res.* **5**, 3281–3286.

Wu, Y. L., Matsueda, G. R., and Bernatowicz, M. (1993). An efficient method for the preparation of omega,omega(')-bis-urethane protected arginine derivatives. *Synth. Commun.* **23**, 3055–3060.

Xiao, T., Takagi, J., Coller, B. S., Wang, J. S., and Springer, T. A. (2004). Structural basis for allostery in integrins and binding to fibrinogen-mimetic therapeutics. *Nature* **432**, 59–67.

Xiong, J. P., Stehle, T., Diefenbach, B., Zhang, R., Dunker, R., Scott, D., Joachimiak, A., Goodman, S. L., and Arnaout, M. A. (2001). Crystal structure of the extracellular segment of integrin $\alpha v\beta 3$. *Science* **294**, 339–345.

Xiong, J. P., Stehle, T., Zhang, R., Joachimiak, A., Frech, M., Goodman, S. L., and Arnaout, M. A. (2002). Crystal structure of the extracellular segment of integrin avb3 in complex with an Arg-Gly-Asp ligand. *Science* **296**, 151–155.

Yang, C. (2001). The effect of calcium phosphate implant coating on osteoconduction. *Oral Surg. Oral Med. Oral Pathol. Oral Radiol. Endod.* **92**, 606–609.

Yusuf-Makagiansar, H., Anderson, M. E., Yakovleva, T. V., Murray, J. S., and Siahaan, T. J. (2002). Inhibition of LFA-1/ICAM-1 and VLA-4/VCAM-1 as a therapeutic approach to inflammation and autoimmune disease. *Med. Res. Rev.* **22**, 146–167.

CHAPTER TWENTY-ONE

Evaluating Integrin Function in Models of Angiogenesis and Vascular Permeability

Sara M. Weis

Contents

Moores UCSD Cancer Center, University of California, San Diego, La Jolla, California

Methods in Enzymology, Volume 426
ISSN 0076-6879, DOI: 10.1016/S0076-6879(07)26021-5

© 2007 Elsevier Inc.
All rights reserved.

Abstract

All vascular biological processes are influenced to some degree by integrins expressed on endothelial cells, vascular smooth muscle cells, fibroblasts, platelets, or other circulating cells. In particular, angiogenesis requires cells to process signals from their microenvironment and respond by altering their cell–cell and cell–matrix adhesion, events which allow migration and vascular remodeling over the period of days to weeks. On the other hand, endothelial cells can respond to a permeability stimulus and alter their junctional adhesion molecules or vesicular transport machinery within seconds or minutes. This chapter will discuss the current understanding of how integrins participate in these processes, and explore the *in vitro* and *in vivo* models available to study the role of integrin function during angiogenesis and vascular leak.

1. INTRODUCTION

Integrins transmit signals from "outside-in" that help a cell decide whether to adhere, move, or undergo apoptosis, depending on the extracellular matrix (ECM) components it encounters. At the same time, integrins can transmit "inside-out" signals to influence their local environment. Thus, integrins functioning as both mechanosensors and signaling molecules play a critical role in vascular remodeling events. The integrin expression profile of a given cell type combined with the ECM ligands in its microenvironment governs the types of signals that the cell can sense and transmit.

The integrin expression profile on endothelial cells undergoing pathological angiogenesis is different from quiescent cells or cells undergoing vasculogenesis during development. In particular, expression of the $\beta 1$ and αv integrin subunits is often increased on cells during angiogenesis. Several decades ago, Erkki Ruoslahti and coworkers discovered a common sequence shared by matrix proteins that can bind the integrins that are generally overexpressed on angiogenic blood vessels (Ruoslahti, 2003). This arginine-glycine-aspartic acid (or RGD) sequence is expressed on fibronectin, vitronectin, osteopontin, collagens, thrombospondin, fibrinogen, and von Willebrand factor, and has since been shown to represent the ligand-binding region for a class of RGD-binding integrins including $\alpha 5 \beta 1$, $\alpha 8 \beta 1$, $\alpha v \beta 1$, $\alpha v \beta 3$, $\alpha v \beta 5$, $\alpha v \beta 6$, $\alpha v \beta 8$, and $\alpha IIb \beta 3$ (Ruoslahti and Pierschbacher, 1986, 1987). This discovery eventually led to the development of integrin antagonists that disrupt the binding between RGD-binding integrins and their ligands, such as RGD-containing synthetic peptides and snake venom disintegrins. Disruption of integrin ligation can lead a cell to undergo apoptosis, and thus represents a powerful tool to disrupt angiogenesis.

1.1. Selective expression of $\alpha v \beta 3$ on angiogenic blood vessels

Endothelial cells grown in culture can bind vitronectin via $\alpha v \beta 3$ or $\alpha v \beta 5$, fibronectin via $\alpha 5 \beta 1$ or $\alpha v \beta 3$, and collagen via $\alpha 1 \beta 1$ or $\alpha 2 \beta 1$. However, $\alpha v \beta 3$ is not detectable on endothelial cells lining normal blood vessels *in vivo*. Instead, $\alpha v \beta 3$ is highly expressed on blood vessels near a tumor (Brooks *et al.*, 1994b), wound tissue (Brooks *et al.*, 1994a), rheumatoid arthritis (Storgard *et al.*, 1999), or focal ischemia (Okada *et al.*, 1996). Accordingly, anti-$\alpha v \beta 3$ treatment blocks pathological angiogenesis but has no effect on preexisting vessels (Brooks *et al.*, 1994a). The selective expression of $\alpha v \beta 3$ on tumor-associated blood vessels has proven useful for targeted delivery of antitumor or angiogenic agents and for noninvasively detecting and imaging tumors. Delivery of $\alpha v \beta 3$-targeted nanoparticles carrying a mutant Raf gene blocks angiogenesis and induces apoptosis of tumor-associated blood vessels, leading to a sustained regression of established primary and metastatic tumors (Hood *et al.*, 2002). Noninvasive imaging is able to detect tumors by the expression of $\alpha v \beta 3$ on angiogenic vessels using a highly potent tetrameric RGD peptide near-infrared fluorescent probe (Wu *et al.*, 2006) or near-infrared quantum dots (Cai *et al.*, 2006).

Vascular endothelial growth factor (VEGF, or VEGF-A165) is a potent inducer of both angiogenesis and vascular leak. Immediately following ischemic stroke in the brain, microvessels show an early expression of both VEGF and $\alpha v \beta 3$ (Abumiya *et al.*, 1999). In tumor cells, it appears that $\alpha v \beta 3$ can directly regulate the production of VEGF, which promotes extensive neovascularization to support tumor growth *in vivo* (De *et al.*, 2005). In cultured endothelial cells, treatment with VEGF induces tyrosine phosphorylation of VEGFR2 and its co-association with $\beta 3$ (but not $\beta 1$ or $\beta 5$), suggesting that $\alpha v \beta 3$ may participate in the full activation of VEGFR2 (Soldi *et al.*, 1999).

Another ligand for $\alpha v \beta 3$ is the RGD site on the sixth Ig-like domain of the cell adhesion molecule, L1-CAM (Montgomery *et al.*, 1996). L1-CAM forms homophilic or heterophilic interactions, and can be immobilized in the ECM. For example, melanoma cells can shed and deposit L1 into the ECM such that $\alpha v \beta 3$ (on either tumor cells or endothelial cells) can recognize L1 as a cell–cell or cell-substrate interaction (Montgomery *et al.*, 1996). Accordingly, matrix-immobilized L1-Ig6 promotes angiogenesis by ligating and phosphorylating $\alpha v \beta 3$. Furthermore, treatment with either VEGF or L1-Ig6 induces phosphorylation and coassociation of VEGFR2 and $\alpha v \beta 3$ (Hall and Hubbell, 2004). This situation is relevant to cancer, since tumor cells can shed L1-CAM onto the local ECM, which could recruit endothelial cells and guide the angiogenic or vascular permeability response by influencing both $\alpha v \beta 3$ and VEGFR2. L1 also associates with the extracellular domain of the VEGF receptor, neuropilin-1 (NRP-1) (Castellani, 2002). Ovarian carcinoma cells overexpress L1, but not NRP-1.

Instead, it is thought that L1 on carcinoma cells facilitates their binding to other cell types expressing NRP-1, such as the cells lining the peritoneum where ovarian carcinoma cells tend to metastasize (Stoeck *et al.*, 2006) or potentially endothelial cells during the tumor-induced angiogenic response. Taken together, these studies suggest that $\alpha v\beta 3$ binding to fibronectin, vitronectin, or L1-CAM within the matrix or on other cell types can influence VEGF signaling, angiogenesis, and vascular leak.

1.2. Distinct functions for $\alpha v\beta 3$ and $\alpha v\beta 5$ during angiogenesis and vascular leak

While $\alpha v\beta 3$ recognizes several different ligands, $\alpha v\beta 5$ primarily binds to vitronectin. In carcinoma cells, $\alpha v\beta 3$ is detected in focal contacts colocalizing with vinculin, talin, and the ends of actin filaments, while $\alpha v\beta 5$ shows a distinct, nonfocal contact distribution on the cell surface (Wayner *et al.*, 1991). Thus, two homologous vitronectin-binding integrins can be distributed differentially on the cell surface and likely direct distinct cellular processes. In fact, studies using function-blocking antibodies or other integrin antagonists have revealed that $\alpha v\beta 3$ is required for angiogenesis induced by bFGF or TNF-α, while $\alpha v\beta 5$ is required for angiogenesis induced by VEGF, TGF-α, or phorbol ester (Friedlander *et al.*, 1995).

Cyclic RGD peptides or fragments of type IV collagen (tumstatin) are soluble ligands for $\alpha v\beta 3$ that transmit antiangiogenic signals, while $\alpha v\beta 3$ binding to vitronectin promotes angiogenesis. Fastatin is an endogenous angiogenesis regulator that inhibits endothelial cell migration, growth, and angiogenesis by binding to $\alpha v\beta 3$ (Nam *et al.*, 2005). In normal mice, blocking $\alpha v\beta 3$ function with the cyclic-RGDfV peptide reduces the extent of injury and edema in a model of ischemic stroke by reducing VEGFR2 activity and VEGF production, thereby preserving the blood–brain barrier (Shimamura *et al.*, 2006). Therefore, $\alpha v\beta 3$ function on endothelial cells depends on whether it is occupied by a soluble or insoluble ligand.

However, lack of $\beta 3$ expression does not produce a similar outcome compared with $\beta 3$ antagonism. Mice lacking $\beta 3$ surprisingly show enhanced angiogenesis and vascular leak in response to VEGF (Reynolds *et al.*, 2002; Taverna *et al.*, 2004). In $\beta 3$-deficient mice, enhanced expression and activity of VEGFR2 appears to be a compensatory mechanism accounting for their phenotype (Reynolds *et al.*, 2004; Robinson *et al.*, 2004). As $\beta 3$-deficient mice age, they develop moderate spontaneous cardiac hypertrophy and mild cardiac inflammation, which is exacerbated by transverse aortic constriction, and can partially be attributed to knockout of $\beta 3$ on blood-borne cells (Ren *et al.*, 2007). Capillaries in the hearts of $\beta 3$-deficient mice show an immature phenotype which resembles angiogenic tumor-associated vessels (Weis *et al.*, 2007). This phenotype can be normalized by systemic treatment with inhibitors of VEGF or VEGFR2, again supporting the notion that

β3-deficient mice are especially sensitive to VEGF signaling (Weis *et al.*, 2007). To analyze the role of integrin phosphorylation and signaling in angiogenesis, Byzova and colleagues generated knock-in mice that express a mutant β3 unable to undergo tyrosine phosphorylation which, importantly, does not result in compensatory increases in VEGFR2 expression (Mahabeleshwar *et al.*, 2006). Unlike the global β3 knockout mice, the "DiYF" mice showed impaired neovascularization in response to VEGF, consistent with studies utilizing integrin antagonists. In the DiYF knock-in mice, VEGF stimulation did not induce a complex between VEGFR2 and β3, VEGFR2 phosphorylation was significantly reduced, and integrin activation was impaired (Mahabeleshwar *et al.*, 2006). Together these studies suggest that β3 expression may influence VEGFR2/VEGF expression and sensitivity during angiogenesis and vascular leak.

Unlike β3-deficient mice, mice lacking β5 show no vascular leak in response to VEGF (Eliceiri *et al.*, 2002) or a variety of inflammatory-induced permeability agents including TGF-β and thrombin (Su *et al.*, 2007). Actin stress fiber formation induced during the permeability response can be attenuated by blocking αvβ5, suggesting that αvβ5 regulates vascular permeability by facilitating interactions with the actin cytoskeleton (Su *et al.*, 2007). Accordingly, β5-deficient mice show less edema in models of ischemic stroke (Eliceiri *et al.*, 2002) and pulmonary ischemia-reperfusion or ventilator-induced lung injury (Su *et al.*, 2007). In fact, a function-blocking antibody for β5 (ALULA) blocks edema and provides protection in these experimental models (Su *et al.*, 2007). These studies suggest that β5 may represent a therapeutic target to block vascular leak.

1.3. Role of β1 integrins in angiogenesis and vascular permeability

The impact of specific integrins during angiogenesis depends on developmental stage. While αvβ3 or αvβ5 are not essential for development, vasculogenesis absolutely requires α5β1 and its ligand fibronectin. In addition, β1 integrins also participate in postnatal, pathological angiogenesis. α5β1 integrin binds fibronectin and contributes to an angiogenesis pathway that is distinct from VEGF-mediated angiogenesis, yet important for the growth of tumors (Kim *et al.*, 2000). In contrast, antagonism of α1β1 and α2β1 with function-blocking antibodies blocks VEGF-dependent, tumor-associated angiogenesis (Senger *et al.*, 2002). Blocking α6β1 function inhibits VEGF-induced adhesion, migration, and tumor angiogenesis (Lee *et al.*, 2006). Similarly, α9β1 binds VEGF and influences VEGF-induced angiogenesis (Vlahakis *et al.*, 2007). These studies illustrate the specificity of integrin-mediated signaling, and highlight the need to understand how each integrin–ligand binding event within each microenvironment influences angiogenesis.

Integrin $\alpha 4\beta 1$ is selectively expressed on proliferating endothelial cells, while its ligand VCAM-1 is expressed only on proliferating mural cells. Varner and colleagues have established that antagonists of this integrin-ligand pair block the adhesion of mural cells to proliferating endothelial cells, induce apoptosis of both endothelial cells and pericytes, and thereby inhibit neovascularization (Garmy-Susini *et al.*, 2005). $\alpha 4\beta 1$ promotes angiogenesis by allowing the invasion of myeloid cells into tumors, and $\alpha 4\beta 1$ antagonists prevent monocyte-induced angiogenesis, macrophage colonization of tumors, and tumor angiogenesis (Jin *et al.*, 2006b). Varner's group has extended these findings to implicate $\alpha 4\beta 1$ in the homing of circulating progenitor cells to tumor neovasculature that expresses VCAM-1 and cellular fibronectin (Jin *et al.*, 2006a).

$\beta 1$ integrins can also influence the disruption of vascular permeability, particularly during inflammation since the recruitment of circulating cells into inflammatory tissue involves recognition of cell adhesion molecules. Specifically, $\beta 1$ integrins can influence the ability of circulating cells to arrest, adhere, and extravasate at sites of injury and vascular leak (Sixt *et al.*, 2006). Expression of $\beta 1$ integrins on lymphocytes, monocytes, leukocytes, and neutrophils can regulate their binding and transit through the endothelial layer. Accordingly, blocking $\beta 1$-integrin function reduces transmigration of circulating cells through the endothelial cell barrier. However, in the lung, treatment with cyclic RGD peptides or antibodies to $\alpha 5\beta 1$ disrupts cell–matrix adhesion and increases endothelial permeability (Curtis *et al.*, 1995).

1.4. Role of $\alpha 6\beta 4$ in angiogenesis

The laminin-5 receptor $\alpha 6\beta 4$ is expressed on endothelial cells comprising mature blood vessels, and plays a somewhat unique role during tumor-associated angiogenesis. Unlike the other integrins implicated in angiogenesis, $\alpha 6\beta 4$ is unique since overexpression of $\beta 4$ is sufficient to transform fibroblasts, enhance anchorage-independent growth of carcinoma cells, and induce tumorigenesis (Bertotti *et al.*, 2005). While $\alpha 6\beta 4$ influences tumor invasion and metastasis, it appears to function as an adapter protein or "signaling accomplice" rather than a cell adhesion molecule (Bertotti *et al.*, 2005; Trusolino *et al.*, 2001).

1.5. Endogenous negative regulators of integrin function

Bussolino and colleagues have recently described semaphorins as endogenous negative regulators of integrin function. Expressed on both vascular and neuronal cells, semaphorin family members can modulate the affinity of certain integrin receptors toward their matrix ligands (Bussolino *et al.*, 2006). In addition, a number of ECM fragments can act as endogenous regulators of integrins. Tumstatin, a cleavage fragment of the $\alpha 3$ chain of type IV collagen, acts as an antiangiogenic agent that can block angiogenesis induced by VEGF or tumors *in vivo* (Hamano *et al.*, 2003). The suppressive effects of tumstatin require $\alpha v \beta 3$ expressed on pathological angiogenic blood vessels to selectively induce endothelial cell apoptosis. However, the presence of cyclic RGD peptides does not affect the $\alpha v \beta 3$-mediated activity of tumstatin (Maeshima *et al.*, 2000). Endostatin is a similar cleaved fragment of the $\alpha 1$ chain of type XVIII collagen, and it acts as an antiangiogenic agent via its relationship with $\alpha 5 \beta 1$ (Sudhakar *et al.*, 2003). Angiostatin is a cleaved fragment of plasminogen which also binds integrins and acts as an antiangiogenic agent by inhibiting endothelial cell migration and proliferation while inducing apoptosis. Angiostatin can bind to integrins $\alpha v \beta 3$, $\alpha 9 \beta 1$, and $\alpha 4 \beta 1$, although its activity can be attributed to its specific interaction with $\alpha v \beta 3$ on angiogenic endothelial cells (Tarui *et al.*, 2001).

1.6. Clinical testing of integrin antagonists as antiangiogenic agents

Many preclinical studies have established the antiangiogenic properties of integrin antagonists. As a result, clinical trials are currently underway to evaluate melanoma or prostate cancer patients treated with Abegrin (etaracizumab; previously known as Vitaxin or MEDI-522), which is a humanized version of the LM609 anti-$\alpha v \beta 3$ monoclonal antibody. The $\alpha v \beta 3$-targeting cyclic RGD peptide cilengitide (EMD 121974) is also currently being tested as an antiangiogenic agent to treat glioblastoma multiforme, prostate cancer, or lymphoma. CNTO 95 is a monoclonal antibody that inhibits αv integrins is being tested in patients with advanced Stage IV melanoma. ATN-161, a non-RGD–based small peptide antagonist of $\alpha 5 \beta 1$, is undergoing Phase I/II trials for patients with recurrent intracranial malignant glioma or advanced solid tumors. Volociximab (M200; anti-a5$\beta 1$) is being tested in patients with metastatic renal cell carcinoma. Adenoviral or human recombinant endostatin is being tested in patients with advanced refractory solid tumors, while angiostatin has been tested in combination therapies for patients with non–small-cell lung cancer and solid tumors.

2. ASSESSING ROLE OF INTEGRINS DURING ANGIOGENESIS

Manipulation of integrin function can impact angiogenesis and vascular leak during many pathological situations. Several factors are important to consider while evaluating the role of specific integrins or developing anti-integrin therapeutic strategies. For example, it is important to verify whether a potential drug specifically targets one or more integrins, to monitor what cell types express these integrins *in vivo* in normal tissues or during the pathology in question, and to determine whether this drug acts as an agonist or antagonist. Strategies that target tumor-associated angiogenesis must ultimately be successful *in vivo*. However, tumor angiogenesis is a complex process that involves integrins on both tumor cells and the host tissues, complicating evaluation of anti-integrin strategies. *In vitro* and *in vivo* angiogenesis models are therefore useful to test the role of particular integrins in this vascular process.

The process of angiogenesis *in vivo* has been studied using a number of well-defined methods. Mouse models are disadvantaged in terms of studying integrins, since reagents for testing and identifying integrins in mice are generally lacking. Most antibodies used to block integrin function do not recognize mouse integrins, and few anti-integrin antibodies for immuno-histochemistry, immunoblotting, or immunoprecipitation are commercially available. However, gene-targeted mouse models are available to study the impact of deletion of individual integrin subunits or knock-in of mutant integrin genes.

2.1. *In vitro* models of angiogenesis

In vitro angiogenesis assays have historically utilized different combinations of ECM components and growth factors to induce endothelial cells to form tube-like structures in two-dimensional (2D) culture. This type of experiment has evolved more recently to the assessment of tube formation within a three-dimensional (3D) gel. Davis and colleagues have used 3D culture experiments to demonstrate how $\alpha v\beta 3$ and $\alpha 5\beta 1$ regulate vacuolation and lumen formation by endothelial cells cultured in 3D fibrin matrices. These events can be inhibited or regressed by including function blocking antibodies to $\alpha v\beta 3/\alpha v\beta 5$, $\alpha 5$, or a cyclic RGD peptide (Bayless *et al.*, 2000).

Hughes and coworkers have established an innovative variation on this model in which endothelial cells are cultured on beads and then embedded within fibrin gels (Nakatsu *et al.*, 2003a, 2003b). This process can be tracked using time-lapse microscopy to visualize endothelial cells sprouting from the surface of the beads into the gel, including tip cells that extend processes similar to angiogenesis *in vivo*. Behind the tip cells, a single layer of endothelial cells forms a patent lumen. Hughes and colleagues had previously identified genes upregulated in endothelial cells forming tubes in 3D collagen gels, compared to migrating and proliferating cells in 2D cultures (Aitkenhead *et al.*, 2002). These 3D culture methods provide a robust system that can be employed to test the antiangiogenic activity of integrin antagonists and together with genetic manipulation of integrins and related proteins can be used to better establish relevant molecular mechanisms.

Differentiation of embryonic stem (ES) cells *in vitro* has become a powerful model to track the formation of a vascular network. The endothelial cells that differentiate form primitive blood vessels as a vascular plexus develops. This relative 2D culture model allows observation of the process of mammalian vascular development, using real-time imaging (Kearney and Bautch, 2003; Wang *et al.*, 1992). Embryonic Stem cells can be isolated from mice with endothelial-specific expression of green fluorescent protein (GFP) or mice with targeted deletion or mutant knock-in of genes of interest (Kearney *et al.*, 2004). Detailed methods for this model have been published in a previous issue of *Methods in Enzymology* (Kearney and Bautch, 2003). Hynes and coworkers have employed this model to recapitulate the vascular defects observed in mice lacking the vascular integrins using homozygous null ES cells (Francis, 2006; Francis *et al.*, 2002; Taverna and Hynes, 2001).

2.2. Chick chorioallantoic membrane angiogenesis assay

The chorioallantoic membrane (CAM) in the chicken embryo has been used since the early 1900s to study angiogenesis *in vivo* (see the short commentary by Ribatti, 2004). Many years later, Cheresh and colleagues employed this model to demonstrate the requirement of $\alpha v\beta 3$ for angiogenesis induced by VEGF, while that induced by bFGF requires $\alpha v\beta 5$ (Brooks *et al.*, 1994a; Friedlander *et al.*, 1995). This model is still being used routinely as a screen for integrin antagonists and antiangiogenic agents. Variations of this model include injection of human tumor cells or implantation of a tumor mass intravenously or onto the CAM to induce angiogenesis and monitor tumor growth or metastatic spread.

The basic method to test growth factor-induced angiogenesis involves the placement of a filter disc containing an angiogenic growth factor (e.g., VEGF or bFGF) onto the exposed CAM of a 10-day-old fertilized chicken egg. The disc is gently placed on the CAM near a major blood vessel, but not directly on a group of tiny vessels. Several days later, the filter disc is excised with some CAM tissue and placed on a small Petri dish, CAM side down. The angiogenic response is assessed by counting the number of branch points using a dissecting microscope. This CAM angiogenesis assay provides a useful tool to study integrin function, since most function-blocking integrin antibodies are active in the chick embryo (compared with the mouse). This method is relatively inexpensive, does not require special equipment or training, and can be used to screen a large number of samples.

2.3. Matrigel plug angiogenesis assay in the mouse

A number of different rodent models have been used to study angiogenesis in an adult mammalian system. The murine skin model of angiogenesis using subcutaneous injection of Matrigel provides a useful *in vivo* model.

This method can provide information about the antiangiogenic properties of integrin antagonists in combination with other potential antiangiogenic compounds. Each plug can contain a combination of growth factors, integrin antagonists, or test agents, while the mouse can be a gene-targeted strain or receive systemic treatment with additional agents. Variation between subjects is minimal, so that groups as small as four to six mice are sufficient to detect statistically significant differences. The basic technique is given below, along with several methods to assess the angiogenic response.

1. Inject Matrigel plugs containing growth factors and/or test compounds. Shave the mice several days before the experiment to avoid skin irritation. Prepare Matrigel mixture to inject 400 μl per plug (one containing growth factor and one control plug per mouse). Growth factor-reduced Matrigel (BD Biosciences) should be thawed for several h in an ice bucket. Prepare angiogenic Matrigel solution containing 400 ng/ml recombinant human bFGF or VEGF and a control Matrigel solution containing an equivalent volume of sterile 0.9% saline. Inject 400 μl subcutaneously to each flank and allow 5 to 7 days for angiogenesis to occur. The optimal time for this experiment varies considerably depending upon mouse strain and growth factors used.

2. Harvest plugs and assess angiogenic response using one of several methods:
   a. Quantify angiogenesis using FITC-lectin. At the end of the experiment, inject 20 μg of a fluorescently labeled lectin that binds mouse endothelial cells (such as FITC-GSL I/BSL I, or FITC-GSL I-B$_4$) via the tail vein and allow it to circulate for approximately 10 min before euthanizing the mouse. Carefully dissect and remove the Matrigel plugs, keeping all tubes containing FITC-Lectin in the dark. If desired, visualize perfused blood vessels in whole plugs using confocal microscopy. Place plugs in 300 to 500 μl of a suitable lysate buffer (e.g., RIPA) and grind briefly using a tissue homogenizer. Spin at high speed for 10 min in refrigerated centrifuge. Dilute supernatant 1:20 and read on fluorescent plate reader.
   b. Estimate angiogenic response from hemoglobin content (Passaniti et al., 1992). Remove plugs, mince, and briefly microfuge. Analyze the supernatant colorimetrically for its hemoglobin concentration using the Drabkin method (Sigma).
   c. Process plugs for immunohistochemistry to count vessel density. Embed plugs in freezing medium (e.g., OCT) and cryosection to obtain several sections through the thickness of each plug. Perform immunostaining for endothelial-specific markers (e.g., CD31, VE-cadherin, or VEGFR2). Count the number of vessels per unit area plug.
   d. Perform whole-mount immunofluorescence staining and examine using confocal microscopy. Fix plugs in methanol-acetone (1:1) for 30 min, wash, block in 3% normal serum for 1 h, incubate in primary

antibodies overnight at 4°, wash, label with fluorescently-tagged secondary antibodies, wash, and post-fix in 4% paraformaldehyde for 10 min. Confocal microscopy can be used to visualize the 3D vascular network and expression of any proteins or phosphoproteins of interest.

An adaptation of the Matrigel plug assay has been developed by Trevigen, Inc. The Directed In Vivo Angiogenesis Assay (DIVAA) kit utilizes a small proprietary angioreactor that contains only 20 μl of basement membrane gel within a small cylindrical tube. This angioreactor is implanted under the skin and induces vascular ingrowth similar to the Matrigel plug model. This system has the advantage that the volume is well-controlled and the vascular content within the angioreactor can easily be quantified upon removal using a smaller amount of FITC-lectin that is added to the explant rather than injected intravenously to the mouse. The expense of this kit is roughly equivalent to the cost of Matrigel and FITC-lectin required per plug using the conventional Matrigel Plug assay. Other similar assays involve the implantation of a sponge or nylon mesh containing growth factors and/or test agents. Together, these *in vivo* models allow testing of several conditions, doses, or treatments using four to six mice per group.

2.4. Assessing angiogenesis using intravital microscopy

The dorsal window chamber model of angiogenesis is technically more difficult, but allows for time-lapse intravital imaging to assess the dynamic angiogenic response. This model is described in detail in Chapter 16 (makale, this volume). Another model involves monitoring the angiogenic response on the rat (or mouse) mesentery developed by Norrby *et al.* (1986, 1990). The advantages of this model include the ability to induce a robust angiogenic response in a relatively sparsely vascularized adult tissue, and the ability to clearly visualize all stages of angiogenesis within a relatively planar environment. Details of this method were recently reviewed (Norrby, 2006).

2.5. Ocular angiogenesis models

A number of ocular angiogenesis models have been used with great success, and have been used to elucidate integrin function (Friedlander, 2007) or assess integrin antagonists such as endostatin or angiostatin (Campochiaro and Hackett, 2003). The corneal micropocket assay was originally developed using rabbits, and has since been routinely used in mice allowing for smaller volumes of systemically-delivered test agents and the use of gene-targeted models (Kenyon *et al.*, 1996). Briefly, a small (0.4 × 0.4 mm) slow-release pellet containing a sufulcrate/Hydron gel and angiogenic test agents is implanted into a small slit within the cornea. The angiogenic response within the normally avascular cornea can be easily visualized noninvasively using slit-lamp microscopy. In addition, the cornea can be harvested at the termination of the study for mRNA and protein analysis. The benefit of this model is that any growth factor and/or test agent can be incorporated into the pellet for slow release during the experiment.

Unspecific wounding can be considered a more relevant pathological model to investigate the complexity of factors involved in ocular angiogenesis compared to the application of a single factor in the corneal micropocket assay. Wounding of the cornea produces angiogenesis that is highly dependent on VEGF/$\alpha v \beta 5$ pathways, but does not require bFGF/$\alpha v \beta 3$ signaling (da Costa Pinto and Malucelli, 2002). The corneal wound angiogenesis assay is technically less difficult to induce compared with the micropocket assay, only requiring the direct application of a silver nitrate cauterization stick (Schonherr et al., 2004), a filter containing alkali solution (Zhang et al., 2002), or a single loop of silk suture (Toyofuku et al., 2007) to the center of the cornea. These wounds induce a reproducible inflammatory and angiogenic response that is easily monitored noninvasively over time, as for the corneal micropocket model. The initial angiogenic response is visible within 24 to 48 h after the injury as the limbal vessels begin to extend into the cornea, with vascular sprouts directed toward the site of injury in the center of the cornea. After 3 to 4 days, the vessels elongate evenly from all sides. After 6 to 7 days, regression and remodeling of these new blood vessels occurs as redundant vessels are pruned. Thus, this experimental model can be used to assess integrin function in gene-targeted mice or using integrin antagonists during either the initial angiogenic phase or the subsequent vascular regression.

Patients with macular degeneration suffer from choriodal neovascularization associated with increased $\alpha v \beta 3$ expression and a dependence on bFGF (but not VEGF) signaling (Friedlander, 2007). In mice, laser-induced coagulation models this aspect of macular degeneration. The pupils are dilated and a laser is used to disrupt Bruch's membrane at several locations, inducing an angiogenic response within the choroidal layer (Campochiaro and Hackett, 2003). To assess choroidal neovascularization, fluorescein angiograms can be performed noninvasively at several time points following laser treatment. At the end of the experiment, the eyes can be harvested to quantify the angiogenic response using immunohistochemistry or to process the tissue for mRNA/protein analysis.

Oxygen-induced retinopathy (OIR) is a model for human retinopathy of prematurity (ROP). OIR is induced by exposing neonatal mice to high oxygen (75%) from postnatal days 7 through 12, which halts vascular development and induces vascular regression in the retina (Ritter et al., 2005). When animals are returned to room air, the retina is hypoxic and thus vascular proliferation is induced. Therefore, this model of OIR first induces extensive vaso-obliteration followed by extensive neovascularization. Ocular tissues taken from diabetic patients who suffer from ROP express both $\alpha v \beta 3$ and

$\alpha v\beta 5$ integrins, and the angiogenic response in this model depends on the activities of both VEGF and bFGF (Friedlander, 2007).

2.6. Evaluating angiogenesis in the zebrafish

Development of small molecule inhibitors requires innovative biological screens for selection of lead compounds. The zebrafish model is becoming widely used as a screening tool for drug discovery for several reasons. The statistical power is much greater due to the high numbers of embryos which can be generated with each mating. Assays are rapid (finished within one week) and toxicity data are accumulated quickly due to the sensitivity of the developing embryo. Importantly, screening compounds can be done *in vivo*, thereby eliminating all of the problems associated with typical enzymatic screening approaches.

1. Intersegmental vessel formation. The metamerically repeating system of angiogenic trunk intersegmental vessels represents an easily imaged model of angiogenesis. Each intersegmental vessel is initially formed from three endothelial cells that emerge from the dorsal aorta and migrate as a chain along myotomal boundaries. This process initiates at 20-h postfertilization (20 hpf), and is completed by 36 hpf. The *fli1*-EGFP transgenic zebrafish (in which GFP is expressed in endothelial cells) can be utilized to study angiogenesis in real time with confocal or multiphoton microscopy to visualize the formation of GFP-labeled vasculature. Since the formation of intersegmental vessels will be completed by 36 hpf in the normal zebrafish, 48 hpf is the main endpoint of the assay by which time the entire angiogenic process of intersegmental vessel formation is complete.

2. Delivery of therapeutic agents. Integrin antagonists can be assessed for their ability to disrupt angiogenesis (assessed as the formation of intersegmental vessels) in the developing *fli1*-EGFP embryo. Drugs or vehicle are simply added to the water at 20 hpf.

3. Imaging live embryos. Time-lapse multiphoton confocal microscopy can be performed to evaluate real-time intersegmental vessel formation between 20 and 48 hpf. The zebrafish is anesthetized with a standard solution of 0.02% tricaine for 2 to 4 min, until the gills stop moving. When fish are calm and insensitive to touch, the fish body (but not the gills) is placed in agarose gel. It takes approximately 3 to 10 min for agarose gel to solidify. During imaging, fish are kept in a small, humidified, sealed chamber under a thin layer of fish system water.

4. Endpoint for angiogenesis. For routine drug screening, fluorescence microscopy can be used to determine the effect of each inhibitor on the extent of intersegmental vessel formation in the *fli1*-EGFP fish at the endpoint of the experiment (48 hpf). In addition, some embryos can be

fixed and whole-mount immunohistochemistry using antibodies for specific signaling proteins will be used to verify the effects of the inhibitors. Alternatively, 10 to 15 embryos can be homogenized and the protein lysate used for Western blotting to detect various signaling cascades.

Alternatively, angiogenesis can be assessed in the adult zebrafish during tail fin regeneration after amputation. Specifically, the *fli1*-EGFP zebrafish can be used to visualize the regeneration of GFP-labeled vasculature during regeneration of the caudal (tail) fin after amputation. The ability of inhibitors to disrupt the formation of blood vessels during tail fin regeneration in the adult *fli1*-EGFP fish can be assayed in a moderate throughput system. Generally, test agents can be added for 3 days after tail amputation and imaged between day 1 and day 7 to assess the antiangiogenic response over time. Tail fin regeneration can be examined using a fluorescent dissection scope and a color camera. In addition, images of endothelial tip cells can be acquired with a multiphoton microscopy system.

3. Assessing Role of Integrins in Vascular Permeability

3.1. *In vitro* models to assess vascular permeability

The process of vascular permeability has extreme consequences *in vivo*, since leak of blood and plasma proteins can quickly damage tissues at sites of leak. *In vitro*, permeability can be assessed by examining the integrity of an endothelial cell monolayer in the presence or absence of permeability factors and/or test agents. Commonly used *in vitro* models include the "transwell assay," in which an endothelial cell monolayer is grown to confluency on a Boyden chamber (or transwell). Tracers of various sizes are added to the media in the upper chamber, and then permeability of the monolayer is assessed as movement of tracer from the upper to lower chambers over time (Esser *et al.*, 1998).

In another *in vitro* model, the impedance of a cell-covered electrode can be partitioned into a cell–matrix resistance, a cell–cell resistance, and a membrane capacitance (Winter *et al.*, 1999). When a permeability agent is added to the cell layer, the impedance is continuously measured to assess changes in cell–cell adhesion. This cell-based system offers continuous monitoring of impedance of sensor electrodes and facilitates screening of agents that induce or block the permeability response. Similar results are obtained by monitoring electrical resistance of cells plated on a Boyden chamber using an Ag/AgCl-electrode (Gotsch *et al.*, 1997).

The physical appearance of individual cells within an endothelial cell monolayer during a permeability response can be assessed by immunostaining. Work by Dejana and colleagues have used these methods to assess the role of vascular integrins during vascular leak and maintenance of the endothelial cell barrier. For example, $\alpha2\beta1$ and $\alpha5\beta1$ are located at cell–cell contacts in endothelial cells. Integrin αv, but not the $\beta3$ chain, is also located to cell junctions. Accordingly, antibodies to $\alpha5\beta1$ (but not $\beta3$) or cyclic RGD peptides can disrupt cell–cell junctions while maintaining initial cell–matrix adhesion (Lampugnani et al., 1991).

Haselton and colleagues developed an in vitro method for measuring the diffusional permeability of an endothelial monolayer cultured on porous microcarrier beads. To determine permeability, tracers of different molecular weights are applied to a chromatography column containing cell-covered beads, and the permeability of the endothelial monolayer is computed from the tracer elution profiles. This model has been used to characterize the permeability of endothelial monolayers and to investigate permeability changes produced by various agents (Haselton et al., 1989). A similar model was developed by Killackey and coworkers (1986), in which endothelial cells are cultured to confluence on denatured collagen-coated dextran microcarriers or gelatin microcarriers. Confluent cells prevent staining of the microcarriers with Evans blue dye, while increases in staining as determined by the spectrophotometric quantitation of the dye represent monolayer permeability after treatment with permeability agents (Killackey et al., 1986).

3.2. The "Miles assay" to assess leak in the skin

Vascular leak in the skin can be evaluated using the Miles assay, developed in 1952 (Miles and Miles, 1952). The vascular permeability properties of permeability agents (such as VEGF) can be determined by leakage of Evan's blue dye (a common marker for vascular leakage) after local intradermal injection. This response can be assessed in normal mice, mice with gene-targeted deletions, or mice treated with potential inhibitors of the vascular leak response (Eliceiri et al., 1999). Importantly, this method also allows measurement of the time course of vascular leak in vivo.

1. *Inducing leak.* Evan's blue dye (1 mg/ml in 100 μl sterile saline) is injected intravenously through the tail vein of a conscious mouse restrained briefly in a rodent restrainer. The torso of the mouse is shaved, and then saline (as a control) or VEGF (400 ng recombinant VEGF in 10 μl sterile saline) is injected subcutaneously in the torso skin on opposite sides of the mouse.
2. *Evaluating leak.* Skin patches can be photographed using a digital camera to evaluate the time course of vascular permeability by monitoring the leakage of Evan's blue dye at each location over time up to 1 h. At the end of the experiment, the regions surrounding the injection sites are dissected using a dermatological skin punch (8-mm diameter) and the

extent of permeability quantified by elution of the Evan's blue dye in formamide and measuring absorbance at 600 nm using a spectrophotometer. This model has been adapted to assess vascular leak in the lung or other organs (Liao *et al.*, 2002).

3.3. Evaluation of biochemical signals during the vascular leak response

Signaling events mediated by angiogenic cytokines *in vivo* differ from studies performed *in vitro* in many respects. To address this, we have developed an *in vivo* signaling assay to assess the regulation of downstream mediators of vascular leak within the intact mouse.

1. Inducing a systemic permeability response *in vivo*. Adult mice are injected via the tail vein with a permeability agent (e.g., 2 to 5 μg recombinant VEGF) diluted in 100 μl sterile saline. This solution can also contain test agents to assess their ability to block vascular leak. After 2, 5, 10, 30, or 60 min, the heart, lung, brain, and liver are rapidly excised, minced on a glass plate using razor blades, and homogenized in 3 ml RIPA buffer on ice. After 10 min of refrigerated centrifugation, the supernatant is used for biochemical analysis or aliquots can be frozen at $-80°$ and stored long term. This method typically yields several milligrams of protein per organ, which can be used for many immunoprecipitation and immunoblotting experiments.
2. Biochemical analysis. To examine the biochemical signaling events induced by growth factor administration, immunoprecipitation is performed to pull-down endothelial specific complexes (e.g., containing VE–cadherin or VEGFR2). Subsequent Western blotting can detect phosphorylation events or co-association between proteins of interest. Using this technique we have shown that VEGF stimulation induces the rapid and transient dissociation between VEGFR2 and VE–cadherin, an event that requires Src kinase activity (Weis *et al.*, 2004b). We have also shown that VEGF induces a transient dissociation between VE–cadherin and β-catenin that also requires Src kinase activity *in vivo* (Weis *et al.*, 2004a). Phosphoprotein analysis by 1D- or 2D-PAGE and Western blotting provides a time course of biochemical signals relevant to the rapid onset of vascular leak *in vivo*.

3.4. Evaluation of structural and ultrastructural changes during the vascular leak response

McDonald and coworkers have utilized the binding properties of various lectins to investigate the permeability response *in vivo*. Lectin binding is localized by peroxidase histochemistry and viewed in tissue whole mounts by differential-interference contrast microscopy (Thurston *et al.*, 1996). They identified lectins that bound uniformly to the endothelium of normal and inflamed venules and other lectins that bound weakly or not at all to venules. A third group of lectins bound selectively to focal patches of inflamed venules, but bound weakly to normal venules. Finally, a fourth group bound preferentially to focal patches in inflamed venules and also bound uniformly to normal venules. Sites of plasma leakage can also be marked by extravasation of a particulate tracer (monastral blue). In combination, the patterns of lectin binding and tracer extravasation can reveal sites of vascular leak *in vivo*.

Dvorak and colleagues have performed a long series of elegant studies to examine vascular leak at the ultrastructural level (Feng *et al.*, 1996). Using electron microscopy, they have established several pathways by which molecules can be transported through individual endothelial cells. In particular, their work has revealed a network of vesiculo-vacuolar organelles, transendothelial cell pores, and fenestrations which together can allow vascular permeability through an intracellular route (Feng *et al.*, 1999).

The consequences of vascular leak *in vivo* depend on the tissue microenvironment. In the heart or brain, leak from capillaries quickly causes damage to the surrounding tissue. Following induction of leak (e.g., intravenous VEGF injection), the ultrastructural changes can be monitored using transmission electron microscopy. In the hearts of mice just 2 min following systemic VEGF injection, changes in the capillary endothelial cells are visible. By 10 min, gaps between adjacent endothelial cells are visible, sometimes plugged by platelets. By 30 to 60 min following a single VEGF injection, the endothelial cells return to a normal phenotype and vascular integrity has been restored. However, multiple systemic treatments with VEGF or experimentally induced myocardial infarction create long-lasting effects including myocyte damage and regions of interstitial edema (Weis *et al.*, 2004b).

4. Concluding Remarks

4.1. Studying angiogenesis and vascular leak in models of cancer

Methods for assessing the requirement of specific integrins in angiogenesis and vascular leak have been discussed above. Extension of these concepts to the situation during cancer requires careful consideration. Genetic manipulation of integrin function in carcinoma cells is easily achieved *in vitro*, and can certainly influence the vascular response to the tumor *in vivo*. Hynes and colleagues have tested the requirement for $\beta3$ and $\beta5$ integrins during tumor-associated angiogenesis mice lacking these integrin subunits

(Taverna *et al.*, 2004, 2005). However, angiogenesis and vascular leak in tumor-bearing mice treated with cyclic RGD peptides or chick CAMs treated with integrin function–blocking antibodies involves integrin disruption on the host vasculature and stromal tissue as well as the tumor cells. Likewise, $\alpha v \beta 3$-targeted drug or gene delivery methods may demonstrate more benefit in models with highly $\beta 3$-positive carcinoma cells.

Like other cancers, $\alpha v \beta 3$ is overexpressed on angiogenic blood vessels in glioblastoma multiform (GBM). Kanamori and colleagues have shown that tumor-specific $\alpha v \beta 3$ overexpression has growth-suppressive effects in gliomas (Kanamori *et al.*, 2004). To directly test the role of host versus tumor cell $\beta 3$ expression, Kanamori and colleagues implanted $\beta 3$-expressing GBM cells into wild-type and $\beta 3$-null mice. As reported previously by Hynes and colleagues for other cancer models (Taverna *et al.*, 2004), $\beta 3$-null mice showed enhanced angiogenesis in the GBM model (Kanamori *et al.*, 2006). However, despite enhanced angiogenesis in the $\beta 3$-null mice, GBM tumor size was similar between genotypes. Furthermore, the tumors in $\beta 3$-knockout mice were infiltrated with more apoptosis-inducing macrophages than the tumors in wild-type mice. These results demonstrate that host $\alpha v \beta 3$ expression has opposing actions, enhancing tumor vascularization and growth while independently enhancing macrophage-mediated tumor cell apoptosis (Kanamori *et al.*, 2006).

4.2. Acknowledging and understanding diversity among subjects

The D'Amato laboratory has published a series of studies investigating the effect of genetic diversity on angiogenesis (recently reviewed by Rogers and D'Amato, 2006). Using the corneal micropocket angiogenesis assay, D'Amato and colleagues demonstrated a $\sim$10-fold range of angiogenic response to bFGF and VEGF depending on strain that correlated well with the *in vitro* migratory activity of endothelial cells isolated from these mice. Furthermore, a differential sensitivity to angiogenesis inhibitors was reported between strains (Rohan *et al.*, 2000). In fact, the genetic loci responsible for sensitivity to bFGF (Rogers *et al.*, 2004) and VEGF (Rogers *et al.*, 2003) were identified. In collaboration with the Kerbel laboratory, D'Amato's group has also demonstrated a striking correlation between bFGF- or VEGF-induced angiogenesis and intrinsic circulating endothelial cell or endothelial progenitor cell levels between various strains of mice (Shaked *et al.*, 2005). These findings suggest that integrin-mediated vascular processes might vary considerably among mouse background strains. Moreover, the experimental procedures used to test integrin function may be more or less effective on a particular mouse background. Finally, careful analysis of integrin-targeted mouse models on several background strains may be necessary to fully understand integrin function.

Gender can also influence experimental findings. We recently reported that male (but not female) β3-null mice show an immature capillary phenotype, which could be induced to a similar extent in male (but not female) mice by several intravenous treatments with VEGF (Weis *et al.*, 2007). This finding suggests that growth factor sensitivity is influenced both by integrin expression and gender-specific hormones. In fact, steroid hormones can regulate the expression of many genes, including integrins (Cid *et al.*, 1999). While our finding in the β3-null mice represents the first account of a gender-specific phenotype in an integrin-null mouse, previous studies have generally not specified which gender was studied and/or did not include mice of each gender in the results. The relationship between integrins, hormones, and growth factors likely plays a role in the results obtained in mouse models of angiogenesis and vascular permeability, and thus should be acknowledged in the design of experiments and reported more reliably in publication.

4.3. Necessity for better tools to study integrin function

What tools would enable us to better understand the role of integrins in vascular biology? Despite the preponderance of mouse models, the current arsenal of integrin reagents is lacking. Antibodies to block integrin function in mice are just recently being developed. Sheppard and colleagues have recently described a function-blocking antibody against $\alpha v \beta 5$, which is active in mice (Su *et al.*, 2007). Reliable and specific antibodies for protein analysis and immunostaining of individual integrin subunits as well as specific $\alpha \beta$ pairs would enhance our ability to detect changes in integrin expression in mice under various experimental conditions. Finally, inducible and conditional knockin/knockout mouse models of specific integrins would allow study of selective integrin null expression in particular tissues in the adult, and could possibly minimize the compensatory mechanisms that arise during development.

REFERENCES

Abumiya, T., Lucero, J., Heo, J. H., Tagaya, M., Koziol, J. A., Copeland, B. R., and del Zoppo, G. J. (1999). Activated microvessels express vascular endothelial growth factor and integrin alpha(v)beta3 during focal cerebral ischemia. *J. Cereb. Blood Flow Metab.* **19,** 1038–1050.

Aitkenhead, M., Wang, S. J., Nakatsu, M. N., Mestas, J., Heard, C., and Hughes, C. C. (2002). Identification of endothelial cell genes expressed in an *in vitro* model of angiogenesis: Induction of ESM-1, (beta)ig-h3, and NrCAM. *Microvasc. Res.* **63,** 159–171.

Bayless, K. J., Salazar, R., and Davis, G. E. (2000). RGD-dependent vacuolation and lumen formation observed during endothelial cell morphogenesis in three-dimensional fibrin matrices involves the $\alpha v \beta 3$ and $\alpha 5 \beta 1$ integrins. *Am. J. Pathol.* **156,** 1673–1683.

Bertotti, A., Comoglio, P. M., and Trusolino, L. (2005). Beta4 integrin is a transforming molecule that unleashes Met tyrosine kinase tumorigenesis. *Cancer Res.* **65,** 10674–10679.

Brooks, P. C., Clark, R. A., and Cheresh, D. A. (1994a). Requirement of vascular integrin alpha v beta 3 for angiogenesis. *Science* **264,** 569–571.

Brooks, P. C., Montgomery, A. M., Rosenfeld, M., Reisfeld, R. A., Hu, T., Klier, G., and Cheresh, D. A. (1994b). Integrin alpha v beta 3 antagonists promote tumor regression by inducing apoptosis of angiogenic blood vessels. *Cell* **79,** 1157–1164.

Bussolino, F., Valdembri, D., Caccavari, F., and Serini, G. (2006). Semaphoring vascular morphogenesis. *Endothelium* **13,** 81–91.

Cai, W., Shin, D. W., Chen, K., Gheysens, O., Cao, Q., Wang, S. X., Gambhir, S. S., and Chen, X. (2006). Peptide-labeled near-infrared quantum dots for imaging tumor vasculature in living subjects. *Nano Lett.* **6,** 669–676.

Campochiaro, P. A., and Hackett, S. F. (2003). Ocular neovascularization: A valuable model system. *Oncogene* **22,** 6537–6548.

Castellani, V. (2002). The function of neuropilin/L1 complex. *Adv. Exp. Med. Biol.* **515,** 91–102.

Cid, M. C., Esparza, J., Schnaper, H. W., Juan, M., Yague, J., Grant, D. S., Urbano-Márquez, A., Hoffman, G. S., and Kleinman, H. K. (1999). Estradiol enhances endothelial cell interactions with extracellular matrix proteins via an increase in integrin expression and function. *Angiogenesis* **3,** 271–280.

Curtis, T. M., McKeown-Longo, P. J., Vincent, P. A., Homan, S. M., Wheatley, E. M., and Saba, T. M. (1995). Fibronectin attenuates increased endothelial monolayer permeability after RGD peptide, anti-alpha 5 beta 1, or TNF-alpha exposure. *Am. J. Physiol.* **269,** L248–L260.

da Costa Pinto, F. A., and Malucelli, B. E. (2002). Inflammatory infiltrate, VEGF and FGF–2 contents during corneal angiogenesis in STZ-diabetic rats. *Angiogenesis* **5,** 67–74.

De, S., Razorenova, O., McCabe, N. P., O'Toole, T., Qin, J., and Byzova, T. V. (2005). VEGF-integrin interplay controls tumor growth and vascularization. *Proc. Natl. Acad. Sci. USA* **102,** 7589–7594.

Eliceiri, B. P., Paul, R., Schwartzberg, P. L., Hood, J. D., Leng, J., and Cheresh, D. A. (1999). Selective requirement for Src kinases during VEGF-induced angiogenesis and vascular permeability. *Mol. Cell* **4,** 915–924.

Eliceiri, B. P., Puente, X. S., Hood, J. D., Stupack, D. G., Schlaepfer, D. D., Huang, X. Z., Sheppard, D., and Cheresh, D. A. (2002). Src-mediated coupling of focal adhesion kinase to integrin alpha(v)beta5 in vascular endothelial growth factor signaling. *J. Cell Biol.* **157,** 149–160.

Esser, S., Lampugnani, M. G., Corada, M., Dejana, E., and Risau, W. (1998). Vascular endothelial growth factor induces VE-cadherin tyrosine phosphorylation in endothelial cells. *J. Cell Sci.* **111,** 1853–1865.

Feng, D., Nagy, J. A., Hipp, J., Dvorak, H. F., and Dvorak, A. M. (1996). Vesiculo-vacuolar organelles and the regulation of venule permeability to macromolecules by vascular permeability factor, histamine, and serotonin. *J. Exp. Med.* **183,** 1981–1986.

Feng, D., Nagy, J. A., Pyne, K., Hammel, I., Dvorak, H. F., and Dvorak, A. M. (1999). Pathways of macromolecular extravasation across microvascular endothelium in response to VPF/VEGF and other vasoactive mediators. *Microcirculation* **6,** 23–44.

Francis, S. E. (2006). Integrins and vascular development in differentiated embryonic stem cells *in vitro. Methods Mol. Biol.* **330,** 331–340.

Francis, S. E., Goh, K. L., Hodivala-Dilke, K., Bader, B. L., Stark, M., Davidson, D., and Hynes, R. O. (2002). Central roles of alpha5beta1 integrin and fibronectin in vascular development in mouse embryos and embryoid bodies. *Arterioscler. Thromb. Vasc. Biol.* **22,** 927–933.

Friedlander, M. (2007). Fibrosis and diseases of the eye. *J. Clin. Invest.* **117,** 576–586.

Friedlander, M., Brooks, P. C., Shaffer, R. W., Kincaid, C. M., Varner, J. A., and Cheresh, D. A. (1995). Definition of two angiogenic pathways by distinct alpha v integrins. *Science* **270,** 1500–1502.

Garmy-Susini, B., Jin, H., Zhu, Y., Sung, R. J., Hwang, R., and Varner, J. (2005). Integrin alpha4beta1-VCAM-1-mediated adhesion between endothelial and mural cells is required for blood vessel maturation. *J. Clin. Invest.* **115,** 1542–1551.

Gotsch, U., Borges, E., Bosse, R., Boggemeyer, E., Simon, M., Mossmann, H., and Vestweber, D. (1997). VE-cadherin antibody accelerates neutrophil recruitment *in vivo*. *J. Cell Sci.* **110,** 583–588.

Hall, H., and Hubbell, J. A. (2004). Matrix-bound sixth Ig-like domain of cell adhesion molecule L1 acts as an angiogenic factor by ligating alphavbeta3-integrin and activating VEGF-R2. *Microvasc. Res.* **68,** 169–178.

Hamano, Y., Zeisberg, M., Sugimoto, H., Lively, J. C., Maeshima, Y., Yang, C., Hynes, R. O., Werb, Z., Sudhakar, A., and Kalluri, R. (2003). Physiological levels of tumstatin, a fragment of collagen IV alpha3 chain, are generated by MMP–9 proteolysis and suppress angiogenesis via alphaV beta3 integrin. *Cancer Cell* **3,** 589–601.

Haselton, F. R., Mueller, S. N., Howell, R. E., Levine, E. M., and Fishman, A. P. (1989). Chromatographic demonstration of reversible changes in endothelial permeability. *J. Appl. Physiol.* **67,** 2032–2048.

Hood, J. D., Bednarski, M., Frausto, R., Guccione, S., Reisfeld, R. A., Xiang, R., and Cheresh, D. A. (2002). Tumor regression by targeted gene delivery to the neovasculature. *Science* **296,** 2404–2407.

Jin, H., Aiyer, A., Su, J., Borgstrom, P., Stupack, D., Friedlander, M., and Varner, J. (2006a). A homing mechanism for bone marrow-derived progenitor cell recruitment to the neovasculature. *J. Clin. Invest.* **116,** 652–662.

Jin, H., Su, J., Garmy-Susini, B., Kleeman, J., and Varner, J. (2006b). Integrin alpha4beta1 romotes monocyte trafficking and angiogenesis in tumors. *Cancer Res.* **66,** 2146–2152.

Kanamori, M., Kawaguchi, T., Berger, M. S., and Pieper, R. O. (2006). Intracranial microenvironment reveals independent opposing functions of host $\alpha V\beta 3$ expression on glioma growth and angiogenesis. *J. Biol. Chem.* **281,** 37256–37264.

Kanamori, M., Vanden Berg, S. R., Bergers, G., Berger, M. S., and Pieper, R. O. (2004). Integrin beta3 overexpression suppresses tumor growth in a human model of gliomagenesis: Implications for the role of beta3 overexpression in glioblastoma multiforme. *Cancer Res.* **64,** 2751–2758.

Kearney, J. B., and Bautch, V. L. (2003). *In vitro* differentiation of mouse ES cells: Hematopoietic and vascular development. *Methods Enzymol.* **365,** 83–98.

Kearney, J. B., Kappas, N. C., Ellerstrom, C., DiPaola, F. W., and Bautch, V. L. (2004). The VEGF receptor flt-1 (VEGFR-1) is a positive modulator of vascular sprout formation and branching morphogenesis. *Blood* **103,** 4527–4535.

Kenyon, B. M., Voest, E. E., Chen, C. C., Flynn, E., Folkman, J., and D'Amato, R. J. (1996). A model of angiogenesis in the mouse cornea. *Invest. Ophthalmol. Vis. Sci.* **37,** 1625–1632.

Killackey, J. J., Johnston, M. G., and Movat, H. Z. (1986). Increased permeability of microcarrier-cultured endothelial monolayers in response to histamine and thrombin. A model for the *in vitro* study of increased vasopermeability. *Am. J. Pathol.* **122,** 50–61.

Kim, S., Bell, K., Mousa, S. A., and Varner, J. A. (2000). Regulation of angiogenesis *in vivo* by ligation of integrin alpha5beta1 with the central cell-binding domain of fibronectin. *Am. J. Pathol.* **156,** 1345–1362.

Lampugnani, M. G., Resnati, M., Dejana, E., and Marchisio, P. C. (1991). The role of integrins in the maintenance of endothelial monolayer integrity. *J. Cell Biol.* **112,** 479–490.

Lee, T. H., Seng, S., Li, H., Kennel, S. J., Avraham, H. K., and Avraham, S. (2006). Integrin regulation by vascular endothelial growth factor in human brain microvascular endothelial cells: role of alpha6beta1 integrin in angiogenesis. *J. Biol. Chem.* **281,** 40450–40460.

Liao, F., Doody, J. F., Overholser, J., Finnerty, B., Bassi, R., Wu, Y., Dejana, E., Kussie, P., Bohlen, P., and Hicklin, D. J. (2002). Selective targeting of angiogenic tumor vasculature by vascular endothelial-cadherin antibody inhibits tumor growth without affecting vascular permeability. *Cancer Res.* **62,** 2567–2575.

Maeshima, Y., Colorado, P. C., and Kalluri, R. (2000). Two RGD-independent alpha vbeta 3 integrin binding sites on tumstatin regulate distinct anti-tumor properties. *J. Biol. Chem.* **275,** 23745–23750.

Mahabeleshwar, G. H., Feng, W., Phillips, D. R., and Byzova, T. V. (2006). Integrin signaling is critical for pathological angiogenesis. *J. Exp. Med.* **203,** 2495–2507.

Miles, A. A., and Miles, E. M. (1952). Vascular reactions to histamine, histamine-liberator and leukotaxine in the skin of guinea pigs. *J. Physiol.* **118,** 228–257.

Montgomery, A. M., Becker, J. C., Siu, C. H., Lemmon, V. P., Cheresh, D. A., Pancook, J. D., Zhao, X., and Reisfeld, R. A. (1996). Human neural cell adhesion molecule L1 and rat homologue NILE are ligands for integrin alpha v beta 3. *J. Cell Biol.* **132,** 475–485.

Nakatsu, M. N., Sainson, R. C., Aoto, J. N., Taylor, K. L., Aitkenhead, M., Perez-del-Pulgar, S., Carpenter, P. M., and Hughes, C. C. (2003a). Angiogenic sprouting and capillary lumen formation modeled by human umbilical vein endothelial cells (HUVEC) in fibrin gels: The role of fibroblasts and Angiopoietin–1. *Microvasc. Res.* **66,** 102–112.

Nakatsu, M. N., Sainson, R. C., Perez-del-Pulgar, S., Aoto, J. N., Aitkenhead, M., Taylor, K. L., Carpenter, P. M., and Hughes, C. C. (2003b). VEGF(121) and VEGF (165) regulate blood vessel diameter through vascular endothelial growth factor receptor 2 in an *in vitro* angiogenesis model. *Lab Invest.* **83,** 1873–1885.

Nam, J. O., Jeong, H. W., Lee, B. H., Park, R. W., and Kim, I. S. (2005). Regulation of tumor angiogenesis by fastatin, the fourth FAS1 domain of betaig-h3, via alphavbeta3 integrin. *Cancer Res.* **65,** 4153–4161.

Norrby, K. (2006). *In vivo* models of angiogenesis. *J. Cell. Mol. Med.* **10,** 588–612.

Norrby, K., Jakobsson, A., and Sorbo, J. (1986). Mast-cell-mediated angiogenesis: A novel experimental model using the rat mesentery. *Virchows Arch. B Cell Pathol. Mol. Pathol.* **52,** 195–206.

Norrby, K., Jakobsson, A., and Sorbo, J. (1990). Quantitative angiogenesis in spreads of intact rat mesenteric windows. *Microvasc. Res.* **39,** 341–348.

Okada, Y., Copeland, B. R., Hamann, G. F., Koziol, J. A., Cheresh, D. A., and del Zoppo, G. J. (1996). Integrin alphavbeta3 is expressed in selected microvessels after focal cerebral ischemia. *Am. J. Pathol.* **149,** 37–44.

Passaniti, A., Taylor, R. M., Pili, R., Guo, Y., Long, P. V., Haney, J. A., Pauly, R. R., Grant, D. S., and Martin, G. R. (1992). A simple, quantitative method for assessing angiogenesis and antiangiogenic agents using reconstituted basement membrane, heparin, and fibroblast growth factor. *Lab. Invest.* **67,** 519–528.

Ren, J., Avery, J., Zhao, H., Schneider, J. G., Ross, F. P., and Muslin, A. J. (2007). β3 Integrin deficiency promotes cardiac hypertrophy and inflammation. *J. Mol. Cell. Cardiol.* **42,** 367–377.

Reynolds, A. R., Reynolds, L. E., Nagel, T. E., Lively, J. C., Robinson, S. D., Hicklin, D. J., Bodary, S. C., and Hodivala-Dilke, K. M. (2004). Elevated Flk1 (Vascular Endothelial Growth Factor Receptor 2). Signaling Mediates Enhanced Angiogenesis in β3-Integrin-Deficient Mice. *Cancer Res.* **64,** 8643–8650.

Reynolds, L. E., Wyder, L., Lively, J. C., Taverna, D., Robinson, S. D., Huang, X., Sheppard, D., Hynes, R. O., and Hodivala-Dilke, K. M. (2002). Enhanced pathological angiogenesis in mice lacking beta3 integrin or beta3 and beta5 integrins. *Nat. Med.* **8,** 27–34.

Ribatti, D. (2004). The first evidence of the tumor-induced angiogenesis *in vivo* by using the chorioallantoic membrane assay dated 1913. *Leukemia* **18,** 1350–1351.

Ritter, M. R., Aguilar, E., Banin, E., Scheppke, L., Uusitalo-Jarvinen, H., and Friedlander, M. (2005). Three-dimensional *in vivo* imaging of the mouse intraocular vasculature during development and disease. *Invest. Ophthalmol. Vis. Sci.* **46**, 3021–3026.

Robinson, S. D., Reynolds, L. E., Wyder, L., Hicklin, D. J., and Hodivala-Dilke, K. M. (2004). {beta}3-Integrin Regulates Vascular Endothelial Growth Factor-A-Dependent Permeability. *Arterioscler. Thromb. Vasc. Biol.* **24**, 2108–2114.

Rogers, M. S., and D'Amato, R. J. (2006). The effect of genetic diversity on angiogenesis. *Exp. Cell Res.* **312**, 561.

Rogers, M. S., Rohan, R. M., Birsner, A. E., and D'Amato, R. J. (2003). Genetic loci that control vascular endothelial growth factor-induced angiogenesis. *FASEB J.* **17**, 2112–2114.

Rogers, M. S., Rohan, R. M., Birsner, A. E., and D'Amato, R. J. (2004). Genetic loci that control the angiogenic response to basic fibroblast growth factor. *FASEB J.* **18**, 1050–1059.

Rohan, R. M., Fernandez, A., Udagawa, T., Yuan, J., and D'Amato, R. J. (2000). Genetic heterogeneity of angiogenesis in mice. *FASEB J.* **14**, 871–876.

Ruoslahti, E. (2003). The RGD story: A personal account. *Matrix Biol.* **22**, 459–465.

Ruoslahti, E., and Pierschbacher, M. D. (1986). Arg-Gly-Asp: A versatile cell recognition signal. *Cell* **44**, 517–518.

Ruoslahti, E., and Pierschbacher, M. D. (1987). New perspectives in cell adhesion: RGD and integrins. *Science* **238**, 491–497.

Schonherr, E., Sunderkotter, C., Schaefer, L., Thanos, S., Grassel, S., Oldberg, A., Iozzo, R. V., Young, M. F., and Kresse, H. (2004). Decorin deficiency leads to impaired angiogenesis in injured mouse cornea. *J. Vasc. Res.* **41**, 499–508.

Senger, D. R., Perruzzi, C. A., Streit, M., Koteliansky, V. E., de Fougerolles, A. R., and Detmar, M. (2002). The alpha(1)beta(1) and alpha(2)beta(1) integrins provide critical support for vascular endothelial growth factor signaling, endothelial cell migration, and tumor angiogenesis. *Am. J. Pathol.* **160**, 195–204.

Shaked, Y., Bertolini, F., Man, S., Rogers, M. S., Cervi, D., Foutz, T., Rawn, K., Voskas, D., Dumont, D. J., Ben-David, Y., Lawler, J., Henkin, J., *et al.* (2005). Genetic heterogeneity of the vasculogenic phenotype parallels angiogenesis: Implications for cellular surrogate marker analysis of antiangiogenesis. *Cancer Cell* **7**, 101.

Shimamura, N., Matchett, G., Yatsushige, H., Calvert, J. W., Ohkuma, H., and Zhang, J. (2006). Inhibition of integrin $\alpha v\beta$ 3 ameliorates focal cerebral ischemic damage in the rat middle cerebral artery occlusion model. *Stroke* **37**, 1902–1909.

Sixt, M., Bauer, M., Lammermann, T., and Fassler, R. (2006). Beta1 integrins: Zip codes and signaling relay for blood cells. *Curr. Opin. Cell Biol.* **18**, 482–490.

Soldi, R., Mitola, S., Strasly, M., Defilippi, P., Tarone, G., and Bussolino, F. (1999). Role of alphavbeta3 integrin in the activation of vascular endothelial growth factor receptor-2. *EMBO J.* **18**, 882–892.

Stoeck, A., Schlich, S., Issa, Y., Gschwend, V., Wenger, T., Herr, I., Marme, A., Bourbie, S., Altevogt, P., and Gutwein, P. (2006). L1 on ovarian carcinoma cells is a binding partner for Neuropilin-1 on mesothelial cells. *Cancer Lett.* **239**, 212–226.

Storgard, C. M., Stupack, D. G., Jonczyk, A., Goodman, S. L., Fox, R. I., and Cheresh, D. A. (1999). Decreased angiogenesis and arthritic disease in rabbits treated with an alphavbeta3 antagonist. *J. Clin. Invest.* **103**, 47–54.

Su, G., Hodnett, M., Wu, N., Atakilit, A., Kosinski, C., Godzich, M., Huang, X. Z., Kim, J. K., Frank, J. A., Matthay, M. A., Sheppard, D., and Pittet, J.-F. (2007). Integrin $\alpha v\beta5$ regulates lung vascular permeability and pulmonary endothelial barrier function. *Am. J. Respir. Cell Mol. Biol.* **36**, 377–386.

Sudhakar, A., Sugimoto, H., Yang, C., Lively, J., Zeisberg, M., and Kalluri, R. (2003). Human tumstatin and human endostatin exhibit distinct antiangiogenic activities mediated by alpha v beta 3 and alpha 5 beta 1 integrins. *Proc. Natl. Acad. Sci. USA* **100**, 4766–4771.

Tarui, T., Miles, L. A., and Takada, Y. (2001). Specific interaction of angiostatin with integrin alpha vbeta 3 in endothelial cells. *J. Biol. Chem.* **276,** 39562–39568.

Taverna, D., Crowley, D., Connolly, M., Bronson, R. T., and Hynes, R. O. (2005). A direct test of potential roles for beta3 and beta5 integrins in growth and metastasis of murine mammary carcinomas. *Cancer Res.* **65,** 10324–10329.

Taverna, D., and Hynes, R. O. (2001). Reduced blood vessel formation and tumor growth in alpha5-integrin–negative teratocarcinomas and embryoid bodies. *Cancer Res.* **61,** 5255–5261.

Taverna, D., Moher, H., Crowley, D., Borsig, L., Varki, A., and Hynes, R. O. (2004). Increased primary tumor growth in mice null for beta3- or beta3/beta5-integrins or selectins. *Proc. Natl. Acad. Sci. USA* **101,** 763–768.

Thurston, G., Baluk, P., Hirata, A., and McDonald, D. M. (1996). Permeability-related changes revealed at endothelial cell borders in inflamed venules by lectin binding. *Am. J. Physiol.* **271,** H2547–H2562.

Toyofuku, T., Yabuki, M., Kamei, J., Kamei, M., Makino, N., Kumanogoh, A., and Hori, M. (2007). Semaphorin-4A, an activator for T-cell–mediated immunity, suppresses angiogenesis via Plexin-D1. *EMBO J.* **26,** 1373–1384.

Trusolino, L., Bertotti, A., and Comoglio, P. M. (2001). A signaling adapter function for alpha6beta4 integrin in the control of HGF-dependent invasive growth. *Cell* **107,** 643–654.

Vlahakis, N. E., Young, B. A., Atakilit, A., Hawkridge, A. E., Issaka, R. B., Boudreau, N., and Sheppard, D. (2007). Integrin alpha 9beta 1 directly binds to vascular endothelial growth factor (VEGF)-A and contributes to VEGF-A induced angiogenesis. *J. Biol. Chem.* **282,** 15187–15196.

Wang, R., Clark, R., and Bautch, V. L. (1992). Embryonic stem cell-derived cystic embryoid bodies form vascular channels: An *in vitro* model of blood vessel development. *Development* **114,** 303–316.

Wayner, E. A., Orlando, R. A., and Cheresh, D. A. (1991). Integrins alpha v beta 3 and alpha v beta 5 contribute to cell attachment to vitronectin but differentially distribute on the cell surface. *J. Cell Biol.* **113,** 919–929.

Weis, S., Cui, J., Barnes, L., and Cheresh, D. (2004a). Endothelial barrier disruption by VEGF-mediated Src activity potentiates tumor cell extravasation and metastasis. *J. Cell Biol.* **167,** 223–229.

Weis, S., Shintani, S., Weber, A., Kirchmair, R., Wood, M., Cravens, A., McSharry, H., Iwakura, A., Yoon, Y. S., Himes, N., Burstein, D., Doukas, J., *et al.* (2004b). Src blockade stabilizes a Flk/cadherin complex, reducing edema and tissue injury following myocardial infarction. *J. Clin. Invest.* **113,** 885–894.

Weis, S. M., Lindquist, J. N., Barnes, L. A., Lutu-Fuga, K. M., Cui, J., Wood, M. R., and Cheresh, D. A. (2007). Cooperation between VEGF and β3 integrin during cardiac vascular development. *Blood* **109,** 1962–1970.

Winter, M. C., Kamath, A. M., Ries, D. R., Shasby, S. S., Chen, Y. T., and Shasby, D. M. (1999). Histamine alters cadherin-mediated sites of endothelial adhesion. *Am. J. Physiol.* **277,** L988–L995.

Wu, Y., Cai, W., and Chen, X. (2006). Near-infrared fluorescence imaging of tumor integrin alpha v beta 3 expression with Cy7-labeled RGD multimers. *Mol. Imaging Biol.* **8,** 226–236.

Zhang, H., Li, C., and Baciu, P. C. (2002). Expression of integrins and MMPs during alkaline-burn-induced corneal angiogenesis. *Invest. Ophthalmol. Vis. Sci.* **43,** 955–962.

Author Index

Subject Index

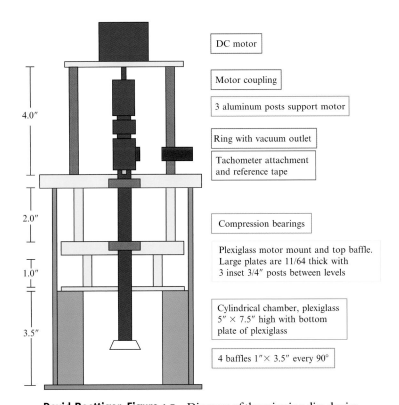

4.0″

2.0″

1.0″

3.5″

DC motor

Motor coupling

3 aluminum posts support motor

Ring with vacuum outlet

Tachometer attachment
and reference tape

Compression bearings

Plexiglass motor mount and top baffle.
Large plates are 11/64 thick with
3 inset 3/4″ posts between levels

Cylindrical chamber, plexiglass
5″ × 7.5″ high with bottom
plate of plexiglass

4 baffles 1″× 3.5″ every 90°

David Boettiger, Figure 1.5 Diagram of the spinning disc device.

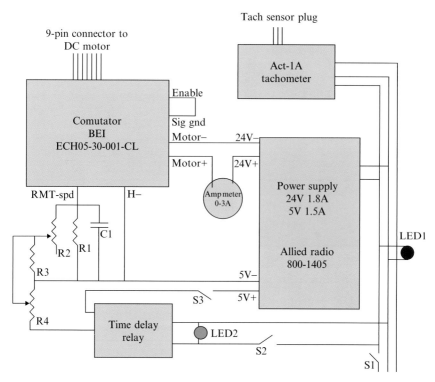

David Boettiger, Figure 1.6   Wiring diagram for motor controller and tachometer. For specific components, see parts list.

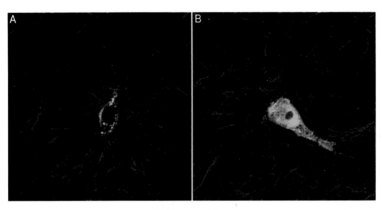

Patricia J. Keely *et al.*, Figures 2.3(A) and 2.4(B)   Merged MPE and SHG images from Fig. 3(A) and Fig. 4(B). MPE of GFP-vinculin isolated with a 480-to-550–nm band-pass filter pseudocolored in green. SHG image of collagen isolated with a 445-nm, narrow band-pass filter pseudocolored in red.)

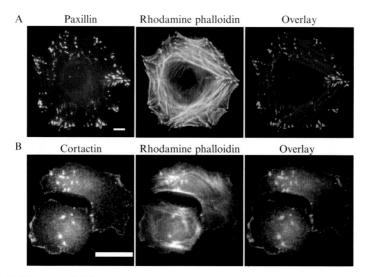

| A | Paxillin | Rhodamine phalloidin | Overlay |
| B | Cortactin | Rhodamine phalloidin | Overlay |

Keefe T. Chan et al., Figure 3.1 (A) Immunofluorescent image of a HeLa cell costained with an anti-paxillin antibody (green) and rhodamine phalloidin (red). Bar, 10 microns. (B) Immunofluorescent image of an MTLn3 cell co-stained with an anti-cortactin antibody (green) and rhodamine phalloidin (red). Bar, 10 microns.

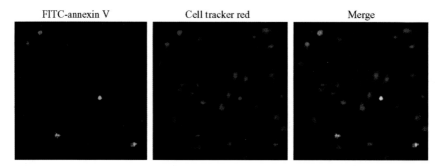

FITC-annexin V · Cell tracker red · Merge

Alireza Alavi and Dwayne G. Stupack, Figure 5.4 Scoring of apoptosis among endothelial cells spread on 3D collagen gels. Endothelial cells labeled with Cell Tracker Orange (red channel) were plated on 35-mm dishes on top of a 3D collagen matrix, and 0.6% agarose was layered on top of the cells. Apoptosis was detected by incubating cells in FITC-annexin V (green channel) and confocal imaging. Apoptotic cells appear yellow in the merge.

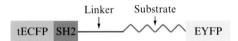

Yingxiao Wang and Shu Chien, Figure 9.1   A schematic drawing of the Src reporter composition.

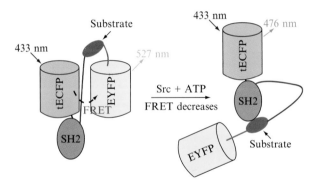

Yingxiao Wang and Shu Chien, Figure 9.2   A schematic cartoon of the activation mechanism of the Src reporter.

Yingxiao Wang and Shu Chien, Figure 9.3   A schematic drawing of the Src reporter with tECFP and EYFP positioned between the SH2 domain and the substrate peptide and flanking the linker.

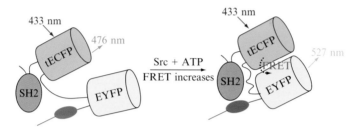

Yingxiao Wang and Shu Chien, Figure 9.4 A schematic cartoon of the activation mechanism of the Src reporter with tECFP and EYFP positioned between the SH2 domain and the substrate peptide and flanking the linker.

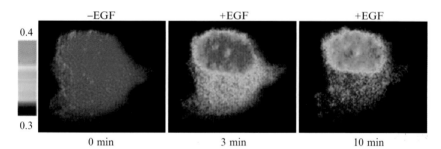

Yingxiao Wang and Shu Chien, Figure 9.9 EGF induced a FRET change of the Src biosensor. HeLa cells were transfected with the Src biosensor and stimulated with EGF (50 ng/ml) for various time periods as indicated. The color scale bar on the left represents the tECFP/citrine emission ratio, with cold colors indicating low Src activity and hot colors indicating high levels of Src activation. The color images represent the emission ratio maps of the Src biosensor in the cell.

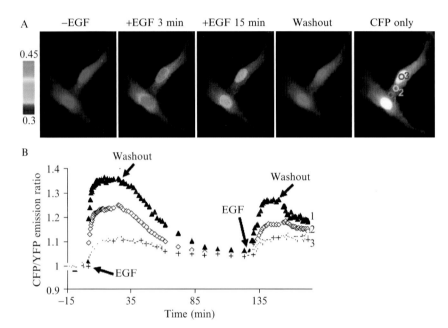

Yingxiao Wang and Shu Chien, Figure 9.11 The FRET response of the monomeric Src biosensor is reversible. HeLa cells transfected with the monomerized Src biosensor were stimulated with EGF (50 ng/ml), washed with serum-free medium (washout), and subsequently subjected to EGF and washout for a second time. (A) The tECFP/citrine emission ratio images of the monomeric Src biosensor. The color scale bar on the left represents the tECFP/citrine emission ratio, with cold colors indicating low Src activity and hot colors indicating high levels of Src activation. The representative emission ratio images (control, 3 and 15 min after EGF, and after washout) are shown in color on the left, and the CFP-only image is shown in black and white on the far right. (B) Time courses of the average tECFP/citrine emission ratios of the monomeric Src biosensor from different regions of cells in (A).

A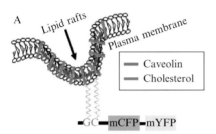

B

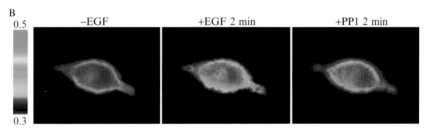

Yingxiao Wang and Shu Chien, Figure 9.12   (A) A schematic drawing of the monomeric Src biosensor targeted to the plasma membrane. (B) The tECFP/citrine emission
ratio images. HeLa cells transfected with the membrane-targeted Src biosensor were stimulated with EGF (50 ng/ml) and subsequently incubated with PP1 for the time periods
as indicated. The color scale bar on the left represents the tECFP/citrine emission ratio,
with cold colors indicating low Src activity and hot colors indicating high levels of Src
activation.

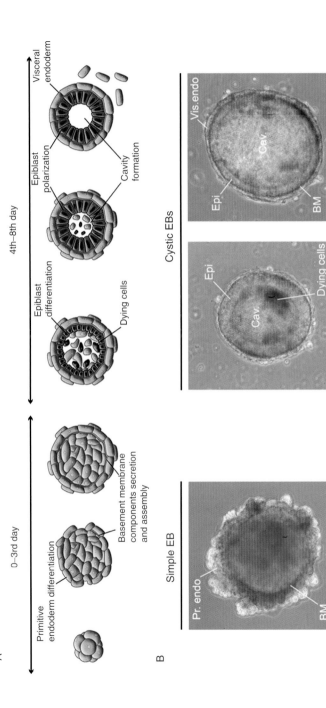

Eloi Montanez *et al.*, Figure 12.1 Stage of EB development. (A) Schematic representation of the major events that take place during EB differentiation. (B) Phase–contrast pictures of simple and cystic EBs. BM, basement membrane; cav, cavity; EB, embryoid bodies; Epi, epiblast; Pr. endo, primitive endoderm; Vis. endo, visceral endoderm.

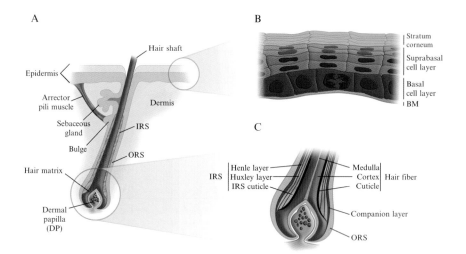

Eloi Montanez *et al.*, Figure 12.6 (A) Scheme of the mammalian skin and anagen hair follicle. (B) Close-up of the epidermis and the underlying BM separating the epidermal from the dermal compartment. The basal cell layer contains stem cells and transiently amplifying cells, which ensure the self-renewal of the epidermis. After the transiently amplifying cells cease proliferation, they differentiate and mature into suprabasal keratinocytes (the suprabasal cell layer). During this transition, the keratinocytes lose their contact with the BM and undergo an apoptosis–related process termed terminal differentiation. Finally, they die and become cornified squames (the stratum corneum) and are shed from the surface of the skin. (C) The hair follicle is made up of eight concentric epithelial sheaths: the ORS, which is continuous with the basal keratinocyte layer of the epidermis, the companion layer, three layers of the IRS, and three layers of hair-producing cells. BM, basement membrane; IRS, inner root sheath; ORS, outer root sheath.

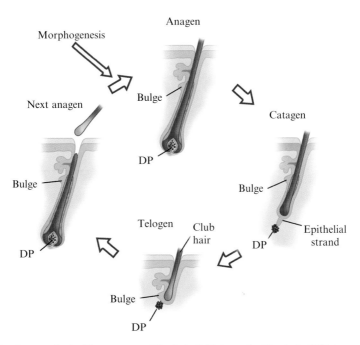

Eloi Montanez *et al.*, Figure 12.7 The hair follicle cycle. The hair-follicle morphogenesis begins during embryogenesis and is completed with the formation of the anagen hair follicle, which is made of an upper permanent part starting above the arrector pili muscle insertion, and a lower part, which undergoes cycling changes with a phase of growth (anagen), regression (catagen), and rest (telogen). The first postnatal anagen (and every subsequent one) starts when the mesenchymal cells of the DP induce the stem cells located in the bulge region of the hair follicle to proliferate and migrate into the hair matrix region of the follicle, where they further differentiate into the eight epithelial lineages. DP, dermal papilla.

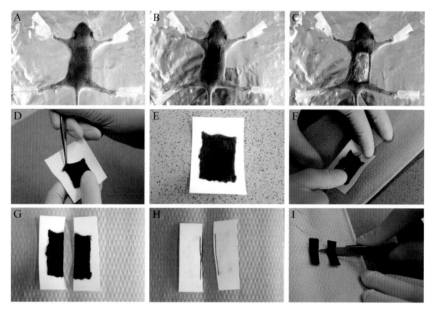

Eloi Montanez *et al.*, Figure 12.8   Isolation of mouse skin. For a detailed description, see Section 4.1.

Eloi Montanez *et al.*, Figure 12.10   Preparation of paraffin blocks. For a detailed description, see Section 4.5.

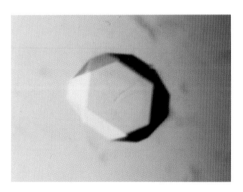

Jian-Ping Xiong *et al.*, Figure 14.3   A hexagonal crystal of ectodomain of human integrin $\alpha V \beta 3$ grown at 4°.

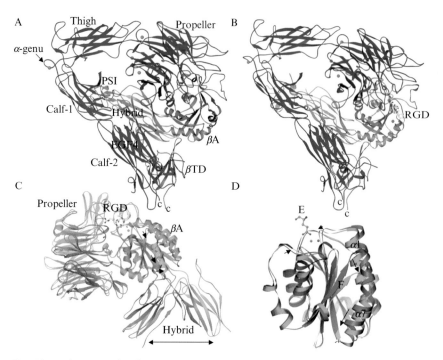

Jian-Ping Xiong *et al.*, Figure 14.7 Integrin crystal structures and conformational states. All diagrams in this figure were made using ribbon models (Carson, 1987). Ribbon diagrams showing the crystal structure of the ectodomain of integrin αVβ3 in its unliganded (A) and RGD-bound (B) (Xiong *et al.*, 2001, 2002, 2004) states. In (A), the protein is bent at a flexible region (α-genu, arrow), with an adjacent metal ion (orange sphere) (the β-genu is not shown). The four metal ions at the bottom of the propeller (orange spheres) and the ADMIDAS ion (purple sphere) are shown. In (B), two additional metal ions at MIDAS (cyan sphere) and LIMBS (gray sphere) are found, the former directly coordinating the Asp residue from the cyclic pentapeptide cilengitide, Arg-Gly-Asp-[D-Phe]-[N-methyl-Val-] (carbon, nitrogen, and oxygen atoms are in yellow, blue, and red, respectively). (C) Ribbon diagram of the propeller, βA, and hybrid domains from the liganded bent αVβ3 ectodomain (red) and the αIIbβ3 integrin fragment (green) (Xiao *et al.*, 2004). Nonmoving parts of the Cα tracings are in gray. (D) Ribbon diagram of the crystal structures of open (red) and closed (green) αA domain from integrin CD11b (Lee *et al.*, 1995a,b). Major conformational differences are indicated by arrows that also show the direction of movement. The ligand Glu is in gold.

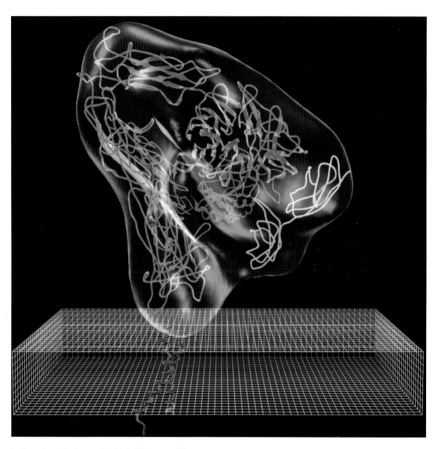

Brian D. Adair and Mark Yeager, Figure 15.13 The 3D density map (grayscale transparency) of the integrin $\alpha_V\beta_3$ in a complex with fibronectin determined by EM and single-particle image analysis. Fibronectin domains 9 and 10 were localized by examining a difference map between the integrin with and without a bound fragment of fibronectin. The X-ray structures of the α_V (blue) and β_3 (red) chains (Xiong *et al.*, 2001) have been docked into the EM density envelope. Additional density (lower right) can accommodate fibronectin domain 10 adjacent to the ligand binding site (green) as well as domain 9 at the synergy site (yellow). The complex is shown adjacent to the white grid box, which represents the 30-Å–thick hydrophobic portion of the cellular membrane across which signals are transmitted. An R-handed α-helical coiled-coil (purple) is shown spanning the lipid bilayer, as we have proposed for the inactive conformation of $\alpha_{IIb}\beta_3$ (Adair and Yeager, 2002). (Modified from Adair and Yeager, 2005.) Graphics by Michael E. Pique and Mark Yeager.

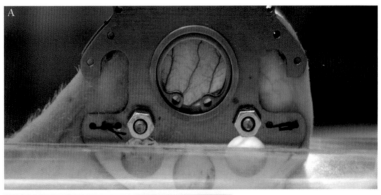

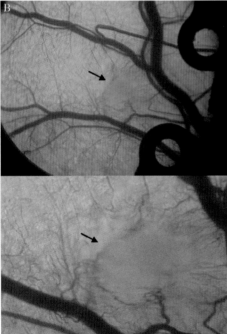

Milan Makale, Figure 16.1 (A) Dorsal skinfold window chamber is shown implanted on a hairless athymic (*nu/nu*) mouse. The subcutaneous vasculature is visible. (B) Human pancreatic FG tumor (800-micron diameter) growing within the chamber. The top panel is a general view at about 1.5×, and the bottom image is a closer view of the same tumor at 2.5×.

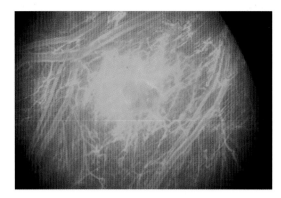

Milan Makale, Figure 16.7 Fluorescence microscope image of window chamber with human pancreatic tumor after intravenous FITC-dextran (2 M mol. wt.) and FITC-lectin. Leakage around the tumor can be seen; tumor-associated vessels are permeable. The tumor cells are expressing dSRed and growing in a *nu/nu* mouse window chamber.

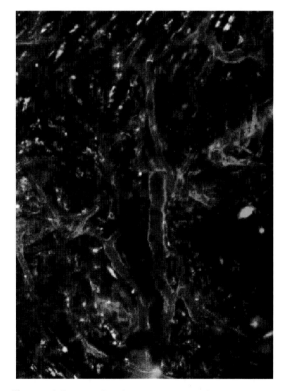

Milan Makale, Figure 16.8 Confocal microscope image of a human M21 (avβ3 integrin positive) melanoma tumor growing in the window chamber. Tumor cells (blue) are labeled with a fluorophore-linked nanoparticle that recognizes avβ3 on the tumor cells. The typically chaotic tumor blood vessels are labeled with rhodamine-lectin, which appears red in the image.

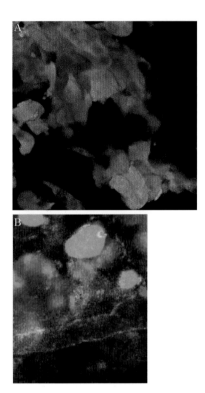

Milan Makale, Figure 16.9 (A) Scanning multiphoton 3D rendering. The invading periphery of a human FGM tumor expressing dSRed was imaged at 40×. (B) Multiphoton image taken at 40×; the vessels, labeled with rhodamine-lectin show up in red, and the human melanoma (M21L, $av\beta3$ integrin negative) cells in green (expressing GFP), surround the microvessels. (Images acquired in collaboration with Benjamin Migliori, and Philbert Tsai, Ph.D., and David Kleinfeld, Ph.D., Department of Physics, University of California, San Diego.)

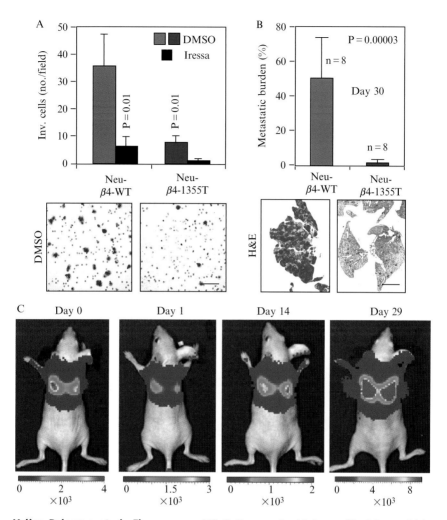

Yuliya Pylayeva _et al._, Figure 19.2 (A) Cells treated with Iressa (10 μM) or vehicle alone (DMSO) were subjected to matrigel invasion assay in response to FBS. The graph shows the mean number of invaded cells ($\pm$ standard deviation [SD]) per microscopic field from triplicate samples. The bottom panel shows representative fields. Scale bar = 50 μm. (B) Cells were injected in the tail vein of nude mice. Percentages of lung section areas occupied by metastases ($\pm$SD) were quantified 30 days later by image analysis. Bottom panels show representative images of sagittal lung sections. Scale bar = 2 mm. (C) Primary MMTV-polyoma MT–derived tumor cells (1×10^6 cells/mouse) were infected with TGL vector and injected in the tail vein of nude mice. Representative images taken at four time points (day 0, day 1, day 14, and day 29) are shown. Color scale represents number of photons per second (photon flux). (With permission, panels A and B from Guo, W., Pylayeva, Y., Pepe, A., Yoshioka, T., Muller, W. J., Inghirami, G., and Giancotti, F. G. (2006). Beta 4 integrin amplifies ErbB2 signaling to promote mammary tumorigenesis. _Cell_ 126, 489–502.)

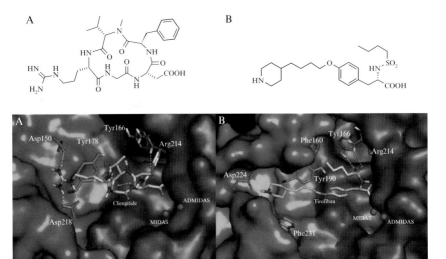

Dominik Heckmann and Horst Kessler, Figure 20.7 Structure of two ligand-integrin complexes based on X-ray structures. (A) Cilengitide binding $\alpha v \beta 3$. (B) Tirofiban binding $\alpha IIb \beta 3$. The receptors are shown as Connolly-surface representation, (α-subunits in blue, β-subunits in red). Interacting residues and metal ions are highlighted, hydrogen bonds shown as dot lines.

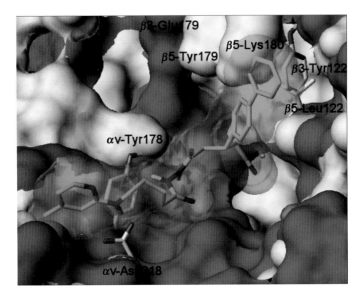

Dominik Heckmann and Horst Kessler, Figure 20.8 Superposition of the $\alpha v \beta 3$ receptor (grey) and the homology model of the $\alpha v \beta 5$ receptor (white) both represented as Connolly surface. The $\alpha v \beta 3$-selective ligand (4) is docked into the $\alpha v \beta 3$ receptor (green transparent), the MIDAS cation is shown as golden sphere.

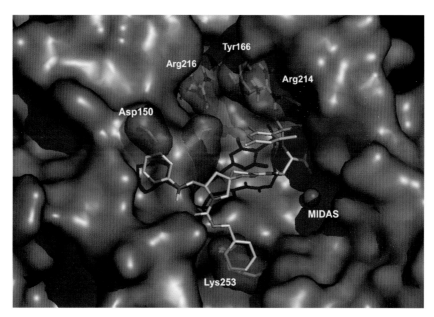

Dominik Heckmann and Horst Kessler, Figure 20.9 Connolly surface of α5β1 with SJ749 modelled into. For comparison, mutated residues in αvβ3 are pictured as sticks, labeled and shown with transparent Connolly surface.

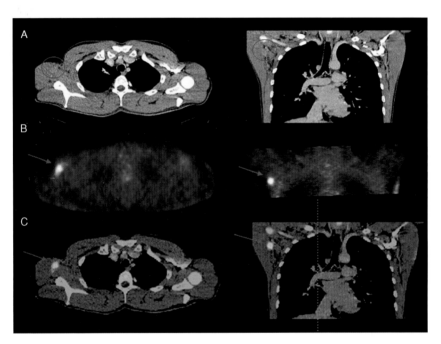

Dominik Heckmann and Horst Kessler, Figure 20.14 Lateral (left) and longitudinal (right) sections from a patient with axillary lymph node metastasis. (A) represents the CT-image, (B) the PET image after application of 220 MBq. Galacto–RGD (15), 2 h p.i., and (C) a superposition of both images. The site of the tumor is marked by a circle or arrows (Haubner *et al.*, 2005).